Graduate Texts in Mathematics 76

Shigeru Iitaka

Algebraic Geometry

An Introduction to Birational Geometry of Algebraic Varieties

Springer-Verlag
New York Heidelberg Berlin

Shigeru Iitaka
Department of Mathematics
Faculty of Science
University of Tokyo
Tokyo
JAPAN

AMS Subject Classification (1981): 14-01

Library of Congress Cataloging in Publication Data
Iitaka, Shigeru.
Algebraic geometry.
(Graduate texts in mathematics; 76)
"This volume grew out of the author's book in Japanese published in 3 volumes by Iwanami, Tokyo in 1977."
Includes index.
1. Geometry, Algebraic. 2. Algebraic varieties.
I. Title. II. Series.
QA564.I36 516.3′5 80-28195

Printed in the United States of America.

9 8 7 6 5 4 3 2 1

ISBN 0-387-90546-4 Springer-Verlag New York Heidelberg Berlin
ISBN 3-540-90546-4 Springer-Verlag Berlin Heidelberg New York

Preface

The aim of this book is to introduce the reader to the geometric theory of algebraic varieties, in particular to the birational geometry of algebraic varieties.

This volume grew out of the author's book in Japanese published in 3 volumes by Iwanami, Tokyo, in 1977. While writing this English version, the author has tried to rearrange and rewrite the original material so that even beginners can read it easily without referring to other books, such as textbooks on commutative algebra. The reader is only expected to know the definition of Noetherin rings and the statement of the Hilbert basis theorem.

The new chapters 1, 2, and 10 have been expanded. In particular, the exposition of D-dimension theory, although shorter, is more complete than in the old version. However, to keep the book of manageable size, the latter parts of Chapters 6, 9, and 11 have been removed.

I thank Mr. A. Sevenster for encouraging me to write this new version, and Professors K. K. Kubota in Kentucky and P. M. H. Wilson in Cambridge for their careful and critical reading of the English manuscripts and typescripts. I held seminars based on the material in this book at The University of Tokyo, where a large number of valuable comments and suggestions were given by students Iwamiya, Kawamata, Norimatsu, Tobita, Tsushima, Maeda, Sakamoto, Tsunoda, Chou, Fujiwara, Suzuki, and Matsuda.

Fall 1981 Shigeru Iitaka

Contents

Introduction

The purpose of algebraic geometry is to study comprehensively varieties defined by a set of polynomial equations in many variables

$$f_1(X_1, \ldots, X_n) = \cdots = f_r(X_1, \ldots, X_n) = 0.$$

Properties of varieties should be independent of the choice of coordinate systems. For example, the variety defined by $X_2 = 0$ ($r = 1, n = 2$) is equivalent to that defined by $Y_1 - Y_2^2 = 0$ ($r = 1, n = 2$) under the invertible transformation $X_1 = Y_2$, $X_2 = Y_1 - Y_2^2$. This equivalence is interpreted as the existence of an isomorphism of rings

$$\frac{k[X_1, X_2]}{(X_2)} \cong \frac{k[Y_1, Y_2]}{(Y_1 - Y_2^2)}.$$

Thus, the study of a set of polynomial equations can be reduced to the study of a commutative ring $k[X_1, \ldots, X_n]/\mathfrak{a}$, where $\mathfrak{a}$ is an ideal generated by $f_1, \ldots, f_r$. From this viewpoint, one arrives naturally at the concepts of affine schemes and then of schemes.

However, ever since the last century, it has been believed that the more essential properties of varieties are those which are birationally invariant.

A plane curve is defined by an irreducible polynomial $\varphi(X, Y)$. The degree of φ is said to be the degree of the curve. Plane curves defined by irreducible polynomials f_1 and f_2 are birationally equivalent if the field $Q(k[X, Y]/(f_1))$ is isomorphic to $Q(k[X, Y]/(f_2))$, where $Q(R)$ denotes the field of fractions of the integral domain R.

The degree is not, however, a birational invariant. As Abel noted, the number of linearly independent Abelian differentials of the first kind (also called regular 1-forms) on a given curve C is more important than the degree, since it is a birational invariant. This number is called the genus of

C, denoted by $g(C)$. Curves can be classified into the following three classes according to their genera:

$$\text{The class I:}\quad g(C) = 0.$$

$$\text{The class II:}\quad g(C) = 1.$$

$$\text{The class III:}\quad g(C) \geq 2.$$

A similar birational classification for 2-dimensional varieties (called surfaces) was obtained by Italian algebraic geometers around the beginning of our own century.

Given a variety V of dimension n, many birational invariants can be defined, such as the plurigenera, the i-th irregularity, and the Kodaira dimension. Let $\kappa(V)$ denote the Kodaira dimension of V, which can take the values $-\infty, 0, 1, \ldots, n$. By means of the Kodaira dimension, varieties of dimension n can be classified into $n + 2$ classes. When $n = 1$, this classification agrees with that given by the genus.

Many fundamental properties of the Kodaira dimension have been found, giving some basic information about the structure of varieties.

Let V be a variety of dimension n and suppose that $\kappa(V) \geq 0$. Then by Theorem 10.3 (fibering theorem), there exists a dominating morphism $f\colon V^* \to W$ such that (1) V^* is birationally equivalent to V, (2) $\dim W = \kappa(V)$, (3) general fibers $f^{-1}(x)$ are irreducible, and (4) for a (strictly) general point x of W, $\kappa(f^{-1}(x)) = 0$.

Varieties V with $\kappa(V) = n$ are said to be of general type or of hyperbolic type. Roughly speaking, almost all varieties are of hyperbolic type and these have rather general properties in common. For example, if V is of hyperbolic type, then the automorphism group $\mathrm{Aut}(V)$ of V is a finite group.

The number of linearly independent regular 1-forms on a complete nonsingular variety V is also a birational invariant, denoted by $q(V)$. In particular, if $\dim V = 1$, then $q(V)$ turns out to be the genus of V. In general, an Abelian variety of dimension $q(V)$, the Albanese variety $\mathrm{Alb}(V)$, is associated with V, together with the Albanese map $\alpha_V\colon V \to \mathrm{Alb}(V)$.

Recently Kawamata proved that if $\kappa(V) = 0$, then α_V is surjective and general fibers $\alpha_V^{-1}(x)$ are irreducible. Thus in the case where $\kappa(V) = 0$ and $q(V) > 0$, the structure of V can be studied using Albanese maps. However, nothing is known about V when $\kappa(V) = q(V) = 0$. If $\dim V = 2$, it has been shown that such a V is birationally equivalent to a $K3$ surface or an Enriques surface.

In the case where $\kappa(V) = -\infty$ and $q(V) > 0$, consider again the Albanese map $\alpha_V\colon V \to \mathrm{Alb}(V)$. One has a morphism $\psi\colon V \to Z$ obtained from the Stein factorization of $\alpha_V\colon V \to \alpha_V(V)$. Then it is conjectured that $\kappa(\psi^{-1}(x)) = -\infty$ for a general point x of Z. Actually, this has been proved for $n \leq 3$ (by Enriques for $n = 2$; by Viehweg for $n = 3$). The case where $\kappa(V) = -\infty$ and $q(V) = 0$ seems the most difficult case to study. When $n = 2$, such a V is a rational surface, i.e., a surface birationally equivalent to $\mathbf{P}^1 \times \mathbf{P}^1$. This fact, discovered by Castelnuovo, was the starting point of the classification theory of algebraic surfaces by the Italian school. But in the higher dimensional case, nothing is known about such V.

Chapter 10 may serve as a guide to this rapidly developing theory of birational classification of varieties.

It is unreasonable to say that only birationally invariant properties are worth studying. For instance, the affine line $\mathbf{A}^1$ is quite different from $G_m = \mathbf{A}^1 - \{0\}$, which are both very important. However, they are birationally equivalent.

Any variety is birationally equivalent to a complete variety. Thus, when considering noncomplete varieties and studying their properties, we can no longer use birational equivalence. However, in this case, a more delicate equivalence relation, called proper birational equivalence, is introduced (see Chapter 2). One can find many proper-birational invariants such as the logarithmic genera, logarithmic irregularities, and logarithmic Kodaira dimension, which are defined by making use of logarithmic forms. A proper birational equivalence between affine normal varieties is just an isomorphism between them; hence the corresponding normal rings are isomorphic. Thus, in our study of proper birational properties of varieties, the theory of (normal) rings and birational geometry are unified; thus the theorems on Kodaira dimension could be translated into ring theory and so on.

Algebraic geometry should be a synthesis of algebra and geometry. But, in practice, it has been an algebraic approach to geometry. Our new birational geometry (e.g., proper birational geometry) is not only a revival of old birational geometry but is also a beginning of some grand unified theory of algebra and geometry.

Chapter 1

Schemes

§1.1 Spectra of Rings

a. We begin by defining spectra of commutative rings with identity, which are the base spaces of the affine schemes introduced in §1.11.

In all that follows, commutative rings A, B, ... with identity elements 1_A, 1_B, ... are referred to simply as *rings*, and ring homomorphisms $\varphi: A \to B$ are assumed to satisfy $\varphi(1_A) = 1_B$.

Definition. The *spectrum* of a ring A is the set of all prime ideals of A, denoted by Spec A.

Note that the ring A itself is not considered to be a prime ideal.

If $\varphi : A \to B$ is a ring homomorphism and $\mathfrak{p}$ is a prime ideal of B, then $\varphi^{-1}(\mathfrak{p})$ is also a prime ideal of A. We note that $1_A \notin \varphi^{-1}(\mathfrak{p})$, since $1_B \notin \mathfrak{p}$ and $\varphi(1_A) = 1_B$.

Definition. The mapping ${}^a\varphi$: Spec $B \to$ Spec A defined by ${}^a\varphi(\mathfrak{p}) = \varphi^{-1}(\mathfrak{p})$ is said to be the *mapping associated with* φ.

EXAMPLE 1.1. (i) For the trivial ring 0, we have Spec $0 = \varnothing$.

(ii) If k is a field (a field is always assumed to be nontrivial, i.e., $k \neq \{0\}$), then Spec $k = \{(0)\}$.

(iii) Spec $\mathbb{Z} = \{(0)\} \bigcup \{(p) \mid p \text{ is a prime number}\}$.

(iv) If $k[\mathsf{X}]$ is a ring of polynomials over an algebraically closed field k, then

$$\text{Spec } k[\mathsf{X}] = \{(0)\} \bigcup \{(\mathsf{X} - \alpha) \mid \alpha \in k\},$$

which can be written as Spec $k[\mathsf{X}] = \{*\} \bigcup k$ with the abbreviations $* = (0)$, $\alpha = (\mathsf{X} - \alpha)$.

(v) If $k[\mathsf{X}, \mathsf{Y}]$ is a polynomial ring in two variables over an algebraically closed field k, then

$$\text{Spec } k[\mathsf{X}, \mathsf{Y}] = \{(0)\} \bigcup \{(f) \mid f \text{ is a nonconstant irreducible polynomial in } \mathsf{X} \text{ and } \mathsf{Y}\} \bigcup \{(\mathsf{X} - \alpha, \mathsf{Y} - \beta) \mid (\alpha, \beta) \in k^2\}.$$

Proof of (v). Clearly, it suffices to show that every nonprincipal prime ideal $\mathfrak{p}$ is of the form $(\mathsf{X} - \alpha, \mathsf{Y} - \beta)$. Let $f \in \mathfrak{p}\backslash\{0\}$ be a polynomial with $\deg_{\mathsf{Y}} f$ minimal such that f is irreducible. Since $\mathfrak{p}$ is nonprincipal, there is an element $g \in \mathfrak{p}\backslash(f)$. By the Euclidean algorithm in the ring $k(\mathsf{X})[\mathsf{Y}]$, there is a $p \in k[\mathsf{X}]\backslash(0)$ and $q, r \in k[\mathsf{X}, \mathsf{Y}]$ satisfying $pg = qf + r$, where either $r = 0$ or $\deg_{\mathsf{Y}} r < \deg_{\mathsf{Y}} f$. It follows that $r = 0$ by the choice of f and because $r = pg - qf \in \mathfrak{p}$. But then $p \in (f)$ because (f) is a prime ideal, $pg = qf \in (f)$, and $g \notin (f)$. Thus $\deg_{\mathsf{Y}} f \leq \deg_{\mathsf{Y}} p = 0$; i.e., $f \in k[\mathsf{X}]$. Since k is algebraically closed and $f \in k[\mathsf{X}]$ is irreducible, f must be linear and so $\mathfrak{p}$ contains a polynomial of the form $\mathsf{X} - \alpha$ with $\alpha \in k$. Interchanging the roles of X and Y, one also sees that $\mathfrak{p}$ contains a polynomial of the form $\mathsf{Y} - \beta$ with $\beta \in k$. But then $\mathfrak{p}$ contains the maximal ideal $(\mathsf{X} - \alpha, \mathsf{Y} - \beta)$ and hence must be equal to it (cf. Exercise 1.1).

The reader can easily verify that the maximal ideals of $k[\mathsf{X}, \mathsf{Y}]$ are precisely those of the form $(\mathsf{X} - \alpha, \mathsf{Y} - \beta)$ with $(\alpha, \beta) \in k^2$. □

b. We shall introduce a topology on Spec A. For any ideal $\mathfrak{a}$ of A, define the set $V(\mathfrak{a})$ to be $\{\mathfrak{p} \in \text{Spec } A \mid \mathfrak{p} \supseteq \mathfrak{a}\}$, and for any $f \in A$, define $V(f)$ to be $\{\mathfrak{p} \in \text{Spec } A \mid \mathfrak{p} \ni f\}$. Then, $V(f) = V(fA)$ and the following properties are easily verified:

(i) $V(0) = \text{Spec } A$, $V(1) = \varnothing$.
(ii) If $\mathfrak{a}$ and $\mathfrak{b}$ are ideals such that $\mathfrak{a} \subseteq \mathfrak{b}$, then $V(\mathfrak{a}) \supseteq V(\mathfrak{b})$.
(iii) $V(\mathfrak{a} \cap \mathfrak{b}) = V(\mathfrak{a}\mathfrak{b}) = V(\mathfrak{a}) \cup V(\mathfrak{b})$. In particular, $V(fg) = V(f) \cup V(g)$ for all $f, g \in A$.
(iv) If $\{\mathfrak{a}_\lambda \mid \lambda \in \Lambda\}$ is a set of ideals of A, then $V(\sum_{\lambda \in \Lambda} \mathfrak{a}_\lambda) = \bigcap_{\lambda \in \Lambda} V(\mathfrak{a}_\lambda)$.
(v) For any ideal $\mathfrak{a}$ of A, if $\varphi: A \to A/\mathfrak{a}$ is the natural homomorphism then ${}^a\varphi: \text{Spec}(A/\mathfrak{a}) \to \text{Spec } A$ is one-to-one and $\text{Im } {}^a\varphi = V(\mathfrak{a})$.

Definition. The topology on Spec A is introduced by taking the sets in $\{V(\mathfrak{a}) \mid \mathfrak{a} \text{ ideals of } A\}$ as the closed sets.

The open sets of this topology are just those of the form

$$D(\mathfrak{a}) = \text{Spec } A\backslash V(\mathfrak{a}) = \{\mathfrak{p} \in \text{Spec } A \mid \mathfrak{p} \not\supseteq \mathfrak{a}\}.$$

When expressed in terms of the $D(\mathfrak{a})$, properties (i) through (v) will be referred to as properties (i′) through (v′), respectively. For example, in view of the property (iv′), if $\mathfrak{a}$ is the ideal generated by $\{f_\lambda \mid \lambda \in \Lambda\}$, then

$$D(\mathfrak{a}) = \bigcup_{\lambda \in \Lambda} D(f_\lambda),$$

where $D(f) = \operatorname{Spec} A \backslash V(f) = \{\mathfrak{p} \in \operatorname{Spec} A \mid \mathfrak{p} \not\ni f\}$.

Note that the sets $D(f)$ form an open base for the topology of Spec A.

The following properties of the mapping ${}^a\varphi$: Spec $B \to$ Spec A associated with $\varphi: A \to B$ are easily verified.

Proposition 1.1

(i) *${}^a\varphi$ is continuous. More precisely, if $f \in A$, then $({}^a\varphi)^{-1}(D(f)) = D(\varphi(f))$, and if $\mathfrak{a}$ is an ideal of A, then $({}^a\varphi)^{-1}(V(\mathfrak{a})) = V(\mathfrak{a}B)$, where $\mathfrak{a}B$ is the ideal of B generated by $\varphi(\mathfrak{a})$.*

(ii) *For any ideal $\mathfrak{b}$ of B, one has*

$${}^a\varphi(V(\mathfrak{b})) \subseteq V(\varphi^{-1}(\mathfrak{b})).$$

(iii) *If $\varphi: A \to A/\mathfrak{a}$ is the natural homomorphism, then ${}^a\varphi$:* Spec $A/\mathfrak{a} \to$ Spec A *is a homeomorphism of* Spec $A/\mathfrak{a}$ *onto $V(\mathfrak{a})$.*

(iv) *If $\varphi: A \to B$ is surjective, then ${}^a\varphi$ is a homeomorphism of* Spec B *onto the closed subset $V(\mathrm{Ker}\ \varphi)$ of* Spec A.

(v) *On the other hand, if ${}^a\varphi$:* Spec $B \to$ Spec A *is surjective, then $V(\mathrm{Ker}\ \varphi) = V(\varphi^{-1}(0)) =$* Spec A. *Hence* Ker φ *is a subset of every prime ideal of A.*

For the proof of the next propositions, we need a result from ring theory.

Definition. If $\mathfrak{a}$ is an ideal of A, the *radical* $\sqrt{\mathfrak{a}}$ is $\{a \in A \mid a^m \in \mathfrak{a}$ for some integer $m > 0\}$. $\sqrt{(0_A)}$ is said to be the *nilradical* of A, which consists of all nilpotent elements of A.

$\sqrt{\mathfrak{a}}$ is an ideal containing $\mathfrak{a}$.

c. The following result is a key lemma in the first stage of the theory of spectra.

Lemma 1.1. *If $\mathfrak{a}$ is a proper ideal (i.e., $\mathfrak{a} \neq A$) of A, then there exists a maximal ideal containing $\mathfrak{a}$.*

PROOF. Let $\mathfrak{F}$ be the set of all proper ideals of A containing $\mathfrak{a}$. Then $\mathfrak{F}$ is not empty, since $\mathfrak{a} \in \mathfrak{F}$. The set $\mathfrak{F}$ is naturally ordered by set inclusion, i.e., $\mathfrak{a}_\lambda \leq \mathfrak{a}_\mu$ if and only if $\mathfrak{a}_\lambda \subseteq \mathfrak{a}_\mu$. We shall show that $\mathfrak{F}$ is an inductively ordered set. In fact, letting $\{\mathfrak{a}_\lambda \mid \lambda \in \Lambda\}$ be an arbitrary linearly ordered subset of $\mathfrak{F}$, define $\mathfrak{a}_* = \bigcup_{\lambda \in \Lambda} \mathfrak{a}_\lambda$, which becomes a proper ideal, i.e., $\mathfrak{a}_* \in \mathfrak{F}$ and $\mathfrak{a}_\lambda \leq \mathfrak{a}_*$ for any $\lambda \in \Lambda$. Hence, the ordered set $\mathfrak{F}$ is inductive.

By Zorn's lemma, $\mathfrak{F}$ has a maximal element $\mathfrak{m}$, which is a maximal ideal containing $\mathfrak{a}$. □

Corollary

(*i*) Spec $A = \varnothing$ *if and only if* $A = 0$.
(*ii*) $V(\mathfrak{a}) = \varnothing$ *if and only if* $\mathfrak{a} = A$.

EXAMPLE 1.2. Let $a_1, \ldots, a_r \in A$. If there is no prime ideal containing $a_1, \ldots, a_r$, then there exist $b_1, \ldots, b_r$ in A such that $a_1 b_1 + \cdots + a_r b_r = 1$.

PROOF. Let $\mathfrak{a} = \sum_{j=1}^{r} a_j A$. If $\mathfrak{a} \neq A$, then there exists a prime ideal $\mathfrak{p}$ containing $\mathfrak{a}$ by Lemma 1.1. □

§1.2 Examples of Spectra as Topological Spaces

a. In $X = \text{Spec } \mathbb{Z}$, the closed sets are X, $\varnothing$, and the sets of the form $\{(p_1), \ldots, (p_s)\}$, i.e., the finite sets of prime numbers.

b. Let k be an algebraically closed field and $k[\mathsf{X}]$ be the corresponding polynomial ring. Then as was seen in Example 1.1.(iv), Spec $k[\mathsf{X}]$ can be written as $\{*\} \cup k$. Since any ideal of $k[\mathsf{X}]$ is either (0), (1), or (f), where $f \in k[\mathsf{X}]\backslash k$, the closed sets are $\{*\} \cup k$, $\varnothing$, and the sets of the form $\{\lambda_1, \ldots, \lambda_s\}$ where the λ_i are the roots of $f(x) = 0$ for such an f.

Note that in this case a union of finitely many closed sets $F_1, F_2, \ldots, F_r$ none of which is the whole space is not the whole space. In other words, if $U_1, \ldots, U_r$ are nonempty open sets, then $U_1 \cap \cdots \cap U_r$ is not empty.

Now, let $A = k[\mathsf{X}, \mathsf{Y}]$ as in Example 1.1.(v). Then

$$\text{Spec } A = \{*\} \cup \{(f) \mid f \in k[\mathsf{X}, \mathsf{Y}]\backslash k \text{ is irreducible}\} \cup k^2.$$

If we consider k^2 as the topological space with the topology induced from Spec A, then the closed sets are k^2, $\varnothing$, and finite unions of the finite sets and the sets of the form $\{(a, b) \in k^2 \mid \varphi(a, b) = 0\}$ for $\varphi \in k[\mathsf{X}, \mathsf{Y}]$. This is easily checked, since A is Noetherian.

c. Let $A = k[[\mathsf{X}]]$, the formal power series ring over a field k. Then Spec $A = \{(0), (\mathsf{X})\}$. The closed sets are $\{(0), (\mathsf{X})\}$, $\{(\mathsf{X})\}$, and $\varnothing$. Hence, the closure of (0) is the whole space. Spec A is a topological space with two points which is not discrete.

§1.3 Rings of Fractions, the Case A_f

Let A be a ring. For any element f of A, define A_f to be $A[\mathsf{X}]/(f\mathsf{X} - 1)$, where $A[\mathsf{X}]$ is the ring of polynomials over A. Letting $\psi_f(a) = a \bmod (f\mathsf{X} - 1)$ for $a \in A$, and $\xi = \mathsf{X} \bmod (f\mathsf{X} - 1)$, one has the ring homomorphism

$\psi_f\colon A \to A_f$ and ξ satisfies $\psi_f(f) \cdot \xi = 1$. Thus we denote ξ by $1/f$. A_f is then generated by $1/f$ as an A-algebra, i.e., $A_f = A[1/f]$. For simplicity, we write $a/1$ instead of $\psi_f(a)$.

Proposition 1.2.

(i) *$a/1 = b/1$ if and only if $f^n a = f^n b$ for some $n \geq 0$.*
(ii) *$A_f = 0$ if and only if f is a nilpotent element.*
(iii) (Universal mapping property). *For every ring homomorphism $\varphi\colon A \to B$ such that $\varphi(f)$ has a multiplicative inverse, there is a unique ring homomorphism $\varphi^{\#}\colon A_f \to B$ such that $\varphi^{\#} \circ \psi_f = \varphi$. (This means that among all pairs (B, φ) of rings B and homomorphisms $\varphi\colon A \to B$ such that $\varphi(f)$ is invertible, the pair (A_f, ψ_f) is the universal one.)*
(iv) *${}^a\psi_f\colon$ Spec $A_f \to$ Spec A is a homeomorphism onto $D(f)$, i.e., Spec $A_f \approx D(f)$.*

Proof. (i) It suffices to prove this for $b = 0$. $a/1 = 0$ if and only if $a \in (f\mathsf{X} - 1)$, i.e., there exist $n \geq 0$ and $b_0, \ldots, b_n \in A$ such that $a = (f\mathsf{X} - 1)(b_0 + b_1\mathsf{X} + \cdots + b_n\mathsf{X}^n)$, i.e.,

$$a = -b_0, fb_0 - b_1 = 0, \ldots, fb_{n-1} - b_n = 0, fb_n = 0. \qquad (*)$$

If $(*)$ holds, then $f^{n+1}a = -f^{n+1}b_0 = \cdots = fb_n = 0$. Conversely, if $f^{n+1}a = 0$, by taking $b_i = -f^i a$ for $0 \leq i \leq n$, then $(*)$ holds. So there exists $n \geq 0$ such that $f^{n+1}a = 0$ if and only if $(*)$ holds, which is equivalent to $a/1 = 0$.

(ii) This follows from (i) and the fact that $A_f = 0$ if and only if $1/1 = 1_{A_f} = 0$.

(iii) Define a ring homomorphism $\Phi\colon A[\mathsf{X}] \to B$ by $\Phi|_A = \varphi$ and $\Phi(\mathsf{X}) = 1/\varphi(f)$. Then $\Phi(f\mathsf{X} - 1) = \Phi(f)\Phi(\mathsf{X}) - 1 = \varphi(f) \cdot (1/\varphi(f)) - 1 = 0$; hence Φ determines a ring homomorphism $\varphi^{\#}\colon A_f \to B$ such that $\varphi^{\#}(a/f^r) = \varphi(a)/\varphi(f)^r$, i.e., $\varphi^{\#} \circ \psi_f = \varphi$. Since $A_f = A[1/f]$, any ring homomorphism $\varphi'\colon A_f \to B$ such that $\varphi' \circ \psi_f = \varphi$ becomes $\varphi^{\#}$.

(iv) The proof is left to the reader, since the proof of the more general case will be given in §1.4 (cf. Lemma 1.3). □

Corollary. *The kernel of ψ_f is the ideal $I(f) = \{a \in A \mid af^m = 0$ for some $m \geq 1\}$.*

Proposition 1.3. *Let $\mathfrak{a}$ be a proper ideal of a ring A. Then*

$$\sqrt{\mathfrak{a}} = \bigcap_{\mathfrak{p} \in V(\mathfrak{a})} \mathfrak{p}.$$

Proof. Since $\sqrt{\mathfrak{p}} = \mathfrak{p}$ for any prime ideal $\mathfrak{p}$, it follows that $\sqrt{\mathfrak{a}} \subseteq \bigcap_{\mathfrak{p} \in V(\mathfrak{a})} \mathfrak{p}$. Let $f \notin \sqrt{\mathfrak{a}}$. Then $a \equiv f \bmod \mathfrak{a}$ is not a nilpotent element of $A/\mathfrak{a}$; hence, Spec$(A/\mathfrak{a})_a$ is not empty by Lemma 1.1. By $V(\mathfrak{a}) \cap D(f) \approx D(a) \approx$ Spec$(A/\mathfrak{a})_a$, we have $\mathfrak{p}_1 \in V(\mathfrak{a}) \cap D(f)$; hence $\mathfrak{p}_1 \supseteq \mathfrak{a}$ and $f \notin \mathfrak{p}_1$. Thus $f \notin \bigcap_{\mathfrak{p} \in V(\mathfrak{a})} \mathfrak{p}$. □

Corollary. Let $\mathfrak{a}$ and $\mathfrak{b}$ be ideals of A.

(i) *$V(\mathfrak{a}) \subseteq V(\mathfrak{b})$ if and only if $\sqrt{\mathfrak{a}} \supseteq \sqrt{\mathfrak{b}}$.*
(ii) *$V(\mathfrak{a}) = V(\mathfrak{b})$ if and only if $\sqrt{\mathfrak{a}} = \sqrt{\mathfrak{b}}$.*
(iii) *$V(\mathfrak{a}) = \operatorname{Spec} A$ if and only if $\mathfrak{a} \subseteq \sqrt{(0_A)}$.*
(iv) *$D(f) = \varnothing$ if and only if f is nilpotent.*

PROOF. All the assertions follow immediately from Proposition 1.3. □

Proposition 1.4. *Let $\varphi: A \to B$ be a homomorphism of rings. Then for any ideal $\mathfrak{b}$ of B,*

(i) *The closure of ${}^a\varphi(V(\mathfrak{b}))$ is $V(\varphi^{-1}(\mathfrak{b}))$.*
(ii) *${}^a\varphi(\operatorname{Spec} B)$ is dense in $V(\operatorname{Ker} \varphi)$.*
(iii) *${}^a\varphi$ is dominating (i.e., ${}^a\varphi(\operatorname{Spec} B)$ is dense in $\operatorname{Spec} A$) if and only if $\operatorname{Ker} \varphi \subseteq \sqrt{(0_A)}$.*

PROOF. (i) If ${}^a\varphi(V(\mathfrak{b})) \subseteq V(f)$ for some $f \in A$, then any $\mathfrak{q} \in V(\mathfrak{b})$ satisfies ${}^a\varphi(\mathfrak{q}) = \varphi^{-1}(\mathfrak{q}) \ni f$; hence $\mathfrak{q} \ni \varphi(f)$. But since $\sqrt{\mathfrak{b}} = \bigcap_{\mathfrak{q} \in V(\mathfrak{b})} \mathfrak{q}$ by Proposition 1.3, one has $\sqrt{\mathfrak{b}} \ni \varphi(f)$ and so $\varphi^{-1}(\sqrt{\mathfrak{b}}) \ni f$; hence

$$V(f) \supseteq V(\varphi^{-1}(\sqrt{\mathfrak{b}})) = V(\varphi^{-1}(\mathfrak{b})).$$

This implies that the closure of ${}^a\varphi(V(\mathfrak{b}))$ includes $V(\varphi^{-1}(\mathfrak{b}))$. By Proposition 1.1.(ii), we obtain the assertion.

(ii) This follows immediately from (i).

(iii) By assertion (ii), ${}^a\varphi$ is dominating if and only if $V(\operatorname{Ker} \varphi) = \operatorname{Spec} A$. By Corollary (iii) to Proposition 1.3, $V(\operatorname{Ker} \varphi) = V(0_A)$ if and only if $\operatorname{Ker} \varphi \subseteq \sqrt{(0_A)}$. □

Corollary. *For any $f \in A$, $D(f)$ is dense in $\operatorname{Spec} A$ if and only if $I(f) \subseteq \sqrt{(0_A)}$.*

PROOF. Since ${}^a\psi_f: \operatorname{Spec} A_f \to \operatorname{Spec} A$ is a homeomorphism onto $D(f)$, we can apply Proposition 1.4. □

§1.4 Rings and Modules of Fractions

a. Let M be an A-module and S be a multiplicative subset of A (i.e., $1 \in S$, and $st \in S$ whenever $s, t \in S$). We want to construct the most general A-module N such that for any $s \in S$ and $b \in N$, $sx = b$ is solvable with $x \in N$.

On the Cartesian product $S \times M$, we define the following relation:

$$(s, m) \sim (s', m') \Leftrightarrow t(s'm - sm') = 0 \quad \text{for some } t \in S.$$

It is easy to check that this is an equivalence relation. (If we define $(s, m) \sim (s', m')$ as $s'm = sm'$, transitivity does not hold in general.) Writing $\frac{m}{s}$ or m/s for the equivalence class of (s, m), and defining addition by

$$\frac{m}{s} + \frac{m'}{s'} = \frac{s'm + sm'}{ss'},$$

the set of equivalence classes becomes an additive group. If we define scalar multiplication by an element of A by

$$a \cdot \frac{m}{s} = \frac{am}{s} \quad \text{for any } a \in A,$$

then this additive group becomes an A-module.

Definition. The A-module defined above is the *module of fractions* obtained from M and S, and denoted by $S^{-1}M$. In case $M = A$, the A-module $S^{-1}A$ can be made into a ring with multiplication defined by $(a/s)\cdot(a'/s') = (aa')/(ss')$, where $a, a' \in A$ and $s, s' \in S$.

Define $\psi_{A,S}\colon A \to S^{-1}A$ by $\psi_{A,S}(a) = a/1$. Then $\psi_{A,S}$ is a ring homomorphism and is called the *canonical homomorphism.* Clearly, $S^{-1}M$ can be regarded as an $S^{-1}A$-module.

For any ring homomorphism $\varphi\colon A \to B$ such that every element of $\varphi(S)$ is invertible, define $\Phi\colon S \times A \to B$ by $\Phi(s, a) = \varphi(a)/\varphi(s)$. Then if $(s, a) \sim (s', a')$, one has $\varphi(a)/\varphi(s) = \varphi(a')/\varphi(s')$; hence $\Phi(s, a) = \Phi(s', a')$. Thus Φ defines a ring homomorphism $\varphi^{\#}\colon S^{-1}A \to B$ such that $\Phi(s, a) = \varphi^{\#}(a/s)$. $\varphi^{\#}$ is uniquely determined by φ, since $\varphi^{\#}(a/1) = \varphi(a)$, i.e., $\varphi = \varphi^{\#} \circ \psi_{A,S}$. This means that $(S^{-1}A, \psi_{A,S})$ is universal among all pairs (B, φ) such that all elements of $\varphi(S)$ are invertible.

Furthermore, define the A-homomorphism $\psi_{M,S}\colon M \to S^{-1}M$ by

$$\psi_{M,S}(m) = m/1.$$

If N is an $S^{-1}A$-module and $\varphi\colon M \to N$ is an A-homomorphism, then there is a unique $S^{-1}A$-homomorphism $\varphi^{\#}\colon S^{-1}M \to N$ such that $\varphi = \varphi^{\#} \circ \psi_{M,S}$. $\varphi^{\#}$ is defined by $\varphi^{\#}(m/s) = \varphi(m)/s$ for any $m \in M$ and $s \in S$.

If $f \in A$, then $S_f = \{f^n \mid n \in \mathbf{N} \cup \{0\}\}$ is a multiplicative subset of A and $S_f^{-1}A$ enjoys the same universal mapping property as that of A_f stated in Proposition 1.4. Hence, $S_f^{-1}A \cong A_f$ by the map Ψ defined by $\Psi(a/f^n) = a\mathsf{X}^n \bmod (f\mathsf{X} - 1)$ for any a and $n \geq 0$.

Definition. Let A be a ring and let Q be the set of elements of A which are not zero divisors. Then Q is a multiplicative subset of A and $Q^{-1}A$ is said to be the *total quotient ring* of A. If A is an integral domain, $Q^{-1}A$ is said to be the *field of fractions* of A (or *quotient field* of A), denoted by $Q(A)$.

If S is a multiplicative subset of A contained in Q, then $\psi_{A,S}: A \to S^{-1}A$ is injective. Furthermore, $S^{-1}A$ is isomorphic to $\psi_{A,Q}(A)[1/s \mid s \in S]$, which is often denoted by $A[1/s \mid s \in S]$.

b. With the notation of subsection **a**, one has the following result.

Lemma 1.2

(i) *$S^{-1}A \otimes_A M \cong S^{-1}M$ both as A-modules and $S^{-1}A$-modules.*
(ii) *If $0 \to M' \to M \to M'' \to 0$ is an exact sequence of A-modules, then $0 \to S^{-1}M' \to S^{-1}M \to S^{-1}M'' \to 0$ is an exact sequence of $S^{-1}A$-modules.*
(iii) *If S_1 and S_2 are multiplicative subsets of A and if S_1' is the image of S_1 in $S_2^{-1}A$, then there is a natural isomorphism*

$$(S_1S_2)^{-1}M \cong S_1'^{-1}(S_2^{-1}M).$$

PROOF. (i). By the universal mapping property applied to the homomorphism

$$\varphi_1: M \to S^{-1}A \otimes_A M$$

defined by $\varphi_1(m) = 1 \otimes m$, there is an $S^{-1}A$-homomorphism

$$\varphi_1^{\#}: S^{-1}M \to S^{-1}A \otimes_A M$$

which satisfies $\varphi_1^{\#}(m/s) = (1/s) \otimes m$. By the universal mapping property of tensor products applied to a bilinear map

$$\varphi_2: S^{-1}A \times M \to S^{-1}M$$

defined by $\varphi_2(a, m) = am$, there is an A-homomorphism

$$\varphi_2^{\#}: S^{-1}A \otimes_A M \to S^{-1}M$$

defined by $\varphi_2^{\#}((a/s) \otimes m) = am/s$ for all $a/s \in S^{-1}A$ and $m \in M$. Since $\varphi_1^{\#} \circ \varphi_2^{\#} = \mathrm{id}$ and $\varphi_2^{\#} \circ \varphi_1^{\#} = \mathrm{id}$, these maps are A-isomorphisms between $S^{-1}A \otimes_A M$ and $S^{-1}M$. $\varphi_1^{\#}$ and $\varphi_2^{\#}$ are $S^{-1}A$-homomorphisms and so the two modules are $S^{-1}A$-isomorphic.

The proofs of assertions (ii) and (iii) are left to the reader. □

EXAMPLE 1.3. $S^{-1}A \otimes_A S^{-1}A \cong S^{-1}A$. In particular, $\mathbb{Q} \otimes_{\mathbb{Z}} \mathbb{Q} \cong \mathbb{Q}$.

Lemma 1.3. *The associated map* ${}^a\psi_{A,S}$: Spec $S^{-1}A \to$ Spec A *is a homeomorphism onto the subspace* $\sum_S = \{\mathfrak{p} \mid \mathfrak{p} \cap S = \varnothing\}$ *of* Spec A.

PROOF. We use the following notation: $B = S^{-1}A$, $\mathfrak{p} \in \operatorname{Spec} A$; $a, a_1, a_2 \in A$; $s, s_1, s_2, t \in S$; $p \in \mathfrak{p}$. We first show that if $\mathfrak{p} \in \sum_S$, then $\mathfrak{p}B = \{p/s \mid p \in \mathfrak{p}, s \in S\}$ is a prime ideal. In fact, supposing that $a_1/s_1 \cdot a_2/s_2 = p/s \in \mathfrak{p}B$, we have $tsa_1a_2 = tps_1s_2$ for some t. Since $\mathfrak{p} \cap S = \varnothing$, it follows that $a_1 a_2 \in \mathfrak{p}$; hence either a_1 or a_2 is in $\mathfrak{p}$, i.e., a_1/s_1 or a_2/s_2 is in $\mathfrak{p}B$. A similar argument shows that $\mathfrak{p}B \neq B$ and so $\mathfrak{p}B \in \operatorname{Spec} B$.

Define $h: \sum_S \to \operatorname{Spec} B$ by $h(\mathfrak{p}) = \mathfrak{p}B$. Then ${}^a\psi_{A,S} \circ h = \mathrm{id}$ and $h \circ {}^a\psi_{A,S} = \mathrm{id}$. In fact, the first assertion is equivalent to $\psi_{A,S}^{-1}(\mathfrak{p}B) = \mathfrak{p}$. Clearly, $\mathfrak{p} \subseteq \psi_{A,S}^{-1}(\mathfrak{p}B)$. On the other hand, if $a/1 = p/s$, then there is t with $tas = tp \in \mathfrak{p}$; hence $a \in \mathfrak{p}$, and the assertion is proved. The second assertion is proved similarly. By Proposition 1.1.(i), ${}^a\psi_{A,S}$ is continuous. Since $h^{-1}(D(a/s)) = {}^a\psi_{A,S}(D(a/s)) = D(a) \cap \sum_S$ for all a and s, h is also continuous, i.e., ${}^a\psi_{A,S}: \operatorname{Spec} S^{-1}A \to \sum_S$ is a homeomorphism. □

Example 1.4. ${}^a\psi_f: \operatorname{Spec} A_f \to D(f)$ is a homeomorphism, since $A_f \cong S_f^{-1}A$.

Remark. $\sum_S$ is the intersection of all $D(s)$, $s \in S$.

c. If $\mathfrak{p}$ is a prime ideal of A, then $S = A \backslash \mathfrak{p}$ is a multiplicative subset.

Definition. We define $A_{\mathfrak{p}}$ to be $S^{-1}A$, which is called the localization of A (or the local ring of Spec A) at $\mathfrak{p}$.

By Lemma 1.3, $\operatorname{Spec} A_{\mathfrak{p}} \approx \{\mathfrak{q} \in \operatorname{Spec} A \mid \mathfrak{q} \subseteq \mathfrak{p}\}$. Thus $A_{\mathfrak{p}}$ has a unique maximal ideal $\mathfrak{p}A_{\mathfrak{p}}$; hence $A_{\mathfrak{p}}$ is a local ring.

Definition. The field $A_{\mathfrak{p}}/\mathfrak{p}A_{\mathfrak{p}}$ is said to be the *residue class field* of Spec A at $\mathfrak{p}$, denoted $k(\mathfrak{p})$.

Note that $k(\mathfrak{p})$ is the field of fractions of $A/\mathfrak{p}$, i.e., $k(\mathfrak{p}) = Q(A/\mathfrak{p})$.

Definition. For any $f \in A$, $f/1 \bmod \mathfrak{p}A_{\mathfrak{p}}$ is denoted by $\bar{f}(\mathfrak{p})$. We call $\mathfrak{p}$ a *zero* of f, if $\bar{f}(\mathfrak{p}) = 0$, i.e., $\mathfrak{p} \ni f$.

Example 1.5. Let $k[\mathsf{X}_1, \ldots, \mathsf{X}_n]$ be a polynomial ring over a field k and $\mathfrak{p} = (\mathsf{X}_1 - \alpha_1, \ldots, \mathsf{X}_n - \alpha_n)$, where the $\alpha_i \in k$. Then $k \cong k(\mathfrak{p})$, and for a polynomial f, $\bar{f}(\mathfrak{p}) = f(\alpha_1, \ldots, \alpha_n)$ with the natural identification $k = k(\mathfrak{p})$. Note that with this identification, k can be regarded as a $k[\mathsf{X}_1, \ldots, \mathsf{X}_n]$-module. For example, one has

$$\mathsf{X}_i \cdot 1_k = \alpha_i \quad \text{for all } i.$$

Example 1.6. Let $f_1, \ldots, f_r$ be elements of a ring A. If $f_1, \ldots, f_r$ do not have common zeros, then there exist $g_1, \ldots, g_r \in A$ such that $1 = g_1 f_1 + \cdots + g_r f_r$. This is the same situation as in Example 1.2, since $\bar{f}(\mathfrak{p}) = 0$ is equivalent to $f \in \mathfrak{p}$.

Remark. $V(f)$ is the set of zeros of f, i.e., $V(f) = \{\mathfrak{p} \in \operatorname{Spec} A \mid \bar{f}(\mathfrak{p}) = 0\}$. Equivalently, $D(f) = \{\mathfrak{p} \in \operatorname{Spec} A \mid \bar{f}(\mathfrak{p}) \neq 0\}$.

Example 1.7. Let M be an A-module and let $x \in M$. Suppose that $x/1 = 0$ in $M \otimes_A A_{\mathfrak{m}}$ for all maximal ideals $\mathfrak{m}$ of A. Then define $\mathfrak{a}$ to be $\{a \in A \mid ax = 0\}$. If $\mathfrak{a} \neq A$, $\mathfrak{a}$ would be contained in a maximal ideal $\mathfrak{m}$ by Lemma 1.1. By hypothesis, $bx = 0$ for some $b \notin \mathfrak{m}$; hence $b \in \mathfrak{a} \subseteq \mathfrak{m}$, a contradiction. Thus $\mathfrak{a} = A$; i.e., $x = 0$.

Now let A be an integral domain. Then $A \subseteq A_{\mathfrak{m}} \subseteq Q(A)$ for any maximal ideal $\mathfrak{m}$ of A, and hence $A \subseteq \bigcap_{\mathfrak{m}} A_{\mathfrak{m}}$. $M = (\bigcap_{\mathfrak{m}} A_{\mathfrak{m}})/A$ satisfies $M \otimes_A A_{\mathfrak{m}} = 0$ for any $\mathfrak{m}$ and so $M = 0$. Thus $A = \bigcap_{\mathfrak{m}} A_{\mathfrak{m}}$.

§1.5 Nullstellensatz

a. Let E be a subset of Spec A and let $f \in A$. By $\bar{f}|_E = 0$ we mean that $\bar{f}(\mathfrak{p}) = 0$ for any $\mathfrak{p} \in E$. For example, $\bar{f}|_{V(f)} = 0$. The following result is an abstract form of the Nullstellensatz.

Theorem 1.1. *Let $\mathfrak{a}$ be an ideal of A and let $f \in A$. Then*

$$\bar{f}|_{V(\mathfrak{a})} = 0 \quad \textit{if and only if} \quad f \in \sqrt{\mathfrak{a}}.$$

Proof. $\bar{f}|_{V(\mathfrak{a})} = 0 \Leftrightarrow \bar{f}(\mathfrak{p}) = 0$ for all $\mathfrak{p} \in V(\mathfrak{a}) \Leftrightarrow f/1 \in \mathfrak{p}A_{\mathfrak{p}}$ for all $\mathfrak{p} \supseteq \mathfrak{a} \Leftrightarrow f \in \bigcap_{\mathfrak{p} \in V(\mathfrak{a})} \mathfrak{p} \Leftrightarrow f \in \sqrt{\mathfrak{a}}$ by Proposition 1.3. □

b. Definition. A (not necessarily Hausdorff) topological space X is said to be *quasi-compact*, if every open cover of X admits a finite subcover.

Theorem 1.2. *Let $f \in A$. Then $D(f)$ is a quasi-compact space.*

Proof. Since $D(f) \approx$ Spec A_f by Proposition 1.2.(iv), it suffices to prove the result for $X =$ Spec A. Suppose that $\{D(f_\lambda) \mid \lambda \in \Lambda\}$ is an open cover of X. Letting $\mathfrak{a}$ be the ideal generated by $\{f_\lambda \mid \lambda \in \Lambda\}$, we have $X = \bigcup_{\lambda \in \Lambda} D(f_\lambda) = D(\mathfrak{a})$; hence $V(\mathfrak{a}) = \varnothing$. From Corollary (ii) to Lemma 1.1, it follows that $\mathfrak{a} \ni 1$, i.e., $1 = \sum_{i=1}^{s} g_i f_{\lambda_i}$ for some $g_i \in A$. Then $X = D(f_{\lambda_1}) \cup \cdots \cup D(f_{\lambda_s})$. Since $\{D(f) \mid f \in A\}$ is an open base, we obtain the result. □

c. Definition. A topological space X is said to be a T_0-*space* if at least one member of any pair of distinct points has an open neighborhood not containing the other point.

Proposition 1.5. Spec *A is a T_0-space.*

Proof. Let $\{\mathfrak{p}, \mathfrak{q}\}$ be a pair of distinct points. Then as subsets of A, $\mathfrak{p}\backslash\mathfrak{q} \neq \varnothing$ or $\mathfrak{q}\backslash\mathfrak{p} \neq \varnothing$. If $\mathfrak{p}\backslash\mathfrak{q} \neq \varnothing$, then for any $f \in \mathfrak{p}\backslash\mathfrak{q}$, one has $\mathfrak{p} \notin D(f)$ and $\mathfrak{q} \in D(f)$. The other case is similar. □

§1.6 Irreducible Spaces

a. Definition. A topological space X is said to be *irreducible* if it cannot be written as a finite union of closed proper subsets of X. A subset V of X is called *irreducible* if the topology on V induced by that of X is irreducible.

Definition. A maximal irreducible subset of X is called an *irreducible component* of X.

The following properties are easily proved.

Proposition 1.6. *Let X be a topological space and let V be a subset of X.*

(i) *If V is irreducible, then so is the closure of V.*
(ii) *The closure $\overline{\{x\}}$ of a point x is irreducible.*
(iii) *X is irreducible if and only if every nonempty open subset is dense.*
(iv) *Every irreducible subset of X is contained in an irreducible component of X.*
(v) *The irreducible components of X are closed in X and together form a closed cover of X.*

Proof. The proofs of all the statements except (iv) are easy and left to the reader. One can prove (iv) by Zorn's Lemma. □

Definition. If $\{X_i | i \in I\}$ is the set of irreducible components of X, then $X = \bigcup_{i \in I} X_i$ is called the *irreducible decomposition* of X.

Definition. A topological space is said to be *Noetherian* if every descending chain $F_1 \supseteq F_2 \supseteq \cdots$ of closed subsets stabilizes.

Clearly, spectra of Noetherian rings are Noetherian spaces. Furthermore, every closed subspace of a Noetherian space is also Noetherian.

Lemma 1.4. *A Noetherian space X has a finite number of irreducible components.*

Proof. Suppose that X is a Noetherian space with infinitely many irreducible components. Then the set $\mathfrak{F}$ of closed subspaces of X with infinitely many irreducible components is nonempty. Since X is Noetherian, $\mathfrak{F}$ has a minimal element V. Clearly, V is reducible, say $V = W_1 \cup W_2$ where both W_1 and W_2 are closed proper subsets of V. By the choice of V, the W_i have a finite number of irreducible components, i.e., $W_i = \bigcup_{j=1}^{a(i)} W_{ij}$. It is easy to see that all the W_{ij} are irreducible closed subsets of V and $V = \bigcup_{i,j} W_{ij}$; hence V has a finite number of irreducible components. Hence, $V \notin \mathfrak{F}$, which is absurd. □

b. Definition. Let X be a topological space. If $X = \overline{\{x\}}$ for some $x \in X$, then X is irreducible and x is said to be a *generic point* of X. If $\{x\}$ is a closed subset, then x is said to be a *closed point* of X.

Lemma 1.5. *Let X be a T_0-space and let V be a subset of X.*

(i) *If V is nonempty, quasi-compact, and closed, then V contains a closed point.*
(ii) *If $V = \overline{\{x\}}$ for some x, then x is the unique generic point of V.*

PROOF. (i) Let $\mathfrak{F} = \{W | W$ is a closed and nonempty subset of $V\}$. Clearly, $V \in \mathfrak{F}$. Define the order $\leq$ in $\mathfrak{F}$ by $W_1 \leq W_2$ if and only if $W_1 \supseteq W_2$. We claim that the ordered set $\mathfrak{F}$ is inductively ordered. In fact, let $\{W_\lambda | \lambda \in \Lambda\}$ be a linearly ordered subset. The set $K = \bigcap_{\lambda \in \Lambda} W_\lambda$ is closed, and nonempty. To prove the last assertion, assume $K = \varnothing$. Then, since V is quasi-compact, there are $\lambda_1, \ldots, \lambda_m \in \Lambda$ such that $W_{\lambda_1} \cap \cdots \cap W_{\lambda_m} = \varnothing$. Take $W_\mu = \max\{W_{\lambda_1}, \ldots, W_{\lambda_m}\}$. Then $W_\mu \subseteq \bigcap_i W_{\lambda_i} = \varnothing$, which contradicts the hypothesis $W_\mu \in \mathfrak{F}$.

Thus by Zorn's Lemma, $\mathfrak{F}$ has a maximal element P. If P is not a set consisting of a single point, there exist distinct points p, q of P, and so there is an open subset U of X such that $p \in U$ and $q \notin U$ (or $p \notin U$ and $q \in U$), since X is a T_0-space. But then $W = P \cap (V \backslash U) \in \mathfrak{F}$ and $W \subset P$. This contradicts the maximality of P. Thus P is a closed point.

(ii) By the definition of T_0-space, at least one member of any pair of distinct points is not contained in the closure of the other point. Therefore, we have $x = y$, whenever $\overline{\{x\}} = \overline{\{y\}}$. □

Proposition 1.7. *Suppose that a topological space X has a finite number of irreducible components $X_1, \ldots, X_m$ and that each X_i has a generic point x_i. Then an open subset U of X is dense if and only if U contains all the x_i.*

PROOF. If an open subset U_0 does not contain x_i, then $x_i \in X \backslash U_0$ and so $X_i = \overline{\{x_i\}} \subseteq X \backslash U$. Since X has a finite number of irreducible components, $X'_i = \bigcup_{j \neq i} X_j$ is closed and $X = X_i \cup X'_i$. Hence, $U_0 \subseteq X \backslash X_i \subseteq X'_i \neq X$; i.e., U_0 is not dense. Thus if U is dense, then U contains all the x_i. The converse is obvious. □

c. Let A be a ring. Then $X = \operatorname{Spec} A$ is a T_0-space, by Proposition 1.5.

For $\mathfrak{p} \in X$, $\overline{\{\mathfrak{p}\}}$ is the intersection of all $V(f)$ such that $\mathfrak{p} \in V(f)$, i.e., $f \in \mathfrak{p}$. Hence, $\overline{\{\mathfrak{p}\}} = V(\sum_{f \in \mathfrak{p}} fA) = V(\mathfrak{p})$. Therefore, $V(\mathfrak{p})$ is irreducible. In particular, $\mathfrak{p}$ is a closed point if and only if $\mathfrak{p}$ is a maximal ideal. If $V(\mathfrak{p})$ is an irreducible component of X, then $\mathfrak{p}$ is a minimal prime ideal of A. The converse is also true.

Proposition 1.8. *$V(\mathfrak{a})$ is irreducible if and only if $\sqrt{\mathfrak{a}}$ is a prime ideal. In this case, $\sqrt{\mathfrak{a}}$ is the generic point of $V(\mathfrak{a})$.*

PROOF. We may assume $\mathfrak{a}$ is proper. Replacing A by $A/\mathfrak{a}$, we can assume $\mathfrak{a} = (0_A)$. For any f, $g \in A$ with $fg \in \sqrt{(0_A)}$, one has $X = V(fg) = V(f) \cup V(g)$. If X is irreducible, then $X = V(f)$ or $X = V(g)$. In the first case, by Corollary (iii) to Proposition 1.3, one has $f \in \sqrt{(0_A)}$. In the second case, one also has $g \in \sqrt{(0_A)}$. This implies that $\sqrt{(0_A)}$ is prime. The converse is obvious. □

d. Since $\mathbb{C}$ is a subspace of Spec $\mathbb{C}[\mathsf{X}]$ (cf. Example 1.1.(iv)), we regard $\mathbb{C}$ as a topological subspace of Spec $\mathbb{C}[\mathsf{X}]$. Every point of $\mathbb{C}$ is closed; hence the irreducible space $\mathbb{C}$ has no generic point. In order to introduce a generic point of $\mathbb{C}$, we have to define a weaker topology on $\mathbb{C}$. We take as an open base for the new topology $\{D(f) \cap \mathbb{C} \mid f \in \mathbb{Q}[\mathsf{X}]\}$. This topology is called the $\mathbb{Q}$-topology on $\mathbb{C}$. For example, the closure of $\{\sqrt{2}\}$ in the $\mathbb{Q}$-topology is $\{\sqrt{2}, -\sqrt{2}\}$, which has two generic points. In general, if α is an algebraic number, then $\overline{\{\alpha\}}$ is the set of all conjugates of α over $\mathbb{Q}$. If α is a transcendental number, then α is a generic point of $\mathbb{C}$. Note that $\mathbb{C}$ with the $\mathbb{Q}$-topology is not a T_0-space.

§1.7 Integral Extension of Rings

a. Definition. Let B be a subring of a ring A. An element of A is said to be *integral* over B if it is a root of a monic polynomial with coefficients in B. If every element of A is integral over B, then we say that A is *integral over* B, or A/B is an integral extension of rings.

Proposition 1.9. *Let α be an element of A. The following properties are equivalent:*

(i) *α is integral over B.*
(ii) *$B[\alpha]$ is finitely generated as a B-module, where $B[\alpha]$ is the subring of A generated by B and α.*
(iii) *There exists a subring B_1 of A containing B and α which is finitely generated as a B-module.*

PROOF. (i) ⇒ (ii) Since α is integral, there is a monic polynomial $F(\mathsf{X}) = \mathsf{X}^n + b_1\mathsf{X}^{n-1} + \cdots + b_n$, with $b_i \in B$ such that $F(\alpha) = 0$. It is easy to check that $B[\alpha] = \sum_{i=0}^{n-1} B\alpha^i$.

(ii) ⇒ (iii) Obvious.

(iii) $\Rightarrow$ (i) By hypothesis, there exist $b_1, \ldots, b_n \in B_1$ such that $B_1 = \sum_{i=1}^n Bb_i$. Since $\alpha b_i \in B_1$ for $1 \leq i \leq n$, we have $\beta_{ij} \in B$ such that

$$\alpha b_i = \sum_{j=1}^{n} \beta_{ij} b_j \quad \text{for} \quad 1 \leq i \leq n.$$

Hence, letting 1_n be the identity matrix, $[\beta_{ij}]$ be the matrix whose components are $\beta_{i,j}$ for $1 \leq i, j \leq n$, and $[b_i]$ be the vector with components b_i for $1 \leq i \leq n$, we can write

$$(\alpha \cdot 1_n - [\beta_{ij}]) \cdot [b_i] = 0.$$

Putting $F(\mathsf{X}) = \det(\mathsf{X} \cdot 1_n - [\beta_{ij}])$, we have $F(\alpha)b_i = 0$ for all i; hence $F(\alpha) = 0$. Since $F(\mathsf{X})$ is a monic polynomial with coefficients in B, α is integral over B. □

Corollary 1. *If $\alpha_1, \ldots, \alpha_n$ are integral over B, then $B[\alpha_1, \ldots, \alpha_n]$ is finitely generated as a B-module and is integral over B.*

Proof. If $n = 1$, this is the implication (i) $\Rightarrow$ (ii) of Proposition 1.9. Assuming the corollary holds for $n - 1$, $B_1 = B[\alpha_1, \ldots, \alpha_{n-1}]$ is finitely generated and so $B_1 = \sum_{i=1}^s B\beta_i$ for some β_i. Since α_n is integral, one has $B[\alpha_n] = \sum_{j=0}^{m-1} B\alpha_n^j$ for some m. Hence, $B[\alpha_1, \ldots, \alpha_n] = B_1[\alpha_n] = \sum_{j=0}^{m-1} B_1 \alpha_n^j = \sum_{i,j} B\beta_i \alpha_n^j$, which is finitely generated as a B-module. □

Corollary 2. *The subset $B'_A = \{\alpha \in A \mid \alpha$ is integral over $B\}$ is a subring of A.*

Proof. Take α and $\beta \in B'_A$. By the above corollary, $B[\alpha, \beta]$ is integral over B. Since $\alpha \pm \beta, \alpha\beta \in B[\alpha, \beta]$, these are integral over B, i.e., $\alpha \pm \beta, \alpha\beta \in B'_A$. □

Corollary 3. *Let C be a subring of B. If A is integral over B and if B is integral over C, then A is integral over C.*

Proof. Take an arbitrary element α of A. By hypothesis, α is integral over B, i.e., there exist $b_1, \ldots, b_n \in B$ such that $\alpha^n + b_1\alpha^{n-1} + \cdots + b_n = 0$. Now α is integral over a subring $C_1 = C[b_1, \ldots, b_n]$ which is finitely generated as a C-module by Corollary 1. Hence, $C_1[\alpha]$ is finitely generated as a C_1-module; so is finitely generated as a C-module as well. This implies that α is integral over C by the implication (ii) $\Rightarrow$ (i) of Proposition 1.9. □

Corollary 4. *Any element α of A integral over B'_A belongs to B'_A.*

Proof. This follows immediately from Corollary 3. □

b. The next two lemmas will be used in the proof of Noether's Normalization Theorem (Theorem 1.3).

Lemma 1.6 (Nagata). *Let A be a polynomial ring $k[\mathsf{X}_1, \ldots, \mathsf{X}_n]$ over a field k and let F be a nonconstant polynomial. Then there exist polynomials $Y_2 = \mathsf{X}_2 - \mathsf{X}_1^{m_2}, \ldots, Y_n = \mathsf{X}_n - \mathsf{X}_1^{m_n}$, where the m_i are positive integers, such that A is integral over a subring $B = k[F, Y_2, \ldots, Y_n]$.*

PROOF. It suffices to show that X_1 is integral over B for appropriately chosen $m_2, \ldots, m_n$, since $A = B[\mathsf{X}_1]$ by Corollary 2 of Proposition 1.9. Let $m_2, \ldots, m_n$ be positive integers and define $Y_2 = \mathsf{X}_2 - \mathsf{X}_1^{m_2}, \ldots, Y_n = \mathsf{X}_n - \mathsf{X}_1^{m_n}$. If one puts

$$H(\mathsf{T}) = F(\mathsf{T}, Y_2 + \mathsf{T}^{m_2}, \ldots, Y_n + \mathsf{T}^{m_n}) - F \in B[\mathsf{T}],$$

where $B[\mathsf{T}]$ is the polynomial ring over B, then one has $H(\mathsf{X}_1) = 0$. Thus it suffices to show that the m_i can be chosen in such a way that the leading coefficient of H is independent of the Y_i for $2 \leq i \leq n$.

To do this, let $m_1 = 1$ and let δ be the total degree of F. The polynomial F is written as a linear combination of monomials $M_\alpha(\mathsf{X}_1, \ldots, \mathsf{X}_n) = \mathsf{X}_1^{\alpha_1} \cdots \mathsf{X}_n^{\alpha_n}$, where $\alpha = (\alpha_1, \ldots, \alpha_n)$ with $\alpha_i \geq 0$, i.e., $F = \sum_{|\alpha| \leq \delta} a_\alpha M_\alpha(\mathsf{X}_1, \ldots, \mathsf{X}_n)$, with $|\alpha| = \alpha_1 + \cdots + \alpha_n$ and $a_\alpha \in k$. Hence,

$$H(\mathsf{T}) = \sum_\alpha a_\alpha M_\alpha(\mathsf{T}^{m_1}, Y_2 + \mathsf{T}^{m_2}, \ldots, Y_n + \mathsf{T}^{m_n}) = \sum_\alpha a_\alpha(\mathsf{T}^{\sum_{i=1}^n \alpha_i m_i} + H_\alpha(\mathsf{T})),$$

where the $H_\alpha(\mathsf{T})$ are polynomials with $\deg_{\mathsf{T}} H_\alpha(\mathsf{T}) < \sum_{i=1}^n \alpha_i m_i$. Thus one needs only to choose the m_i so that the sums $\sum_{i=1}^n \alpha_i m_i$ for α satisfying $|\alpha| \leq \delta$ are all different; for example, one can take $m_i = (\delta + 1)^{i-1}$ for $i = 2, \ldots, n$. □

Lemma 1.7. *Let B be a subring of a ring A and let $\mathfrak{a}$ be an ideal of A, so that $B/(\mathfrak{a} \cap B)$ can be identified with a subring of $A/\mathfrak{a}$. If A is integral over B, then so is $A/\mathfrak{a}$ over $B/(\mathfrak{a} \cap B)$.*

PROOF. Obvious. □

Theorem 1.3 (Noether's Normalization Theorem). *Let A be an integral domain finitely generated over a field k, i.e., $A = k[x_1, \ldots, x_n]$. Then there exist elements $y_1, \ldots, y_r$ of A such that*

(i) a subring $B = k[y_1, \ldots, y_r]$ is isomorphic as a k-algebra to the ring of polynomials in r variables over k, i.e., $y_1, \ldots, y_r$ are algebraically independent over k, and
(ii) A is integral over B.

PROOF. We prove this by induction on n. The case $n = 0$ is trivial. There is a surjective homomorphism φ of a polynomial ring $R = k[\mathsf{X}_1, \ldots, \mathsf{X}_n]$ onto A such that $\varphi(\mathsf{X}_i) = x_i$. Since A is an integral domain, $\mathfrak{p} = \operatorname{Ker} \varphi$ is a prime ideal. Since the result is trivial for $\mathfrak{p} = 0$, we may suppose that there is an element $F \in \mathfrak{p} \backslash (0)$. Since F is necessarily nonconstant, Lemma 1.6 shows that there are polynomials $Y_2, \ldots, Y_n$ such that $R = k[\mathsf{X}_1, \ldots, \mathsf{X}_n]$ is integral over $S = k[F, Y_2, \ldots, Y_n]$. Then $A = R/\mathfrak{p}$ is integral over $A_1 = S/(\mathfrak{p} \cap S)$ by Lemma 1.7. Since $F \in \mathfrak{p} \cap S$, the subring A_1 is generated by Y_i mod $\mathfrak{p} \cap S$, $i = 2, \ldots, n$. By the induction hypothesis, there exist elements $y_1, \ldots, y_r \in A_1$ such that $y_1, \ldots, y_r$ are algebraically independent over k and A_1 is integral over $B = k[y_1, \ldots, y_r]$. Since A is integral over A_1, A is also integral over B by Corollary 3 to Proposition 1.9. □

Remark. In Theorem 1.3, the set $\{y_1, \ldots, y_r\}$ is a transcendence basis of $Q(A)$ over k.

§1.8 Hilbert Nullstellensatz

a. In Example 1.1.(v), we found that if k is algebraically closed, then the maximal ideals of $k[\mathsf{X}, \mathsf{Y}]$ are of the form $(\mathsf{X} - \alpha, \mathsf{Y} - \beta)$. This suggests that the "classical" points of Spec A might be the maximal ideals of A. To substantiate this, we begin with the following result.

Theorem 1.4 (Weak Hilbert Nullstellensatz). *Let R be a ring finitely generated over a field k. If $\mathfrak{m}$ is a maximal ideal of R, then $R/\mathfrak{m}$ is a field algebraic over k.*

PROOF. Since $A = R/\mathfrak{m}$ is an integral domain finitely generated over a field k, by Theorem 1.3, there exist $y_1, \ldots., y_r \in A$ such that both conclusions (i) and (ii) of Theorem 1.3 hold. If $r > 0$, then $\eta = y_1^{-1} \in A$, since A is a field. Hence, η is integral over a polynomial ring $k[y_1, \ldots, y_r]$, i.e.,

$$\eta^m + F_1\eta^{m-1} + \cdots + F_m = 0$$

for some $F_i \in k[y_1, \ldots, y_r]$. But then

$$1 + F_1 y_1 + \cdots + F_m y_1^m = 0,$$

which contradicts the fact that $y_1, \ldots, y_r$ are algebraically independent. Therefore, $r = 0$ and so A is algebraic over k. □

If, in addition, k is algebraically closed, the natural map $\psi: k \to R/\mathfrak{m}$ is bijective. Letting $x_1, \ldots, x_n$ be generators of R, we have $\alpha_i = \psi^{-1}$ (x_i mod $\mathfrak{m}$) for all i. Then $x_i - \alpha_i \in \mathfrak{m}$ and so $(x_1 - \alpha_1, \ldots, x_n - \alpha_n) \subseteq \mathfrak{m}$. Since the ideal $(x_1 - \alpha_1, \ldots, x_n - \alpha_n)$ is maximal, we conclude that $\mathfrak{m} = (x_1 - \alpha_1, \ldots, x_n - \alpha_n)$. Hence $k(\mathfrak{m}) \cong k$. With this identification, for any $\varphi(x_1, \ldots, x_n) \in R$, we have $\bar{\varphi}(\mathfrak{m}) = \varphi(\alpha_1, \ldots, \alpha_n)$.

b. Definition. Let A be a k-algebra and let $\mathfrak{p} \in$ Spec A. $\mathfrak{p}$ is said to be *k-algebraic*, if $k(\mathfrak{p})$ is algebraic over k.

Since $k(\mathfrak{p}) = Q(A/\mathfrak{p}) \supseteq A/\mathfrak{p} \supseteq k$, it follows that if $\mathfrak{p}$ is k-algebraic, then $A/\mathfrak{p}$ is a field, i.e., $\mathfrak{p}$ is the closed point of Spec A, by the next lemma.

Lemma 1.8. *Let A be an integral domain which is integral over a subring B. Then A is a field if and only if B is a field.*

PROOF. Suppose first that A is a field and $x \in B\backslash(0)$. Since A is a field, one has $1/x \in A$ and so

$$\left(\frac{1}{x}\right)^n + b_1\left(\frac{1}{x}\right)^{n-1} + \cdots + b_n = 0$$

for some $b_i \in B$. From this equation, one has

$$\frac{1}{x} = -b_1 - b_2 x - \cdots - b_n x^{n-1} \in B.$$

Hence, B is a field. The converse is proved similarly. □

Theorem 1.5. *If A is a ring finitely generated over a field k, then the set of all closed points is dense in* Spec A.

PROOF. Since $\{D(f) | f \in A\}$ is an open base of the topology on Spec A, it suffices to prove that any nonempty $D(f)$ contains a closed point of Spec A. Since $A_f \neq 0$, Spec A_f has a closed point y, which is k-algebraic by Theorem 1.4. In fact, $A_f = A[1/f]$ is finitely generated over k. Letting ${}^a\psi_f(y) = x$, one clearly has $k(x) \cong k(y)$. Thus x belongs to $D(f)$ and is k-algebraic. Hence, x is a closed point of Spec A. □

EXAMPLE 1.8. Let R be a localization of the polynomial ring $k[X_1, \ldots, X_n]$ at $\mathfrak{m} = (X_1, \ldots, X_n)$. Then an open subset $V_0 = D(X_1) \cup \cdots \cup D(X_n) =$ Spec $R\backslash\{\mathfrak{m}\}$ is quasi-compact, and thus contains closed points of V_0 by Lemma 1.5.(ii). But these are not closed points of Spec R if $n > 0$.

c. Definition. Let $\mathfrak{a}$ be an ideal of a ring A. The intersection of all maximal ideals containing $\mathfrak{a}$ is denoted by $\mathfrak{R}(\mathfrak{a})$. The *Jacobson radical* of A is defined to be $\mathfrak{R}((0_A))$.

The next result is the Nullstellensatz of Hilbert.

Theorem 1.6 (*Hilbert Nullstellensatz*). *Let A be a ring finitely generated over a field k and let $\mathfrak{a}$ be a proper ideal of A. Then $\sqrt{\mathfrak{a}} = \mathfrak{R}(\mathfrak{a})$.*

PROOF. We use the same notation as in the proof of Proposition 1.3. By Theorem 1.5, $D(a)$ contains closed points of Spec $A/\mathfrak{a}$; hence there exists a maximal ideal $\mathfrak{m}_1$ containing $\mathfrak{a}$ such that $f \notin \mathfrak{m}_1$. □

d. Let k be an algebraically closed field. Then the closed points of Spec $k[\mathsf{X}_1, \ldots, \mathsf{X}_n]$ are the maximal ideals of the form $(\mathsf{X}_1 - \alpha_1, \ldots, \mathsf{X}_n - \alpha_n)$. Thus, we identify k^n with the set of all closed points, i.e., $(\alpha_1, \ldots, \alpha_n) \leftrightarrow (\mathsf{X}_1 - \alpha_1, \ldots, \mathsf{X}_n - \alpha_n)$. For an ideal I, we put $\mathbf{V}(I) = V(I) \cap k^n$, which consists of the points $p = (\alpha_1, \ldots, \alpha_n)$ such that $\varphi(\alpha_1, \ldots, \alpha_n) = 0$ for all $\varphi \in I$. The condition $p = (\alpha_1, \ldots, \alpha_n) \in \mathbf{V}(I)$ is equivalent to $(\mathsf{X}_1 - \alpha_1, \ldots, \mathsf{X}_n - \alpha_n) \supseteq I$; hence,

$$\mathfrak{N}(I) = \bigcap_{(\alpha_1, \ldots, \alpha_n) \in \mathbf{V}(I)} (\mathsf{X}_1 - \alpha_1, \ldots, \mathsf{X}_n - \alpha_n).$$

In other words, $\psi \in \mathfrak{N}(I)$ if and only if $\bar{\psi}|_{\mathbf{V}(I)} = 0$. By Theorem 1.6, $\sqrt{I} = \mathfrak{N}(I)$. Thus, $\bar{\psi}|_{\mathbf{V}(I)} = 0$ *if and only if* $\psi \in \sqrt{I}$. This is the classical result due to Hilbert.

Proposition 1.10. *There is a one-to-one correspondence* Ψ *between the closed subsets of* $X =$ Spec $k[\mathsf{X}_1, \ldots, \mathsf{X}_n]$ *and the closed subsets of* k^n *defined by* $\Psi(F) = F \cap k^n$, *if* k *is an algebraically closed field.*

PROOF. A closed subset F can be written as $V(I)$ for an ideal I such that $\sqrt{I} = I$. Let F_1 be the closure of $\Psi(F)$ in X. Then $F_1 \subseteq F$ and $F_1 = V(I_1)$ for some ideal $I_1 \supseteq I$. Any element f of I_1 vanishes on $\Psi(F) = \mathbf{V}(I)$. By the above result, $f \in \sqrt{I} = I$; hence $I_1 = I$ and so $F_1 = F$. □

Note that the one-to-one correspondence Ψ is order-preserving for the ordering by inclusion.

Proposition 1.11. *Let* k *be an algebraically closed field and let* L/k *be any field extension. An irreducible polynomial* $f \in k[\mathsf{X}_1, \ldots, \mathsf{X}_n]$ *remains irreducible in* $L[\mathsf{X}_1, \ldots, \mathsf{X}_n]$.

PROOF. Suppose that $f = gh$, where g and h belong to $L[\mathsf{X}_1, \ldots, \mathsf{X}_n] \backslash L$. Writing $g = \sum_\alpha a_\alpha M_\alpha(\mathsf{X})$ and $h = \sum_\alpha b_\alpha M_\alpha(\mathsf{X})$, where the $M_\alpha(\mathsf{X})$ are monomials and $a_\alpha, b_\alpha \in L$, we let $A = k[a_\alpha, b_\alpha \mid \alpha \text{ multi-indices}]$ and let a^* and b^* be the products of all nonzero a_α and all nonzero b_α, respectively. Choose a closed point p of $D(a^*) \cap D(b^*) \subseteq$ Spec A. Then $k(p) \cong k$ by Theorem 1.4 and we have the surjection $\eta\colon A \to k(p) \cong k$. Thus $f = \eta(g) \cdot \eta(h)$, where $\eta(g) = \sum_\alpha \eta(a_\alpha) \cdot M_\alpha(\mathsf{X})$ and $\eta(h) = \sum_\alpha \eta(b_\alpha) \cdot M_\alpha(\mathsf{X})$ are not constants. This implies that f is reducible. □

EXAMPLE 1.9. Let $k(\mathsf{X})$ be a purely transcendental extension of a field k. Then $f(\mathsf{T}) = \mathsf{T}^n + \mathsf{X}$ is irreducible in $k(\mathsf{X})[\mathsf{T}]$, but reducible in $k(\mathsf{X})(\mathsf{X}^{1/n})[\mathsf{T}]$.

§1.9 Dimension of Spec A

a. Definition. Let X be a nonempty topological space. The *dimension* $\dim X$ of X is the supremum of all the nonnegative integers l for which there are closed irreducible subsets $F_0, \ldots, F_l$ of X such that

$$\varnothing \neq F_0 \subset F_1 \subset \cdots \subset F_l \subseteq X.$$

If $x \in X$, we define the *dimension* of X at x by

$$\dim_x X = \inf\{\dim U \mid U \text{ is an open neighborhood of } x\}.$$

Clearly, if $\{X_i \mid i \in I\}$ is the set of all (closed) irreducible components of X, then $\dim X = \sup_{i \in I} \dim X_i$. Moreover, if $\{U_\lambda \mid \lambda \in \Lambda\}$ is an open cover of X, then $\dim X = \sup\{\dim U_\lambda \mid \lambda \in \Lambda\}$.

Definition. If Y is a closed irreducible subset of X, the *codimension* codim (Y, X) of Y in X is defined to be the supremum of all the nonnegative integers l for which there are closed irreducible subsets $F_0, \ldots, F_l$ of X with

$$Y = F_0 \subset F_1 \subset \cdots \subset F_l \subseteq X.$$

If Z is an arbitrary closed subset of X, then the *codimension* of Z in X is defined by

$$\operatorname{codim}(Z, X) = \inf\{\operatorname{codim}(Y, X) \mid Y \text{ is a closed irreducible subset of } X \text{ contained in } Z\},$$

and denoted by codim(Z), when there is no danger of confusion.

EXAMPLE 1.10. (i) A space consisting of a single point has dimension 0.

(ii) $\dim \operatorname{Spec} k[X] = 1$.

(iii) Let $\mathbb{R}^n$ be the Euclidean space with the usual metric topology. Then, irreducible subsets of $\mathbb{R}^n$ are the single point sets. Hence, $\dim \mathbb{R}^n = 0$. Thus, the notion of dimension defined here is useless for the usual metric spaces.

b. Definition. Let A be a ring. The (Krull) *dimension* $\dim A$ of A is defined to be the supremum of all the nonnegative integers l for which there are prime ideals $\mathfrak{p}_0, \ldots, \mathfrak{p}_l$ of A such that

$$\mathfrak{p}_l \subset \mathfrak{p}_{l-1} \subset \cdots \subset \mathfrak{p}_0 \subset A.$$

For such prime ideals, one has

$$\varnothing \neq V(\mathfrak{p}_0) \subset V(\mathfrak{p}_1) \subset \cdots \subset V(\mathfrak{p}_l) \subseteq X = \operatorname{Spec} A.$$

Conversely, if $F_0 \subset F_1 \subset \cdots \subset F_l \subseteq X$ is a sequence of closed irreducible subsets, then $F_j = V(\mathfrak{p}_j)$ for certain prime ideals $\mathfrak{p}_j$ which satisfy the above inclusion relation by Corollary (i) to Proposition 1.3. Hence, one obtains the following result.

Lemma 1.9

$$\dim A = \dim \operatorname{Spec} A.$$

If $\mathfrak{p} \in \operatorname{Spec} A$, then $\operatorname{codim}(V(\mathfrak{p}), \operatorname{Spec} A)$ is also called the *height* of $\mathfrak{p}$, and denoted by $\operatorname{ht}(\mathfrak{p})$. Further, the height of a proper ideal $\mathfrak{a}$ is defined to be $\operatorname{codim}(V(\mathfrak{a}), \operatorname{Spec} A)$, denoted $\operatorname{ht}(\mathfrak{a})$.

Note that $\operatorname{ht}(\mathfrak{p}) = 0$ if and only if $\mathfrak{p}$ is a minimal prime ideal. Furthermore, $\operatorname{ht}(\mathfrak{p}) + \dim(A/\mathfrak{p}) \leq \dim A$, by definition.

EXAMPLE 1.11. Let $A = \mathbb{Z}_{(2)}[\mathsf{X}]$ and $\mathfrak{p} = (2\mathsf{X} - 1)$. Then $A/\mathfrak{p}$ is a field $\mathbb{Q}$ and $\operatorname{ht}(\mathfrak{p}) = 1$. But $\dim A \geq 2$, since there exists a sequence of prime ideals $(0) \subset (\mathsf{X}) \subset (2, \mathsf{X})$.

Thus, $\operatorname{ht}(\mathfrak{p}) + \dim(A/\mathfrak{p}) = \dim A$ does not hold in general. However, one can prove the equality for an integral domain finitely generated over a field (see the Corollary to Theorem 1.9).

c. Now, we want to compute the dimension of polynomial rings. For this computation, we begin by studying prime ideals of integral extensions of rings.

Proposition 1.12. *Let A be a ring integral over a subring B, let $\mathfrak{p}$ be a prime ideal of B, and let $\mathfrak{q}$ be a prime ideal of A. Then $\mathfrak{q} \cap B = \mathfrak{p}$ if and only if $\mathfrak{q}(B\backslash\mathfrak{p})^{-1}A$ is a maximal ideal of $(B\backslash\mathfrak{p})^{-1}A$.*

PROOF. Consider first the case where B is a local ring and $\mathfrak{p}$ is the unique maximal ideal. Then $B\backslash\mathfrak{p}$ consists of units and so the assertion is reduced to the following statement: $\mathfrak{q} \cap B = \mathfrak{p}$ *if and only if* $\mathfrak{q}$ *is a maximal ideal of* A. To show this, suppose $\mathfrak{q} \cap B = \mathfrak{p}$. Then $A/\mathfrak{q}$ is integral over $B/\mathfrak{p}$ by Lemma 1.7 and so is a field by Lemma 1.8, i.e., $\mathfrak{q}$ is a maximal ideal. Conversely, if $\mathfrak{q}$ is maximal, then $A/\mathfrak{q}$ is a field which is integral over a subring $B/(\mathfrak{q} \cap B)$. Again by Lemma 1.8, we conclude that $B/(\mathfrak{q} \cap B)$ is a field; hence $\mathfrak{q} \cap B$ is a maximal ideal, which is $\mathfrak{p}$. Thus $\mathfrak{p} = \mathfrak{q} \cap B$.

To treat the general case, let $S = B\backslash\mathfrak{p}$. Then, clearly, $S^{-1}A$ is integral over $S^{-1}B = B_{\mathfrak{p}}$ and $\mathfrak{p}B_{\mathfrak{p}}$ is the unique maximal ideal. Hence,

$$\mathfrak{q} \cap B = \mathfrak{p} \Leftrightarrow (\mathfrak{q}S^{-1}A) \cap B_{\mathfrak{p}} = \mathfrak{p}B_{\mathfrak{p}} \Leftrightarrow \mathfrak{q}S^{-1}A \text{ is a maximal ideal.} \qquad \square$$

Theorem 1.7 (Cohen, Seidenberg). *If A is integral over a subring B and $\mathfrak{p} \in \operatorname{Spec} B$, then there exists $\mathfrak{q} \in \operatorname{Spec} A$ such that $\mathfrak{q} \cap B = \mathfrak{p}$. Moreover, if $\mathfrak{q}' \in \operatorname{Spec} A$ satisfies $\mathfrak{q}' \supseteq \mathfrak{q}$ and $\mathfrak{q}' \cap B = \mathfrak{q} \cap B$, then $\mathfrak{q}' = \mathfrak{q}$.*

PROOF. Take a maximal ideal $\mathfrak{m}$ of $S^{-1}A$ by Lemma 1.1, where S is $B\backslash\mathfrak{p}$. Then $\mathfrak{q} = \psi_{A,S}^{-1}(\mathfrak{m})$ satisfies $\mathfrak{m} = \mathfrak{q}S^{-1}A$ by the proof of Lemma 1.3. Hence by Proposition 1.12, $\mathfrak{q} \cap B = \mathfrak{p}$. If $\mathfrak{q}' \cap B = \mathfrak{p}$ with $\mathfrak{q}' \supseteq \mathfrak{q}$, then $\mathfrak{q}'S^{-1}A$ is maximal and so $\mathfrak{q}'S^{-1}A \supseteq \mathfrak{q}S^{-1}A$ implies that $\mathfrak{q}'S^{-1}A = \mathfrak{q}S^{-1}A$. Then $\mathfrak{q}' = \mathfrak{q}$ by the proof of Lemma 1.3. □

Corollary. *Under the same hypothesis as above*, $\dim A = \dim B$.

PROOF. Suppose that $\mathfrak{q}_0, \ldots, \mathfrak{q}_n$ are prime ideals of A such that

$$\mathfrak{q}_n \subset \mathfrak{q}_{n-1} \subset \cdots \subset \mathfrak{q}_0.$$

Then the $\mathfrak{q}_i \cap B$ are prime ideals of B with $\mathfrak{q}_n \cap B \subseteq \mathfrak{q}_{n-1} \cap B \subseteq \cdots \subseteq \mathfrak{q}_0 \cap B$. Furthermore, the inclusions are strict by the last theorem. Thus $\dim B \geq \dim A$.

Next, suppose that $\mathfrak{p}_0, \ldots, \mathfrak{p}_m$ are prime ideals of B with $\mathfrak{p}_m \subset \mathfrak{p}_{m-1} \subset \cdots \subset \mathfrak{p}_0$. By the last theorem, there is a prime ideal $\mathfrak{q}_m$ of A with $\mathfrak{q}_m \cap B = \mathfrak{p}_m$. Furthermore, since $A/\mathfrak{q}_m$ is integral over $B/\mathfrak{p}_m = B/(\mathfrak{q}_m \cap B)$ by Lemma 1.7, there exists a prime ideal $\mathfrak{q}_{m-1}$ containing $\mathfrak{q}_m$ such that $(\mathfrak{q}_{m-1}/\mathfrak{q}_m) \cap (B/\mathfrak{p}_m) = \mathfrak{p}_{m-1}/\mathfrak{p}_m$; hence $\mathfrak{q}_{m-1} \cap B = \mathfrak{p}_{m-1}$ and $\mathfrak{q}_{m-1} \neq \mathfrak{q}_m$ by the last theorem. Continuing in this manner, we can construct a sequence of prime ideals of A with $\mathfrak{q}_m \subset \mathfrak{q}_{m-1} \subset \cdots \subset \mathfrak{q}_0$. Hence, $\dim A \geq \dim B$ and so the dimensions are equal. □

Theorem 1.8. *If k is a field, then* $\dim k[\mathsf{X}_1, \ldots, \mathsf{X}_n] = n$.

PROOF. The proof is by induction on n. The case $n = 0$ is trivial. Let $A = k[\mathsf{X}_1, \ldots, \mathsf{X}_n]$ and choose a sequence of prime ideals $\mathfrak{p}_0, \ldots, \mathfrak{p}_{l-1}$, $\mathfrak{p}_l = (0)$ with $\mathfrak{p}_l \subset \mathfrak{p}_{l-1} \subset \cdots \subset \mathfrak{p}_0$. Let $x_i = \mathsf{X}_i \bmod \mathfrak{p}_{l-1}$. Then $x_1, \ldots, x_n$ are algebraically dependent; hence $r = \text{tr. deg}_k\, Q(A/\mathfrak{p}_{l-1}) < n$. By Theorem 1.3, there exist $y_1, \ldots, y_r \in A/\mathfrak{p}_{l-1}$ such that $\bar{A} = A/\mathfrak{p}_{l-1}$ is integral over $\bar{B} = k[y_1, \ldots, y_r]$. By the induction hypothesis, one has $\dim \bar{B} = r$, and from the Corollary to Theorem 1.7, it follows that $\dim \bar{A} = \dim \bar{B} = r$. Since $l - 1 \leq \dim \bar{A}$, we have $l \leq \dim \bar{A} + 1 = r + 1 \leq n$. Hence $\dim A = \sup\{l\} \leq n$. Finally, since we have the sequence of prime ideals

$$0 \subset (\mathsf{X}_1) \subset (\mathsf{X}_1, \mathsf{X}_2) \subset \cdots \subset (\mathsf{X}_1, \ldots, \mathsf{X}_n),$$

we conclude that $\dim A = n$. □

Corollary 1. *If A is an integral domain finitely generated over a field k, then* $\dim A = \text{tr. deg}_k\, Q(A)$.

PROOF. By Theorem 1.3, there exist $y_1, \ldots, y_n$ with $n = \text{tr. deg}_k\, Q(A)$ such that A is integral over $B = k[y_1, \ldots, y_n]$. By the last theorem, $\dim B = n$, and by the Corollary to Theorem 1.7, we have $\dim A = \dim B = n$. □

Corollary 2. *Let* $f \in k[X_1, \ldots, X_n]$ *be an irreducible polynomial. Then* $\mathrm{ht}(f) = 1$ *and* $\dim(k[X_1, \ldots, X_n]/(f)) = n - 1$.

PROOF. The first assertion is obvious, since (f) is a prime ideal by [N1, p. 37]. To prove the second assertion, we have only to note that tr. deg $Q(k[X_1, \ldots, X_n]/(f)) = n - 1$ by Corollary 1. □

Theorem 1.9. *Let* A *be an integral domain finitely generated over a field* k *and* $\mathfrak{p}_l = 0, \mathfrak{p}_{l-1}, \ldots, \mathfrak{p}_0$ *be prime ideals such that* $\mathfrak{p}_l \subset \cdots \subset \mathfrak{p}_i \subset \mathfrak{p}_{i-1} \subset \cdots \subset \mathfrak{p}_0$. *Suppose that there are no prime ideals between* $\mathfrak{p}_i$ *and* $\mathfrak{p}_{i-1}$ *for* $1 \leq i \leq l$ *and* $\mathfrak{p}_0$ *is maximal. Then* $l = \dim A$.

PROOF. The proof is by induction on $n = \dim A$. The case $n = 0$ is trivial. Let $\bar{A} = A/\mathfrak{p}_{l-1}$. Then we have a sequence of prime ideals $0 = \mathfrak{p}_{l-1}\bar{A} \subset \cdots \subset \mathfrak{p}_0 \bar{A}$. Since there are no prime ideals between $\mathfrak{p}_i \bar{A}$ and $\mathfrak{p}_{i-1}\bar{A}$, and since $\mathfrak{p}_0 \bar{A}$ is maximal, we have $l - 1 = \dim \bar{A}$ by the induction hypothesis. We claim that $\dim \bar{A} = \dim A - 1$. To show this, applying Theorem 1.3, we have $y_1, \ldots, y_n \in A$ such that A is integral over $B = k[y_1, \ldots, y_n]$. By Theorem 1.7, $\mathfrak{p}_{l-1} \cap B \neq \mathfrak{p}_l \cap B = 0$. Hence, we take a nonzero element $F \in \mathfrak{p}_{l-1} \cap B$ and choose $Y_2, \ldots, Y_n$ such that B is integral over $C = k[F, Y_2, \ldots, Y_n]$ by Lemma 1.6. By Theorem 1.7, $\mathfrak{p}_{l-1} \cap C$ is obviously (F) and so $\bar{A} = A/\mathfrak{p}_{l-1}$ is integral over $C/(F) \cong k[Y_2, \ldots, Y_n]$. Thus $\dim \bar{A} = n - 1$ has been shown and the proof is complete, since $\dim A = \dim \bar{A} + 1 = n$.

□

Corollary. If $\mathfrak{p}$ *is a prime ideal of* A, *then* $\dim A/\mathfrak{p} + \mathrm{ht}(\mathfrak{p}) = \dim A$.

PROOF. This follows immediately from Theorem 1.9. □

Remark. The condition that A is finitely generated is indispensable in Theorems 1.8 and 1.9.

EXAMPLE 1.12. Let $A = \mathbb{C}[[X]]$ be the ring of formal power series. Then $\dim A = 1$ and tr. $\deg_{\mathbb{C}} Q(A) = \infty$.

d. We conclude this section with a simple result on 0-dimensional spectra.

Lemma 1.10. *Let* X *be a Noetherian* T_0*-space. Then* $\dim X = 0$ *if and only if* X *is a finite set with the discrete topology.*

PROOF. Since X is Noetherian, X is a union of a finite number of irreducible components X_i $(1 \leq i \leq m)$. If $\dim X = 0$, then $\dim X_i = 0$; hence $X_i = \{x_i\}$, where the x_i are closed points by Lemma 1.5.(i). Therefore, X is a finite set with the discrete topology. The converse is evidently true. □

Let A be a Noetherian ring such that $X = \operatorname{Spec} A$ is 0-dimensional. Then $\operatorname{Spec} A = \{\mathfrak{p}_1, \ldots, \mathfrak{p}_m\}$ where the $\mathfrak{p}_i$ are maximal ideals. These are minimal prime ideals of A; hence, by Proposition 1.3, $\sqrt{(0_A)} = \mathfrak{p}_1 \cap \cdots \cap \mathfrak{p}_m$, denoted by I. Since A is Noetherian, there exists r such that $I^r = 0$.

Putting $\mathfrak{a}_i = \mathfrak{p}_i^r$, one has $\mathfrak{a}_1 \cap \cdots \cap \mathfrak{a}_m = 0$ and $\mathfrak{a}_i + \mathfrak{a}_j = A$ for any $i \neq j$. To prove the last claim, suppose that $\mathfrak{a}_i + \mathfrak{a}_j$ is proper; there exists a maximal ideal $\mathfrak{m}$ containing $\mathfrak{a}_i$ and $\mathfrak{a}_j$. Then $\mathfrak{m} \supseteq \mathfrak{p}_i + \mathfrak{p}_j = A$, which is a contradiction. Thus, one has by the Chinese remainder theorem,

$$A = \frac{A}{(0_A)} \cong \frac{A}{\mathfrak{a}_1} \oplus \cdots \oplus \frac{A}{\mathfrak{a}_m}.$$

Each $A/\mathfrak{a}_i$ has a unique prime ideal $\mathfrak{m}_i = \mathfrak{p}_i/\mathfrak{a}_i$. We now use the next result.

Lemma 1.11. *Let A be a Noetherian ring with* $\operatorname{Spec} A = \{\mathfrak{p}\}$. *Then any ideal $\mathfrak{a}$ of A has a descending chain of ideals $\mathfrak{a}_0 \supseteq \mathfrak{a}_1 \supseteq \mathfrak{a}_2 \supseteq \cdots \supseteq \mathfrak{a}_t$ such that $\mathfrak{a}_0 = \mathfrak{a}$, $\mathfrak{a}_t = 0$, and $\mathfrak{a}_i = \mathfrak{a}_{i+1}$ or $\mathfrak{a}_i/\mathfrak{a}_{i+1} \cong A/\mathfrak{p}$ for all i.*

Proof. Since $\sqrt{(0_A)} = \mathfrak{p}$ by Proposition 1.3, there exists $r > 0$ such that $\mathfrak{p}^r = 0$. Since $\mathfrak{p}^{i-1}/\mathfrak{p}^i$ is an $A/\mathfrak{p}$-module with a finite base, there exist ideals $\mathfrak{q}_{i-1,0}, \ldots, \mathfrak{q}_{i-1,s}$ such that

$$\mathfrak{p}^{i-1} = \mathfrak{q}_{i-1,0} \supset \cdots \supset \mathfrak{q}_{i-1,s} = \mathfrak{p}^i$$

with $\mathfrak{q}_{i-1,j}/\mathfrak{q}_{i-1,j+1} \cong A/\mathfrak{p}$. Thus, we have a descending chain of ideals $\{I_i \mid 1 \leq i \leq t\}$ such that

$$I_1 = \mathfrak{p} \supset I_2 \supset \cdots \supset I_t = 0$$

with $I_i/I_{i+1} \cong A/\mathfrak{p}$ for all i. For any ideal $\mathfrak{a}$ of A, we let $\mathfrak{a}_i = \mathfrak{a} \cap I_i$. Then $\mathfrak{a}_i = \mathfrak{a}_{i+1}$ or $\mathfrak{a}_i/\mathfrak{a}_{i+1} \cong A/\mathfrak{p}$. □

Definition. Let A be a ring. If every descending chain of ideals stabilizes, then A is said to be an *Artinian ring*.

The ring A in Lemma 1.11 is an Artinian local ring. Hence, if a Noetherian ring A has only a finite number of prime ideals, which are also maximal ideals, then A is a direct sum of Artinian local rings, and so A is an Artinian ring.

Theorem 1.10. *Let A be a Noetherian ring with* $\dim A = 0$. *Then A is an Artinian ring. Furthermore, if A is finitely generated over a field k, then A is a finite-dimensional vector space over k.*

Proof. The first part was proved in the above discussion. If A is finitely generated over k, then each $A/\mathfrak{p}_i = K_i$ is algebraic over k by Theorem 1.4, i.e., $\dim_k K_i < \infty$. Hence,

$$\dim_k A = \sum_{i=1}^{m} \dim_k(A/\mathfrak{a}_i) = \sum_{i=1}^{m} \sum_{j=0}^{r-1} \dim_{K_i}(\mathfrak{p}_i^j/\mathfrak{p}_i^{j+1}) \dim_k K_i < \infty.$$

□

EXAMPLE 1.13. Let $f, g \in k[\mathsf{X}, \mathsf{Y}]$ be polynomials such that f and g have no common polynomial factors. Then $A = k[\mathsf{X}, \mathsf{Y}]/(f, g)$ is 0-dimensional; hence $\dim_k A$ is finite, whenever $A \neq 0$.

§1.10 Sheaves

In this section, we fix a topological space X and let Ω denote the set of all open subsets of X. The reader who is familiar with the general theory of sheaves can skip this section.

a. Definition. A *presheaf* $\mathscr{F}$ of Abelian groups on X is a pair of collections

$$\mathscr{F} = (\{\mathscr{F}(U) \mid U \in \Omega\}, \{r_{V,U} \mid U, V \in \Omega \text{ and } U \supseteq V\}),$$

where every $\mathscr{F}(U)$ is an Abelian group and every $r_{V,U}$ is a group homomorphism, from $\mathscr{F}(U)$ to $\mathscr{F}(V)$ satisfying the following conditions:

(1) $\mathscr{F}(\varnothing) = \{0\}$.
(2) $r_{U,U}: \mathscr{F}(U) \to \mathscr{F}(U)$ is the identity for all $U \in \Omega$.
(3) $r_{W,U} = r_{W,V} \circ r_{V,U}$ for all $U, V, W \in \Omega$ with $U \supseteq V \supseteq W$.

The $r_{V,U}$ are called restriction maps and so $r_{V,U}(\alpha)$ for $\alpha \in \mathscr{F}(U)$ is often denoted $\alpha|_V$. Thus condition (3) can be rewritten as

$$(\alpha|_V)|_W = \alpha|_W.$$

Definition. The presheaf $\mathscr{F}$ is called a *sheaf of Abelian groups* on X, if the following two additional conditions are satisfied:

(4) Let $\{U_j \mid j \in J\}$ be an arbitrary open cover of U, i.e., $U_j \in \Omega$ and $U = \bigcup_{j \in J} U_j$. If $\alpha \in \mathscr{F}(U)$, then $\alpha = 0$ whenever $\alpha|_{U_j} = 0$ for all $j \in J$.
(5) Let $\{U_j \mid j \in J\}$ be an arbitrary open cover of U. If $\alpha_j \in \mathscr{F}(U_j)$ are given for all j such that $\alpha_i|_{U_i \cap U_j} = \alpha_j|_{U_i \cap U_j}$ for all i, j, then there exists an $\alpha \in \mathscr{F}(U)$ such that $\alpha|_{U_j} = \alpha_j$ for all j.

Remarks. (1) Property (1) follows from (4). In fact, it suffices to consider the open cover $\{U_j \mid j \in J\}$ of $\varnothing$ where J is also empty.

(2) By property (4), the α in (5) is uniquely determined.

(3) If the $\mathscr{F}(U)$ are sets, or rings, or k-vector spaces, and the $r_{V,U}$ are maps, or ring homomorphisms, or k-linear maps, then $\mathscr{F}$ is said to be a

presheaf (or sheaf) of sets, or rings, or k-vector spaces, respectively (For a sheaf of sets, see subsection **d** below).

(4) $\mathscr{F} = (\{\mathscr{F}(U) | U \in \Omega\}, \{r_{V,U} | U, V \in \Omega, U \supseteq V\})$ is often written as $(\{\mathscr{F}(U)\}, \{r_{V,U}\})$.

Definition. Let $\mathscr{O} = (\{\mathscr{O}(U)\}, \{r_{V,U}\})$ be a (pre)sheaf of rings and $\mathscr{F} = (\{\mathscr{F}(U)\}, \{s_{V,U}\})$ be a (pre)sheaf of Abelian groups on X such that the $\mathscr{F}(U)$ are $\mathscr{O}(U)$-modules for all $U \in \Omega$, and $s_{V,U}(af) = r_{V,U}(a)s_{V,U}(f)$ for all $a \in \mathscr{O}(U)$, $f \in \mathscr{F}(U)$ and $U, V \in \Omega$ with $U \supseteq V$. Then $\mathscr{F}$ is said to be a *(pre)sheaf of $\mathscr{O}$-modules* on X. A sheaf of $\mathscr{O}$-modules is also simply called an *$\mathscr{O}$-module.*

Remark. By "presheaf" we mean a presheaf of Abelian groups.

Definition. If X_0 is an open subset of X and $\mathscr{F}$ is a (pre)sheaf on X. Then $(\{\mathscr{F}(U) \mid U \in \Omega \text{ and } U \subseteq X_0\}, \{r_{V,U} \mid V, U \in \Omega, U \supseteq V, \text{ and } V, U \subseteq X_0\})$ is a (pre)sheaf on X_0, which is called the *restriction* of $\mathscr{F}$ to X_0, denoted $\mathscr{F}|_{X_0}$.

Clearly, if $\mathscr{F}$ is an $\mathscr{O}$-module, then $\mathscr{F}|_{X_0}$ is an $\mathscr{O}|_{X_0}$-module.

EXAMPLE 1.14. Assume X is irreducible and let G be an Abelian group. Define $\mathscr{F}(U) = G$ if $U \neq \varnothing$, and $\mathscr{F}(\varnothing) = \{0\}$, and further define $r_{V,U} = \mathrm{id}_G$ if $V \neq \varnothing$ and $r_{\varnothing,U} = 0$. Then $(\{\mathscr{F}(U)\}, \{r_{V,U}\})$ is a sheaf, which is called the *constant sheaf* with coefficients in G on X. $\mathscr{F}$ is often denoted by G_X.

Remark. Let $U, V \in \Omega$ with $U \cap V = \varnothing$. If $\mathscr{F}$ is a sheaf on X, then $\mathscr{F}(U \cup V) \cong \mathscr{F}(U) \oplus \mathscr{F}(V)$.

b. Definition. For $x \in X$, we let Ω_x be $\{U \in \Omega \mid U \ni x\}$. If $\mathscr{F}$ is a presheaf on X, the *stalk* $\mathscr{F}_x$ of $\mathscr{F}$ at x is defined by

$$\mathscr{F}_x = \operatorname*{inj\,lim}_{U \in \Omega_x} \mathscr{F}(U).$$

Here, the inj lim denotes the direct (or injective) limit and so $\mathscr{F}_x$ is isomorphic to $\coprod_{U \in \Omega_x} \mathscr{F}(U)/\sim$, where $\sim$ is the equivalence relation defined by

$$\alpha \sim \beta \Leftrightarrow \alpha \in \mathscr{F}(U), \beta \in \mathscr{F}(V),$$

where $U, V \in \Omega_x$ and there is $W \subseteq U \cap V$ in Ω_x with $\alpha|_W = \beta|_W$.

Since the $\mathscr{F}(U)$ are Abelian groups and the $r_{V,U}$ are homomorphisms, the direct limit is also an Abelian group in a natural way. If $\alpha \in \mathscr{F}(U)$, the equivalence class of α in $\mathscr{F}_x$ is denoted by α_x and is called the *germ* of α at x. By definition, $(\alpha \pm \beta)_x = \alpha_x \pm \beta_x$ for all $\alpha, \beta \in \mathscr{F}(U)$ and $U \in \Omega_x$.

If $\mathscr{O}$ is a presheaf of rings, the stalks $\mathscr{O}_x$ of $\mathscr{O}$ are rings. If $\mathscr{F}$ is a presheaf of $\mathscr{O}$-modules, the stalks $\mathscr{F}_x$ of $\mathscr{F}$ are $\mathscr{O}_x$-modules.

EXAMPLE 1.15. Let X be the complex analytic affine space $\mathbb{C}^n$ with the usual metric topology. For an open set U, let $\mathscr{F}(U)$ be the set of holomorphic functions on U. Then $\{\mathscr{F}(U)\}$ with the natural restriction maps is a sheaf, denoted by $\mathcal{O}^{\text{hol}}$. Then for any $p = (a_1, \ldots, a_n) \in \mathbb{C}^n$, $\mathcal{O}_p^{\text{hol}}$ is isomorphic to the ring of convergent power series $\mathbb{C}\{z_1 - a_1, \ldots, z_n - a_n\}$.

c. Definition. For presheaves $\mathscr{F}$ and $\mathscr{G}$ on X, a *homomorphism* $\theta\colon \mathscr{F} \to \mathscr{G}$ is defined to be a collection of group homomorphisms $\{\theta_U\colon \mathscr{F}(U) \to \mathscr{G}(U) \mid U \in \Omega\}$ such that $\theta_U(\alpha)|_V = \theta_V(\alpha|_V)$ for all $\alpha \in \mathscr{F}(U)$, i.e., the square

$$\begin{array}{ccc} \mathscr{F}(U) & \xrightarrow{\theta_U} & \mathscr{G}(U) \\ {\scriptstyle\text{restriction}}\big\downarrow & & \big\downarrow{\scriptstyle\text{restriction}} \\ \mathscr{F}(V) & \xrightarrow{\theta_V} & \mathscr{G}(V) \end{array}$$

is commutative, whenever $U, V \in \Omega$ and $U \supseteq V$.

If, in addition, $\mathscr{F}$ and $\mathscr{G}$ are sheaves, then θ is said to be a *homomorphism of sheaves*. When there is no danger of confusion, homomorphisms of presheaves (and of sheaves) are referred to simply as *homomorphisms*.

Definition. If $\mathscr{F}$ and $\mathscr{G}$ are $\mathcal{O}$-modules and $\theta\colon \mathscr{F} \to \mathscr{G}$ is a homomorphism such that $\theta_U\colon \mathscr{F}(U) \to \mathscr{G}(U)$ are $\mathcal{O}(U)$-homomorphisms for all $U \in \Omega$, then θ is said to be an *$\mathcal{O}$-homomorphism.*

Definition. The set of all $\mathcal{O}$-homomorphisms from $\mathscr{F}$ to $\mathscr{G}$ is naturally endowed with the structure of an Abelian group, denoted by $\operatorname{Hom}_{\mathcal{O}}(\mathscr{F}, \mathscr{G})$.

Note that there exists a natural identification $\operatorname{Hom}_{\mathcal{O}}(\mathcal{O}, \mathscr{F}) = \mathscr{F}(X)$ by $\theta \leftrightarrow \theta_X(1)$.

Definition. If $\theta_U = \mathrm{id}_U\colon \mathscr{F}(U) \to \mathscr{F}(U)$ for all $U \in \Omega$, then $\{\theta_U\}$ is said to be the *identity*, denoted by id. If $\theta\colon \mathscr{F} \to \mathscr{G}$ and $\eta\colon \mathscr{G} \to \mathscr{H}$ are homomorphisms, then the composition $\eta \circ \theta\colon \mathscr{F} \to \mathscr{H}$ is defined by $(\eta \circ \theta)_U = \eta_U \circ \theta_U$ for all $U \in \Omega$. If $\mathscr{F} = \mathscr{H}$, $\eta \circ \theta = \mathrm{id}$ and $\theta \circ \eta = \mathrm{id}$, then θ and η are called *isomorphisms*, and the notations $\mathscr{F} \cong \mathscr{G}$ or $\theta\colon \mathscr{F} \xrightarrow{\sim} \mathscr{G}$ will be used.

Definition. Let $\theta\colon \mathscr{F} \to \mathscr{G}$ be a homomorphism of presheaves. For any $x \in X$ and $\alpha \in \mathscr{F}(U)$ such that $U \in \Omega_x$, $V \in \Omega_x$, and $U \supseteq V$, one has $\theta_U(\alpha)|_V = \theta_V(\alpha|_V)$; hence $\theta_U(\alpha)_x = \theta_U(\alpha')_x$, if $\alpha_x = \alpha'_x$ for $\alpha' \in \mathscr{F}(U)$. Thus, the mapping $\alpha_x \mapsto \theta_U(\alpha)_x$ is well defined, and will be denoted by θ_x, i.e., $\theta_x(\alpha_x) = \theta_U(\alpha)_x$. θ_x is said to be a *germ defined* by θ.

It is easy to see that θ_x is a group homomorphism. If $\mathscr{F}$ and $\mathscr{G}$ are $\mathcal{O}$-modules and θ is an $\mathcal{O}$-homomorphism, then θ_x is an $\mathcal{O}_x$-homomorphism.

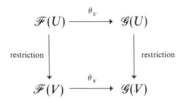

If $\mathscr{H}$ is a presheaf and $\eta: \mathscr{G} \to \mathscr{H}$ is a homomorphism, then $(\eta \circ \theta)_x = \eta_x \circ \theta_x$ for any $x \in X$.

d. To construct the sheafification of a presheaf, we need the following concept.

Definition. An *étale space* is defined to be a triple $(\mathbf{F}, \pi, X)$ where $\mathbf{F}$ and X are topological spaces, and $\pi: \mathbf{F} \to X$ is a surjective continuous map which is a local homeomorphism, i.e., every point $p \in \mathbf{F}$ has an open neighborhood V such that $\pi(V)$ is an open subset of X and $\pi|_V: V \to \pi(V)$ is a homeomorphism.

When there is no risk of confusion, we write simply $\mathbf{F}$ instead of $(\mathbf{F}, \pi, X)$.

Definition. Let $(\mathbf{F}, \pi, X)$ be an étale space. For any $x \in X$, $\mathbf{F}_x = \pi^{-1}(x)$ is said to be the *stalk* of $\mathbf{F}$ at x. For any open subset U of X, we define $\Gamma(U, \mathbf{F})$ to be the set of all continuous maps $\sigma: U \to \mathbf{F}$ such that $\pi \circ \sigma = \mathrm{id}_U$. Any element of $\Gamma(U, \mathbf{F})$ is said to be a *section* of $\mathbf{F}$ over U.

Note that the topology on the stalk $\mathbf{F}_x$ induced from $\mathbf{F}$ is discrete.

Definition. Let $(\mathbf{F}, \pi, X)$ and $(\mathbf{G}, \pi', X)$ be étale spaces with the same topological space X. A continuous map $\varphi: \mathbf{F} \to \mathbf{G}$ is said to be an *étale space map*, if $\pi' \circ \varphi = \pi$.

Given an étale space $(\mathbf{F}, \pi, X)$, we shall construct a *presheaf of sets* by letting $\mathscr{F}^*(U)$ be $\Gamma(U, \mathbf{F})$ for $U \in \Omega$, and letting $r_{V,U}$ be the natural restriction map whenever $U, V \in \Omega$ and $U \supseteq V$.

Furthermore, $\mathscr{F}^*$ is a sheaf of sets, i.e., $\mathscr{F}^*$ satisfies the following conditions:

(4)′ Let $\{U_j | j \in J\}$ be an arbitrary open cover of an open set U. If α and $\beta \in \mathscr{F}^*(U)$ satisfy $\alpha|_{U_j} = \beta|_{U_j}$ for all $j \in J$, then $\alpha = \beta$.
(5)′ is the same as (5) in subsection **a**.

Definition. $\mathscr{F}^*$ is denoted by $\mathbf{F}^b$, and is called the *sheaf associated with the étale space* $\mathbf{F}$.

Given an étale space map $\varphi: (\mathbf{F}, \pi, X) \to (\mathbf{G}, \pi', X)$, we have maps $\varphi_U: \Gamma(U, \mathbf{F}) \to \Gamma(U, \mathbf{G})$ for all $U \in \Omega$ defined by $\varphi_U(\alpha) = \varphi \circ \alpha$, since $\varphi_U(\alpha)|_V = \varphi_V(\alpha|_V)$ for all $U, V \in \Omega$ such that $U \supseteq V$. $\{\varphi_U\}$ is a homomorphism between the sheaves of sets $\mathbf{F}^b$ and $\mathbf{G}^b$, and is denoted by φ^b.

Lemma 1.12. *Let $(\mathbf{F}, \pi, X)$ be an étale space. For any $x \in X$, there is a canonical bijection between the stalk $\mathbf{F}^b_x$ and $\mathbf{F}_x = \pi^{-1}(x)$.*

Proof. For any $\sigma \in \Gamma(U, \mathbf{F})$ with $U \in \Omega_x$, define j_U by $j_U(\sigma) = \sigma(x) \in \mathbf{F}_x$. Since $j_V(\sigma|_V) = j_U(\sigma)$ for $V \subseteq U$ such that $V \in \Omega_x$, we know that $\{j_U | U \in \Omega_x\}$ determines $j_{\{x\}}: \mathbf{F}_x^b \to \mathbf{F}_x$. We claim that $j_{\{x\}}$ is a bijection. To prove that $j_{\{x\}}$ is onto, take an arbitrary point p of $\mathbf{F}_x$. There is an open neighborhood W of p such that $\pi|_W: W \to \pi(W)$ is a homeomorphism. Let β be $(\pi|_W)^{-1}$, which is a section of $\mathbf{F}$ over $\pi(W)$. $j_{\pi(W)}(\beta) = \beta(x) = (\pi|_W)^{-1}(x) = p$; hence $j_{\{x\}}(\beta_x) = p$. Now, suppose that $j_U(\sigma) = p$ for some $U \in \Omega_x$. Then p has an open neighborhood W such that $\pi(W) \subseteq U$ and $\pi|_W: W \to \pi(W)$ is a homeomorphism. Hence, $\sigma \sim (\pi|_W)^{-1}$ ($\sim$ was defined in subsection **b**). But since $(\pi|_W)^{-1}$ is determined by p and $W \in \Omega_p$, this implies that $j_{\{x\}}$ is one-to-one. □

e. Given a presheaf $\mathscr{F}$ on X, we construct an étale space $(\mathscr{F}^+, \pi, X)$ as follows. First, let $\mathscr{F}^+ = \coprod_{x \in X} \mathscr{F}_x$ as a set. To introduce the topology on $\mathscr{F}^+$, define $\mathbf{W}(\alpha, U) = \{\alpha_x | x \in U\}$ for $U \in \Omega$ and $\alpha \in \mathscr{F}(U)$. The topology of $\mathscr{F}^+$ is defined by taking $\{\mathbf{W}(\alpha, U) | U \in \Omega, \alpha \in \mathscr{F}(U)\}$ as an open base. To do so, we have to check that if $p \in \mathbf{W}(\alpha, U) \cap \mathbf{W}(\alpha', U')$, then there exist W and β such that $p \in \mathbf{W}(\beta, W) \subseteq \mathbf{W}(\alpha, U) \cap \mathbf{W}(\alpha', U')$. In fact, $p = \alpha_x = \alpha'_x$ for some $x \in U \cap U'$, and so $\alpha|_W = \alpha'|_W$ for some $W \in \Omega_x$. But then, taking $\beta = \alpha|_W$, one has $\mathbf{W}(\beta, W) \subseteq \mathbf{W}(\alpha, U) \cap \mathbf{W}(\alpha', U')$.

Finally, we define $\pi: \mathscr{F}^+ \to X$ by $\pi(p) = x$ if $p \in \mathscr{F}_x$. Then π is a local homeomorphism. In fact, if $\alpha_x \in \mathscr{F}_x$ is represented by $\alpha \in \mathscr{F}(U)$ for some $U \in \Omega_x$, then $\pi: \mathbf{W}(\alpha, U) \to U$ is a homeomorphism. This completes the construction of $\mathscr{F}^+$.

For any $\alpha \in \mathscr{F}(U)$ and $U \in \Omega$, one can obtain a section $\alpha^+ \in \Gamma(U, \mathscr{F}^+)$ by $\alpha^+(x) = \alpha_x$ for all $x \in U$.

Now, let $\sigma \in \Gamma(U, \mathscr{F}^+)$. For any $x \in U$, $\sigma(x)$ is a germ of some $\beta \in \mathscr{F}(V(x))$, $V(x) \in \Omega_x$. Hence, we have two sections σ and β^+ such that $\sigma(x) = \beta^+(x)$. Thus, there exists $W \in \Omega_x$ such that $\sigma|_W = \beta^+|_W$. In other words, there exist an open cover $\{U_j | j \in J\}$ of U and $\beta_j \in \mathscr{F}(U_j)$ for all j such that $\sigma(x) = \beta_{jx}$ whenever $x \in U_j$.

Now, since all the $\mathscr{F}(U)$ are Abelian groups, all the $\Gamma(U, \mathscr{F}^+)$ become Abelian groups in such a way that $\alpha \mapsto \alpha^+$ is a group homomorphism. In particular, 0^+ is the zero element of $\Gamma(U, \mathscr{F}^+)$. Clearly, the natural restriction maps $\Gamma(U, \mathscr{F}^+) \to \Gamma(V, \mathscr{F}^+)$, where $U, V \in \Omega$ and $U \supseteq V$, are homomorphisms; hence, $\mathscr{F}^{+b}$ is a sheaf of Abelian groups.

f. Definition. If $\mathscr{F}$ is a presheaf, then $\mathscr{F}^{+b}$ is said to be the *sheafification* of $\mathscr{F}$, denoted by ${}^a\mathscr{F}$.

One has the homomorphism $\zeta_{\mathscr{F}, U}: \mathscr{F}(U) \to {}^a\mathscr{F}(U) = \Gamma(U, \mathscr{F}^+)$ defined by $\zeta_{\mathscr{F}, U}(\alpha) = \alpha^+$. Since $\zeta_{\mathscr{F}, V}(\alpha|_V) = \zeta_{\mathscr{F}, U}(\alpha)|_V$ for all $\alpha \in \mathscr{F}(U)$ and $V \in \Omega$ with $U \supseteq V$, $\{\zeta_{\mathscr{F}, U}\}$ is a homomorphism of presheaves, denoted by $\zeta_{\mathscr{F}}$.

Lemma 1.13. *If $\mathscr{F}$ is a sheaf, then $\zeta_{\mathscr{F}}$ is an isomorphism of sheaves.*

Proof. Since $\mathscr{F}$ satisfies condition (5) (respectively (4)) of sheaves in §1.10.**a**, $\zeta_{\mathscr{F},U}$ is surjective (respectively injective) for any $U \in \Omega$, by the argument in subsection **e**. □

Let $\theta\colon \mathscr{F} \to \mathscr{G}$ be a homomorphism of presheaves. Then the étale space map $\theta^+ : \mathscr{F}^+ \to \mathscr{G}^+$ is defined as follows: $\theta^+(\alpha_x) = \theta_x(\alpha_x)$, for $\alpha_x \in \mathscr{F}_x$. Then θ^{+b} is a homomorphism from ${}^a\mathscr{F}$ to ${}^a\mathscr{G}$, denoted by ${}^a\theta$. Clearly, one has $\zeta_{\mathscr{G}} \circ \theta = {}^a\theta \circ \zeta_{\mathscr{F}}$, i.e., the following square is commutative:

$$\begin{array}{ccc} \mathscr{F} & \xrightarrow{\theta} & \mathscr{G} \\ {\scriptstyle \zeta_{\mathscr{F}}}\downarrow & & \downarrow{\scriptstyle \zeta_{\mathscr{G}}} \\ {}^a\mathscr{F} & \xrightarrow{{}^a\theta} & {}^a\mathscr{G} \end{array}$$

If $\mathscr{G}$ is a sheaf, then $\zeta_{\mathscr{G}}$ is an isomorphism by Lemma 1.13. In this case, any homomorphism $\eta\colon {}^a\mathscr{F} \to \mathscr{G}$ gives rise to $\theta = \eta \circ \zeta_{\mathscr{F}}$, and so $\eta = \zeta_{\mathscr{G}}^{-1} \circ {}^a\theta$. Moreover, if ${}^a\theta = 0$, then $\theta = 0$. Thus we obtain the next result.

Lemma 1.14. *If $\mathscr{G}$ is a sheaf, then there exists an isomorphism h:* $\mathrm{Hom}(\mathscr{F}, \mathscr{G}) \to \mathrm{Hom}({}^a\mathscr{F}, \mathscr{G})$ *defined by $h(\theta) = \zeta_{\mathscr{G}}^{-1} \circ {}^a\theta$.*

Lemma 1.15. *Let $\mathscr{F}$ and $\mathscr{G}$ be sheaves and let $\theta\colon \mathscr{F} \to \mathscr{G}$ be a homomorphism. If θ_x is an isomorphism (respectively zero) for all $x \in X$, then θ is an isomorphism (respectively zero). In particular, if $\mathscr{F}_x = 0$ for all $x \in X$, then $\mathscr{F} = 0$.*

Proof. By hypothesis, θ^+ is an isomorphism (respectively zero), and hence so is ${}^a\theta$. Since $\zeta_{\mathscr{F}}$ and $\zeta_{\mathscr{G}}$ are isomorphisms, θ is also an isomorphism (respectively zero). □

Lemma 1.16. *If $\mathscr{F}$ is a presheaf, then $\mathscr{F}_x$ is canonically isomorphic to $({}^a\mathscr{F})_x$ for every point x of X.*

Proof. By Lemma 1.12, $({}^a\mathscr{F})_x = (\mathscr{F}^{+b})_x \cong \pi^{-1}(x)$, where $\mathscr{F}^+ = ({}^+\mathscr{F}, \pi, X)$. But since $\pi^{-1}(x) = \mathscr{F}_x$, one obtains the assertion. □

Proposition 1.13. *Let $\mathscr{F}$ and $\mathscr{G}$ be presheaves and let $\theta\colon \mathscr{F} \to \mathscr{G}$ be a homomorphism. If $\theta_x\colon \mathscr{F}_x \to \mathscr{G}_x$ is an isomorphism for all $x \in X$, then ${}^a\theta\colon {}^a\mathscr{F} \to {}^a\mathscr{G}$ is an isomorphism.*

Proof. This follows immediately from Lemmas 1.15 and 1.16. □

Definition. Let $\mathscr{F}$ be a presheaf on X, and $U \in \Omega$. Define $\Gamma(U, \mathscr{F})$ to be $\Gamma(U, \mathscr{F}^+) = {}^a\mathscr{F}(U)$.

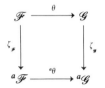

By Lemma 1.13, if $\mathscr{F}$ is a sheaf, then $\mathscr{F}(U)$ is isomorphic to $\Gamma(U, \mathscr{F})$ via $\zeta_{\mathscr{F}, U}$. In this case, we shall identify $\mathscr{F}(U)$ with $\Gamma(U, \mathscr{F})$, and call elements of $\mathscr{F}(U)$ *sections* of $\mathscr{F}$ over U.

If $\mathscr{O}$ is a presheaf of rings, then ${}^a\mathscr{O}$ is a sheaf of rings. If $\mathscr{F}$ is a presheaf of $\mathscr{O}$-modules, then ${}^a\mathscr{F}$ is a sheaf of ${}^a\mathscr{O}$-modules.

g. Definition. Let $\mathscr{F}$ and $\mathscr{G}$ be presheaves on X. $\mathscr{G}$ is said to be a *subpresheaf* of $\mathscr{F}$, if $\mathscr{G}(U)$ is a subgroup of $\mathscr{F}(U)$ for all $U \in \Omega$ and if the square

$$\begin{array}{ccc} \mathscr{G}(U) & \hookrightarrow & \mathscr{F}(U) \\ {\scriptstyle \text{restriction}} \downarrow & & \downarrow {\scriptstyle \text{restriction}} \\ \mathscr{G}(V) & \hookrightarrow & \mathscr{F}(V) \end{array}$$

is commutative for all $U, V \in \Omega$ with $U \supseteq V$.

If, in addition, both $\mathscr{F}$ and $\mathscr{G}$ are sheaves, then $\mathscr{G}$ is said to be a *subsheaf* of $\mathscr{F}$.

It is clear that if $\mathscr{G}$ is a subpresheaf of a sheaf $\mathscr{F}$, then ${}^a\mathscr{G}$ can be identified with a subsheaf of $\mathscr{F}$.

Definition. Let $\theta: \mathscr{F} \to \mathscr{G}$ be a homomorphism of presheaves. Then the *kernel* of θ is the unique subpresheaf Ker θ of $\mathscr{F}$ with $(\operatorname{Ker} \theta)(U) = \operatorname{Ker} \theta_U$ for all $U \in \Omega$.

If $\mathscr{F}$ and $\mathscr{G}$ are sheaves, then Ker θ is also a sheaf. Furthermore, for any $x \in X$, one has $(\operatorname{Ker} \theta)_x = \operatorname{Ker} \theta_x$.

Definition. Let $\theta: \mathscr{F} \to \mathscr{G}$ be a homomorphism of presheaves. Then the *presheaf image* preim θ of θ is the unique subpresheaf of $\mathscr{G}$ with $(\operatorname{preim} \theta)(U) = \operatorname{Im}(\theta_U)$ for all $U \in \Omega$.

Definition. If $\mathscr{G}$ is a sheaf, the sheafification of preim θ is a subsheaf of $\mathscr{G}$, which is said to be the sheaf image of θ, denoted by Im θ.

Note that even if $\mathscr{F}$ and $\mathscr{G}$ are sheaves, preim θ need not be a sheaf. But for every point $x \in X$,

$$(\operatorname{Im} \theta)_x = \operatorname{Im}(\theta_x) = (\operatorname{preim} \theta)_x.$$

Let $\theta: \mathscr{F} \to \mathscr{G}$ be a homomorphism of sheaves. One says that θ is *injective* or *one-to-one* if Ker $\theta = 0$, i.e., θ_U is injective for all $U \in \Omega$. The homomorphism θ is said to be *surjective* if Im $\theta = \mathscr{G}$, i.e., $\theta_x: \mathscr{F}_x \to \mathscr{G}_x$ is surjective for every $x \in X$. By Lemma 1.15, if θ is injective and surjective, then it is an isomorphism (cf. Lemma 1.19 and the remark immediately following).

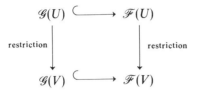

EXAMPLE 1.16. Let X be a topological space $\{*, p_1, p_2\}$, where $*$ is the generic point of X, and p_1, p_2 are closed points. For example, letting $A = \{f(\mathsf{X})/g(\mathsf{X}) | f, g \in \mathbb{Q}[\mathsf{X}]$ such that $g(0) \neq 0, g(1) \neq 0\}$, Spec A is an example of X. The open subsets of X are $U_1 = \{*, p_1\}$, $U_2 = \{*, p_2\}$, $U = U_1 \cap U_2 = \{*\}$, X, and $\varnothing$. Let $\mathbb{Q}[\mathsf{T}_1, \mathsf{T}_2]$ be a polynomial ring in two variables. Define sheaves $\mathscr{F}$ and $\mathscr{G}$ by $\mathscr{F}(X) = \mathbb{Q}$, $\mathscr{F}(U_i) = \mathbb{Q}[\mathsf{T}_i]$ $(i = 1, 2)$, $\mathscr{F}(U) = \mathbb{Q}[\mathsf{T}_1, \mathsf{T}_2]$, with the inclusion maps as restriction maps; $\mathscr{G}(X) = \mathscr{G}(U_1) = \mathscr{G}(U_2) = \mathscr{G}(U) = \mathbb{Q}[\mathsf{T}_1]$ with the identity maps as restriction maps. Furthermore, define a homomorphism $\theta\colon \mathscr{F} \to \mathscr{G}$ by $\theta(\mathsf{T}_i) = \mathsf{T}_1$ for $i = 1, 2$. Then, $\mathsf{T}_1 \in (\text{preim } \theta)(U_1) = (\text{preim } \theta)(U_2) = \mathbb{Q}[\mathsf{T}_1]$, but $\mathsf{T}_1 \notin (\text{preim } \theta)(X) = \mathbb{Q}$. Hence, preim θ is not a sheaf.

Let $\mathscr{G}$ be a subpresheaf of a presheaf $\mathscr{F}$. Then a presheaf $\mathscr{H}$ is defined as follows: $\mathscr{H}(U) = \mathscr{F}(U)/\mathscr{G}(U)$ for any $U \in \Omega$ and the restriction maps $\bar{r}_{V,U}\colon \mathscr{H}(U) \to \mathscr{H}(V)$ are derived from the $r_{V,U}\colon \mathscr{F}(U) \to \mathscr{F}(V)$ of $\mathscr{F}$, where U, $V \in \Omega$ with $U \supseteq V$.

Definition. The presheaf $\mathscr{H}$ defined above is denoted by $\mathscr{F}//\mathscr{G}$, and is called the *quotient presheaf* of $\mathscr{F}$ by $\mathscr{G}$.

Even if $\mathscr{F}$ and $\mathscr{G}$ are sheaves, $\mathscr{F}//\mathscr{G}$ need not be a sheaf. For example, in Example 1.16, $\mathscr{F}//\text{Ker } \theta \cong \text{preim } \theta$, which is not a sheaf.

Definition. If $\mathscr{G}$ is a subsheaf of a sheaf $\mathscr{F}$, the quotient sheaf of $\mathscr{F}$ by $\mathscr{G}$ is defined to be the sheafification of $\mathscr{F}//\mathscr{G}$, i.e., ${}^a(\mathscr{F}//\mathscr{G})$, which is denoted $\mathscr{F}/\mathscr{G}$.

Let $\pi_U\colon \mathscr{F}(U) \to \mathscr{F}(U)/\mathscr{G}(U)$ denote the natural homomorphism for all $U \in \Omega$. Then $\pi = \{\pi_U \mid U \in \Omega\}$ is a homomorphism from $\mathscr{F}$ to $\mathscr{F}//\mathscr{G}$. If $\mathscr{F}$ and $\mathscr{G}$ are sheaves, then a homomorphism ${}^a\pi\colon {}^a\mathscr{F} \to \mathscr{F}/\mathscr{G}$ is associated.

Definition. ${}^a\pi \circ \zeta_{\mathscr{F}}\colon \mathscr{F} \to \mathscr{F}/\mathscr{G}$ is said to be the *canonical homomorphism* of the quotient sheaf $\mathscr{F}/\mathscr{G}$, again denoted by π.

Remark. Since preim $\pi = \mathscr{F}//\mathscr{G}$, one has Im $\pi = \mathscr{F}/\mathscr{G}$, where $\mathscr{G}$ is a subsheaf of the sheaf $\mathscr{F}$.

Lemma 1.17. *Let $\mathscr{G}$ be a subsheaf of a sheaf $\mathscr{F}$. If $\mathscr{F}/\mathscr{G} = 0$, then $\mathscr{G} = \mathscr{F}$.*

PROOF. Since $\mathscr{G}_x = \mathscr{F}_x$ from $(\mathscr{F}/\mathscr{G})_x = 0$ for all $x \in X$, it follows that $\mathscr{G} = \mathscr{F}$ by Lemma 1.15. □

h. Definition. A sequence of presheaves $\mathscr{F}_i$ and homomorphisms θ_i for $i \in \mathbb{Z}$

$$\cdots \longrightarrow \mathscr{F}_i \xrightarrow{\theta_i} \mathscr{F}_{i+1} \xrightarrow{\theta_{i+1}} \mathscr{F}_{i+2} \longrightarrow \cdots \qquad (*)$$

is said to be *presheaf-exact*, if the sequence of Abelian groups

$$\cdots \longrightarrow \mathscr{F}_i(U) \xrightarrow{\theta_{iU}} \mathscr{F}_{i+1}(U) \xrightarrow{\theta_{i+1,U}} \mathscr{F}_{i+2}(U) \longrightarrow \cdots \quad (*)(U)$$

is exact for all U; i.e., preim $\theta_i =$ Ker θ_{i+1} for all i.

Definition. A sequence of sheaves $\mathscr{F}_i$ and homomorphisms θ_i

$$\cdots \longrightarrow \mathscr{F}_i \xrightarrow{\theta_i} \mathscr{F}_{i+1} \xrightarrow{\theta_{i+1}} \mathscr{F}_{i+2} \longrightarrow \cdots \quad (*)$$

is said to be *sheaf-exact*, if the sequence

$$\cdots \longrightarrow \mathscr{F}_{i,x} \xrightarrow{\theta_{ix}} \mathscr{F}_{i+1,x} \xrightarrow{\theta_{i+1,x}} \mathscr{F}_{i+2,x} \longrightarrow \cdots \quad (*)_x$$

is exact for all $x \in X$.

One can easily prove the next result.

Lemma 1.18. *If a sequence of presheaves*

$$\cdots \longrightarrow \mathscr{F}_i \xrightarrow{\theta_i} \mathscr{F}_{i+1} \xrightarrow{\theta_{i+1}} \mathscr{F}_{i+2} \longrightarrow \cdots$$

is presheaf-exact, then the sequence of sheafifications of the $\mathscr{F}_i$

$$\cdots \longrightarrow {}^a\mathscr{F}_i \xrightarrow{{}^a\theta_i} {}^a\mathscr{F}_{i+1} \xrightarrow{{}^a\theta_{i+1}} {}^a\mathscr{F}_{i+2} \longrightarrow \cdots$$

is sheaf-exact.

Proof. This is obtained from the fact that direct limit is an exact functor. □

The converse of the above lemma does not hold in general. However, one can prove a slightly weaker result.

Lemma 1.19. *If a sequence of sheaves*

$$0 \longrightarrow \mathscr{G} \xrightarrow{\theta} \mathscr{F} \xrightarrow{\eta} \mathscr{H}$$

is sheaf-exact, then it is also presheaf-exact.

Proof. Since Ker θ is a sheaf and (Ker $\theta)_x =$ Ker $\theta_x = 0$ for all $x \in X$, it follows that Ker $\theta = 0$, i.e., Ker $\theta_U = 0$ for all $U \in \Omega$. Similarly, one has $\eta \circ \theta = 0$; hence Im $\theta \subseteq$ Ker η. But since Im $\theta_x =$ Ker η_x for all $x \in X$, one has Im $\theta =$ Ker η by Lemma 1.17. In view of the isomorphism $\mathscr{G} \cong$ preim θ, preim θ is a sheaf Im θ. Hence, one has preim $\theta =$ Ker η. □

Remark. By applying the last lemma to $\mathscr{H} = 0$, one sees that if $0 \to \mathscr{G} \xrightarrow{\theta} \mathscr{F} \to 0$ is sheaf-exact, then θ is an isomorphism, i.e., θ_U is an isomorphism for all $U \in \Omega$. This is generalized as follows.

Proposition 1.14. *If a sequence of sheaves*

$$0 \to \mathscr{G} \to \mathscr{F} \to \mathscr{H} \to 0$$

is sheaf-exact, then $\mathscr{F}/\mathscr{G} \cong \mathscr{H}$ *canonically.*

PROOF. Since one has the exact sequences

$$0 \to \mathscr{G}(U) \to \mathscr{F}(U) \to \mathscr{H}(U)$$

for all $U \in \Omega$ by Lemma 1.19, the homomorphisms $(\mathscr{F}//\mathscr{G})(U) \to \mathscr{H}(U)$ are obtained, i.e. the homomorphism $\mathscr{F}//\mathscr{G} \to \mathscr{H}$ is obtained. Denoting this last homomorphism by π, we have $\eta = {}^a\pi\colon \mathscr{F}/\mathscr{G} \to \mathscr{H}$ is defined. Since η_x: $(\mathscr{F}/\mathscr{G})_x \cong \mathscr{F}_x/\mathscr{G}_x \to \mathscr{H}_x$ is an isomorphism for every $x \in X$, η is an isomorphism by Lemma 1.15. □

For any sheaf homomorphism $\theta\colon \mathscr{G} \to \mathscr{F}$, one has the sheaf-exact sequence

$$0 \to \operatorname{Ker}\theta \to \mathscr{G} \to \operatorname{Im}\theta \to 0.$$

Hence, $\mathscr{G}/\operatorname{Ker}\theta \cong \operatorname{Im}\theta$.

Definition. The *cokernel* of θ is defined to be the sheaf $\mathscr{F}/\operatorname{Im}\theta$, denoted by $\operatorname{Coker}\theta$.

Then one has the sheaf-exact sequence

$$0 \to \operatorname{Im}\theta \to \mathscr{F} \to \operatorname{Coker}\theta \to 0.$$

Proposition 1.15. *A sequence of sheaves*

$$\cdots \longrightarrow \mathscr{F}_i \xrightarrow{\theta_i} \mathscr{F}_{i+1} \xrightarrow{\theta_{i+1}} \mathscr{F}_{i+2} \longrightarrow \cdots$$

is sheaf-exact if and only if $\operatorname{Ker}\theta_{i+1} = \operatorname{Im}\theta_i$ *for all i.*

PROOF. If the sequence is exact, then $(\theta_{i+1} \circ \theta_i)_x = \theta_{i+1,x} \circ \theta_{i,x} = 0$ for all $x \in X$; hence $\theta_{i+1} \circ \theta_i = 0$ by Lemma 1.15. Thus $\operatorname{Im}\theta_i \subseteq \operatorname{Ker}\theta_{i+1}$, and $(\operatorname{Ker}\theta_{i+1}/\operatorname{Im}\theta_i)_x = 0$ for all $x \in X$. Hence, one has $\operatorname{Im}\theta_i = \operatorname{Ker}\theta_{i+1}$ by Lemma 1.17. The converse is obvious. □

Let $\mathscr{O}$ be a sheaf of rings on X, and $\mathscr{F}$, $\mathscr{G}$ be $\mathscr{O}$-modules. If $\theta\colon \mathscr{G} \to \mathscr{F}$ is an $\mathscr{O}$-homomorphism, then $\operatorname{Ker}\theta$, $\operatorname{Im}\theta$, and $\operatorname{Coker}\theta$ are naturally $\mathscr{O}$-modules. Lemmas 1.14 through 1.19 and Propositions 1.13, 1.14, and 1.15 hold even for $\mathscr{O}$-modules and $\mathscr{O}$-homomorphisms.

Since the $\mathcal{O}$-modules and sheaf-exact sequences are the main objects in our study, a sheaf-exact sequence of $\mathcal{O}$-modules is referred to as an exact sequence of $\mathcal{O}$-modules.

i. Definition. Let $\mathcal{O}$ be a presheaf of rings on X and let $\mathscr{F} = (\{\mathscr{F}(U)\}, \{r_{V,U}\})$ and $\mathscr{G} = (\{\mathscr{G}(U)\}, \{s_{V,U}\})$ be presheaves of $\mathcal{O}$-modules. Then $(\{\mathscr{F}(U)\otimes_{\mathcal{O}(U)}\mathscr{G}(U)\}, \{r_{V,U}\otimes s_{V,U}\})$ is also a presheaf of $\mathcal{O}$-modules, called the *presheaf tensor product* of $\mathscr{F}$ and $\mathscr{G}$ over $\mathcal{O}$.

Even if $\mathcal{O}$, $\mathscr{F}$, and $\mathscr{G}$ are sheaves, this tensor product need not be a sheaf, as will be shown in the next example.

EXAMPLE 1.17. We use the notation of Example 1.16. Define $\mathcal{O}$ by $\mathcal{O}(U) = \mathbb{Q}[\mathsf{T}_1, \mathsf{T}_2]$ and $\mathcal{O}(U_1) = \mathcal{O}(U_2) = \mathcal{O}(X) = \mathbb{Q}$, and let $r_{V,W}$ be the inclusion map if $V \neq \varnothing$ and let $r_{\varnothing,W} = 0$. Furthermore, define $\mathscr{H}$ by $\mathscr{H}(U) = \mathbb{Q}[\mathsf{T}_1, \mathsf{T}_2]$, $\mathscr{H}(U_1) = \mathbb{Q}[\mathsf{T}_2]$, $\mathscr{H}(U_2) = \mathbb{Q}[\mathsf{T}_1]$, $\mathscr{H}(X) = \mathbb{Q}$, and let $r_{V,W}$ be the inclusion map if $V \neq \varnothing$ and let $r_{\varnothing,W} = 0$. $\mathscr{F}$ and $\mathscr{H}$ are $\mathcal{O}$-modules in the natural way. If $\mathscr{P}$ is the presheaf tensor product of $\mathscr{F}$ and $\mathscr{H}$ over $\mathcal{O}$, then $\mathscr{P}(U) = \mathbb{Q}[\mathsf{T}_1, \mathsf{T}_2]$, $\mathscr{P}(U_1) = \mathbb{Q}[\mathsf{T}_1\otimes 1, 1\otimes\mathsf{T}_2]$, $\mathscr{P}(U_2) = \mathbb{Q}[\mathsf{T}_2\otimes 1, 1\otimes\mathsf{T}_1]$, and $\mathscr{P}(X) = \mathbb{Q}$. Furthermore, one has $(\mathsf{T}_1\otimes 1)|_U = \mathsf{T}_1$, $(\mathsf{T}_2\otimes 1)|_U = \mathsf{T}_2$, $(1\otimes\mathsf{T}_1)|_U = \mathsf{T}_1$, and $(1\otimes\mathsf{T}_2)|_U = \mathsf{T}_2$. Hence, $(\mathsf{T}_i\otimes 1)|_U = (1\otimes\mathsf{T}_i)|_U = \mathsf{T}_i$ for $i = 1, 2$, but $\mathscr{P}(U_1\cup U_2) = \mathscr{P}(X) = \mathbb{Q}$.

Definition. If $\mathcal{O}$ is a sheaf and $\mathscr{F}$ and $\mathscr{G}$ are (sheaves of) $\mathcal{O}$-modules, the *tensor product* of $\mathscr{F}$ and $\mathscr{G}$ over $\mathcal{O}$ is the sheafification of the presheaf tensor product of $\mathscr{F}$ and $\mathscr{G}$ over $\mathcal{O}$, which is denoted by $\mathscr{F}\otimes_{\mathcal{O}}\mathscr{G}$ or simply $\mathscr{F}\otimes\mathscr{G}$. $\mathscr{F}\otimes_{\mathcal{O}}\mathscr{G}$ is an $\mathcal{O}$-module.

Proposition 1.16

(i) *$(\mathscr{F}\otimes_{\mathcal{O}}\mathscr{G})_x \cong \mathscr{F}_x\otimes_{\mathcal{O}_x}\mathscr{G}_x$ for any $x\in X$.*
(ii) *$\mathcal{O}\otimes_{\mathcal{O}}\mathscr{F}\cong\mathscr{F}$.*
(iii) *$(\mathscr{F}\otimes_{\mathcal{O}}\mathscr{G})\otimes_{\mathcal{O}}\mathscr{H}\cong\mathscr{F}\otimes_{\mathcal{O}}(\mathscr{G}\otimes_{\mathcal{O}}\mathscr{H})$, where $\mathscr{F}$, $\mathscr{G}$, and $\mathscr{H}$ are $\mathcal{O}$-modules.*
(iv) *$(\mathscr{F}\otimes_{\mathcal{O}}\mathscr{G})|_U\cong(\mathscr{F}|_U)\otimes_{\mathcal{O}|_U}(\mathscr{G}|_U)$ for any open subset U of X.*
(v) *If a sequence of $\mathcal{O}$-modules*

$$\mathscr{F}'\to\mathscr{F}\to\mathscr{F}''\to 0$$

is exact, then for any $\mathcal{O}$-module $\mathscr{G}$, the sequence

$$\mathscr{F}'\otimes_{\mathcal{O}}\mathscr{G}\to\mathscr{F}\otimes_{\mathcal{O}}\mathscr{G}\to\mathscr{F}''\otimes_{\mathcal{O}}\mathscr{G}\to 0$$

is exact.

PROOF. (i) By Lemma 1.16,

$$(\mathscr{F} \otimes_{\mathcal{O}} \mathscr{G})_x \cong \operatorname*{inj\,lim}_{U \in \Omega_x} \mathscr{F}(U) \otimes_{\mathcal{O}(U)} \mathscr{G}(U).$$

It is easy to check that the right-hand side is isomorphic to $\mathscr{F}_x \otimes_{\mathcal{O}_x} \mathscr{G}_x$.

(ii) This follows from $\mathcal{O}(U) \otimes_{\mathcal{O}(U)} \mathscr{F}(U) \cong \mathscr{F}(U)$ for any $U \in \Omega$.

(iii) This follows from

$$(\mathscr{F}(U) \otimes_{\mathcal{O}(U)} \mathscr{G}(U)) \otimes_{\mathcal{O}(U)} \mathscr{H}(U) \cong \mathscr{F}(U) \otimes_{\mathcal{O}(U)} (\mathscr{G}(U) \otimes_{\mathcal{O}(U)} \mathscr{H}(U))$$

for any $U \in \Omega$.

(iv) Let $\mathscr{P}$ be the presheaf tensor product of $\mathscr{F}$ and $\mathscr{G}$ over $\mathcal{O}$. Then $({}^a\mathscr{P})|_U \rightrightarrows {}^a(\mathscr{P}|_U)$. Clearly, we have ${}^a(\mathscr{P}|_U) \cong \mathscr{F}|_U \otimes_{\mathcal{O}|_U} \mathscr{G}|_U$ and ${}^a\mathscr{P}|_U \cong (\mathscr{F} \otimes_{\mathcal{O}} \mathscr{G})|_U$.

(v) This follows from (i). □

j. Definition. Let $\mathcal{O}$ be a sheaf of rings, and let $\mathscr{F}$, $\mathscr{G}$ be $\mathcal{O}$-modules. The presheaf $(\{\mathrm{Hom}_{\mathcal{O}|_U}(\mathscr{F}|_U, \mathscr{G}|_U) \mid U \in \Omega\}, \{\text{natural restriction maps}\})$ is a sheaf, denoted by $\mathscr{H}om_{\mathcal{O}}(\mathscr{F}, \mathscr{G})$.

We shall prove the following properties of this sheaf.

Proposition 1.17.

(i) *There is a canonical homomorphism*

$$u_{\langle x \rangle}\colon \mathscr{H}om_{\mathcal{O}}(\mathscr{F}, \mathscr{G})_x \to \mathrm{Hom}_{\mathcal{O}_x}(\mathscr{F}_x, \mathscr{G}_x)$$

for every point $x \in X$.

(ii) $\mathscr{H}om_{\mathcal{O}}(\mathcal{O}, \mathscr{F}) \cong \mathscr{F}$.

(iii) *If a sequence of $\mathcal{O}$-modules*

$$0 \to \mathscr{F}' \to \mathscr{F} \to \mathscr{F}'' \to 0$$

is exact, then both sequences of $\mathcal{O}$-modules

$$0 \to \mathscr{H}om_{\mathcal{O}}(\mathscr{F}'', \mathscr{G}) \to \mathscr{H}om_{\mathcal{O}}(\mathscr{F}, \mathscr{G}) \to \mathscr{H}om_{\mathcal{O}}(\mathscr{F}', \mathscr{G}),$$

and

$$0 \to \mathscr{H}om_{\mathcal{O}}(\mathscr{G}, \mathscr{F}') \to \mathscr{H}om_{\mathcal{O}}(\mathscr{G}, \mathscr{F}) \to \mathscr{H}om_{\mathcal{O}}(\mathscr{G}, \mathscr{F}'')$$

are exact for any $\mathcal{O}$-module $\mathscr{G}$.

PROOF. For any $U \in \Omega_x$, we have the natural homomorphism

$$\mathrm{Hom}_{\mathcal{O}|_U}(\mathscr{F}|_U, \mathscr{G}|_U) \to \mathrm{Hom}_{\mathcal{O}_x}(\mathscr{F}_x, \mathscr{G}_x)$$

defined by $\theta \mapsto \theta_x$ (see §1.10.c), and so we obtain

$$u_{\langle x \rangle}\colon \mathscr{H}om_{\mathcal{O}}(\mathscr{F}, \mathscr{G})_x \to \mathrm{Hom}_{\mathcal{O}_x}(\mathscr{F}_x, \mathscr{G}_x)$$

by $u_{\langle x \rangle}(\theta) = \theta_x$. The proofs of (ii) and (iii) are quite easy. □

Remark. Note that $u_{\langle x \rangle}$ in (i) is neither injective nor surjective in general.

EXAMPLE 1.18. Let X be a topological space $\{*, p\}$ such that $*$ is the generic point and p is a closed point (see §1.2.**c**). If $\mathscr{F}$ and $\mathscr{G}$ are $\mathcal{O}$ $(=\mathbb{Z}_X)$-modules, then $\mathscr{F}_p = \mathscr{F}(X)$ and $\mathscr{G}_p = \mathscr{G}(X)$. Furthermore, $\mathscr{H}om_{\mathcal{O}}(\mathscr{F}, \mathscr{G})_p = \mathrm{Hom}_{\mathcal{O}}(\mathscr{F}, \mathscr{G}) = \{(\varphi, \psi) \mid \varphi\colon \mathscr{F}(X) \to \mathscr{G}(X),\ \psi\colon \mathscr{F}(U) \to \mathscr{G}(U)$ with $\psi(\alpha|_U) = \varphi(\alpha)|_U$ for $\alpha \in \mathscr{F}(X)\}$, U being $\{*\}$. Hence, $u_{\langle p \rangle}(\varphi, \psi) = \varphi$.

§1.11 Structure Sheaves on Spectra

a. Let A be a ring. We wish to define a sheaf $\tilde{A}$ of rings on the spectrum Spec A such that $\tilde{A}(D(f)) = A_f$ for all $f \in A$. It is easy to see that if $D(f) = D(g)$, then the A-algebras A_f and A_g are A-isomorphic. However, in order to define a sheaf $\tilde{A}$, one has to associate a uniquely determined ring A_f with every open set of the type $D(f)$.

Lemma 1.20. *Let f and $g \in A$. If $D(f) = D(g)$, then $I(f) = I(g)$, where $I(f)$ is the kernel of the canonical homomorphism $\psi_f\colon A \to A_f$ (cf. Corollary to Proposition* 1.2).

PROOF. By Corollary (ii) to Proposition 1.3, there exist $m, n \in \mathbb{N}$ and $a, b \in A$ such that $f^m = ag$ and $g^n = bf$. If $\alpha \in I(f)$, then $f^r\alpha = 0$ for some $r \in \mathbb{N}$; hence $g^{nr}\alpha = b^r f^r \alpha = 0$. Thus $I(f) \subseteq I(g)$. Similarly one can prove $I(g) \subseteq I(f)$ and so $I(f) = I(g)$. □

Therefore, if $D(f) = D(g)$, one has the ring $\bar{A} = A/I(f) = A/I(g)$, in which $\bar{f} = f \bmod I(f)$ is not a zero divisor. In fact, if $\bar{\alpha} \cdot \bar{f} = 0$, where $\bar{\alpha}$ is α mod $I(f)$, then $\alpha f \in I(f)$, i.e., $\alpha f \cdot f^s = 0$ for some $s > 0$, i.e., $\alpha \cdot f^{s+1} = 0$. This implies $\alpha \in I(f)$, i.e., $\bar{\alpha} = 0$. Similarly, $\bar{g} = g \bmod I(f)$ is not a zero divisor. Hence, $\bar{f}$ and $\bar{g}$ have inverses $1/\bar{f}$ and $1/\bar{g}$ in the total quotient ring $Q(\bar{A})$ of $\bar{A}$ (§1.4.**a**). Furthermore, from $\bar{f}^m = \bar{a} \cdot \bar{g}$ and $\bar{g}^n = \bar{b} \cdot \bar{f}$, we conclude $\bar{A}[1/\bar{f}] = \bar{A}[1/\bar{g}]$ which is an $\bar{A}$-subalgebra of $Q(\bar{A})$. Clearly, $\bar{A}[1/\bar{f}] \cong A[\mathsf{X}]/(f\mathsf{X} - 1)$ as an A-algebra. Throughout this section, we denote $\bar{A}[1/\bar{f}]$ by A_f. Thus by the above argument, if $D(f) = D(g)$, then $A_f = A_g$.

Now, if $D(f) \supseteq D(g)$, then $D(f) \supseteq D(g) = D(f) \cap D(g) = D(fg)$ and so $I(f) \subseteq I(fg) = I(g)$ by Lemma 1.20. The natural homomorphism $A/I(f) \to A/I(g)$ induces an A-homomorphism $A_f \to A_g = A_{fg}$, which is denoted by $r_{g,f}$. If $D(f) \supseteq D(g) \supseteq D(h)$, one can prove that $r_{h,f} = r_{h,g} \circ r_{g,f}$. Furthermore, $r_{f,1} = \psi_f\colon A \to A_f$. Since the family of open subsets of the form $D(f), f \in A$, is an open base for the topology of Spec A, the following lemma shows that $\mathscr{F} = (\{A_f \text{ for } D(f) \mid f \in A\}, \{r_{g,f} \mid D(f) \supseteq D(g)\})$ determines a sheaf of rings.

Lemma 1.21. *Let X be a topological space and let $\mathbb{B}$ be an open base for the topology on X. Suppose that $(\{\mathscr{F}(U) \mid U \in \mathbb{B}\}, \{r_{V,U} \mid U, V \in \mathbb{B}$ and $U \supseteq V\})$ is a pair consisting of a collection of Abelian groups (respectively rings) and a collection of group homomorphisms (respectively ring homomorphisms) satisfying conditions (1), (2), and (3) in §1.10.**a** with $\mathbb{B}$ in place of Ω. Then there exists a sheaf $\mathscr{F}^* = (\{\mathscr{F}^*(U)\}, \{r^*_{V,U}\})$ of Abelian groups (respectively rings) such that $\mathscr{F}^*_x = \mathscr{F}_x = \text{injlim}_{U \in \mathbb{B}_x} \mathscr{F}(U)$, where $\mathbb{B}_x = \{U \in \mathbb{B} \cap \Omega_x\}$ for all $x \in X$. Furthermore, if $\mathscr{F}$ satisfies (4) and (5) with $\mathbb{B}$ in place of Ω, then $\mathscr{F}^*(U)$ can be identified with $\mathscr{F}(U)$ and then $r^*_{V,U} = r_{V,U}$, where $U, V \in \mathbb{B}$ and $U \supseteq V$. Such a $\mathscr{F}^*$ is unique up to isomorphism.*

PROOF. First form an étale space $\mathscr{F}^+ = (\mathscr{F}^+, \pi, X)$ with a topology on $\mathscr{F}^+ = \coprod_{x \in X} \mathscr{F}_x$ defined by taking $\{\mathbf{W}(\alpha, U) \mid \alpha \in \mathscr{F}(U), U \in \mathbb{B}\}$ as an open base. $\mathscr{F}^{+b}$ is a sheaf and $\zeta_{\mathscr{F}}: \mathscr{F} \to \mathscr{F}^{+b}$ is canonically defined as in §1.10.**e**,**f**. For $x \in X$, one has $\mathscr{F}^{+b}_x \cong \text{injlim}_{U \in \mathbb{B}_x} \mathscr{F}(U)$. Now suppose that $\mathscr{F}$ satisfies (4) and (5). Then by the similar argument to the proof of Lemma 1.13, one can prove $\mathscr{F}^*(U) \cong \mathscr{F}(U)$ for any $U \in \mathbb{B}$. $\mathscr{F}^*$ defined as $\mathscr{F}^{+b}$ has the required property. □

b. In order to prove that $(\{A_f\}, \{r_{g,f}\})$ satisfies conditions (4) and (5), we will first prove the next result.

Lemma 1.22. *If $g, f_j \in A$, for $j \in J$, satisfy $\bigcup_{j \in J} D(f_j) = D(g)$, then there exists a finite subset F of J such that for any given m there exist $a_i \in A$, $i \in F$,* with $\sum_{i \in F} a_i f_i^m = g^n$ *for some $n > 0$.*

PROOF. Since $D(g)$ is quasi-compact by Theorem 1.2, we have a finite subset F of J such that $D(g) = \bigcup_{i \in F} D(f_i) = \bigcup_{i \in F} D(f_i^m) = D(\sum_{i \in F} Af_i^m)$. Hence, $g^n \in \sum_{i \in F} Af_i^m$ for some n by Corollary (ii) to Proposition 1.3. □

Proposition 1.18. *Suppose that $\{D(f_j) \mid j \in J\}$ is an open cover of $D(g)$.*

(i) If $\alpha \in A_g$ satisfies $r_{f_j, g}(\alpha) = 0$ for all $j \in J$, then $\alpha = 0$.
(ii) If $\alpha_j \in A_{f_j}$, $j \in J$, are given such that $r_{f_i f_j, f_j}(\alpha_j) = r_{f_i f_j, f_i}(\alpha_i)$ for all i, $j \in J$, then there exists $\alpha \in A_g$ such that $r_{f_i, g}(\alpha) = \alpha_i$ for all i.

PROOF. Clearly, we may assume that $g = 1$ and J is a finite set F.

(i) Since $\psi_{f_i}(\alpha) = 0$, there exists m_i such that $f_i^{m_i} \alpha = 0$. Since F is finite, we can choose m_i, $i \in F$, independently of i, i.e., $m = m_i$. By Lemma 1.22, there exist $a_i \in A$ for all $i \in F$, such that $1 = \sum_{i \in F} a_i f_i^m$ and so $\alpha = \sum_{i \in F} \alpha a_i f_i^m = 0$.

(ii) Each α_i can be written in the form b_i/f_i^l where $b_i \in A$ and l can be chosen independently of i. Since $r_{f_i f_j, f_i}(\alpha_i) = r_{f_i f_j, f_j}(\alpha_j)$, there exists m such that $f_i^m f_j^{m+l} b_i = f_j^m f_i^{m+l} b_j$, where m can be chosen independently of i, j. By

Lemma 1.22, one has $c_i \in A (i \in F)$ such that $1 = \sum_{i \in F} c_i f_i^{m+l}$. Now, $\alpha = \sum_{i \in F} c_i f_i^m b_i$ satisfies

$$\alpha f_j^{m+l} = \sum_{i \in F} c_i f_i^m b_i f_j^{m+l} = \sum_{i \in F} c_i f_j^m b_j f_i^{m+l} = f_j^m b_j.$$

Hence, $r_{f_j, 1}(\alpha) = b_j / f_j^l = \alpha_j$ for all $j \in F$. □

Thus, by the last proposition and Lemma 1.21, we have a sheaf $\mathscr{F}^*$ of rings on Spec A such that $\mathscr{F}^*(D(f)) = A_f$ for any $f \in A$. The sheaf $\mathscr{F}^*$ is denoted by $\tilde{A}$.

c. Definition. The pair (Spec A, $\tilde{A}$) is said to be an *affine scheme* associated with the ring A. The sheaf $\tilde{A}$ is said to be the *structure sheaf* of the affine scheme.

Proposition 1.19. *If* $\mathfrak{q} \in$ Spec A, *the stalk* $(\tilde{A})_\mathfrak{q}$ *is naturally isomorphic as an* A*-algebra to the local ring* $A_\mathfrak{q}$.

Proof. By definition, one has

$$\tilde{A}_\mathfrak{q} = \operatorname*{injlim}_{D(f) \ni \mathfrak{q}} \tilde{A}(D(f)) \cong \operatorname*{injlim}_{f \notin \mathfrak{q}} A_f.$$

The next lemma shows that the right-hand side is isomorphic to $A_\mathfrak{q}$. □

Lemma 1.23. *Let* S *be a multiplicative subset of* A *and let* M *be an* A*-module. Then* $\operatorname{injlim}_{f \in S} A_f \cong S^{-1} A$ *and* $\operatorname{injlim}_{f \in S} M \otimes_A A_f \cong S^{-1} M$.

Proof. Any $\alpha \in \operatorname{injlim}_{f \in S} A_f$ is represented by $a/f^n \in A_f$, and in the direct limit, $a/f^n \sim b/g^m$, where $b/g^m \in A_g$, if and only if $ag^m h = bf^n h$ for some $h \in S$. Hence, $\alpha \mapsto a/f^n \in S^{-1}A$ gives rise to the required isomorphism. One can prove the second assertion similarly. □

d. Let X be a topological space and let $\mathcal{O}$ be a sheaf of rings on X.

Definition. The pair $(X, \mathcal{O})$ is said to be a *ringed space* with base space X and structure sheaf $\mathcal{O}$.

Remarks (1) When no confusion can arise, $(X, \mathcal{O})$ is referred to simply as X. Then the structure sheaf $\mathcal{O}$ is denoted by $\mathcal{O}_X$.

(2) If U is an open subset of X, then U becomes a ringed space with $\mathcal{O}_U = \mathcal{O}_X|_U$.

Definition. If all stalks $\mathcal{O}_{X,x}$ of a ringed space $(X, \mathcal{O}_X)$ are local rings, then $(X, \mathcal{O}_X)$ is said to be a *local ringed space*.

By Proposition 1.19, any affine scheme is a local ringed space. $(\mathbb{C}^n, \mathcal{O}^{\text{hol}})$ in Example 1.16 is also a local ringed space.

The affine scheme (Spec A, $\tilde{A}$) is often referred to as Spec A. For example, if k is a field, Spec k is not just a topological space $\{p\}$, but a pair consisting of a point p and a field k. Spec 0 is a pair ($\emptyset$, 0), called an empty scheme.

Definition. Let R be a ring, and let A be the polynomial ring $R[\mathsf{X}_1, \ldots, \mathsf{X}_n]$. Then (Spec A, $\tilde{A}$) is said to be the *affine n-space* over R, denoted by $\mathbf{A}_R^n$.

The *dimension* of a ringed space $(X, \mathcal{O}_X)$ is defined to be the dimension of X. Hence, dim Spec A = dim A. For example, if R is a field k, then dim $\mathbf{A}_k^n = \dim k[\mathsf{X}_1, \ldots, \mathsf{X}_n] = n$, by Theorem 1.8.

e. Let M be an A-module. For any $f \in A$, we define M_f to be $M \otimes_A A_f$, and $\psi_{M,f}\colon M \to M_f$ to be $\psi_{M,S}$, where $S = S_f = \{f^n \mid n \geq 0\}$. Then, by the same argument as in §1.11.**b**, one obtains the sheaf $\mathscr{F}$ such that $\mathscr{F}(D(f)) \cong M_f$ for any $f \in A$. Since M_f is naturally an A_f-module, $\mathscr{F}$ becomes an $\tilde{A}$-module, which is denoted $\tilde{M}$. Note that $M \cong \tilde{M}(X) \cong \mathscr{F}(X)$.

Definition. The $\tilde{A}$-module $\tilde{M}$ is said to be the *$\tilde{A}$-module associated with M*.

Remark. By Lemma 1.23, one obtains a result similar to Proposition 1.19, i.e., for any $\mathfrak{q} \in$ Spec A, the stalk $(\tilde{M})_{\mathfrak{q}}$ is $A_{\mathfrak{q}}$-isomorphic to $M_{\mathfrak{q}} = M \otimes_A A_{\mathfrak{q}}$.

Let M and N be A-modules and let $\varphi\colon M \to N$ be an A-homomorphism. For any $f \in A$, one has the A_f-homomorphism $\varphi_f\colon M_f \to N_f$ defined by $\varphi_f(x/f^m) = \varphi(x)/f^m$ for any $x \in M$ and $m \geq 0$. Then, if $D(f) \supseteq D(g)$, one has the commutative square:

$$\begin{array}{ccc} M_f & \xrightarrow{\varphi_f} & N_f \\ {\scriptstyle r_{g,f}}\downarrow & & \downarrow{\scriptstyle r_{g,f}} \\ M_g & \xrightarrow{\varphi_g} & N_g \end{array}$$

Since $\mathbb{B} = \{D(f) \mid f \in A\}$ is an open base, $\{\varphi_f \mid f \in A\}$ induces an $\tilde{A}$-homomorphism $\tilde{M} \to \tilde{N}$, denoted by $\tilde{\varphi}$, i.e., $\tilde{\varphi}_{D(f)} = \varphi_f$.

Lemma 1.24. *Let M,N be A-modules and let $\theta\colon \tilde{M} \to \tilde{N}$ be an $\tilde{A}$-homomorphism over $X =$ Spec A. Then $\theta = \tilde{\varphi}$, where $\varphi = \theta_X$.*

Proof. One has the canonical homomorphisms $\psi_{M,f}\colon M \to M_f$ and $\psi_{N,f}\colon N \to N_f$, for any $f \in A$. Then by the definition of θ, one has the commutative square:

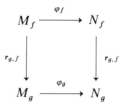

$$\begin{array}{ccc} M & \xrightarrow{\theta_X} & N \\ {\scriptstyle \psi_{M,f}}\downarrow & & \downarrow{\scriptstyle \psi_{N,f}} \\ M_f & \xrightarrow{\theta_{D(f)}} & N_f \end{array}$$

By the universal property of $(M_f, \psi_{M,f})$, $\theta_{D(f)}$ coincides with φ_f. Hence, $\theta = \tilde{\varphi}$. □

Thus one has the isomorphism $\mathrm{Hom}_A(M, N) \cong \mathrm{Hom}_{\tilde{A}}(\tilde{M}, \tilde{N})$.

f. For any $f \in A$, $D(f)$ is an open subset of Spec A; hence one has an affine scheme $(D(f), \tilde{A}|_{D(f)})$. Then identifying $D(f)$ with Spec A_f (Example 1.4), one has an isomorphism $\tilde{A}|_{D(f)} \cong \tilde{A}_f$. More generally, if M is an A-module, then $\tilde{M}|_{D(f)} \cong \tilde{M}_f$. Since this is easily verified, we leave it to the reader as an exercise. Thus, in particular, $\mathrm{Hom}_{\tilde{A}|D(f)}(\tilde{M}|_{D(f)}, \tilde{N}|_{D(f)}) \cong \mathrm{Hom}_{\tilde{A}_f}(\tilde{M}_f, \tilde{N}_f) \cong \mathrm{Hom}_{A_f}(M_f, N_f)$. Similarly, one checks that $\tilde{M} \otimes_{\tilde{A}} \tilde{N} \cong (M \otimes_A N)^{\sim}$, since $M_f \otimes_{A_f} N_f \cong (M \otimes_A N) \otimes_A A_f$.

Proposition 1.20. *If a sequence of A-modules*

$$0 \longrightarrow M' \xrightarrow{\varphi} M \xrightarrow{\psi} M'' \longrightarrow 0$$

is exact, then the associated sequence

$$0 \longrightarrow \tilde{M}' \xrightarrow{\tilde{\varphi}} \tilde{M} \xrightarrow{\tilde{\psi}} \tilde{M}'' \longrightarrow 0$$

is also an exact sequence of $\tilde{A}$-modules

PROOF. By Lemma 1.2.(ii), one has the exact sequences of $A_{\mathfrak{p}}$-modules for any point $\mathfrak{p}$ of Spec A:

$$0 \longrightarrow M'_{\mathfrak{p}} \longrightarrow M_{\mathfrak{p}} \longrightarrow M''_{\mathfrak{p}} \longrightarrow 0.$$

But since $M_{\mathfrak{p}} = \tilde{M}_{\mathfrak{p}}$ by the remark in §1.11.**e**, one obtains the exactness of the sequence of $\tilde{A}$-modules. □

Corollary. *If a sequence of $\tilde{A}$-modules*

$$0 \longrightarrow \tilde{M}' \xrightarrow{\theta} \tilde{M} \xrightarrow{\eta} \tilde{M}'' \longrightarrow 0$$

is exact, then the sequence

$$0 \longrightarrow M' \xrightarrow{\theta_X} M \xrightarrow{\eta_X} M'' \longrightarrow 0,$$

is also exact, where $X =$ Spec A.

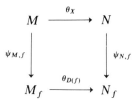

PROOF. According to Lemma 1.19, it suffices to prove that η_X is surjective. One has an exact sequence $M \to M'' \to \text{Coker } \eta_X \to 0$. By the last proposition, the sequence

$$\tilde{M} \longrightarrow \tilde{M}'' \longrightarrow (\text{Coker } \eta_X)^{\sim} \longrightarrow 0$$

is exact. Since η is surjective, one has $(\text{Coker } \eta_X)^{\sim} = 0$. Then $\text{Coker } \eta_X = \Gamma(X, (\text{Coker } \eta_X)^{\sim}) = 0$ (cf. subsection **e**). □

§1.12 Quasi-coherent Sheaves and Coherent Sheaves

a. Let $(X, \mathcal{O}_X)$ be a ringed space and let $\{\mathscr{F}_i \mid i \in I\}$ be a collection of $\mathcal{O}_X$-modules. $\mathscr{F}_{(*)} = (\{\bigoplus_{i \in I} \mathscr{F}_i(U) \mid U \in \Omega\}, \{\text{natural restriction maps}\})$ is a presheaf of $\mathcal{O}_X$-modules. If I is a finite set, one can prove easily that $\mathscr{F}_{(*)}$ is a sheaf, and more precisely, an $\mathcal{O}_X$-module.

Definition. For any index set I, $\bigoplus_{i \in I} \mathscr{F}_i$ denotes the sheafification of $\mathscr{F}_{(*)}$, called the *direct sum* of $\mathscr{F}_i$, $i \in I$. If $\mathscr{F}_i = \mathscr{F}$ for all $i \in I$, we write $\mathscr{F}^{(I)}$ instead of $\bigoplus_{i \in I} \mathscr{F}$. If I is a finite set of n elements, we write $\mathscr{F}^n$ for $\mathscr{F}^{(I)}$. Furthermore, $\mathcal{O}_X^{(I)}$ is said to be a *free* $\mathcal{O}_X$-module.

Remark. The presheaf $\mathscr{F}_{(*)}$ satisfies condition (4) for a sheaf (§1.10.**a**).

Lemma 1.25. *If U is a quasi-compact open subset, then $\mathscr{F}_{(*)}$ satisfies condition* (5) *for any open cover $\{U_\lambda \mid \lambda \in \Lambda\}$ of U.*

PROOF. If $\alpha_\lambda = (\ldots, \alpha_\lambda^{(i)}, \ldots) \in \bigoplus_{i \in I} \mathscr{F}_i(U_\lambda)$ are given such that $\alpha_\lambda|_{U_\lambda \cap U_\mu} = \alpha_\mu|_{U_\lambda \cap U_\mu}$ for any λ, μ, where $\alpha_\lambda^{(i)}$ denotes the i-th component of α_λ, then $\alpha_\lambda^{(i)}|_{U_\lambda \cap U_\mu} = \alpha_\mu^{(i)}|_{U_\lambda \cap U_\mu}$ and so there exist $\alpha^{(i)} \in \mathscr{F}_i(U)$ such that $\alpha^{(i)}|_{U_\lambda} = \alpha_\lambda^{(i)}$ for all $\lambda \in \Lambda$. If $\alpha^{(i)} = 0$ except for finitely many $i \in I$, then $\alpha = (\ldots, \alpha^{(i)}, \ldots)$ belongs to $\bigoplus_{i \in I} \mathscr{F}_i(U)$ and $\alpha|_{U_\lambda} = \alpha_\lambda$. To show this, take a finite subcover $\{U_\lambda \mid \lambda \in \Phi\}$ indexed by a finite set $\Phi \subseteq \Lambda$. Since $\alpha_\lambda^{(i)} = 0$ if $i \in I \backslash F_\lambda$ for some finite subset F_λ, $G = \bigcup_{\lambda \in \Phi} F_\lambda$ is a finite set such that if $i \notin G$, one has $\alpha_\lambda^{(i)} = 0$ for all $\lambda \in \Phi$ and so $\alpha^{(i)}|_{U_\lambda} = \alpha_\lambda^{(i)} = 0$. This implies $\alpha^{(i)} = 0$, since $\{U_\lambda \mid \lambda \in \Phi\}$ is a cover of U. □

Therefore, if A is a ring and I is an index set, then $\tilde{A}^{(I)}(D(f)) = A_f^{(I)}$ for any $f \in A$. Thus $\tilde{A}^{(I)} \cong (A^{(I)})^{\sim}$.

b. Let $(X, \mathcal{O}_X)$ be a ringed space and let $\mathscr{F}$ be an $\mathcal{O}_X$-module.

Definition. If there exists an exact sequence of $\mathcal{O}_U$-modules of the form

$$\mathcal{O}_U^{(J)} \to \mathcal{O}_U^{(I)} \to \mathscr{F}|_U \to 0,$$

where U is an open neighborhood of x, $\mathcal{O}_U = \mathcal{O}_X|_U$, and I, J are index sets, then $\mathscr{F}$ is said to be *quasi-coherent* at x. If $\mathscr{F}$ is quasi-coherent at every point of X, then $\mathscr{F}$ is said to be *quasi-coherent*.

Proposition 1.21. *Let M be an A-module. Then the $\tilde{A}$-module $\tilde{M}$ is quasi-coherent.*

PROOF. If $\{m_i \mid i \in I\}$ is any set of generators of M as an A-module, then the A-homomorphism $\varphi\colon A^{(I)} \to M$ defined by $\varphi((a_i)_{i \in I}) = \sum_{i \in I} a_i m_i$ is surjective. Applying the same process with $\operatorname{Ker}\varphi$ in place of M, one can build an exact sequence of A-modules:

$$A^{(J)} \to A^{(I)} \to M \to 0.$$

By Proposition 1.20, one has the exact sequence

$$(A^{(J)})^{\sim} \to (A^{(I)})^{\sim} \to \tilde{M} \to 0.$$

Since $(A^{(J)})^{\sim} \cong \tilde{A}^{(J)}$ and $(A^{(I)})^{\sim} \cong \tilde{A}^{(I)}$, one obtains the assertion. □

The main result of this subsection can be stated in the following way.

Theorem 1.11. *Let A be a ring and $\mathscr{F}$ be an $\tilde{A}$-module. Then the following conditions are equivalent:*

(a) *$\mathscr{F}$ is quasi-coherent.*
(b) *There exist elements $g_1, \ldots, g_s \in A$ such that $X = \operatorname{Spec} A$ is covered by $\{D(g_i) \mid 1 \le i \le s\}$ and $\mathscr{F}|_{D(g_i)} \cong \mathscr{F}(D(g_i))^{\sim}$ with the identification $D(g_i) = \operatorname{Spec} A_{g_i}$ for all i.*
(c) *For every $f \in A$, the following two conditions hold:*
  (i) *If $t \in \mathscr{F}(\operatorname{Spec} A)$ satisfies $t|_{D(f)} = 0$, then $f^n t = 0$ for some $n > 0$.*
  (ii) *For any $s \in \mathscr{F}(D(f))$ there exists n such that $f^n s$ is of the form $t|_{D(f)}$ for some $t \in \mathscr{F}(\operatorname{Spec} A)$.*
(d) *$\mathscr{F} \cong \tilde{M}$ for some A-module M.*

PROOF. (a) ⇒ (b) Every point $x \in X = \operatorname{Spec} A$ has a neighborhood of the form $D(g)$ such that there exists an exact sequence of $\tilde{A}|_{D(g)}$-modules

$$(\tilde{A}|_{D(g)})^{(J)} \xrightarrow{\ \theta\ } (\tilde{A}|_{D(g)})^{(I)} \longrightarrow \mathscr{F}|_{D(g)} \longrightarrow 0$$

for some index sets J and I. $\varphi = \theta_{D(g)} : A_g^{(J)} \to A_g^{(I)}$ is an A_g-homomorphism and one obtains an exact sequence

$$A_g^{(J)} \xrightarrow{\ \varphi\ } A_g^{(I)} \longrightarrow \operatorname{Coker}\varphi \longrightarrow 0.$$

Since $\tilde{\varphi} = \theta$ with the identification $D(g) = \operatorname{Spec} A_g$, by Proposition 1.20 one has an exact sequence

$$(A_g^{(J)})^{\sim} \xrightarrow{\ \theta\ } (A_g^{(I)})^{\sim} \longrightarrow (\operatorname{Coker}\varphi)^{\sim} \longrightarrow 0.$$

Recalling that $(A_g^{(J)})^\sim = (\tilde{A}|_{D(g)})^{(J)}$ and $(A_g^{(I)})^\sim = (\tilde{A}|_{D(g)})^{(I)}$, $(\text{Coker } \varphi)^\sim$ is isomorphic to Coker θ, which is isomorphic to $\mathscr{F}|_{D(g)}$ by Proposition 1.14. By Theorem 1.2, X is quasi-compact; hence (b) is now satisfied.

(b) ⇒ (c) To show property (i), suppose $t|_{D(f)} = 0$. Then since $t|_{D(fg_i)} \in \mathscr{F}(D(fg_i))$ and $\mathscr{F}(D(fg_i)) = (\mathscr{F}|_{D(g_i)})(D(fg_i)) = \mathscr{F}(D(g_i))_{f/1}$, there exists l such that $f^l t|_{D(g_i)} = 0$. Since there are a finite number of these g_i, the integer l can be chosen independently of i. By condition (4) for a sheaf, one has $f^l t = 0$.

A similar argument shows that (ii) also holds.

(c) ⇒ (d) For any $f \in A$, one has the restriction map $\mathscr{F}(X) \to \mathscr{F}(D(f))$. By the universal property of $(M_f, \psi_{M,f})$ (cf. §1.4.**a**.), one obtains the A_f-homomorphism $\rho_f : \mathscr{F}(X)_f \to \mathscr{F}(D(f))$. Since the $\rho_f (f \in A)$ commute with restrictions, these give rise to an $\tilde{A}$-homomorphism $\tilde{\rho}: \mathscr{F}(X)^\sim \to \mathscr{F}$. Property (c) shows that ρ_f is an isomorphism for any $f \in A$; hence $\tilde{\rho}$ is an isomorphism. Setting $M = \mathscr{F}(X)$, one sees that $\mathscr{F} \cong \tilde{M}$.

(d) ⇒ (a) This was shown in Proposition 1.21. □

c. Let $(X, \mathcal{O}_X)$ be a ringed space and let $\mathscr{F}$ be an $\mathcal{O}_X$-module.

Definition. Let $x \in X$. If there exists $U \in \Omega_x$ such that for some positive integer m, there exists an exact sequence of $\mathcal{O}_U$-modules

$$\mathcal{O}_U^m \xrightarrow{\theta} \mathscr{F}|_U \longrightarrow 0$$

then $\mathscr{F}$ is said to be *of finite type* at x.

In this case, for any point $y \in U$, one has the commutative diagram with bottom row exact:

$$\begin{array}{ccccc} \mathcal{O}(U)^m & \xrightarrow{\theta_U} & \mathscr{F}(U) & & \\ \downarrow & & \downarrow & & \\ \mathcal{O}_y^m & \xrightarrow{\theta_y} & \mathscr{F}_y & \longrightarrow & 0 \end{array}$$

That is, letting $s_i = \theta_U(\mathbf{e}_i)$, where $\mathbf{e}_1, \ldots, \mathbf{e}_m$ are a basis of $\mathcal{O}(U)^m$, $s_{1y}, \ldots, s_{my}$ generate $\mathscr{F}_y$ as an $\mathcal{O}_y$-module. Conversely, if there exist $s_1, \ldots, s_m \in \mathscr{F}(U)$ such that $\mathscr{F}_y = \sum_{i=1}^m \mathcal{O}_y s_{iy}$ for all $y \in U$, then $\mathscr{F}$ is of finite type at x. Such $s_1, \ldots, s_m$ are called *generators* of $\mathscr{F}$ over U.

Lemma 1.26. *Let $\mathscr{F}$ be an $\mathcal{O}_X$-module of finite type at $x \in X$. Suppose that $\alpha_1, \ldots, \alpha_m \in \mathscr{F}_x$ generate $\mathscr{F}_x$ as an $\mathcal{O}_x$-module. Then there exist $V \in \Omega_x$ and $t_1, \ldots, t_m \in \mathscr{F}(V)$ such that $t_{ix} = \alpha_i$ for all i and $\mathscr{F}_y = \sum_{i=1}^m \mathcal{O}_y t_{iy}$ for all $y \in V$.*

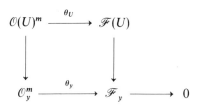

PROOF. By definition, there exist $U \in \Omega_x$ and $s_1, \ldots, s_m \in \mathscr{F}(U)$ such that $\mathscr{F}_y = \sum_{i=1}^m \mathscr{O}_y s_{iy}$ for all $y \in U$. Since $\alpha_1, \ldots, \alpha_m$ are generators of $\mathscr{F}_x$, there exist $\beta_{jl} \in \mathscr{O}_x$ with $s_{jx} = \sum_{l=1}^m \beta_{jl}\alpha_l$ for $1 \le j \le n$. Then there exist an open neighborhood V of x contained in U, sections $t_l \in \mathscr{F}(V)$, and $a_{jl} \in \mathscr{O}_X(V)$ such that $t_{lx} = \alpha_l$ and $a_{jlx} = \beta_{jl}$ for all j, l such that

$$s_j|_V = \sum_{l=1}^m a_{jl} t_l, \quad \text{for all } j.$$

Since the s_{ly} generate $\mathscr{F}_y$, the t_{jy} generate $\mathscr{F}_y$ for all $y \in V$. □

Corollary. *If $\mathscr{F}$ is of finite type at $x \in X$ and $\mathscr{F}_x = 0$, then there exists $U \in \Omega_x$ such that $\mathscr{F}|_U = 0$.*

PROOF. Since 0_x generates $\mathscr{F}_x$, we can apply the last proposition. □

Definition. $\mathscr{F}$ is said to be *of finite type*, if $\mathscr{F}$ is of finite type at every point $x \in X$.

Definition. The *support* of $\mathscr{F}$ is $\{x \in X \mid \mathscr{F}_x \neq 0\}$, denoted by supp $\mathscr{F}$.

EXAMPLE 1.19. Let M be an A-module. Then supp $\tilde{M} = \{\mathfrak{p} \in \operatorname{Spec} A \mid M_\mathfrak{p} \neq 0\}$ by the remark in §1.11.e. This support is also written as $\operatorname{supp}_A M$. If $M = A/\mathfrak{a}$ for some ideal $\mathfrak{a}$ of A, then $\operatorname{supp}(A/\mathfrak{a}) = V(\mathfrak{a})$. Furthermore, if M is finitely generated as an A-module, then $\operatorname{supp}_A M$ is a closed subset.

Proposition 1.22. *$\mathscr{F}$, $\mathscr{G}$ be $\mathscr{O}_X$-modules of finite type and let $\mathscr{H}$ be an $\mathscr{O}_X$-submodule of $\mathscr{F}$. Then,*

(i) *supp $\mathscr{F}$ is a closed subset.*
(ii) *$\mathscr{F} \oplus \mathscr{G}$ and $\mathscr{F} \otimes \mathscr{G}$ are also of finite type.*
(iii) *$\mathscr{F}/\mathscr{H}$ is of finite type.*

PROOF. (i) follows from the Corollary to Lemma 1.26. (ii) and (iii) are easily checked. □

d. Definition. An $\mathscr{O}_X$-module $\mathscr{F}$ is said to be *coherent* if

(1) $\mathscr{F}$ is of finite type, and
(2) for any open subset U and any $\mathscr{O}_U$-homomorphism $\theta\colon \mathscr{O}_U^m \to \mathscr{F}|_U$ where m is a positive integer, Ker θ is an $\mathscr{O}_U$-module of finite type.

Remarks. (1) A coherent $\mathscr{O}_X$-module $\mathscr{F}$ is certainly quasi-coherent. In fact, for any $x \in X$, there exist $U \in \Omega_x$ and an exact sequence

$$\theta\colon \mathscr{O}_U^m \to \mathscr{F}|_U \to 0$$

for some $m > 0$. Since Ker θ is of finite type by condition (2), there exists $V \in \Omega_x$ such that $U \supseteq V$ and $\mathcal{O}_V^n \to \text{Ker}\ \theta|_V \to 0$ for some $n > 0$. Hence, one has the exact sequence

$$\mathcal{O}_V^n \to \mathcal{O}_V^m \to \mathcal{F}|_V \to 0.$$

(2) Coherence is a local property, i.e., $\mathcal{F}$ is coherent if and only if $\mathcal{F}|_{U_\lambda}$ is coherent for every $\lambda \in \Lambda$, where $\{U_\lambda \mid \lambda \in \Lambda\}$ is an open cover of X.

Lemma 1.27. *Let $\mathcal{F}$, $\mathcal{G}$, and $\mathcal{H}$ be $\mathcal{O}_X$-modules such that $\mathcal{F} \subseteq \mathcal{H}$ and $\mathcal{F}$ is a quotient $\mathcal{O}_X$-module of $\mathcal{G}$. If $\mathcal{G}$ is of finite type and $\mathcal{H}$ is coherent, then $\mathcal{F}$ is coherent.*

PROOF. $\mathcal{F}$ satisfies condition (1) for coherence, since $\mathcal{G}$ is of finite type. Condition (2) is easily verified, since $\mathcal{H}$ is coherent. □

The following result due to Serre is quite useful. Although its proof is not hard, it will not be given here.

Theorem 1.12 (Serre). *Let*

$$0 \to \mathcal{F} \to \mathcal{G} \to \mathcal{H} \to 0$$

be an exact sequence of $\mathcal{O}_X$-modules. If any two of $\mathcal{F}$, $\mathcal{G}$ and $\mathcal{H}$ are coherent, so is the third.

PROOF. We refer to Théorème 1 of "Faisceaux algébriques cohérents", by J. P. Serre, *Ann. of Math.* 61, (1955), 197–278. □

Corollary 1.

(i) *If $\mathcal{F}$ and $\mathcal{G}$ are coherent $\mathcal{O}_X$-modules, then so is $\mathcal{F} \oplus \mathcal{G}$. The converse is also true.*
(ii) *Let $\theta: \mathcal{F} \to \mathcal{G}$ be an $\mathcal{O}_X$-homomorphism. If $\mathcal{F}$ and $\mathcal{G}$ are coherent, then* Ker θ, Im θ, *and* Coker θ *are all coherent.*
(iii) *If $\mathcal{F}$ and $\mathcal{G}$ are coherent, then so are $\mathcal{F} \otimes_{\mathcal{O}_X} \mathcal{G}$ and $\mathcal{H}om_{\mathcal{O}_X}(\mathcal{F}, \mathcal{G})$.*

PROOF. (i) The first assertion follows directly from Theorem 1.12 applied to

$$0 \to \mathcal{F} \to \mathcal{F} \oplus \mathcal{G} \to \mathcal{G} \to 0.$$

If $\mathcal{F} \oplus \mathcal{G}$ is coherent, then $\mathcal{F}$ is coherent by Lemma 1.26, since $\mathcal{F} \subseteq \mathcal{F} \oplus \mathcal{G}$ and $\mathcal{F} \cong (\mathcal{F} \oplus \mathcal{G})/\mathcal{G}$.

(ii) Since Im $\theta \subseteq \mathcal{G}$ and $\mathcal{F} \to \text{Im}\ \theta \to 0$ is exact, Im θ is coherent by Lemma 1.26. Applying Theorem 1.12 to the exact sequences

$$0 \to \text{Im}\ \theta \to \mathcal{G} \to \text{Coker}\ \theta \to 0,$$

$$0 \to \text{Ker}\ \theta \to \mathcal{F} \to \text{Im}\ \theta \to 0,$$

one concludes that Coker θ and Ker θ are also coherent.

(iii) For any $x \in X$, there exists $U \in \Omega_x$ and an exact sequence

$$\mathcal{O}_U^n \to \mathcal{O}_U^m \to \mathcal{F}|_U \to 0$$

for some n, m (see Remark (1)). Tensoring by $\mathcal{G}|_U$ and using $\mathcal{O}_U^m \otimes_{\mathcal{O}_U} \mathcal{G}|_U \cong \mathcal{G}^m|_U$, one has by Proposition 1.16.(v) the exact sequence:

$$\mathcal{G}^n|_U \to \mathcal{G}^m|_U \to (\mathcal{F} \otimes_{\mathcal{O}_X} \mathcal{G})|_U (\cong \mathcal{F}|_U \otimes_{\mathcal{O}_U} \mathcal{G}|_U) \to 0.$$

Since $\mathcal{G}$ is coherent, so are $\mathcal{G}^m|_U$ and $\mathcal{G}^n|_U$ by the first assertion. Therefore, $(\mathcal{F} \otimes_{\mathcal{O}_X} \mathcal{G})|_U$ is coherent, and so is $\mathcal{F} \otimes_{\mathcal{O}_X} \mathcal{G}$ by the local property of coherence. The proof for $\mathcal{H}om_{\mathcal{O}_X}(\mathcal{F}, \mathcal{G})$ proceeds similarly using Proposition 1.17. □

Corollary 2. *If $\mathcal{F}$ is coherent and $x \in X$, then the natural $\mathcal{O}_{X,x}$-homomorphism $u_{\langle x \rangle}\colon \mathcal{H}om_{\mathcal{O}_X}(\mathcal{F}, \mathcal{G})_x \to \mathrm{Hom}_{\mathcal{O}_x}(\mathcal{F}_x, \mathcal{G}_x)$ is an isomorphism for any $\mathcal{O}_X$-module $\mathcal{G}$. (The homomorphism $u_{\langle x \rangle}$ was defined in Proposition 1.17.(i).).*

Proof. The proof is similar to that of Corollary 1.(iii). □

e. The following simple result is a key lemma in the theory of schemes.

Lemma 1.28. *Let A be a Noetherian ring and let M be a finitely generated A-module. Then $\tilde{M}$ is coherent. In particular, $\tilde{A}$ is coherent.*

Proof. It suffices to show that every homomorphism $\theta\colon \tilde{A}^m|_{D(f)} \to \tilde{M}|_{D(f)}$ for any f and m has a kernel of finite type. $\varphi = \theta_{D(f)}\colon A_f^m \to M_f$ is an A_f-homomorphism and A_f is also Noetherian by the Hilbert basis theorem, since $A_f \cong A[\mathsf{X}]/(f\mathsf{X} - 1)$. Then $\operatorname{Ker} \varphi (\subseteq A_f^m)$ is finitely generated. Since $\operatorname{Ker} \theta = (\operatorname{Ker} \varphi)^\sim$ by Proposition 1.20, one obtains the assertion. □

Remark. Even if the modules $\tilde{M}$ associated with finitely generated A-modules M are always coherent, A need not be Noetherian. For example, let A be a non-Noetherian valuation ring. Then any kernel of $A^m \to M$ is free; hence $\tilde{A}$ is coherent.

Theorem 1.13. *Let A be a Noetherian ring, and let $\mathcal{F}$ be a quasi-coherent $\tilde{A}$-module. The following conditions are equivalent:*

(a) *$\mathcal{F}$ is coherent.*
(b) *$\mathcal{F}$ is of finite type.*
(c) *$\mathcal{F} \cong \tilde{M}$, where M is finitely generated as an A-module.*

Proof. (c) ⇒ (a) is Lemma 1.28, and (a) ⇒ (b) is trivial. It suffices to show (b) ⇒ (c). By Theorem 1.11, $\mathcal{F} \cong \tilde{M}$, since $\mathcal{F}$ is quasi-coherent. One has an open cover $\{D(f_j) \mid 1 \le j \le r\}$ such that $\tilde{M}|_{D(f_j)}$ is generated by a finite number of elements $\alpha_{j,1}, \ldots, \alpha_{j,s} \in \tilde{M}(D(f_j)) = M_{f_j}$ for all j, where s can be chosen independently of j. Then by the Corollary to Proposition 1.20, $\alpha_{j,1}, \ldots, \alpha_{j,s}$ generate M_{f_j} as an A_{f_j}-module. Writing $\alpha_{j,i}$ in the form $a_{j,i}/f_j^l$, where $a_{j,i} \in M$ and l is independent of j, we have a finitely generated A-module $N = \sum_{j,i} Aa_{j,i}$, which is an A-submodule of M. Further, $\tilde{N}|_{D(f_j)} = \tilde{M}|_{D(f_j)}$ for all j; hence $(M/N)^\sim \cong \tilde{M}/\tilde{N} = 0$ by Proposition 1.20. Thus $M = N$. $\square$

Definition. If A is a Noetherian ring, then the affine scheme Spec A is said to be an *affine Noetherian scheme.*

In this case, $\tilde{A}$ is coherent by Lemma 1.28.

Remark. The sheaf $\mathcal{O}^{\text{hol}}$ defined in Example 1.16 is a coherent sheaf. This result of Oka is a deep and fundamental result in the theory of functions of several complex variables. Lemma 1.28 can be regarded as a ring-theoretic (or affine-schematic) counterpart of Oka's Theorem.

It should be noted that sheaves of germs of differentiable functions are by no means coherent. These facts seem to suggest that coherence is linked with the property of being algebraic or analytic.

§1.13 Reduced Affine Schemes and Integral Affine Schemes

a. Definition. An affine scheme Spec A is said to be *reduced* if A has no nonzero nilpotent elements, i.e., $\sqrt{(0_A)} = (0_A)$. In this case, we say that the ring A is *reduced.*

If A is a ring, $A/\sqrt{(0_A)}$ is a reduced ring, denoted by A_{red}.

Lemma 1.29.

(i) *If A is a reduced ring and S is a multiplicative subset of A, then $S^{-1}A$ is also reduced.*
(ii) *If $A_\mathfrak{m}$ is reduced for every maximal ideal $\mathfrak{m}$ of A, then A is also reduced.*

PROOF. (i) Suppose $(a/s)^n = 0$ for some n, where $a \in A$, $s \in S$. Then $ta^n = 0$ for some $t \in S$; hence $(ta)^n = 0$. This yields $ta = 0$, i.e., $a/s = 0$.

(ii) Let a be a nilpotent element. Put $\mathfrak{a} = (0 : a)$, which is defined as $\{b \in A \mid ab = 0\}$. If $\mathfrak{a}$ is proper, there exists a maximal ideal $\mathfrak{m}$ containing $\mathfrak{a}$ by Lemma 1.1. Then $a/1 \in A_\mathfrak{m}$ is a nilpotent element and so $a/1 = 0$, since $A_\mathfrak{m}$ is reduced. Hence, $ab = 0$ for some $b \notin \mathfrak{m}$. This contradicts the definition of $\mathfrak{m} \supseteq \mathfrak{a}$. Hence $\mathfrak{a} = A$, i.e., $a = 0$. $\square$

Therefore, if A is reduced, every stalk of $\tilde{A}$ is also reduced. Conversely, if $\tilde{A}_\mathfrak{p}$ is reduced for all $\mathfrak{p}$, then A is reduced.

Definition. The affine scheme Spec A_{red} is the *reduced affine scheme* obtained from the affine scheme Spec A.

Note that the base space of Spec A_{red} coincides with that of Spec A.

b. Definition. If A is an integral domain, then Spec A is said to be an *integral affine scheme.*

Proposition 1.23. *Let X be an affine scheme* Spec A.

(i) *If X is irreducible and reduced, then X is integral.*
(ii) *Suppose that X has a finite number of irreducible components $X_1, \ldots, X_u$. If X is connected and every stalk is an integral domain, then X is integral.*

PROOF. (i) Assume that X is irreducible and reduced. Suppose that there exist a and b with $ab = 0$. Then $X = \text{Spec } A = V(a) \cup V(b)$. Since Spec A is irreducible, it follows that $V(a) = X$ or $V(b) = X$; thus $a = 0$ or $b = 0$, since A is reduced.

(ii) By assertion (i), it suffices to derive $u = 1$ from the hypothesis that X is connected and every stalk is an integral domain. Assume $u \geq 2$. Since X is connected, one can assume that $X_1 \cap X_2 \neq \varnothing$. Take $x \in X_1 \cap X_2$ and let x_i denote the generic point of X_i. Corresponding to x and x_i, one has prime ideals $\mathfrak{p}$ and $\mathfrak{p}_i$ of A, respectively. Then $\mathfrak{p}_1$ and $\mathfrak{p}_2$ are minimal prime ideals contained in $\mathfrak{p}$. Since Spec $A_\mathfrak{p} \cong \{\mathfrak{q} \in \text{Spec } A \mid \mathfrak{q} \subseteq \mathfrak{p}\}$ by Lemma 1.3 and $A_\mathfrak{p} = \mathcal{O}_{X,x}$ is an integral domain, it follows that $\mathfrak{p}_1 = \mathfrak{p}_2$; thus $X_1 = X_2$, a contradiction. □

c. Definition. An affine scheme Spec A is said to be *algebraic* if the ring A is finitely generated over a field k. If, in addition, A is an integral domain, then Spec A is said to be an *affine algebraic variety.*

Remark. Let f be a nonconstant irreducible polynomial in X and Y. Then, Spec$(k[\mathsf{X}, \mathsf{Y}]/(f))$ is said to be an *affine algebraic curve* defined by f.

If $f \in k[\mathsf{X}, \mathsf{Y}, \mathsf{Z}]\backslash k$ is irreducible, then Spec$(k[\mathsf{X}, \mathsf{Y}, \mathsf{Z}]/(f))$ is said to be the *affine algebraic surface* defined by f.

§1.14 Morphisms of Affine Schemes

a. First, we introduce direct image sheaves. Let X and Y be topological spaces and let $\psi: X \to Y$ be a continuous map.

Definition. For any sheaf $\mathscr{F}$ on X, the *direct image sheaf* $\psi_* \mathscr{F}$ is defined to be $(\{\mathscr{F}(\psi^{-1}(U)) \mid U$ is an open subset of $Y\}, \{r_{\psi^{-1}(V), \psi^{-1}(U)} \mid U \supseteq V\})$.

It is easy to see that $\psi_* \mathscr{F}$ is a sheaf.

Definition. Let $(X, \mathscr{O}_X)$ and $(Y, \mathscr{O}_Y)$ be ringed spaces. A *morphism* of ringed spaces $(X, \mathscr{O}_X) \to (Y, \mathscr{O}_Y)$ is defined to be a pair (ψ, θ) consisting of a continuous map $\psi: X \to Y$ and a sheaf homomorphism $\theta: \mathscr{O}_Y \to \psi_*(\mathscr{O}_X)$. The homomorphism $(\text{id}, \text{id}): (X, \mathscr{O}_X) \to (X, \mathscr{O}_X)$ is said to be the *identity*, again denoted by id.

Take $x \in X$ and let $y = \psi(x)$. Then a homomorphism $\theta: \mathscr{O}_Y \to \psi_*(\mathscr{O}_X)$ induces $\theta_y: \mathscr{O}_{Y,y} \to \psi_*(\mathscr{O}_X)_y$. Composing θ_y with the canonical homomorphism $\psi_*(\mathscr{O}_X)_y = \text{injlim}_{y \in U} \mathscr{O}_X(\psi^{-1}(U)) \to \text{injlim}_{x \in V} \mathscr{O}_X(V) = \mathscr{O}_{X,x}$, where U and V are open subsets, one has a ring homomorphism: $\mathscr{O}_{Y,\psi(x)} \to \mathscr{O}_{X,x}$, denoted by $\theta_x^\#$.

b. Definition. Let R_1 and R_2 be local rings with maximal ideals $\mathfrak{m}_1$ and $\mathfrak{m}_2$, respectively. A homomorphism $\varphi: R_1 \to R_2$ is said to be *local* if $\varphi(\mathfrak{m}_1) \subseteq \mathfrak{m}_2$, i.e., $\mathfrak{m}_1 R_2 \neq R_2$.

EXAMPLE 1.20. Let $\varphi: A \to B$ be a ring homomorphism and let $\mathfrak{q} \in \text{Spec } B$. Letting $\mathfrak{p} = \varphi^{-1}(\mathfrak{q})$, one has a ring homomorphism $\varphi_\mathfrak{q}: A_\mathfrak{p} \to B_\mathfrak{q}$ defined by $\varphi_\mathfrak{q}(a/s) = \varphi(a)/\varphi(s)$ for any $a \in A$ and $s \notin \mathfrak{p}$. Then $\varphi_\mathfrak{q}$ is a local homomorphism.

Definition. Let $(X, \mathscr{O}_X)$ and $(Y, \mathscr{O}_Y)$ be local ringed spaces. A morphism $(\psi, \theta): (X, \mathscr{O}_X) \to (Y, \mathscr{O}_Y)$ is said to be a *local ringed space morphism* if every $\theta_x^\#: \mathscr{O}_{Y,\psi(x)} \to \mathscr{O}_{X,x}$, for $x \in X$, is local.

Definition. The *composition* of two morphisms $(\psi, \theta): (X, \mathscr{O}_X) \to (Y, \mathscr{O}_Y)$ and $(\rho, \eta): (Y, \mathscr{O}_Y) \to (Z, \mathscr{O}_Z)$ is defined to be the pair $(\rho \circ \psi, \Theta)$, in which $\Theta: \mathscr{O}_Z \to (\rho \circ \psi)_*(\mathscr{O}_X) = \rho_*(\psi_*(\mathscr{O}_X))$ is naturally defined as follows: for any open subset U of Z, $\Theta_U: \mathscr{O}_Z(U) \to \mathscr{O}_X(\psi^{-1}(\rho^{-1}(U)))$ is the composition $\theta_{\rho^{-1}(U)} \circ \eta_U$.

Since $\Theta_x^\# = \theta_x^\# \circ \eta_{\psi(x)}^\#$ for any $x \in X$, the composition is a local ringed space morphism, if both (ψ, θ) and (ρ, η) are local ringed space morphisms.

As (local) ringed spaces $(X, \mathscr{O}_X)$ and $(Y, \mathscr{O}_Y)$ are simply denoted by X and Y, and a morphism $(f, \theta): (X, \mathscr{O}_X) \to (Y, \mathscr{O}_Y)$ is often written as f. In this case, $\theta: \mathscr{O}_Y \to f_*(\mathscr{O}_X)$ is denoted by $\theta(f)$.

EXAMPLE 1.21. If U is an open subset of a (local) ringed space, then U becomes a (local) ringed space with the structure sheaf $\mathscr{O}_X|_U$, i.e., $\mathscr{O}_U = \mathscr{O}_X|_U$. Let $i: U \to X$ be the inclusion map. Define $\theta: \mathscr{O}_X \to i_*(\mathscr{O}_U)$ by letting

$\theta_V: \mathcal{O}_X(V) \to (i_* \mathcal{O}_U)(V) = \mathcal{O}_X(U \cap V)$ be $r_{U \cap V, V}$. Then (i, θ) is a (local ringed space) morphism, again denoted by i.

Definition. Let $f: X \to Y$ be a morphism of ringed spaces, and let U be an open subset of X, which is also regarded as a ringed space. $f \circ i: U \to Y$ is said to be the *restriction* of f to U, and is denoted by $f|_U$.

Definition. Let $f: X \to Y$ be a morphism of ringed spaces. If there exists a morphism $g: Y \to X$ such that $f \circ g$ and $g \circ f$ are the identities, then f is said to be an *isomorphism* and it is said that X is *isomorphic* to Y, or X and Y are *isomorphic*, denoted $X \cong Y$ or $f: X \xrightarrow{\sim} Y$.

c. Let $\varphi: A \to B$ be a ring homomorphism. Then one has the associated continuous map ${}^a\varphi$: Spec $B \to$ Spec A. We shall define a sheaf homomorphism $\theta: \tilde{A} \to {}^a\varphi_*(\tilde{B})$. For any $f \in A$, one has the homomorphism φ_f: $A_f \to B_{\varphi(f)}$ defined by $\varphi_f(a/f^n) = \varphi(a)/\varphi(f)^n$ for $a \in A$ and $n \geq 0$. Then, if $D(f) \supseteq D(g)$, one obtains the commutative square

$$\begin{array}{ccc} A_f & \xrightarrow{\varphi_f} & B_{\varphi(f)} \\ {\scriptstyle r_{g,f}}\downarrow & & \downarrow{\scriptstyle r_{\varphi(g),\varphi(f)}} \\ A_g & \xrightarrow{\varphi_g} & B_{\varphi(g)}. \end{array}$$

Since $\tilde{A}(D(f)) = A_f$ and $({}^a\varphi_* \ \tilde{B})(D(f)) = \tilde{B}({}^a\varphi^{-1}(D(f))) = \tilde{B}(D(\varphi(f))) = B_{\varphi(f)}$, $\{\varphi_f\}$ determine a sheaf homomorphism $\theta: \tilde{A} \to {}^a\varphi_*(\tilde{B})$, denoted again by $\tilde{\varphi}$. Since $\tilde{\varphi}_{\mathfrak{p}}^{\#} = \varphi_{\mathfrak{p}}: A_{{}^a\varphi(\mathfrak{p})} \to B_{\mathfrak{p}}$ for any $\mathfrak{p} \in$ Spec B (see Example 1.20) is a local homomorphism, $({}^a\varphi, \tilde{\varphi})$ is a local ringed space morphism, which is simply denoted by ${}^a\varphi$.

Definition. A *morphism* of affine schemes is defined to be a local ringed space morphism. The set of morphisms from Spec B to Spec A is denoted by Hom(Spec B, Spec A).

Thus, one has the map H: Hom$(A, B) = \{\varphi: A \to B \mid \varphi$ is a ring homomorphism$\} \to$ Hom(Spec B, Spec A) defined by $H(\varphi) = ({}^a\varphi, \tilde{\varphi})$. Then $H(\text{id}) = \text{id}$ and $H(\varphi \circ \psi) = H(\psi) \circ H(\varphi)$.

Theorem 1.14. *Let A and B be rings. Then H*: Hom$(A, B) \to$ Hom(Spec B, Spec A) *is a bijection.*

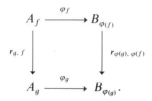

Proof. We write $X =$ Spec B, $\mathcal{O}_X = \tilde{B}$, $Y =$ Spec A, and $\mathcal{O}_Y = \tilde{A}$. Now, if $H(\varphi_1) = H(\varphi_2)$, then $\tilde{\varphi}_1 = \tilde{\varphi}_2$ and $\tilde{\varphi}_{1,Y} = \varphi_1$, $\tilde{\varphi}_{2,Y} = \varphi_2$; hence $\varphi_1 = \varphi_2$. Thus it suffices to show that H is onto. Let $(\psi, \theta): (X, \mathcal{O}_X) \to (Y, \mathcal{O}_Y)$ be a local ringed space morphism. One has $\varphi = \theta_Y: \mathcal{O}_Y(Y) = A \to \mathcal{O}_X(X) = B$. We claim that $\psi = {}^a\varphi$ as a map. For any $\mathfrak{p} \in X$, one has the commutative square

$$\begin{array}{ccc} \mathcal{O}_Y(Y) & \xrightarrow{\theta_Y} & \mathcal{O}_X(X) \\ \downarrow & & \downarrow \\ \mathcal{O}_{Y,\psi(\mathfrak{p})} & \xrightarrow{\theta_{\mathfrak{p}}^{\#}} & \mathcal{O}_{X,\mathfrak{p}}, \end{array}$$

where the vertical arrows are canonical maps of the direct limits. The above square can be rewritten as

$$\begin{array}{ccc} A & \xrightarrow{\varphi} & B \\ \downarrow & & \downarrow \\ A_{\psi(\mathfrak{p})} & \xrightarrow{\theta_{\mathfrak{p}}^{\#}} & B_{\mathfrak{p}} \end{array}$$

Since $\theta_{\mathfrak{p}}^{\#}$ is a local homomorphism, one has

$$f \in \psi(\mathfrak{p}) \Leftrightarrow f/1 \in \psi(\mathfrak{p})A_{\psi(\mathfrak{p})} \Leftrightarrow \theta_{\mathfrak{p}}^{\#}(f/1) \in \mathfrak{p}B_{\mathfrak{p}}.$$

By the above square, $\theta_{\mathfrak{p}}^{\#}(f/1)$ can be written as $\varphi(f)/1$. Hence,

$$\varphi(f)/1 \in \mathfrak{p}B_{\mathfrak{p}} \Leftrightarrow \varphi(f) \in \mathfrak{p} \Leftrightarrow f \in {}^a\varphi(\mathfrak{p}) = \varphi^{-1}(\mathfrak{p}).$$

Therefore,

$$\psi(\mathfrak{p}) = {}^a\varphi(\mathfrak{p}), \text{ i.e., } \psi = {}^a\varphi.$$

For any $g \in A$, one has the commutative square

$$\begin{array}{ccc} A & \xrightarrow{\varphi} & B \\ {\scriptstyle \psi_{A,g}}\downarrow & & \downarrow{\scriptstyle \psi_{B,\varphi(g)}} \\ A_g & \xrightarrow{\theta_{D(g)}} & B_{\varphi(g)}. \end{array}$$

By the universal property of $(A_g, \psi_{A,g})$, one has $\theta_{D(g)} = \varphi_g$. Hence, $\tilde{\varphi} = \theta$ is established. □

§1.15 Definition of Schemes and First Properties

a. Definition. A local ringed space $(X, \mathcal{O}_X)$ is said to be a *scheme* if every point $x \in X$ has an open neighborhood U such that $(U, \mathcal{O}_X|_U)$ is isomorphic as a ringed space to some affine scheme. Such subsets are said to be *affine* *open subsets* or affine open subschemes of X. Any local ringed space morphism between schemes is said to be a *morphism* (*of schemes*).

Proposition 1.24. *Let* $(X, \mathcal{O}_X)$ *be a scheme. Then,*

(i) *The set of all affine open subsets of* X *is an open base for the topology on* X.
(ii) *If* U *is an open subset of* X, $(U, \mathcal{O}_X|_U)$ *is a scheme.*

PROOF. (i) follows from the fact that if $X = \operatorname{Spec} A$, then $(D(f), \tilde{A}_f)$ is an affine scheme and $\{D(f) \mid f \in A\}$ is an open base for the topology on X. (ii) is a consequence of (i). □

As in the cases of ringed spaces and affine schemes, we often use X to denote a scheme $(X, \mathcal{O}_X)$.

Definition. If $\{U_\lambda \mid \lambda \in \Lambda\}$ is an open cover of the base space of a scheme X such that every U_λ is an affine open subset, then it is called an *affine open cover* of X.

Proposition 1.25. *Let* $(X, \mathcal{O}_X)$ *be a scheme and let* $\mathcal{F}$ *be an* $\mathcal{O}_X$*-module.*

(i) X *is a* T_0*-space.*
(ii) *Every nonempty irreducible closed subset* V *of* X *has a unique generic point.*
(iii) $\mathcal{F}$ *is quasi-coherent if and only if there exists an affine open cover* $\{U_\lambda \mid \lambda \in \Lambda\}$ *such that* $\mathcal{F}|_{U_\lambda} \cong \tilde{M}_\lambda$ *for some* $\mathcal{O}_X(U_\lambda)$*-modules* M_λ.

PROOF. (i) This follows from Proposition 1.5, since the property of being a T_0-space is local.

(ii) For a point $x \in V$, take an affine open neighborhood U of x. Then $V \cap U$ is irreducible and closed in U; hence it has the generic point ξ by Proposition 1.8. The closure of ξ in X is V, since $V \cap U$ is dense in V. The uniqueness of the generic point follows from Lemma 1.5.(ii).

(iii) This follows from Theorem 1.11. □

b. Definition. For any point x of a scheme X, the field $\mathcal{O}_{X,x}/\mathfrak{m}_x$ (it is a field because $\mathfrak{m}_x$ is a maximal ideal) is said to be the *residue class field* at x, denoted by $k(x)$.

If U is an affine open neighborhood of x, then $U \cong \operatorname{Spec} A$ and x corresponds to a prime ideal $\mathfrak{p}$ of A. Since $\mathcal{O}_{X,x} \cong A_\mathfrak{p}$ and $\mathfrak{m}_x \cong \mathfrak{p}A_\mathfrak{p}$, it follows that $k(x) \cong k(\mathfrak{p})$.

Definition. Let $(X, \mathcal{O}_X)$ be a scheme and take $g \in \mathcal{O}_X(U)$ for an open subset U. For any point $x \in U$, define $\bar{g}(x)$ to be g_x mod $\mathfrak{m}_x$, i.e., $\bar{g}(x) \in k(x)$. Define U_g to be $\{x \in U \mid \bar{g}(x) \neq 0\}$ and $Z_U(g)$ to be $\{x \in U \mid \bar{g}(x) = 0\}$ (see §1.4.**c**).

If U is an affine scheme Spec A, then $Z_U(g) = V(g)$ and $U_g = D(g)$, for any $g \in \mathcal{O}_X(U) = A$. Therefore, for an arbitrary open subset U of X, U_g becomes an open subset of U.

c. Definition. A scheme X is said to be *reduced* if $\mathcal{O}_X(U)$ is a reduced ring for all open subset U of X. A scheme X is said to be *integral*, if $\mathcal{O}_X(U)$ is an integral domain for all open subsets U of X.

Proposition 1.26. *Let X be a scheme. Then*

(i) *X is reduced if and only if every stalk is reduced.*
(ii) *X is integral if and only if X is irreducible and reduced.*
(iii) *Suppose that X has a finite number of irreducible components. If X is connected and every stalk is an integral domain, then X is integral.*

PROOF. (i) If every stalk is reduced, then any nilpotent element $f \in \mathcal{O}_X(U)$ satisfies $f_x = 0$; hence $f|_{V(x)} = 0$ for some open neighborhood $V(x)$ of x. By sheaf condition (4), this implies that $f = 0$. The converse is clear by Lemma 1.29.(ii).

(ii) Suppose that an integral scheme X is reducible; i.e., $X = X_1 \cup X_2$ for some closed proper subsets X_1 and X_2. Then taking nonempty affine open subsets $U_1 \subseteq X \backslash X_1$ and $U_2 \subseteq X \backslash X_2$, one has $U_1 \cap U_2 = \varnothing$ and $\mathcal{O}_X(U_1 \cup U_2) \cong \mathcal{O}_X(U_1) \oplus \mathcal{O}_X(U_2)$, which has nontrivial zero divisors. The converse is easy, by Proposition 1.23.(i).

(iii) This follows immediately from (ii) and Proposition 1.23.(ii). □

d. Definition. A scheme X is said to be *Noetherian* if there exists a finite affine open cover $\{U_i \mid 1 \leq i \leq r\}$ such that each U_i is an affine Noetherian scheme, i.e., $\mathcal{O}_X(U_i)$ is a Noetherian ring.

If X is a Noetherian scheme, then there exist a finite number of irreducible components $X_1, \ldots, X_m$ of the base space. Thus, each X_i has the generic point x_i; hence $X = \bigcup_{i=1}^m X_i$, and $X_i = \overline{\{x_i\}}$.

e. Let $f: X \to Y$ be a morphism of schemes. Then $\theta(f)_Y: \mathcal{O}_Y(Y) \to \mathcal{O}_X(X)$ is a ring homomorphism. We denote $\mathcal{O}_X(X)$ by $A(X)$ and $\theta(f)_Y$ by $\Gamma(f)$. Thus, to each $f: X \to Y$, the ring homomorphism $\Gamma(f): A(Y) \to A(X)$ is associated. If $g: Y \to Z$ is another morphism of schemes, then $\Gamma(g \circ f) = \Gamma(f) \circ \Gamma(g)$.

Suppose Y is an affine scheme Spec B. Then $\Gamma(f)$ is a homomorphism from B to $A(X)$.

Theorem 1.15. Γ: Hom(X, Spec B) $\to$ Hom(B, $A(X)$) *is a bijection.*

PROOF. First, we shall form the canonical morphism $\Psi: X \to \operatorname{Spec} A(X)$. Letting $\{U_\lambda \mid \lambda \in \Lambda\}$ be an affine open cover of X, one has the inclusion

$i_\lambda\colon U_\lambda \hookrightarrow X$. Then $\psi_\lambda = \Gamma(i_\lambda)\colon A(X) \to A(U_\lambda)$ and $\Psi_\lambda = {}^a\psi_\lambda\colon U_\lambda \to \operatorname{Spec} A(X)$ are obtained for all λ. We claim that $\Psi_\lambda|_{U_\lambda \cap U_\mu} = \Psi_\mu|_{U_\lambda \cap U_\mu}$ for all λ, μ. In fact, let $\{V_\alpha \mid \alpha \in \Phi\}$ be an affine open cover of $U_\lambda \cap U_\mu$. Then the following square of restriction maps becomes commutative:

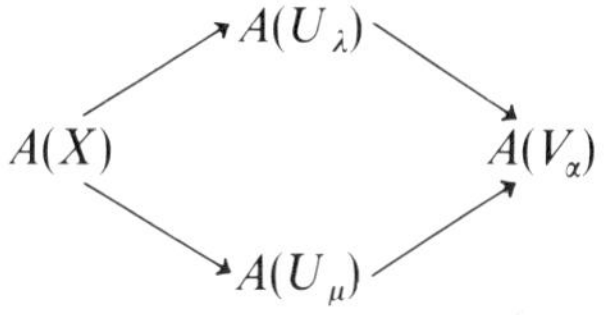

Associated with this cover, one has $\Psi_\lambda|_{V_\alpha} = \Psi_\mu|_{V_\alpha}\colon V_\alpha \to \operatorname{Spec} A(X)$ for all $\alpha \in \Phi$. Hence, $\Psi_\lambda|_{U_\lambda \cap U_\mu} = \Psi_\mu|_{U_\lambda \cap U_\mu}$. $\{\Psi_\lambda \mid \lambda \in \Lambda\}$ determines the morphism $\Psi\colon X \to \operatorname{Spec} A(X)$ by the next lemma.

Lemma 1.30. *Let X and Y be ringed spaces, and let $\{U_\lambda \mid \lambda \in \Lambda\}$ be an open cover of X. If we are given morphisms $\Psi_\lambda\colon U_\lambda \to Y$ such that $\Psi_\lambda|_{U_\lambda \cap U_\mu} = \Psi_\mu|_{U_\lambda \cap U_\mu}$ for all λ, μ, then there exists $\Psi\colon X \to Y$ such that $\Psi|_{U_\lambda} = \Psi_\lambda$ for all $\lambda \in \Lambda$. (In this case, one says that Ψ is obtained by glueing $\{\Psi_\lambda \mid \lambda \in \Lambda\}$).*

Proof. We let $\Psi_\lambda = (\psi_\lambda, \theta_\lambda)$. Then clearly $\{\psi_\lambda \mid \lambda \in \Lambda\}$ determines a continuous map $\psi\colon X \to Y$. Now, we wish to define $\theta\colon \mathcal{O}_Y \to \psi_*(\mathcal{O}_X)$. For any open subset V of Y, and $\alpha \in \mathcal{O}_Y(V)$, one has $\theta_{\lambda, V}(\alpha) \in \Gamma(\psi^{-1}(V) \cap U_\lambda, \mathcal{O}_X)$. Then letting $U'_\lambda = \psi^{-1}(V) \cap U_\lambda$, one has $\theta_{\lambda, V}(\alpha)|_{U_{\lambda'} \cap U_{\mu'}} = \theta_{\mu, V}(\alpha)|_{U_{\lambda'} \cap U_{\mu'}}$. Hence, there exists a unique $\beta \in \Gamma(\psi^{-1}(V), \mathcal{O}_X)$ with $\beta|_{U_{\lambda'}} = \theta_{\lambda, V}(\alpha)$. By letting $\theta_V(\alpha) = \beta$, one obtains $\theta_V\colon \mathcal{O}_Y(V) \to \mathcal{O}_X(\psi^{-1}(V))$. $\{\theta_V\}$ is a homomorphism θ and let $\Psi = (\psi, \theta)$. Clearly, $\Psi|_{U_\lambda} = \Psi_\lambda$ for all λ.

Given a morphism $f\colon X \to \operatorname{Spec} B$, one has the following commutative diagrams by letting i_λ be the inclusion $U_\lambda \to X$ and $f_\lambda = f|_{U_\lambda}$ as follows:

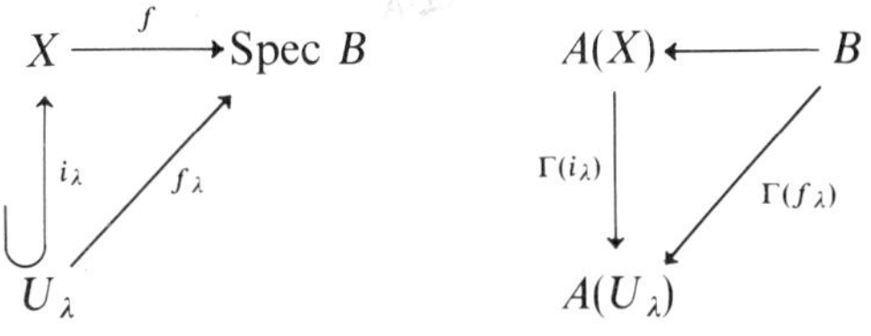

Then, one obtains the commutative diagram

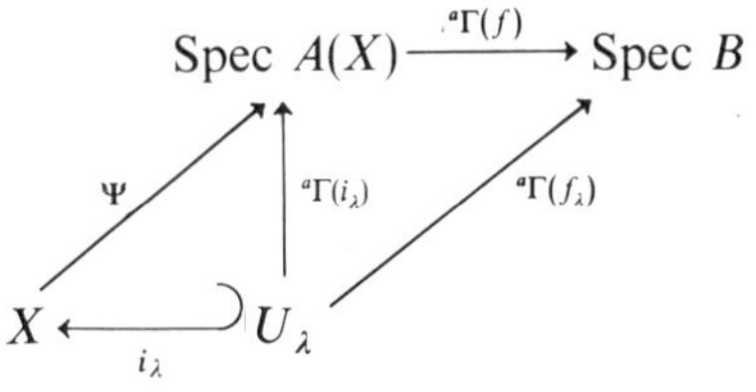

From this, $f_\lambda = f|_{U_\lambda} = {}^a\Gamma(f_\lambda)$ and $({}^a\Gamma(f) \circ \Psi)|_{U_\lambda} = {}^a\Gamma(f_\lambda)$ for all $\lambda \in \Lambda$; hence $f = {}^a\Gamma(f) \circ \Psi$. In particular, if $B = A(X)$ and $f = \Psi$, then $\Gamma(\Psi) = \text{id}$, since

$\Gamma(\Psi)(\alpha)|_{U_\lambda} = \alpha|_{U_\lambda}$ for all $\alpha \in A(X)$ and λ. Suppose that $f = g \circ \Psi$ where g: Spec $A(X) \to$ Spec B is a morphism. Then $\Gamma(f) = \Gamma(\Psi) \circ \Gamma(g) = \Gamma(g)$ and so $g = {}^a\Gamma(g) = {}^a\Gamma(f)$.

If $\varphi \in \mathrm{Hom}(B, A(X))$, then $\Gamma({}^a\varphi \circ \Psi) = \varphi$; thus Γ is bijective. □

This suggests that among all pairs (B, f), $(A(X), \Psi)$ is the universal one.

Given a morphism f: $X \to$ Spec B, there exists a unique morphism g: Spec $A(X) \to$ Spec B with $f = g \circ \Psi$.

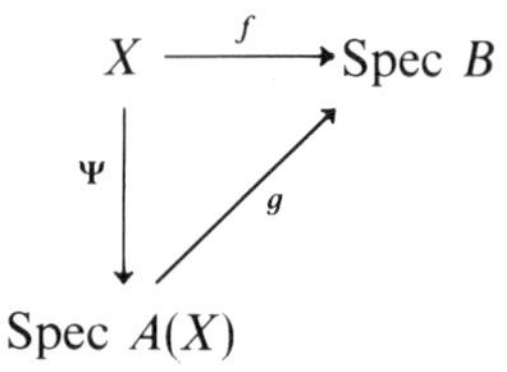

For $s \in A(X)$, $\Psi|_{X_s}$: $X_s \to$ Spec $A(X)$ is derived from the canonical morphism of X_s and a morphism Spec $A(X_s) \to$ Spec $A(X)$, which is associated with the restriction map r: $A(X) \to A(X_s)$. Now we study this map.

Corollary to Theorem 1.15. *r induces the homomorphism φ: $A(X)_s \to A(X_s)$ by $\varphi(a/s^m) = r(a) \cdot r(s)^{-m}$.*

(i) If X is quasi-compact, then φ is injective.
(ii) If X has a finite affine open cover $\{U_j \mid 1 \le j \le n\}$ such that all $U_i \cap U_j$ are quasi-compact, then φ is an isomorphism.

Proof. Since $V(r(s))$ is empty, $r(s)$ is an invertible element. To prove (i), let $\{U_j \mid 1 \le j \le n\}$ be a finite affine open cover of X and $s_j = s|_{U_j}$. If $a \in A(X)$ satisfies $r(a) = 0$, then $r(a)|_{U_j} = 0$ and so $(a|_{U_j}) \cdot s_j^m = 0$, where m can be chosen independently of j. Thus $s^m a = 0$, i.e., $a/1 = 0$ in $A(X)_s$. The proof of (ii) is left to the reader as an exercise. □

§1.16 Subschemes

a. Let X be a ringed space and let $\mathscr{F}$ be an $\mathscr{O}_X$-module of finite type. Then $Y = \mathrm{supp}\,\mathscr{F}$ is a closed subset of X by Proposition 1.22.(i). In this case, one has the following result.

Lemma 1.31. *With the notation as above, if $U, V \in \Omega$ satisfy $U \supseteq V$ and $U \cap Y = V \cap Y$, then $r_{V,U}$: $\mathscr{F}(U) \to \mathscr{F}(V)$ is an isomorphism.*

Proof. One has an open cover $\{V, U \setminus Y\}$ of U. If $a \in \mathscr{F}(U)$ satisfies $a|_V = 0$, then $a = 0$ since $a|_{U \setminus Y} = 0$. For any $b \in \mathscr{F}(V)$, $b|_{V \cap (U \setminus Y)} = 0|_{V \cap (U \setminus Y)}$; hence one has $a \in \mathscr{F}(U)$ such that $a|_V = b$ by the condition (5) for a sheaf in §1.10.**a**. $\square$

Now, we wish to define a sheaf $\mathscr{G}$ on Y such that $\mathscr{G}(V) \cong \mathscr{F}(U)$ where $V = U \cap Y$ for all $U \in \Omega$. First, given an open subset V of Y, we choose an open subset U of X such that $U \cap Y = V$, which we denote by $U\langle V \rangle$. But when $V_1 \supseteq V_2$, one cannot expect in general that $U\langle V_1 \rangle \supseteq U\langle V_2 \rangle$. In this case, there exists U_3 (for example, $U_3 = U\langle V_1 \rangle \cup U\langle V_2 \rangle$) with $U_3 \cap Y = V_1$ and $U_3 \supseteq U\langle V_1 \rangle$, $U\langle V_2 \rangle$. Since the restriction map: $\mathscr{F}(U_3) \to \mathscr{F}(U\langle V_1 \rangle)$, denoted by r_3, is an isomorphism by Lemma 1.31, let $\rho_{2,1} = r_{2,1} \circ r_3^{-1}$, where $r_{2,1}$ is the restriction map from U_3 to $U\langle V_2 \rangle$. Then $(\{\mathscr{F}(U\langle V \rangle) \mid V \text{ is an open subset of } Y\}, \{\rho_{2,1}\})$ is a sheaf, denoted by $\mathscr{F}|_Y$.

Definition. The sheaf $\mathscr{F}|_Y$ is said to be the sheaf obtained from $\mathscr{F}$ by restricting to its support Y, or the *restriction* of $\mathscr{F}$ to Y.

Note that $\mathscr{F}_x \cong (\mathscr{F}|_Y)_x$, for any $x \in Y$.

b. Definition. An $\mathcal{O}_X$-submodule $\mathscr{S}$ of $\mathcal{O}_X$ is said to be a *sheaf* of *$\mathcal{O}_X$-ideals*, or in short, an *$\mathcal{O}_X$-ideal.*

If $\mathscr{S}$ is an $\mathcal{O}_X$-ideal and $x \in X$, then $\mathscr{S}_x$ can be identified with an ideal of $\mathcal{O}_{X,x}$.

Let X be an affine scheme Spec A and let $\mathscr{S}$ be a quasi-coherent $\mathcal{O}_X$-ideal. Then $I = \Gamma(X, \mathscr{S})$ is an ideal of A, and for any $f \in A$, one has the commutative square

$$\begin{array}{ccc} I = \Gamma(X, \mathscr{S}) & \hookrightarrow & A \\ \downarrow r_{D(f),X} & & \downarrow \psi_{A,f} \\ \Gamma(D(f), \mathscr{S}) & \hookrightarrow & A_f. \end{array}$$

Therefore, $IA_f = \Gamma(D(f), \mathscr{S})$. Since $IA_f \cong I \otimes_A A_f$, it follows that $\mathscr{S} = \tilde{I}$, where $\tilde{I}(D(f)) = IA_f$.

Now, let $\mathscr{S}$ be a quasi-coherent $\mathcal{O}_X$-ideal of a scheme X. Since $\mathcal{O}_X/\mathscr{S}$ is of finite type, $Y = \operatorname{supp}(\mathcal{O}_X/\mathscr{S})$ is a closed subset of X.

Proposition 1.27. *$(Y, (\mathcal{O}_X/\mathscr{S})|_Y)$ is a scheme.*

Proof. Let $\{U_\lambda \mid \lambda \in \Lambda\}$ be an affine open cover of X. Then $\mathscr{S}|_{U_\lambda} = \tilde{I}_\lambda$ for some ideal I_λ of $A_\lambda = \mathcal{O}_X(U_\lambda)$. Since $(\mathcal{O}_X/\mathscr{S})|_{U_\lambda \cap Y} \cong (A_\lambda/I_\lambda)^\sim$ one has $(U_\lambda \cap Y, (\mathcal{O}_X/\mathscr{S})|_{U_\lambda \cap Y})$ is isomorphic to an affine scheme $\operatorname{Spec}(A_\lambda/I_\lambda)$, for all λ. $\square$

Definition. $(Y, (\mathcal{O}_X/\mathcal{S})|_Y)$ is said to be the *closed subscheme* of X defined by $\mathcal{S}$.

Remark. If I is an ideal of a ring A, then one can identify the closed subset $V(I)$ of Spec A with Spec(A/I). With this identification, the closed subschemes of an affine scheme Spec A are the schemes of the form Spec (A/I), where I is an ideal of A.

c. Given a scheme X and a closed subset Y of X, we define a sheaf $\mathcal{S}$ by $\mathcal{S}(U) = \{\sigma \in \mathcal{O}_X(U) \mid \bar{\sigma}(x) = 0 \text{ for all } x \in U \cap Y\}$ for any open subset U of X (see §1.16.**b**). Clearly, $\mathcal{S}$ becomes a sheaf. If U is an affine scheme Spec A and $Y \cap U = V(J)$ for some ideal J of A, then $\mathcal{S}(U) = \{\sigma \in A \mid \bar{\sigma}(x) = 0$ for all $x \in V(J)\} = \sqrt{J}$. Hence, $\mathcal{S}|_U = (\sqrt{J})^\sim$ and so $\mathcal{S}$ is quasi-coherent. The support of $\mathcal{O}_X/\mathcal{S}$ is just Y and $(Y, (\mathcal{O}_X/\mathcal{S})|_Y)$ is a reduced scheme.

Definition. If Y coincides with the base space X, the $\mathcal{O}_X$-ideal $\mathcal{S}$ defined above is said to be the *nilradical* of the scheme X. In this case, $(X, \mathcal{O}_X/\mathcal{S})$ is said to be the *reduced scheme obtained from* X, and is often denoted by X_{red}.

Note that if $\mathcal{N}$ is the nilradical of X, then $\mathcal{N}_x$ is the nilradical of $\mathcal{O}_{X,x}$ for all $x \in X$.

d. Let X be a scheme.

Definition. An *open subscheme* of X is defined to be $(U, \mathcal{O}_X|_U)$ where U is an open subset of X. Any closed subscheme Y of some open subscheme of X is said to be a *subscheme* of X.

Proposition 1.28. *Let $f = (\psi, \theta)\colon X \to Y$ be a morphism of schemes. If ψ is a homeomorphism onto a locally closed subset Z of Y and $\theta_x^{\#}\colon \mathcal{O}_{Y,\psi(x)} \to \mathcal{O}_{X,x}$ is surjective for all x of X, then f is the isomorphism from X onto a subscheme Z of Y.*

PROOF. Since Z is a locally closed subset, it is a closed subset of an open subset U of Y. Replacing Y by the open subscheme U, one can assume that Z is a closed subset of Y. Define $\mathcal{S}$ to be the kernel of $\theta\colon \mathcal{O}_Y \to \psi_*(\mathcal{O}_X)$. Since $\psi_*(\mathcal{O}_X)_{\psi(x)} \cong \mathcal{O}_{X,x}$, and $\theta_x^{\#}$ is surjective for all x, the sequence

$$0 \to \mathcal{S} \to \mathcal{O}_Y \to \psi_*(\mathcal{O}_X) \to 0$$

is exact. We shall prove that $\mathcal{S}$ is quasi-coherent. To do so, one can assume that Y is affine, i.e., $Y = \text{Spec } A$. Letting $B = A(X)$ and letting $\Psi\colon X \to \text{Spec } B$ be the canonical morphism defined in §1.15.**e**, f can be expressed as ${}^a\theta_Y \circ \Psi$, by Theorem 1.15. From Ψ, one has a homomorphism $\xi\colon ({}^a\theta_Y)_*(\tilde{B}) \to \psi_*(\mathcal{O}_X)$, which makes the following triangle commutative

$$\begin{array}{ccc} \tilde{A} = \mathcal{O}_Y & \xrightarrow{\theta} & \psi_*(\mathcal{O}_X) \\ \downarrow & \nearrow_{\xi} & \\ ({}^a\theta_Y)_*(\tilde{B}) . & & \end{array}$$

Since θ is surjective, ξ is also surjective. We claim that ξ is injective. For any $a \in A$, $\psi_*(\mathcal{O}_X)(D(a)) = \mathcal{O}_X(\psi^{-1}(D(a))) = A(X_\alpha)$, where $\alpha = \theta_Y(a)$ and furthermore, $({}^a\theta_Y)_*(\tilde{B})(D(a)) = B_\alpha$. Thus by the Corollary to Theorem 1.15, $\xi_{D(a)}$: $B_\alpha \to A(X_\alpha)$ is injective; hence, ξ is an isomorphism. Therefore, $\mathcal{S} = (\mathrm{Ker}\ \theta_Y)^\sim$. □

Definition. A morphism $f: X \to Y$ of schemes is said to be a *closed immersion* (respectively an *open immersion*; respectively an *immersion*), if there exists a closed subscheme Z (respectively an open subscheme, respectively a subscheme) of Y such that $f: X \to Z$ is an isomorphism.

In this case, we denote the subscheme Z by $f(X)$.

Proposition 1.29

(i) *If $f: X \to Y$ and $g: Y \to Z$ are closed immersions (respectively open immersions, respectively immersions), then so is the composition $g \circ f$.*

(ii) *Let $f: X \to Y$ be a morphism and $\{Y_\lambda \mid \lambda \in \Lambda\}$ be an open cover of Y. Then f is a closed immersion (respectively an open immersion, respectively an immersion) if and only if so is f_λ for all λ, where $X_\lambda = f^{-1}(Y_\lambda)$ and $f_\lambda = f|_{X_\lambda}: X_\lambda \to Y_\lambda$.*

PROOF. (i) One may assume that f and g are natural inclusions; i.e., X and Y are closed subschemes (respectively open subschemes, respectively subschemes) of Y and Z, respectively. Then the proof is easy and it is left to the reader. To prove the case of immersion, one can use Proposition 1.28.

(ii) Left to the reader. □

EXAMPLE 1.22. Let $\varphi: A \to B$ be a ring homomorphism. Then the morphism ${}^a\varphi$: Spec $B \to$ Spec A is a closed immersion if and only if φ is surjective.

Proposition 1.30. *Let $f: X \to Y$ be a morphism of schemes, and let Z be a subscheme of Y such that $f(X) \subseteq Z$ as base spaces. Suppose that X is reduced. Then f can be regarded as a morphism into Z.*

PROOF. Clearly, one may assume Z is a closed subscheme. For every point x of X, take affine neighborhoods U_1 of $f(x)$ and U_2 of x such that $f(U_2) \subseteq U_1$. Then, it suffices to prove that $f|_{U_2}$ is a morphism into $U_1 \cap Z$. Thus, one may suppose that $Y =$ Spec A, $X =$ Spec B, $f = {}^a\varphi$, where φ is a homomorphism: $A \to B$ and $Z = V(I)$ for some ideal I of A. Since ${}^a\varphi(\text{Spec } B) \subseteq V(I)$ by hypothesis and since the closure of ${}^a\varphi(\text{Spec } B)$ is written as $V(\text{Ker } \varphi)$ by Proposition 1.4.(i), it follows that $\sqrt{\text{Ker } \varphi} \supseteq \sqrt{I}$. Since B is reduced, one has $\sqrt{\text{Ker } \varphi} = \text{Ker } \varphi$. Hence $\text{Ker } \varphi \supseteq \sqrt{I} \supseteq I$ and φ induces a ring homomorphism ψ: $A/I \to B$; thus ${}^a\varphi$: Spec $B \to$ Spec A can be regarded as the morphism ${}^a\psi$: Spec $B \to V(I) \cong \text{Spec}(A/I)$. □

e. Let $\mathscr{S}_1$ and $\mathscr{S}_2$ be quasi-coherent $\mathscr{O}_X$-ideals, where X is a scheme. Define a presheaf $\mathscr{S}$ by $\mathscr{S}(U) = \mathscr{S}_1(U) + \mathscr{S}_2(U) \subseteq \mathscr{O}_X(U)$ for all $U \in \Omega$. ${}^a\mathscr{S}$ is a quasi-coherent $\mathscr{O}_X$-ideal, denoted by $\mathscr{S}_1 + \mathscr{S}_2$. Then for every point $x \in X$, $(\mathscr{S}_1 + \mathscr{S}_2)_x = \mathscr{S}_{1,x} + \mathscr{S}_{2,x}$. Furthermore, since $\mathscr{O}_{X,x}$ is a local ring, one has

$$1_x \notin (\mathscr{S}_1 + \mathscr{S}_2)_x \quad \text{if and only if} \quad 1_x \notin \mathscr{S}_{1,x} \quad \text{and} \quad 1_x \notin \mathscr{S}_{2,x}.$$

Thus, letting $Y_i = \text{supp}(\mathscr{O}_X/\mathscr{S}_i)$ for $i = 1, 2$, one has $Y_1 \cap Y_2 = \text{supp}\,(\mathscr{O}_X/(\mathscr{S}_1 + \mathscr{S}_2))$. If $X = \text{Spec } A$, $\mathscr{S}_1 = \tilde{I}_1$ and $\mathscr{S}_2 = \tilde{I}_2$, then $\mathscr{S}_1 + \mathscr{S}_2 = (I_1 + I_2)^{\sim}$.

Definition. The subscheme $(Y_1 \cap Y_2, \mathscr{O}_X/(\mathscr{S}_1 + \mathscr{S}_2)|_{Y_1 \cap Y_2})$ is said to be the *scheme-theoretic intersection* of $(Y_1, (\mathscr{O}_X/\mathscr{S}_1)|_{Y_1})$ and $(Y_2, (\mathscr{O}_X/\mathscr{S}_2)|_{Y_2})$.

EXAMPLE 1.23. Let $X = \text{Spec } k[\mathsf{X}_1, \mathsf{X}_2]$ be an affine plane and let $Y_1 = V(\mathsf{X}_1)$, $Y_2 = V(\mathsf{X}_1 + \mathsf{X}_2^m)$. Then $Y_1 \cap Y_2 = V(\mathsf{X}_1, \mathsf{X}_2^m)$. If $m \geq 2$, the intersection $Y_1 \cap Y_2$ is not reduced.

Notice that even in a simple situation like this, nonreduced schemes can arise naturally.

§1.17 Glueing Schemes

a. Let X be a ringed space and let $\{X_i | i \in J\}$ be an open cover of X. Furthermore, let $\{Y_i | i \in J\}$ be a collection of ringed spaces, and let $\{f_i: X_i \to Y_i | i \in J\}$ be a collection of isomorphisms of ringed spaces. Now, the $Y_{ij} = f_i(X_i \cap X_j)$ are open subsets of Y_i and so can be regarded as ringed spaces for all $i, j \in J$. $\eta_{ij} = f_j \circ f_i^{-1}|_{Y_{ij}}$: $Y_{ij} \to Y_{ji}$ are isomorphisms, which satisfy the conditions:

(i) $Y_i = Y_{ii}$ and $\eta_{ii} = \text{id}_{Y_i}$,
(ii) $\eta_{ij} \circ \eta_{ji} = \text{id}_{Y_{ij}}$,
(iii) $\eta_{ij} \circ \eta_{jl}|_{Y_{ijl}} = \eta_{il}|_{Y_{ijl}}$, where $Y_{ijl} = Y_{lj} \cap \eta_{jl}^{-1}(Y_{ji})$.

Lemma 1.32. (Glueing lemma). *Given a collection $\{Y_i | i \in J\}$ of ringed spaces, a collection $\{Y_{ij} | i, j \in J\}$ of open subsets Y_{ij} of Y_i, and a collection $\{\eta_{ij} | i, j \in J\}$ of isomorphisms η_{ji}: $Y_{ij} \to Y_{ji}$ satisfying the above three conditions, there exist a ringed space X, an open cover $\{X_i | i \in J\}$ of X, and a* *collection* $\{f_i: X_i \to Y_i \mid i \in J\}$ *of isomorphisms such that* $\eta_{ji} = f_j \circ f_i^{-1}|_{Y_{ij}}$ *for all* $i, j \in J$.

Proof. First, take the direct sum $\coprod_{i \in J} Y_i$ of topological spaces and introduce the equivalence relation $\sim$ in $\coprod_{i \in J} Y_i$ by

$$y_i \in Y_i \sim y_j \in Y_j \Leftrightarrow y_i \in Y_{ij} \quad \text{and} \quad y_j = \eta_{ji}(y_i).$$

Define X to be the set $\coprod_{i \in J} Y_i/\sim$ and topologize it with the quotient topology. If $p: \coprod_{i \in J} Y_i \to X$ is the projection map, then all $X_i = p(Y_i)$ are open subsets of X. $\{X_i \mid i \in J\}$ is an open cover of X. Let Ω^* be the set of all open subsets U of X such that $U \subseteq X_i$ for some $i \in J$. For every $U \in \Omega^*$, we choose one $i \in J$ such that $U \subseteq X_i$. Thus, one has a function $i: \Omega^* \to J$ defined by $i(U) = i$. We define $\mathscr{F}(U)$ to be $\mathcal{O}_{Y_i}(f_i(U))$, where $i = i(U)$ and $f_i = (p|_{Y_i})^{-1}$. If U and $V \in \Omega^*$ with $U \supseteq V$ are given, then one has $i = i(U)$ and $j = i(V)$; hence $V \subseteq U \cap X_j \subseteq X_i \cap X_j$ and so $V \subseteq X_i$. Since $f_i(V) = \eta_{ji}^{-1}(f_j(V))$, one has an isomorphism

$$\theta_{ji} = \theta(\eta_{ji})_{f_j(V)}: \mathcal{O}_{Y_j}(f_j(V)) \xrightarrow{\sim} \mathcal{O}_{Y_i}(\eta_{ji}^{-1}(f_j(V))) = \mathcal{O}_{Y_i}(f_i(V)).$$

Letting τ_i be the restriction map $\mathcal{O}_{Y_i}(f_i(U)) \to \mathcal{O}_{Y_i}(f_i(V))$, one defines $\rho_{V,U}$ to be $\theta_{ij} \circ \tau_i: \mathcal{O}_{Y_i}(f_i(U)) \to \mathcal{O}_{Y_j}(f_j(V))$, then $\mathscr{F} = (\{\mathcal{O}_{Y_i}(f_i(U)) \mid U \in \Omega^*\}, \{\rho_{V,U} \mid V, U \in \Omega^*\})$ satisfies the conditions in Lemma 1.21. Thus $\mathscr{F}$ defines a sheaf of rings, denoted by $\mathcal{O}_X$. Clearly, $(X, \mathcal{O}_X)$, $\{X_i\}$, $\{f_i\}$ have the required properties. □

Definition. The above X is said to be the *ringed space* obtained by glueing (or pasting) the ringed spaces $\{Y_i \mid i \in J\}$ along $\{Y_{ij} \mid i, j \in J\}$ via the isomorphisms $\{\eta_{ji} \mid i, j \in J\}$.

Note that if all Y_i are schemes, then so is X.

b. Consider two copies of the affine line $\mathbf{A}_k^1$, say $Y_i = \operatorname{Spec} k[\mathsf{X}^{(i)}]$ for $i = 0, 1$. Let $Y_{10} = D(\mathsf{X}^{(1)})$ and $Y_{01} = D(\mathsf{X}^{(0)})$. Define a k-isomorphism $\varphi_{10}: k[\mathsf{X}^{(1)}, 1/\mathsf{X}^{(1)}] \to k[\mathsf{X}^{(0)}, 1/\mathsf{X}^{(0)}]$ by $\varphi_{10}(\mathsf{X}^{(1)}) = 1/\mathsf{X}^{(0)}$. Let $\eta_{00} = \text{id}$, $\eta_{11} = \text{id}$, and $\eta_{10} = {}^a\varphi_{10}: Y_{01} \to Y_{10}$. Glueing Y_0, Y_1 along Y_{01} via $\{\eta_{ij}\}$, one has a scheme X, which is called the *projective line*, denoted by $\mathbf{P}_k^1$.

If one lets $\mathsf{X}^{(0)} = \mathsf{T}_0/\mathsf{T}_1$ where T_0, T_1 are algebraically independent over k, then $\mathsf{X}^{(1)}$ is written as $\mathsf{T}_1/\mathsf{T}_0$. $(\mathsf{T}_0, \mathsf{T}_1)$ is called the *homogeneous coordinate system* of $\mathbf{P}_k^1$.

§1.18 Projective Spaces

a. Let R be a ring and let $R[\mathsf{T}_0, \ldots, \mathsf{T}_n]$ be a polynomial ring over R. Then, for each i, $\mathsf{X}_0^{(i)} = \mathsf{T}_0/\mathsf{T}_i, \ldots, \mathsf{X}_{i-1}^{(i)} = \mathsf{T}_{i-1}/\mathsf{T}_i, \mathsf{X}_{i+1}^{(i)} = \mathsf{T}_{i+1}/\mathsf{T}_i, \ldots, \mathsf{X}_n^{(i)} = \mathsf{T}_n/\mathsf{T}_i$ are algebraically independent over R. Hence, the $A^{(i)} = R[\mathsf{X}_0^{(i)},$

$\ldots, \mathsf{X}_n^{(i)}]$ are polynomial rings over R contained in $Q = R[\mathsf{T}_0, \ldots, \mathsf{T}_n, \mathsf{T}_0^{-1}, \ldots, \mathsf{T}_n^{-1}]$. As subrings of Q, $A^{(i)}[(\mathsf{X}_j^{(i)})^{-1}] = A^{(j)}[(\mathsf{X}_i^{(j)})^{-1}]$ for all $i \neq j$. The identity map of Q induces ξ_{ij}: $A^{(i)}[(\mathsf{X}_j^{(i)})^{-1}] \to A^{(j)}[(\mathsf{X}_i^{(j)})^{-1}]$, which satisfy $\xi_{ij}(\mathsf{X}_j^{(i)}) = (\mathsf{X}_i^{(j)})^{-1}$ and $\xi_{ij}(\mathsf{X}_l^{(i)}) = \mathsf{X}_l^{(j)}/\mathsf{X}_i^{(j)}$ for $l \neq i, j$. If $i = j$, then one puts $\mathsf{X}_j^{(i)} = 1$, $\xi_{ij} = \text{id}$. Setting $Y_i = \operatorname{Spec} A^{(i)} \cong \mathbf{A}_R^n$, $Y_{ij} = D(\mathsf{X}_j^{(i)})$, and $\eta_{ij} = {}^a\xi_{ij}$, one can easily check that $\{\eta_{ij} \mid 0 \leq i, j \leq n\}$ satisfy conditions (i), (ii), (iii) in §1.17.**a**.

Definition. The scheme obtained by glueing $\{Y_i\}$ along $\{Y_{ij}\}$ via $\{\eta_{ij}\}$ is said to be the *projective n-space over R*, denoted by $\mathbf{P}_R^n$. $(\mathsf{T}_0, \ldots, \mathsf{T}_n)$ is said to be the *homogeneous coordinate system* of $\mathbf{P}_R^n$.

Then $\mathbf{P}_R^n$ is covered by the X_i which are isomorphic to the Y_i. In Chapter 3 (§3.2.**a**) projective schemes will be introduced and X_i will be denoted by $D_+(\mathsf{T}_i)$. We use this notation here also.

b. Let W_i be $D(\mathsf{T}_i)$ in $\mathbf{A}_R^{n+1} = \operatorname{Spec} R[\mathsf{T}_0, \ldots, \mathsf{T}_n]$, for all i. Then the inclusion map φ_i: $R[\mathsf{T}_0/\mathsf{T}_i, \ldots, \mathsf{T}_n/\mathsf{T}_i] \to R[\mathsf{T}_0, \ldots, \mathsf{T}_n, 1/\mathsf{T}_i]$ determines $f_i = {}^a\varphi_i$: $W_i \to D_+(\mathsf{T}_i) \subset \mathbf{P}_R^n$. Then since $f_i|_{W_i \cap W_j} = f_j|_{W_i \cap W_j}$ for all i, j, by Lemma 1.30 one has $f: \bigcup_{i=0}^n W_i \to \mathbf{P}_R^n$ such that $f|_{W_i} = f_i$. The morphism f is said to be the *canonical projection*. Note that $\bigcup_{i=0}^n W_i = \mathbf{A}_R^{n+1} \backslash C$, where C is $V(\mathsf{T}_0, \ldots, \mathsf{T}_n)$.

If R is a field k, then $\mathbf{P}_k^n$ is an integral scheme of dimension n. Further, when k is algebraically closed, the set of closed points of $\mathbf{P}_k^n$ is denoted by $\operatorname{spm}(\mathbf{P}_k^n)$ (see §1.25).

Definition. For a field K, define an equivalence relation in $K^{n+1}\backslash\{0\}$ by $(a_0, \ldots, a_n) \sim (b_0, \ldots, b_n)$ if and only if there is a $\lambda \in K$ such that $a_i = \lambda b_i$ for $0 \leq i \leq n$. The equivalence class containing $(a_0, \ldots, a_n)$ is denoted by $(a_0 : \cdots : a_n)$; and is called $(n + 1)$-ratio. The set of $(n + 1)$-ratios is the projective n-space over K in the classical sense.

The proof of the following result is left to the reader.

If k is an algebraically closed field, there is a natural bijection between $\operatorname{spm}(\mathbf{P}_k^n)$ and the classical projective n-space over k.

§1.19 *S*-Schemes and Automorphism Groups of Schemes

a. Definition. Let S be a scheme. An *S-scheme* (or a *scheme over S*) is a pair (X, φ), where X is a scheme and $\varphi: X \to S$ is a morphism. If (Y, ψ) is another S-scheme, a *morphism* $f: (X, \varphi) \to (Y, \psi)$ *of S-schemes* is a morphism f of schemes such that $\psi \circ f = \varphi$, and is often called an *S-morphism*.

If S is an affine scheme Spec R, an S-scheme is often referred to as an *R-scheme*. The *category* of *S-schemes* or *R-schemes* is denoted by (Sch/S) or (Sch/R).

Let X be a scheme. Then there is a unique homomorphism i: $\mathbb{Z} \to A(X) = \Gamma(X, \mathcal{O}_X)$. Hence by Theorem 1.15, one has $\varphi: X \to \mathrm{Spec}\ \mathbb{Z}$ with $\Gamma(\varphi) = i$. Thus every scheme can be considered uniquely as a $\mathbb{Z}$-scheme and the category of schemes can be identified with $(\mathrm{Sch}/\mathbb{Z})$.

b. An S-scheme (X, φ) is often referred to as X/S or simply as X, if there is no danger of confusion. Then φ is said to be the *structure morphism* of X/S, and S is called the *base scheme* of X/S. If X and Y are S-schemes, $\mathrm{Hom}_S(X, Y)$ denotes the set of all S-morphisms from X to Y. If $S = \mathrm{Spec}\ R$, then $\mathrm{Hom}_R(X, Y)$ is often used to denote $\mathrm{Hom}_S(X, Y)$.

Remark. If X is an S-scheme and V is a subscheme of X, then V can be regarded as an S-scheme via $\varphi \circ i$, where i: $V \to X$ is the inclusion map and φ is the structure morphism of X/S.

Let X be a scheme. It is clear that X is an R-scheme if and only if $A(X)$ is an R-algebra. Therefore, one can generalize Theorem 1.15 as follows:

Proposition 1.31. *Let X be an R-scheme and let A be an R-algebra. Then Γ: $\mathrm{Hom}_R(X, \mathrm{Spec}\ A) \cong \mathrm{Hom}_R(A, A(X))$ is bijective.*

Note that $\Psi: X \to \mathrm{Spec}\ A(X)$ is an R-morphism.

c. Let X be an S-scheme.

Definition. $\mathrm{Aut}_S(X)$ denotes the set of all S-isomorphisms of X into itself, which is a group, called the *automorphism group* of X. When S is an affine scheme Spec R, $\mathrm{Aut}_S X$ is often denoted by $\mathrm{Aut}_R(X)$.

Let A be an R-algebra and let $\mathrm{Aut}_R A$ denote the set of all R-isomorphisms of A into itself. By Proposition 1.31, $\mathrm{Aut}_R A \cong \mathrm{Aut}_R(\mathrm{Spec}\ A)$.

EXAMPLE 1.24. Let $k[\mathsf{X}]$ be the polynomial ring over a field k. $\varphi \in \mathrm{Aut}_k\ k[\mathsf{X}]$ is determined by the image of X, which must be of the form $\alpha\mathsf{X} + \beta$, where $\alpha \in k^* = k\backslash\{0\}$ and $\beta \in k$. Thus $\mathrm{Aut}_k\ k[\mathsf{X}]$ is just the group of one-dimensional affine motions, denoted by $\mathrm{Aff}_1(k)$. Hence,

$$\mathrm{Aut}_k\ \mathbf{A}_k^1 \cong \mathrm{Aff}_1(k).$$

Note that this group is a connected two-dimensional affine algebraic group.

EXAMPLE 1.25. Let $k[X_1, X_2]$ be a polynomial ring. Define $\mathrm{Aff}_2(k) = \{\varphi \in \mathrm{Aut}_k(k[X_1, X_2]) \mid$ there exist $[a_{ij}] \in GL_2(k)$ and $(b_1, b_2) \in k^2$ such that

$$\varphi(X_i) = \sum_{j=1}^{2} a_{ji} X_j + b_i \quad \text{for} \quad i = 1, 2\}.$$

Then $\mathrm{Aff}_2(k)$ is an algebraic group of dimension 6. Furthermore, define $J_2(k) = \{\varphi \in \mathrm{Aut}_k(k[X_1, X_2]) \mid \varphi(X_1) = X_1$ and there exist $a_0, \ldots, a_m \in k$ such that

$$\varphi(X_2) = X_2 + a_0 + a_1 X_1 + \cdots + a_m X_1^m\}.$$

Then $J_2(k)$ is also a subgroup of $\mathrm{Aut}_k(k[X_1, X_2])$, but cannot be an algebraic group since it depends on infinitely many parameters $a_0, a_1, a_2, \ldots$.

Remark. We do not develop the theory of algebraic groups in this book (see [*B*]).

§1.20 Product of S-Schemes

Throughout this section, we let S be a scheme and consider products in (Sch/S). If S is affine, we write $R = A(S)$, i.e., $S = \operatorname{Spec} R$.

a. Definition. Let X and Y be S-schemes. The *S-product* of X and Y is a triple (Z, p, q), where Z is an S-scheme and $p: Z \to X$, $q: Z \to Y$ are S-morphisms, such that given any S-scheme P and any S-morphisms $\varphi: P \to X$ and $\psi: P \to Y$, there exists a unique S-morphism $f: P \to Z$ such that $p \circ f = \varphi$ and $q \circ f = \psi$.

By the next lemma, this triple is unique up to S-isomorphism. Hence, Z is said to be the *S-product* (*scheme*), denoted by $X \times_S Y$, and p, q are said to be the *projection morphisms* or projections. Furthermore the f in the definition is denoted by $(\varphi, \psi)_S$. Hence, $p \circ (\varphi, \psi)_S = \varphi$ and $q \circ (\varphi, \psi)_S = \psi$.

Lemma 1.33. *The S-product of X and Y is unique up to S-isomorphism, if it exists.*

PROOF. Suppose (Z', p', q') is another S-product of X and Y. By definition, one has $(p', q')_S: Z' \to Z$ and $(p, q)_S: Z \to Z'$. Since $p' \circ (p, q)_S = p$, $p \circ (p', q')_S = p'$, and $q' \circ (p, q)_S = q$, $q \circ (p', q')_S = q'$, it follows that $p \circ (p', q')_S \circ (p, q)_S = p$, and $q \circ (p', q')_S \circ (p, q)_S = q$; hence $(p', q')_S \circ (p, q)_S = \mathrm{id}$ by the uniqueness of $(\varphi, \psi)_S$ for given φ and ψ. Similarly, one has $(p, q)_S \circ (p', q')_S = \mathrm{id}$; thus $Z \cong Z'$. □

b. To prove the existence of the S-product, we note that the definition of S-product is equivalent to the existence of the bijections $h_P\colon \mathrm{Hom}_S(P, X) \times \mathrm{Hom}_S(P, Y) \cong \mathrm{Hom}_S(P, X \times_S Y)$ for all S-schemes P satisfying $h_P(\varphi, \psi) \circ \sigma = h_Q(\varphi \circ \sigma, \psi \circ \sigma)$ for any φ, ψ, and any S-scheme Q, and any S-morphism $\sigma\colon Q \to P$.

Lemma 1.34. *Let $X = \mathrm{Spec}\ A$ and $Y = \mathrm{Spec}\ B$ be R-affine schemes. Then the S-product $(X \times_S Y, p, q)$ exists, and is obtained as $(\mathrm{Spec}(A \otimes_R B), {}^a\pi_1, {}^a\pi_2)$, where $\pi_1\colon A \to A \otimes_R B$ and $\pi_2\colon B \to A \otimes_R B$ are defined by $\pi_1(a) = a \otimes 1$ and $\pi_2(b) = 1 \otimes b$.*

Proof. For any R-Scheme P and any R-affine scheme W, by Proposition 1.31, one has

$$\Gamma\colon \mathrm{Hom}_R(P, W) \cong \mathrm{Hom}_R(A(W), A(P)).$$

Applying this to $W = X = \mathrm{Spec}\ A$, and once more to $W = Y = \mathrm{Spec}\ B$, one has

$$\mathrm{Hom}_R(P, X) \times \mathrm{Hom}_R(P, Y) \cong \mathrm{Hom}_R(A, A(P)) \times \mathrm{Hom}_R(B, A(P)).$$

But by the universal property of the tensor product of R-algebras, there is a natural bijection from the right-hand side to $\mathrm{Hom}_R(A \otimes_R B, A(P))$, from which there is a natural bijection to $\mathrm{Hom}_R(P, \mathrm{Spec}\ A \otimes_R B)$ by Proposition 1.31. Hence, one has a natural bijection

$$\mathrm{Hom}_R(P, X) \times \mathrm{Hom}_R(P, Y) \cong \mathrm{Hom}_R(P, \mathrm{Spec}\ A \otimes_R B).$$

Therefore, $(\mathrm{Spec}\ A \otimes_R B, {}^a\pi_1, {}^a\pi_2)$ is the R-product of X and Y, since $h_Z({}^a\pi_1, {}^a\pi_2) = \mathrm{id}$, where $Z = \mathrm{Spec}(A \otimes_R B)$. □

Example 1.26. Let $X = \mathrm{Spec}(R/I)$ and $Y = \mathrm{Spec}(R/J)$, which can be regarded as closed subschemes of $\mathrm{Spec}\ R$. Then $X \times_R Y \cong \mathrm{Spec}\ R/(I + J)$. Hence $X \times_R Y$ is R-isomorphic to the scheme-theoretic intersection of X and Y (cf. §1.16.**e**).

Example 1.27. Let $k[\mathsf{X}]$ and $k[\mathsf{X}_1, \ldots, \mathsf{X}_n]$ be polynomial rings over a field k. Then $k[\mathsf{X}] \otimes_k \cdots \otimes_k k[\mathsf{X}] \cong k[\mathsf{X}_1, \ldots, \mathsf{X}_n]$. Hence, the product of n copies of $\mathbf{A}_k^1$ is isomorphic to $\mathbf{A}_k^n$, i.e.,

$$\mathbf{A}_k^1 \times_k \cdots \times_k \mathbf{A}_k^1 \cong \mathbf{A}_k^n.$$

Example 1.28. Let $A = \mathbb{Q}(\sqrt{-1}) \cong \mathbb{Q}[\mathsf{X}]/(\mathsf{X}^2 + 1)$. Then A is integral, but $A \otimes_{\mathbb{Q}} A \cong \mathbb{Q}(\sqrt{-1})[\mathsf{X}]/((\mathsf{X} + \sqrt{-1}) \cdot (\mathsf{X} - \sqrt{-1})) \cong \mathbb{Q}(\sqrt{-1}) \oplus \mathbb{Q}(\sqrt{-1})$. Hence, $\mathrm{Spec}\ A$ consists of a single point, but $(\mathrm{Spec}\ A) \times_{\mathbb{Q}} (\mathrm{Spec}\ A)$ consists of two points.

c. To prove the existence of the S-product of S-schemes (X, ξ) and (Y, η), we need the following simple result.

Lemma 1.35. *Let U, V, and W be open subsets of X, Y, and S, respectively, such that $\xi(U) \subseteq W$ and $\eta(V) \subseteq W$. If $(X \times_S Y, p, q)$ is an S-product of X and Y, then*

$$p^{-1}(U) \cap q^{-1}(V) \cong U \times_W V = U \times_S V \quad \textit{as} \quad S\textit{-schemes.}$$

PROOF. Let $E = p^{-1}(U) \cap q^{-1}(V)$, which is an open subscheme of $X \times_S Y$. We shall prove that $(E, p|_E, q|_E)$ is a W-product of U and V.

Let P be a W-scheme and let $\varphi: P \to U$, $\psi: P \to V$ be W-morphisms. Composing φ, ψ with the natural inclusion maps $U \to X$, $V \to Y$, one has the S-morphisms $\varphi_1: P \to X$, $\psi_1: P \to Y$. Hence, $(\varphi_1, \psi_1)_S: P \to X \times_S Y$ is obtained, and $p \circ (\varphi_1, \psi_1)_S(P) = \varphi_1(P) \subseteq U$ and $q \circ (\varphi_1, \psi_1)_S(P) = \psi_1(P) \subseteq V$. Hence, $(\varphi_1, \psi_1)_S(P) \subseteq p^{-1}(U) \cap q^{-1}(V) = E$; thus $(\varphi_1, \psi_1)_S$ is a morphism into E. On the other hand, $p|_E$ and $q|_E$ are clearly W-morphisms. Thus by the universality of the W-product, $(E, p|_E, q|_E)$ is the W-product of U and V. Now taking $W = S$, one has the isomorphism $p^{-1}(U) \cap q^{-1}(V) \cong U \times_S V$. □

Theorem 1.16. *For any S-schemes X and Y, the S-product of X and Y exists.*

PROOF. First, assume S is an affine scheme Spec R. Let $\{X_\alpha \mid \alpha \in \Phi\}$ and $\{Y_\lambda \mid \lambda \in \Lambda\}$ be affine open covers of X and Y, respectively. By Lemma 1.34, one has the S-products $(X_\alpha \times_S Y_\lambda, p_{\alpha\lambda}, q_{\alpha\lambda})$ for all $\alpha \in \Phi$, $\lambda \in \Lambda$. Glueing these S-schemes, one will construct the S-scheme $X \times_S Y$.

For all α, $\beta \in \Phi$ and λ, $\mu \in \Lambda$, define open subschemes

$$U_{\alpha\lambda,\beta\mu} = p_{\alpha\lambda}^{-1}(X_\alpha \cap X_\beta) \cap q_{\alpha\lambda}^{-1}(Y_\lambda \cap Y_\mu) \subseteq U_{\alpha\lambda} = X_\alpha \times_S Y_\lambda.$$

Then, by Lemma 1.35, one obtains the S-isomorphisms

$$g_{\alpha\lambda,\beta\mu}: U_{\alpha\lambda,\beta\mu} \xrightarrow{\sim} (X_\alpha \cap X_\beta) \times_S (Y_\lambda \cap Y_\mu).$$

Letting $f_{\beta\mu,\alpha\lambda} = g_{\beta\mu,\alpha\lambda}^{-1} \circ g_{\alpha\lambda,\beta\mu}: U_{\alpha\lambda,\beta\mu} \xrightarrow{\sim} U_{\beta\mu,\alpha\lambda}$, one can easily check that $\{f_{\beta\mu,\alpha\lambda} \mid \alpha, \beta \in \Phi; \lambda, \mu \in \Lambda\}$ satisfy the glueing conditions (i), (ii), and (iii) in §1.17. Thus, glueing $\{U_{\alpha\lambda} \mid \alpha \in \Phi, \lambda \in \Lambda\}$ along $\{U_{\alpha\lambda,\beta\mu}\}$ via $\{f_{\alpha\lambda,\beta\mu}\}$, one obtains an S-scheme Z, which admits an affine open cover $\{V_{\alpha\lambda} \mid \alpha \in \Phi, \lambda \in \Lambda\}$ such that S-isomorphisms $h_{\alpha\lambda}: V_{\alpha\lambda} \to U_{\alpha\lambda}$ exist for all $\alpha \in \Phi$, $\lambda \in \Lambda$, satisfying

$$f_{\beta\mu,\alpha\lambda} \circ h_{\alpha\lambda}|_{V_{\alpha\lambda} \cap V_{\beta\mu}} = h_{\beta\mu}|_{V_{\alpha\lambda} \cap V_{\beta\mu}}.$$

Hence, $p_{\alpha\lambda} \circ h_{\alpha\lambda}|_{V_{\alpha\lambda} \cap V_{\beta\mu}} = p_{\beta\mu} \circ h_{\beta\mu}|_{V_{\alpha\lambda} \cap V_{\beta\mu}}$; thus $\{p_{\alpha\lambda} \circ h_{\alpha\lambda}\}$ defines an S-morphism $p: Z \to X$ by Lemma 1.30. Similarly, $\{q_{\alpha\lambda} \circ h_{\alpha\lambda}\}$ defines an S-morphism $q: Z \to Y$. It is easy to see that (Z, p, q) is an S-product of X and Y; Z is denoted by $X \times_S Y$.

Next, consider the case when S is not necessarily affine. Then let $\{S_i \mid i \in J\}$ be an affine open cover of S and let $X_i = \xi^{-1}(S_i)$, $Y_i = \eta^{-1}(S_i)$ for all i. One has the S_i-products $(X_i \times_{S_i} Y_i, p_i, q_i)$ for all $i \in J$. Since S_i is an open subscheme of S, by Lemma 1.35 one has $X_i \times_{S_i} Y_i = X_i \times_S Y_i$. Letting $V_{ij} = p_i^{-1}(X_i \cap X_j) \cap q_i^{-1}(Y_i \cap Y_j) \subseteq X_i \times_S Y_i$, one has the S-isomorphism

$$g_{ij}: V_{ij} \xrightarrow{\sim} (X_i \cap X_j) \times_S (Y_i \cap Y_j).$$

Defining f_{ij} to be $g_{ij}^{-1} \circ g_{ji}: V_{ji} \to V_{ij}$, one can easily check that $\{f_{ij} \mid i, j \in J\}$ satisfies conditions (i), (ii), (iii) in §1.17. Glueing $\{X_i \times_S Y_i\}$ along $\{V_{ij}\}$ with $\{f_{ij}\}$, one obtains an S-scheme Z, which admits an open cover $\{V_i \mid i \in J\}$ with S-isomorphisms $h_i: V_i \xrightarrow{\sim} X_i \times_S Y_i$. $\{p_i \circ h_i\}$ and $\{q_i \circ h_i\}$ define S-morphisms $p: Z \to X$ and $q: Z \to Y$, respectively.

It is rather easy to check that the triple (Z, p, q) is actually the S-product of X and Y. □

d. From the universal property of the S-product, one can easily derive the following elementary properties of S-products.

Proposition 1.32. *Let X, Y, Z be S-schemes and let W be a Z-scheme.*

(i) $X \times_S S \cong X$, *where S denotes* (S, id).
(ii) $X \times_S Y \cong Y \times_S X$; *this isomorphism will be denoted τ.*
(iii) $(X \times_S Y) \times_S Z \cong X \times_S (Y \times_S Z)$.
(iv) $(X \times_S Z) \times_Z W \cong X \times_S W$, *where the W on the right-hand side denotes the S-scheme whose structure morphism is the composition of the structure morphisms $W \to Z$ and $Z \to S$.*
(v) *If $X_Z = X \times_S Z$ and $Y_Z = Y \times_S Z$, then*

$$(X \times_S Y) \times_S Z \cong X_Z \times_Z Y_Z.$$

(vi) *If $\sigma: S \to T$ is an immersion, then*

$$X \times_S Y \cong X \times_T Y.$$

Proof of (vi). Let $\varphi: P \to X$, $\psi: P \to Y$ be T-morphisms, where P is a T-scheme. Then, letting ξ and η be the structure morphisms of the S-schemes X and Y respectively, one has $\sigma \circ \xi \circ \varphi = \sigma \circ \eta \circ \psi$ equal to the structure morphism of the T-scheme P. Since σ is an immersion, it follows that $\xi \circ \varphi = \eta \circ \psi$. Applying the universal property of the S-product, one obtains the isomorphism. □

Note that (vi) was proved in Lemma 1.35, in the case when σ is an open immersion.

Definition. Let X be a scheme over a field k. If K/k is a field extension, $X(K)$ denotes $\mathrm{Hom}_k(\mathrm{Spec}\ K, X)$. Any element of $X(K)$ is said to be a *K-valued point* of X.

Let Y, S be k-schemes and let $\xi: X \to S$, $\eta: Y \to S$ be k-morphisms. Then $(X \times_k Y)(K) = X(K) \times Y(K)$ as sets. Letting α: $X(K) \to S(K)$ and

β: $Y(K) \to S(K)$ be defined by $\alpha(x) = \xi \circ x$ and $\beta(y) = \eta \circ y$, one has $(X \times_S Y)(K) = \{(x, y) \in X(K) \times Y(K) \mid \alpha(x) = \beta(y)\}$.

Note that the right-hand side is the *fiber product of the sets* $X(K)$ and $Y(K)$ over $S(K)$. Thus, in general, the S-product of S-schemes X and Y is also said to be the *fiber product* of X and Y over S.

e. Let f: $X_1 \to X_2$ and g: $Y_1 \to Y_2$ be S-morphisms. Denoting by $(X_i \times_S Y_i, p_i, q_i)$ the S-products of X_i and Y_i for $i = 1, 2$, one obtains the commutative diagram

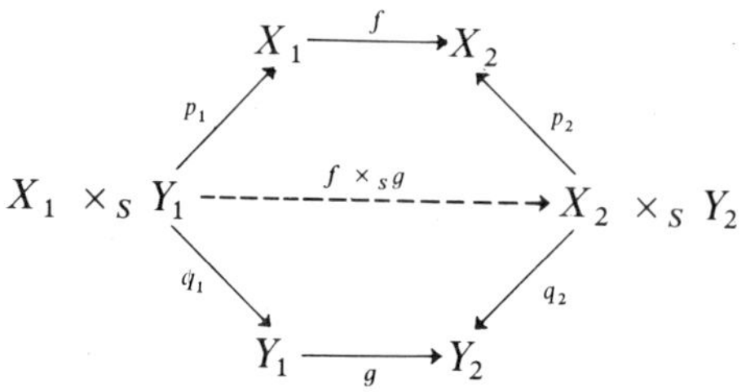

Definition. The *S-product of morphisms* f and g is defined to be $(f \circ p_1, g \circ q_1)_S$, denoted by $f \times_S g$.

Proposition 1.33

(i) *If* f': $X_2 \to X_3$ *and* g': $Y_2 \to Y_3$ *are S-morphisms, then* $(f' \times_S g') \circ (f \times_S g) = f' \circ f \times_S g' \circ g$.

(ii) $f \times_S g = \mathrm{id}_{X_2} \times_S g \circ f \times_S \mathrm{id}_{Y_1}$.

(iii) *Let* ξ: $X \to S$ *and* η: $Y \to S$ *be the structure morphisms. Then identifying* $S \times_S Y$ *with* Y, $\xi \times_S \mathrm{id}_Y$: $X \times_S Y \to S \times_S Y = Y$ *becomes the projection* $X \times_S Y \to Y$.

PROOF. (i) and (ii) are left to the reader. (iii) follows from the commutative diagram

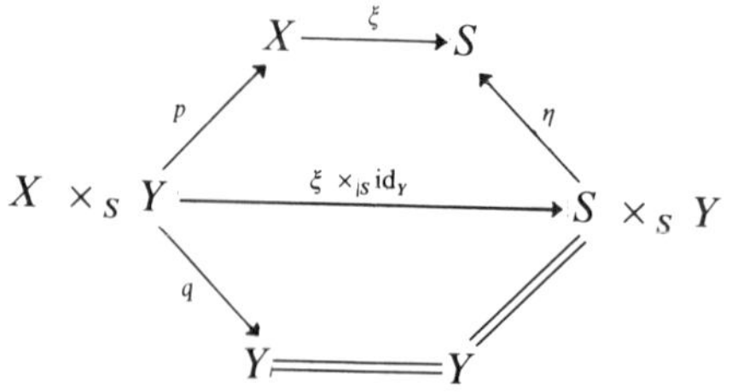

□

Thus, any $f \times_S g$ can be written as a composition of the projection morphisms, which will be studied in the next sections.

If S and Y are affine schemes, say $S = \operatorname{Spec} R$ and $Y = \operatorname{Spec} A$, then $X \otimes_R A$ denotes $X \times_S Y$. Moreover, if A is $T^{-1}R$ for some multiplicative subset T of R, then we let $T^{-1}X$ denote $X \otimes_R T^{-1}R$.

Definition. An S-morphism $f: X \to Y$ is said to be a *dominating morphism* if $f(X)$ is a dense subset of Y.

By Proposition 1.4.(iii), "${}^a\varphi$: Spec $B \to$ Spec A is dominating if and only if $\varphi: A \to B$ has a nilpotent kernel.

§1.21 Base Extension

a. Definition. Let $f: X \to S$ be a morphism of schemes. For a morphism $g: Y \to S$, one has the S-product $(X \times_S Y, p, q)$. Then the Y-scheme $(X \times_S Y, q)$ is denoted by X_Y, and is said to be the *result of extending the base scheme of X from S to Y.*

The projection q is often denoted by f_Y, and is $f \times_S \mathrm{id}_Y$ if one identifies $S \times_S Y$ with Y by Proposition 1.33.(iii).

Lemma 1.36

(i) *Let U be an open subscheme of S. Then $X_U = X \times_S U \cong f^{-1}(U)$, and $f_U = f|_{f^{-1}(U)}$ with the identification $X_U = f^{-1}(U)$.*
(ii) *Furthermore, let V be an open subscheme of Y with $g(V) \subset U$. Then $f_Y^{-1}(V) \cong f^{-1}(U) \times_U V$.*

Proof. (i) Identifying $X \times_S S$ with X by Proposition 1.32.(i), one has $f^{-1}(U) \cong X \times_S U$ by Lemma 1.35, and then $f|_{f^{-1}(U)} = f_U$.
(ii) This is proved similarly. □

b. Let $\mathbb{P}$ be a property of a morphism $f: X \to S$.

Definition. One says that $\mathbb{P}$ is *stable under base extension* if $f_Y: X_Y \to Y$ has the property $\mathbb{P}$ for any morphism $g: Y \to S$.

For instance, if f has a property $\mathbb{P}$ that is stable under base extension, then for any open subset U of S, $f|_{f^{-1}(U)}: f^{-1}(U) \to U$ has the property $\mathbb{P}$ by Lemma 1.36.(i).

Remark. Now, let $f: X \to Y$ be an S-morphism and $g: H \to S$ be a morphism. Then by Proposition 1.32, $X_H = X \times_S H \cong X \times_Y Y \times_S H \cong X \times_Y Y_H$. Hence, X_H can be regarded as the result of extending the base scheme of X from Y to Y_H. Thus one has $f_H = f \times_Y \mathrm{id}_{Y_H}$, which is also written as $f \times_S \mathrm{id}_H$.

c. Proposition 1.34. *The property of being a closed immersion (an open immersion, an immersion) is stable under base extension.*

Proof. Let $f\colon X \to Y$ be a closed immersion and let $g\colon Z \to Y$ be a morphism of schemes. At any point z of Z, one has affine open neighborhoods Z_α of z and Y_λ of $g(z)$ such that $g(Z_\alpha) \subseteq Y_\lambda$. Since $f_Z^{-1}(Z_\alpha) \cong f^{-1}(Y_\lambda) \times_{Y_\lambda} Z_\alpha$ by Lemma 1.36.(ii), one concludes that $f_Z|_{f_Z^{-1}(Z_\alpha)}\colon f_Z^{-1}(Z_\alpha) \to Z_\alpha$ is a closed immersion by Lemma 1.37 below. This implies that f_Z is a closed immersion by Proposition 1.29.

Next, assume f is an open immersion. By Lemma 1.36.(i), $X \times_Y Z \cong g^{-1}(f(X))$, which is an open subscheme of Z; thus f_Z is also an open immersion.

Finally, one can prove the stability of the property of being an immersion by combining the stability of the properties of being a closed immersion or an open immersion. □

In this proof, we used the following lemma.

Lemma 1.37. *Let* ${}^a\varphi$: Spec $B \to$ Spec A *be a closed immersion and let* ${}^a\psi$: Spec $C \to$ Spec A *be a morphism of affine schemes. Then* ${}^a(\varphi \otimes 1_C)$: Spec$(B \otimes_A C) \to$ Spec C *is also a closed immersion.*

Proof. By hypothesis, $\varphi\colon A \to B$ is surjective; hence so is $\varphi \otimes 1_C\colon C = A \otimes_A C \to B \otimes_A C$. By Example 1.22, one obtains the assertion. □

d. Let Y_1 and Y_2 be subschemes of a scheme X. Then by the last proposition, one has the immersion $Y_1 \times_X Y_2 \to X$; thus $Y_1 \times_X Y_2$ can be regarded as a subscheme of X. If Y_1 and Y_2 are open subschemes, then by Lemma 1.35, $Y_1 \times_X Y_2 \cong Y_1 \cap Y_2$. Furthermore, if the Y_i are closed subschemes defined by $\mathcal{O}_X$-ideals $\mathcal{I}_i$, then $Y_1 \times_X Y_2$ is a closed subscheme defined by $\mathcal{I}_1 + \mathcal{I}_2$; thus $Y_1 \times_X Y_2$ is the scheme-theoretic intersection of Y_1 and Y_2 (cf. §1.16.**e**).

§1.22 Graphs of Morphisms

a. Definition. Let $f\colon X \to Y$ be an S-morphism of S-schemes. The morphism $(\mathrm{id}_X, f)_S\colon X \to X \times_S Y$ is called the (S-)*graph* of f, denoted by Γ_f. If $f = \mathrm{id}_X$, then $\Gamma_f = (\mathrm{id}_X, \mathrm{id}_X)_S$ is also said to be the *diagonal morphism*, denoted by $\Delta_{X/S}$ or Δ_φ, where φ is the structure morphism of X.

Proposition 1.35

(i) *If both* Y *and* S *are affine, then* Γ_f *is a closed immersion. In particular, if* X *and* S *are affine, then* $\Delta_{X/S}$ *is a closed immersion.*
(ii) *In general,* Γ_f *is an immersion.*

PROOF. (i) Let $\{X_\lambda \mid \lambda \in \Lambda\}$ be an affine open cover of X, and let $(X \times_S Y, p, q)$ be the S-product. Then $p^{-1}(X_\lambda) \cong X_\lambda \times_S Y$ is affine and $\Gamma_f^{-1}(p^{-1}(X_\lambda)) = X_\lambda$. Hence, to prove (i), it suffices to show (i) in the case when X is also affine; say $S = \operatorname{Spec} R$, $X = \operatorname{Spec} B$, $Y = \operatorname{Spec} A$, and $f = {}^a\varphi$. Γ_f is associated with $\psi: B \otimes_R A \to B$ defined by $\psi(b \otimes a) = b\varphi(a)$. Clearly, ψ is surjective and so Γ_f is a closed immersion.

(ii) First, assume S is affine. Let $\{Y_\alpha \mid \alpha \in \Phi\}$ be an affine open cover of Y. Then, letting $X_\alpha = f^{-1}(Y_\alpha)$, one sees that $\Gamma_f|_{X_\alpha} = \Gamma_{f_\alpha}$, where $f_\alpha = f|_{X_\alpha}$: $X_\alpha \to Y_\alpha$. Hence, $\Gamma_f(X_\alpha) \subseteq X_\alpha \times_S Y_\alpha$, which is an open subset of $X \times_S Y_\alpha$. Therefore, $q^{-1}(Y_\alpha) \cong X \times_S Y_\alpha$ and $\Gamma_f^{-1}(q^{-1}(Y_\alpha)) = X_\alpha$ imply that $\Gamma_f|_{X_\alpha}$: $X_\alpha \to X \times_S Y_\alpha$ is an immersion for all $\alpha \in \Phi$. By Proposition 1.29, one concludes that Γ_f is an immersion.

Finally, if S is not affine, take an affine open cover $\{S_j \mid i \in J\}$ of S. Then, since $(X \times_S Y) \times_S S_j \cong X_j \times_{S_j} Y_j$, where $X_j = X \times_S S_j$ and $Y_j = X \times_S S_j$ (cf. Proposition 1.32.(v)), it follows that $\Gamma_f^{-1}(X_j \times_{S_j} Y_j) = X_j$ and $\Gamma_f|_{X_j}$ is the graph of f_{X_j}: $X_j \to Y_j$, which was proved to be an immersion. Hence, Γ_f is also an immersion. □

Therefore, since Γ_f is an immersion, $\Gamma_f(X)$ is a subscheme of $X \times_S Y$. $\Gamma_f(X)$ is also said to be the *graph* of f.

Remark. If $\Gamma_f(X)$ is a closed subset of $X \times_S Y$, then Γ_f is a closed immersion (see Proposition 1.28).

b. Let $f: X \to Y$ be a morphism and let y be a point of Y. Take an affine open neighborhood $U \cong \operatorname{Spec} A$ of y. The point y corresponds to a prime ideal $\mathfrak{p}$ of A and $\mathcal{O}_{Y,y} \cong A_\mathfrak{p}$. Associated with the natural homomorphism $A \to A_\mathfrak{p}/\mathfrak{p}A_\mathfrak{p} = k(y)$, one has the morphism ι_y : $\operatorname{Spec} k(y) \to U \subseteq Y$. Then $\iota_y((0)) = y$ and ι_y does not depend on the choice of U. We write $\operatorname{Spec} k(y)$ as y.

If y is a closed point, then ι_y is a closed immersion. But in general, ι_y need not be an immersion.

Extending the base scheme of X from Y to $y = \operatorname{Spec} k(y)$, we obtain the commutative square

$$\begin{array}{ccc} X_y = X \times_Y \operatorname{Spec} k(y) & \xrightarrow{\ p\ } & X \\ \Big\downarrow{\scriptstyle f_y = q} & & \Big\downarrow{\scriptstyle f} \\ y = \operatorname{Spec} k(y) & \xrightarrow[\ \iota_y\ ]{} & Y. \end{array}$$

Definition. The $k(y)$-scheme $X \times_Y \operatorname{Spec} k(y)$ is called the *fiber of the morphism f over y*, which is written as X_y, by convention.

Proposition 1.36. *The base space of X_y is homeomorphic to the subspace $f^{-1}(y)$.*

PROOF. From the above diagram, it follows that $X_y = p^{-1}(f^{-1}(y))$; hence $p(X_y) \subseteq f^{-1}(y)$. Taking an affine open subset V of X with $V \cap f^{-1}(y) \neq \emptyset$, one has $p^{-1}(V) \cong X_y \times_X V = X \times_Y \operatorname{Spec} k(y) \times_X V \cong V_y$. Hence, it suffices to prove the result under the assumption that both X and Y are affine, say $X = \operatorname{Spec} B$, $Y = \operatorname{Spec} A$, and y is a prime ideal $\mathfrak{p}$ of A. Then letting $\varphi = \Gamma(f)$, we have the following homeomorphism:

$$f^{-1}(y) = \{\mathfrak{q} \in \operatorname{Spec} B \mid \varphi^{-1}(\mathfrak{q}) = \mathfrak{p}\} \approx \{\mathfrak{q} \in \operatorname{Spec} B_1 \mid \varphi_1^{-1}(\mathfrak{q}) = \mathfrak{p}A_{\mathfrak{p}}\},$$

where $B_1 = \varphi(A \backslash \mathfrak{p})^{-1}B$ and $\varphi_1 \colon A_{\mathfrak{p}} \to B_1$ is the homomorphism obtained from φ. Since $\varphi_1^{-1}(\mathfrak{q}) = \mathfrak{p}A_{\mathfrak{p}}$ is equivalent to $\mathfrak{q} \supseteq \mathfrak{p}A_{\mathfrak{p}} \cdot B_1$, one has

$$\{\mathfrak{q} \in \operatorname{Spec} B_1 \mid \mathfrak{q} \supseteq \mathfrak{p}A_{\mathfrak{p}} \cdot B_1\} \approx \operatorname{Spec}(B_1/\mathfrak{p}A_{\mathfrak{p}} \cdot B_1) = X_y. \qquad \square$$

Remark. The scheme X_y is often denoted by $f^{-1}(y)$.

Definition. If Y is irreducible, $f(X)$ is dense, and $*$ is the generic point of Y, then $f^{-1}(*)$ is said to be the *generic fiber* of f.

c. Definition. Let $f \colon X \to Y$ be a morphism of schemes and let W be a subscheme of Y. The Y-product $X \times_Y W$ is said to be the *inverse image (scheme)* of W by f.

If W is an open subscheme, then $X \times_Y W \cong f^{-1}(W)$. Furthermore, if $W = \{y\}$ is a reduced subscheme consisting of a closed point y, then $X \times_Y W \cong f^{-1}(y)$.

Proposition 1.37. *$X \times_Y W$ is homeomorphic to $f^{-1}(W)$.*

PROOF. One may assume that W is a closed subscheme. Furthermore, as in the proof of the last proposition, one can assume that X and Y are affine, i.e., $X = \operatorname{Spec} B$, $Y = \operatorname{Spec} A$, $W = V(I)$ for some ideal I of A, and $\varphi = \Gamma(f) \colon A \to B$. Then, one has the equalities and the homeomorphism:

$$\begin{aligned} f^{-1}(W) &= \{\mathfrak{q} \in \operatorname{Spec} B \mid \varphi^{-1}(\mathfrak{q}) \in V(I)\} \\ &= \{\mathfrak{q} \in \operatorname{Spec} B \mid \mathfrak{q} \supseteq IB\} \approx \operatorname{Spec}(B/IB) \cong X \times_Y W. \qquad \square \end{aligned}$$

d. Definition. Let $f, g \colon X \to Y$ be S-morphisms. Then the *difference kernel* $\operatorname{Ker}(f, g)$ is defined to be a subscheme $(f, g)_S^{-1}(\Delta_{Y/S}(Y))$ of X.

Letting i be the inclusion of $\operatorname{Ker}(f, g)$, $(f, g)_S \circ i$ is a morphism into $\Delta_{Y/S}(Y)$. Hence, $f \circ i = g \circ i$. We shall prove that $(\operatorname{Ker}(f, g), i)$ is universal among (P, h) where P is an arbitrary S-scheme and $h \colon P \to X$ is an S-morphism.

Proposition 1.38. *For any S-morphism $h \colon P \to X$ such that $f \circ h = g \circ h$, there exists $h^\# \colon P \to \operatorname{Ker}(f, g)$ such that $h = i \circ h^\#$.*

PROOF. Letting $\psi = f \circ h = g \circ h$: $P \to Y$, one has $(f, g)_S \circ h = (f \circ h, g \circ h)_S = (\psi, \psi)_S = j \circ \psi_1$, where j: $\Delta_{Y/S}(Y) \to Z = Y \times_S Y$ is the inclusion map and ψ_1: $P \to \Delta_{Y/S}(Y)$. Hence $h^\# = (h, \psi_1)_Z$: $P \to X \times_Z \Delta_{Y/S}(Y) = \mathrm{Ker}(f, g)$ is obtained, satisfying $h = i \circ h^\#$. □

$$\begin{array}{ccccc} \mathrm{Ker}(f, g) & \xrightarrow{\ i\ } & X & \overset{f}{\underset{g}{\rightrightarrows}} & Y \\ \downarrow & & \downarrow{\scriptstyle (f,g)_S} & & \\ P \xrightarrow{\psi_1} \Delta_{Y/S}(Y) & \overset{j}{\subset} & Y \times_S Y = Z & & \end{array}$$

□

Corollary. *Let f: $X \to Y$ be an S-morphism and let $(X \times_S Y, p, q)$ be the S-product of S-schemes X and Y. Then $\mathrm{Ker}(f \circ p, q)$ coincides with the subscheme $\Gamma_f(X)$, i.e., $\Gamma_f(X) = \varphi^{-1}(\Delta_{Y/S}(Y))$, φ being $f \times_S \mathrm{id}_Y$.*

PROOF. $y \in \psi^{-1}(\varphi(F))$ if and only if $\psi(y) = \varphi(x)$ for some $x \in F$. Hence, by the last proposition, there exists z such that $x = p(z)$ and $y = q(z)$; thus $z \in p^{-1}(F)$ and $y \in q(p^{-1}(F))$. □

e. Proposition 1.39. *Let (X, φ) and (Y, ψ) be nonempty S-schemes, and let $(X \times_S Y, p, q)$ be the S-product. For any two points $x \in X$ and $y \in Y$, $\varphi(x) = \psi(y)$ if and only if $x = p(z)$ and $y = q(z)$ for some $z \in X \times_S Y$.*

PROOF. The if part is clear. Assume that $s = \varphi(x) = \psi(y) \in S$. Then from ι_x and ι_y, one forms the S-morphism δ: $\mathrm{Spec}(k(x) \otimes_{k(s)} k(y)) \to X \times_S Y$. Since $k(x) \otimes_{k(s)} k(y) \neq (0)$, choose any z in the image of $\mathrm{Spec}(k(x) \otimes_{k(s)} k(y))$ under δ. Then $x = p(z)$ and $y = q(z)$. □

Corollary. *Let F be a subset of X. Then $q(p^{-1}(F)) = \psi^{-1}(\varphi(F))$.*

PROOF. $y \in \psi^{-1}(\varphi(F))$ if and only if $\psi(y) = \varphi(x)$ for some $x \in F$. Hence, by the last proposition, there exists z such that $x = p(z)$ and $y = q(z)$; thus $z \in p^{-1}(F)$ and $y \in q(p^{-1}(F))$. □

In particular, if φ is surjective, then so is $q = \varphi_Y$. Hence, the property of being surjective is stable under base extension.

§1.23 Separated Schemes

a. Definition. A morphism f: $X \to Y$ of schemes is said to be *separated* if $\Delta_{X/Y}$: $X \to X \times_Y X$ is a closed immersion. In this case, one also says that the scheme X is *separated over* Y, or is Y*-separated.*

By Proposition 1.35.(i), any morphism of affine schemes is separated.

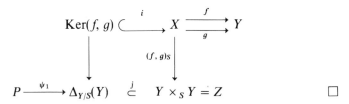

Remark. The property of being separated for schemes is analogous to that of being Hausdorff for topologies.

Proposition 1.40. *Let X be a Y-scheme. Then for any open subsets U and V of X, one has $U \cap V \cong \Delta_{X/Y}(X) \cap (U \times_Y V)$, where $U \times_Y V$ is identified with the open subscheme $p^{-1}(U) \cap q^{-1}(V)$ of $X \times_Y X$. Here, $(X \times_Y X, p, q)$ is the Y-product.*

PROOF. Since $p(\Delta_{X/Y}(U)) = U$ and $q(\Delta_{X/Y}(V)) = V$, it follows that $\Delta_{X/Y}(U \cap V) \subseteq p^{-1}(U) \cap q^{-1}(V) \subset X \times_Y X$. Hence $\Delta_{X/Y}|_{U \cap V}$ is an immersion into $\Delta_{X/Y}(X) \cap p^{-1}(U) \cap q^{-1}(V)$. Furthermore, $\eta = \Delta_{X/Y}^{-1}$: $\Delta_{X/Y}(X) \to X$ satisfies $\eta(\Delta_{X/Y}(X) \cap p^{-1}(U) \cap q^{-1}(V)) \subseteq U \cap V$, since $\eta(p^{-1}(U)) = U$ and $\eta(q^{-1}(V)) = V$. Hence, $\Delta_{X/Y}|_{U \cap V}$ is an isomorphism. □

Proposition 1.41. *Let f: $X \to Y$ be an S-morphism. If Y is separated over S, then the graph Γ_f: $X \to X \times_S Y$ is a closed immersion.*

PROOF. We use the notation in the Corollary to Proposition 1.38. Since $\Delta_{Y/S}(Y)$ is a closed set, so is $\varphi^{-1}(\Delta_{Y/S}(Y)) = \Gamma_f(X)$. □

b. To prove the elementary properties of separated morphisms, we need the next three lemmas.

Lemma 1.38. *Let f: $X \to Y$, g: $Y \to Z$ be morphisms of schemes, and let $(X \times_Y X, p, q)$ be the Y-product. If j denotes $(p, q)_Z$: $X \times_Y X \to X \times_Z X$, the following diagram is commutative:*

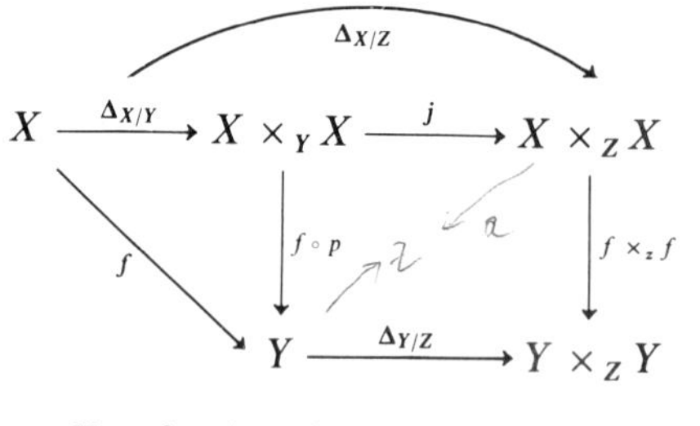

Furthermore, $(X \times_Y X, j, f \circ p)$ is the $Y \times_Z Y$-product of Y and $X \times_Z X$.

PROOF. Left to the reader. □

Lemma 1.39. *Let f: $X \to Y$ and φ: $H \to Y$ be morphisms. Identifying $(X \times_Y X)_H = (X \times_Y X) \times_Y H$ with $X_H \times_H X_H$, one obtains*

$$(\Delta_{X/Y})_H = \Delta_{X_H/H}: X_H \to X_H \times_H X_H$$

and

$$\Delta_{X/Y}(X)_H = \Delta_{X_H/H}(X_H).$$

In particular, if H is an open subset of Y, then letting $p, q\colon X \times_Y X \to X$ be projections, one has $(f \circ p)^{-1}(H) \cong f^{-1}(H) \times_H f^{-1}(H)$, and $\Delta_{X/Y}|_{f^{-1}(H)}\colon f^{-1}(H) \to f^{-1}(H) \times_H f^{-1}(H)$ is written as $\Delta_{f^{-1}(H)/H}$.

Proof. Left to the reader. □

Lemma 1.40. *Let $f\colon X \to Y$ and $g\colon Y \to Z$ be morphisms. Letting $(X \times_Z Y, p, q)$ be the Z product of X and Y, one has $q \circ \Gamma_f = f$ and $q = (g \circ f)_Y$.*

Proof. This follows from the definitions of Γ_f and $(g \circ f)_Y$.

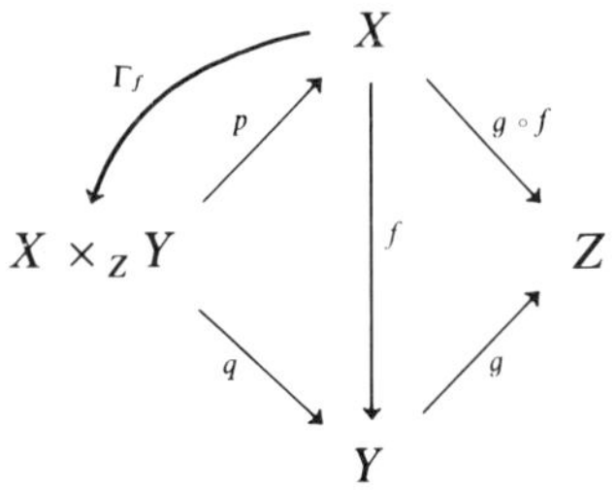

□

Theorem 1.17

(i) *If $f\colon X \to Y$ and $g\colon Y \to Z$ are separated, then so is $g \circ f$.*
(ii) *Any immersion is separated.*
(iii) *The property of being separated is stable under base extension.*
(iv) *If $f\colon X \to Y$ and $g\colon Y \to Z$ are morphisms such that $g \circ f$ is separated, then so is f.*
(v) *If $f\colon X \to Y$ is a morphism and $\{Y_\lambda \mid \lambda \in \Lambda\}$ is an open cover of Y such that $f|_{X_\lambda}\colon X_\lambda \to Y_\lambda$, where X_λ is $f^{-1}(Y_\lambda)$, is separated for all λ, then so is f.*

Proof. (i) By hypothesis, both $\Delta_{X/Y}$ and $\Delta_{Y/Z}$ are closed immersions. With the notation of Lemma 1.38, one has $\Delta_{X/Z} = j \circ \Delta_{X/Y}$. Since j is obtained from $\Delta_{Y/Z}$ by base extension, i.e., $j = (\Delta_{Y/Z})_{X \times_Z X}$, (cf. Lemma 1.38), j is also a closed immersion by Proposition 1.34. Hence $\Delta_{X/Z}$ is a closed immersion by Proposition 1.29.

(ii) If $f\colon X \to Y$ is an immersion, one has the isomorphisms $X \cong X \times_X X \cong X \times_Y X$ by Proposition 1.32.(vi); thus $\Delta_{X/Y}$ is an isomorphism. In particular, f is separated.

(iii) With the notation of Lemma 1.39, $\Delta_{X_H/H} = (\Delta_{X/Y})_H$ is a closed immersion by Proposition 1.34.

(iv) By Lemma 1.40, $q = (g \circ f)_Y$, which is a separated morphism by (iii). Since $f = q \circ \Gamma_f$ and since Γ_f is an immersion, f is a separated morphism by (i) and (ii).

(v) Let $p, q\colon X \times_Y X \to X$ be the projections. Since $(f \circ p)^{-1}(Y_\lambda) = f^{-1}(Y_\lambda) \times_{Y_\lambda} f^{-1}(Y_\lambda) = X_\lambda \times_{Y_\lambda} X_\lambda$ and $\Delta_{X/Y}^{-1}((f \circ p)^{-1}(Y_\lambda)) = X_\lambda$, it suffices to prove that $\Delta_{X/Y}|_{X_\lambda}: X_\lambda \to X_\lambda \times_{Y_\lambda} X_\lambda$ is closed for all λ. But by Lemma 1.39, $\Delta_{X/Y}|_{X_\lambda} = \Delta_{X_\lambda/Y_\lambda}$, which is closed by hypothesis. □

c. Definition. A scheme X is said to be *separated* if X is $\mathbb{Z}$-separated.

If X is separated over a scheme S and S is separated, then so is X by Theorem 1.17.(i).

Proposition 1.42

(*i*) *If an R-scheme is separated, then it is separated over R.*
(*ii*) *Let X be an R-separated R-scheme. If U and V are affine open subsets of X, then so is $U \cap V$.*

PROOF. (i) This is derived from Theorem 1.17.(iv).

(ii) Applying Proposition 1.40 to U and V, one has $U \cap V \cong \Delta_{X/R}(X) \cap (U \times_R V)$, which is a closed subscheme of an affine scheme $U \times_R V$ by hypothesis. Hence, $U \cap V$ is also affine. □

EXAMPLE 1.29. Let $Y_1 = \operatorname{Spec} \mathbb{Q}[\mathsf{X}_1, \mathsf{X}_2]$ and $Y_2 = \operatorname{Spec} \mathbb{Q}[\mathsf{Y}_1, \mathsf{Y}_2]$. Furthermore let $Y_{12} = D(\mathsf{X}_1, \mathsf{X}_2) \cong \mathbf{A}^2_{\mathbb{Q}} \backslash \{0\}$, and $Y_{21} = D(\mathsf{Y}_1, \mathsf{Y}_2)$. Identifying Y_{12} with Y_{21} via $\eta_{12}: Y_{12} \to Y_{21}$ defined by $\eta_{12}(\mathsf{X}_i) = \mathsf{Y}_i$ for $i = 1, 2$, one obtains a scheme Y, which has an affine open cover $\{X_1, X_2\}$ with $X_i \cong Y_i$ for $i = 1, 2$. But $X_1 \cap X_2 \cong Y_{12}$ is not affine, since $A(Y_{12}) = \Gamma(D(\mathsf{X}_1) \cup D(\mathsf{X}_2), \mathcal{O}) = \Gamma(D(\mathsf{X}_1), \mathcal{O}) \cap \Gamma(D(\mathsf{X}_2), \mathcal{O}) = \mathbb{Q}[\mathsf{X}_1, \mathsf{X}_2]$ and $Y_{12} \not\cong \operatorname{Spec} \mathbb{Q}[\mathsf{X}_1, \mathsf{X}_2]$.

EXAMPLE 1.30. Any projective R-space is separated. This fact will be proved in Chapter 3 (§3.5.**a**).

§1.24 Regular Functions and Rational Functions

Throughout this section, let X denote an R-scheme, where R is a ring.

a. Definition. An R-morphism $f: X \to \mathbf{A}^1_R = \operatorname{Spec} R[\mathsf{X}]$ is said to be an (*R*-)*regular function* on X.

By Proposition 1.31,

$$\operatorname{Hom}_R(X, \mathbf{A}^1_R) \cong \operatorname{Hom}_R(R[\mathsf{X}], A(X)) \cong A(X) = \Gamma(X, \mathcal{O}_X),$$
$$f \quad \leftrightarrow \quad \Gamma(f) \qquad \leftrightarrow \quad \Gamma(f)(\mathsf{X})$$

Thus, an (R-)regular function f on X is identified with $\Gamma(f)(\mathsf{X})$, which is a section of $\mathcal{O}_X$ over X.

If R is a field k and $f(x)$ is a closed point $(\mathsf{X} - \alpha)$ of $\mathbf{A}_k^1$ for some $x \in X$, then $\bar{f}(x) = \alpha$.

Proposition 1.43. Let $f \in A(X)$. *If X is reduced and U is an open dense subset, then* $\bar{f}|_U = 0$ *induces* $f = 0$.

PROOF. Since $\mathcal{O}_X$ is a sheaf, we may assume X is an affine scheme Spec A. Since $\bar{f}|_U = 0$, it follows that $V(f) \supseteq U$. But then $V(f) = X$, since U is dense. Hence, f is nilpotent by Corollary (iii) to Proposition 1.3; thus $f = 0$, by hypothesis. □

b. First, we introduce a set $\mathbb{E}$ defined by

$$\mathbb{E} = \{(U, \varphi) \mid U \text{ is an open dense subset of } X, \varphi \in \Gamma(U, \mathcal{O}_X)\}.$$

We define an equivalence relation $\sim$ in $\mathbb{E}$ as follows:

$$(U, \varphi) \sim (V, \psi) \Leftrightarrow \varphi|_W = \psi|_W,$$

for some open dense subset W contained in $U \cap V$.

To prove that $\sim$ is an equivalence relation, one may use the next lemma.

Lemma 1.41

(i) *Let $\varphi: Y \to S$ be a continuous map of topological spaces and let E be a subset of Y. Then $\varphi(\bar{E})^- = \varphi(E)^-$, where $\bar{E}$ denotes the closure of a set E.*
(ii) *If Y is an open subset of S and V is a dense subset of S, then $(V \cap Y)^- = \bar{Y}$.*

PROOF. (i) is well known from general topology. (ii) follows immediately from (i). □

Definition. An *(R-)rational function* on X is defined to be an element of $\mathbb{E}/\sim$. The set of all rational functions of X is denoted $\mathrm{Rat}_R(X)$. The *domain* of a rational function f is defined by

$$\mathrm{dom}(f) = \{x \in X \mid x \in U \text{ for some } (U, \varphi) \in f\}.$$

If $(U, \varphi) \in f$, we say that f is a *rational function defined* by φ.

Let $f, g \in \mathrm{Rat}_R(X)$. Choosing $(U, \varphi) \in f$ and $(V, \psi) \in g$, the rational function defined by $\alpha\varphi|_{U \cap V} + \beta\psi|_{U \cap V}$ depends only on $\alpha, \beta \in R$ and f, g. This rational function is denoted $\alpha f + \beta g$. Furthermore, the rational function defined by $\varphi|_{U \cap V} \cdot \psi|_{U \cap V}$ is denoted by $f \cdot g$. With the operations $+$, $\cdot$, $\mathrm{Rat}_R(X)$ becomes an R-algebra, called the *ring of R-rational functions* of X.

Proposition 1.44. *Assume that X is reduced.*

(i) *If U is an open dense subset of X, then $i: \Gamma(U, \mathcal{O}_X) \to \mathrm{Rat}_R(X)$, defined by $i(\varphi) \ni (U, \varphi)$, is injective. Hence, $\varphi \in \Gamma(U, \mathcal{O}_X)$ is identified with the rational function defined by φ.*

(ii) *If $f \in \mathrm{Rat}_R(X)$, then there exists $\varphi \in \Gamma(\mathrm{dom}(f), \mathcal{O}_X)$ such that $(\mathrm{dom}(f), \varphi) \in f$.*

PROOF. (i) follows from Proposition 1.43.

(ii) Suppose that $\mathrm{dom}(f) = \bigcup_{\lambda \in \Lambda} U_\lambda$, where $(U_\lambda, \varphi_\lambda) \in f$. For any two $(U_\lambda, \varphi_\lambda)$, $(U_\mu, \varphi_\mu) \in f$, there exists an open dense subset W such that $\varphi_\lambda|_W = \varphi_\mu|_W$. Hence $\varphi_\lambda|_{U_\lambda \cap U_\mu} = \varphi_\mu|_{U_\lambda \cap U_\mu}$ by Proposition 1.43. By condition (4) for sheaves, $\{\varphi_\lambda \mid \lambda \in \Lambda\}$ determines $\varphi \in \Gamma(\mathrm{dom}(f), \mathcal{O}_X)$. □

If Y is a subscheme of X and $(U, \varphi) \in f$ is such that $U \cap Y$ is dense in Y, then $(U \cap Y, \varphi|_{U \cap Y})$ determines a rational function of Y, denoted by $f|_Y$.

Thus $f|_{\mathrm{dom}(f)}$ is a regular function if X is reduced.

c. The following result is easily proved.

Lemma 1.42

(i) *If U is an open dense subset of X, then $\mathrm{Rat}_R(U)$ is naturally isomorphic to $\mathrm{Rat}_R(X)$.*

(ii) $\mathrm{Rat}_R(X) = \mathrm{injlim}_U\, A(U)$, *where the U are open dense subsets of X.*

(iii) *If U and V are disjoint subschemes of X, then $\mathrm{Rat}_R(U \coprod V) \cong \mathrm{Rat}_R(U) \oplus \mathrm{Rat}_R(V)$.*

Proposition 1.45

(i) *Suppose that X has a finite number of irreducible components $X_1, \ldots, X_u$. Then $\mathrm{Rat}_R(X) \cong \bigoplus_{i=1}^u \mathrm{Rat}(X_i) \cong \bigoplus_{i=1}^u \mathcal{O}_{X, x_i}$ where the x_i are generic points of the X_i.*

(ii) *If an R-algebra A has a finite number of minimal prime ideals $\mathfrak{p}_1, \ldots, \mathfrak{p}_u$* and if $X = \mathrm{Spec}\, A$, *then* $\mathrm{Rat}(X) \cong T^{-1}A$, *where $T = A \setminus (\mathfrak{p}_1 \cup \cdots \cup \mathfrak{p}_u)$.*

(iii) *Suppose further that A is reduced. Then* $\mathrm{Rat}(X) \cong Q(A)$, *that is the total quotient ring of A.*

PROOF. (i) If X is irreducible with generic point x, then $\mathrm{Rat}(X)$ is isomorphic to $\mathcal{O}_{X,x}$ by definition. In general, let $X_i' = X_i \setminus \bigcup_{j \neq i} X_j$. Then $X' = \bigcup_{i=1}^u X_i'$ is an open dense subset of X; hence $\mathrm{Rat}_R(X') \cong \mathrm{Rat}_R(X)$ by Lemma 1.42.(i). But since $X' = X_1' \coprod \cdots \coprod X_u'$, it follows that

$$\mathrm{Rat}_R(X') \cong \bigoplus_{j=1}^{u} \mathrm{Rat}(X_j') \cong \bigoplus_{j=1}^{u} \mathcal{O}_{X, x_j}.$$

(ii) Since $T = A \backslash \bigcup_{i=1}^{u} \mathfrak{p}_i$, one has

$$f \in T \quad \Leftrightarrow \quad \mathfrak{p}_i \in D(f) \quad \text{for all} \quad i \quad \Leftrightarrow \quad D(f) \quad \text{is dense in} \quad X$$

by Proposition 1.7.

Thus by Lemma 1.42.(ii), $\mathrm{Rat}_R(X) \cong \mathrm{injlim}_{f \in T} \tilde{A}(D(f)) = \mathrm{injlim}_{f \in T} A_f \cong T^{-1}A$ by Lemma 1.23 (see Exercise 1.9).

(iii) Since A is reduced, $I(f) = \mathrm{Ker}\ \psi_f$ coincides with $(0:f)$. If $D(f)$ is dense, then $X = V(I(f)) = V(0:f)$; hence $(0:f) = 0$ by Corollary (ii) to Proposition 1.3. Thus $f \in T$ implies that f is not a zero divisor. Conversely, if f is not a zero divisor, then $I(f) = (0:f) = 0$; hence $f \in T$. Therefore, T coincides with the set of all elements of A which are not zero divisors. Hence, $T^{-1}A$ is the total quotient ring of A. □

d. Let X be an integral R-scheme. For a point $x \in X$, one defines a map j_x: $\mathcal{O}_{X,x} \to \mathrm{Rat}_R(X)$ defined by letting $j_x(\varphi_x)$ be the rational function defined by $\varphi \in \Gamma(U, \mathcal{O}_X)$, $U \ni x$. Then, clearly, j_x is an injective homomorphism. We can identify $\mathcal{O}_{X,x}$ with a subring of $\mathrm{Rat}_R(X)$ via j_x. Furthermore, for any open set U containing x, one has a homomorphism r_x: $\Gamma(U, \mathcal{O}_X) \to \mathcal{O}_{X,x}$ defined by $r_x(\varphi) = \varphi_x$. If $r_x(\varphi) = r_x(\psi)$, then $\varphi|_W = \psi|_W$ for some nonempty open subset W; hence $\varphi = \psi$ by Proposition 1.43. Thus, we identify $\Gamma(U, \mathcal{O}_x)$ with a subring of $\mathcal{O}_{X,x}$. Thus, one has the inclusions

$$\Gamma(U, \mathcal{O}_X) \subseteq \mathcal{O}_{X,x} \subseteq \mathrm{Rat}_R(X).$$

Lemma 1.43. $\Gamma(U, \mathcal{O}_X) = \bigcap_{x \in U} \mathcal{O}_{X,x}$, *for any open subset* U *of* X.

PROOF. Clearly, $\Gamma(U, \mathcal{O}_X) \subseteq \bigcap_{x \in U} \mathcal{O}_{X,x}$. Hence, taking $\varphi \in \bigcap_{x \in U} \mathcal{O}_{X,x}$, one has $\mathrm{dom}(\varphi) \supseteq U$ and so $\varphi|_U$ is regular by Proposition 1.44.(ii); i.e., $\varphi \in \Gamma(U, \mathcal{O}_X)$. □

Lemma 1.44. *Let* R *be an integral domain and let* X *be a quasi-compact integral* R*-scheme such that the structure morphism* $X \to \mathrm{Spec}\ R$ *is dominating. For a multiplicative subset* S *of* R *with* $0 \notin S$, *we have* $A(S^{-1}X) = S^{-1}A(X)$, *where* $S^{-1}X$ *was defined in* §1.20.**e.**

PROOF. Let $\{X_i \mid 1 \leq i \leq m\}$ be a finite open cover of X by nonempty affine subschemes. Then $S^{-1}X$ is covered by $\{S^{-1}X_i \mid 1 \leq i \leq m\}$. By Theorem 1.18, we have inclusions $R \subseteq A(X) \subseteq A(X_i) \subseteq \mathrm{Rat}(X)$ and so $S^{-1}R \subseteq S^{-1}A(X) \subseteq S^{-1}A(X_i) \subseteq \mathrm{Rat}(X)$. Since the $S^{-1}X_i$ are affine, we may suppose that $A(S^{-1}X_i) = S^{-1}A(X_i)$. Thus

$$S^{-1}A(X) \subseteq A(S^{-1}X) = \bigcap_{i=1}^{m} A(S^{-1}X_i) = \bigcap_{i=1}^{m} S^{-1}A(X_i).$$

If $\alpha \in \bigcap_{i=1}^{m} S^{-1}A(X_i)$, then α can be written as a_i/s for $a_i \in A(X_i)$ and $s \in S$. Since $a_i/s = a_j/s$, one has $a_i = a_j$ and so $a_1 = a_2 = \cdots = a_m \in \bigcap_{i=1}^{m} A(X_i) = A(X)$. Therefore, $\alpha \in S^{-1}A(X)$ and so $A(S^{-1}X) = S^{-1}A(X)$. □

Since $A(X) \subseteq A(X_i)$, $X_i \cong \operatorname{Spec} A(X_i) \to \operatorname{Spec} A(X)$ is a dominating morphism and is the restriction of Ψ (defined in §1.15.**d**) to X_i; hence, Ψ: $X \to \operatorname{Spec} A(X)$ is a dominating morphism. Letting $*$ be the generic point of Spec $A(X)$ and S be $A(X)\backslash\{0\}$, one has $S^{-1}X \neq \Psi^{-1}(*)$ and, by the last lemma, $A(\Psi^{-1}(*)) = A(S^{-1}X) \cong S^{-1}A(X) = Q(A(X))$. Hence, we have the following result.

Theorem 1.18. *Let X be a quasi-compact integral scheme. Then* $A(\Psi^{-1}(*)) = Q(A(X))$, *where* Ψ: $X \to \operatorname{Spec} A(X)$ *was introduced in* §1.15.**d**.

Definition. If X is an integral k-scheme such that $\operatorname{Rat}_k(X)$ is finitely generated over k as a field, then tr. $\deg_k QA(X)$ is denoted by $\gamma_k(X)$ or $\gamma(X)$.

The invariant $\gamma_k(X)$ will play a beautiful role in Chapter 10.

§1.25 Rational Maps

a. Rational functions are generalized to rational maps, which are important in birational geometry. Letting X and Y be S-schemes, we define a set $\mathbb{E}^*$ by

$$\mathbb{E}^* = \{(U, \varphi) \mid U \text{ is an open dense subset of } X, \text{ and } \varphi\colon U \to Y \text{ is an } S\text{-morphism}\}.$$

We introduce an equivalence relation $\sim$ in $\mathbb{E}^*$ as follows:

$$(U, \varphi) \sim (V, \psi) \Leftrightarrow \varphi|_W = \psi|_W,$$

for some open dense subset W contained in $U \cap V$.

Definition. An (S-)*rational map* f: $X \to Y$ is defined to be an element of $\mathbb{E}^*/\sim$. The *domain* of an (S-)rational map f is defined by

$$\operatorname{dom}(f) = \{x \in X \mid x \in U \text{ for some } (U, \varphi) \in f\}.$$

If $x \in \operatorname{dom}(f)$, then we say that f is *defined at* x.

Proposition 1.46

(i) *Suppose that X is reduced and Y is S-separated. If two S-morphisms φ, ψ: $X \to Y$ satisfy $\varphi|_U = \psi|_U$ for some open dense subset U, then $\varphi = \psi$.*
(ii) *If two morphisms φ, ψ: $X \to Y$ are equal as rational maps, then they are equal as morphisms.*
(iii) *Let f: $X \to Y$ be a rational map. Then* $(\operatorname{dom}(f), \varphi) \in f$ *for some* φ.

PROOF. (i) Let K be the subscheme $\operatorname{Ker}(\varphi, \psi)$ defined in §1.22.**d**. Since $\varphi|_U = \psi|_U$, U is contained in K by Proposition 1.38. Hence K is dense.

Furthermore, since $\Delta_{Y/S}(Y)$ is closed, and since $K = (\varphi, \psi)_S^{-1}(\Delta_{Y/S}(Y))$, K is a closed subscheme. Thus $K = X$, since X is reduced; hence, $\varphi = \psi$.

(ii) This follows from (i) immediately.

(iii) This can be shown in the same way as the proof of Proposition 1.44. □

In the assertion (iii), if $X = \text{dom}(f)$, then we say that f is a morphism φ.

Definition. Let Y be a subscheme of X. If a rational map $f: X \to Z$ is represented by (U, φ) such that $U \cap Y$ is dense in Y, then the rational map defined by $\varphi|_{U \cap Y}$ is said to be the *restriction* of f to Y, denoted $f|_Y$.

b. As suggested by Proposition 1.46, rational maps are better behaved for separated integral schemes. Thus, we suppose that X and Y are *R-separated integral schemes.*

Definition. Let $f: X \to Y$ be a rational map and let ξ be the generic point of X. The closure of the point $f(\xi)$ has the structure of a reduced closed subscheme of Y, which is said to be the *closed image* of X by f, and denoted by $\text{Im}(f)$.

$\text{Im}(f)$ is an R-separated integral scheme and $f|_{\text{dom}(f)}: \text{dom}(f) \to Y$ can be regarded as a morphism into $\text{Im}(f)$ by Proposition 1.30. Hence, f can be considered as a rational map from X into $\text{Im}(f)$.

Definition. If $\text{Im}(f) = Y$, i.e., $\varphi(U)$ is dense in Y for some $(U, \varphi) \in f$, then we say that f is *dominating.*

Definition. Let $f: X \to Y$ and $g: Y \to Z$ be rational maps, where Z is an R-separated integral scheme.

If $\text{Im}(f) \cap \text{dom}(g) \neq \emptyset$, one says that the *composition* $g \circ f$ *can be defined.* In this case, $g \circ f$ is defined to be the rational map represented by $(U_0, \psi \circ \varphi_0)$, where $\varphi = f|_{\text{dom}(f)}$, $\psi = g|_{\text{dom}(g)}$, $U_0 = \varphi^{-1}(\text{dom}(g))$, and $\varphi_0 = \varphi|_{U_0}$.

In the above definition, note that $U_0 = \varphi^{-1}(\text{dom}(g))$ is not empty.

In particular, if f is dominating, then the composition $g \circ f$ can be defined for any rational map $g: Y \to Z$.

Remark. A rational map $f: X \to \mathbf{A}_R^1$ is just a rational function on X.

EXAMPLE 1.31. If $f: X \to Y$ is dominating, then for any $\varphi \in \text{Rat}_R(Y)$, $\varphi \circ f \in \text{Rat}_R(X)$ is defined, and is called the *pullback* of φ by f.

As in the case of regular functions, one writes $f^*(\varphi)$ instead of $\varphi \circ f$. $f^*\colon \mathrm{Rat}_R(Y) \to \mathrm{Rat}_R(X)$ is a ring homomorphism, which is injective, since $\mathrm{Rat}_R(Y)$ is a field.

Theorem 1.19. *Let $f\colon X \to Y$ be a dominating morphism between R-separated integral schemes. Then $\mathcal{O}_Y$ is identified with a subsheaf of $f_*(\mathcal{O}_X)$.*

Proof. For any nonempty open subset U of Y, one has a commutative square

$$\begin{array}{ccc} \mathcal{O}_Y(U) & \xrightarrow{\theta(f)_U} & (f_*\mathcal{O}_X)(U) = \mathcal{O}_X(f^{-1}(U)) \\ \cap\downarrow & & \cap\downarrow \\ \mathrm{Rat}_R(Y) & \xrightarrow{f^*} & \mathrm{Rat}_R(X). \end{array}$$

Hence, $\theta(f)_U$ is injective for all U; i.e., $\theta(f)$ is injective. □

Thus, $\theta(f)_U$ is often loosely written as f^*.

c. Let X be a scheme algebraic over a field k, i.e., a scheme which admits an open finite cover of affine algebraic schemes.

Definition. Define $\mathrm{spm}(X)$ to be the set of all closed points of X.

Lemma 1.45. $\mathrm{spm}(X)$ *is a dense subset of X. If X is affine, then* $\mathrm{spm}(X)$ *is the set of all maximal ideals of $A(X)$.*

Proof. This follows from Theorems 1.4 and 1.5. □

Let Y be another scheme algebraic over k and let $f\colon X \to Y$ be a k-morphism. If $x \in \mathrm{spm}(X)$, then one has the induced k-homomorphism: $k(f(x)) \to k(x)$. Since the field extension $k(x)/k$ is algebraic, so is $k(f(x))/k$; hence $f(x) \in \mathrm{spm}(Y)$. Thus $\bar{f} = f|_{\mathrm{spm}(X)}$ is a continuous map from $\mathrm{spm}(X)$ to $\mathrm{spm}(Y)$.

§1.26 Morphisms of Finite Type

Throughout this and the following sections, let X, Y, and Z denote schemes, and let f and g denote morphisms.

a. Definition. A morphism $f\colon \mathrm{Spec}\, B \to \mathrm{Spec}\, A$ is provisionally said to be *of f.g. type*, if B regarded as an A-algebra via $\varphi = \Gamma(f)$ is finitely generated, i.e., there exist $b_1, \ldots, b_m \in B$ such that $B = A[b_1, \ldots, b_m]$ (defined to be $\varphi(A)[b_1, \ldots, b_m]$).

Furthermore, $f: X \to Y$ is said to be *of finite type* if there exists an affine open cover $\{Y_\lambda \mid \lambda \in \Lambda\}$ of Y, such that each $X_\lambda = f^{-1}(Y_\lambda)$ can be covered by a finite number of affine open subsets $X_{\lambda 1}, \ldots, X_{\lambda m}$, where all $f|_{X_{\lambda j}}$: $X_{\lambda j} \to Y_\lambda$ are of f.g. type.

Lemma 1.46. *Let $f: X \to Y$ be a morphism of finite type. If V is an affine open subset of Y, then $f^{-1}(V)$ is covered by affine open subsets $U_1, \ldots, U_n$ such that the $f|_{U_j}: U_j \to V$ are of f.g. type.*

PROOF. Since $Y_\lambda \cap V$ is an open subset of the affine scheme V, $Y_\lambda \cap V$ is covered by affine open subsets of the form V_φ $(= D(\varphi)$, see §1.15.**b**) with $\varphi \in A(V)$. Furthermore, $f|_{X_{\lambda j}}$: $X_{\lambda j} \to Y_\lambda$ is of f.g. type and $X_{\lambda j} \cap f^{-1}(V_\varphi) \cong X_{\lambda j} \times_{Y_\lambda} V_\varphi$ by Lemma 1.35. $X_{\lambda j} \cap f^{-1}(V_\varphi) \to V_\varphi$ is a morphism of affine schemes that is of f.g. type, since $A(X_{\lambda j} \cap f^{-1}(V_\varphi)) \cong A(X_{\lambda j}) \otimes_{A(Y_\lambda)} A(V_\varphi)$ is finitely generated over $A(V_\varphi)$.

Thus define a set $\Phi = \{\varphi \in A(V) \mid V_\varphi \subseteq Y_\lambda \cap V$ for some $\lambda \in \Lambda\}$. Then $\{V_\varphi \mid \varphi \in \Phi\}$ is an affine open cover of the affine scheme V; hence by Theorem 1.2, there exists a finite subset $\{\varphi_1, \ldots, \varphi_a\}$ of Φ such that $V = \bigcup_{p=1}^a V_{\varphi_p}$. Thus there exist a finite number of affine open subsets of the form $X_{\lambda jp} = X_{\lambda j} \cap f^{-1}(V_{\varphi_p})$ which cover f^{-1}(V). Since $f|_{X_{\lambda jp}}$: $X_{\lambda jp} \to V$ is of f.g. type, we obtain the assertion. □

Lemma 1.47. *If ${}^a\varphi$: Spec $B \to$ Spec A is of finite type, then it is also of f.g. type, i.e., B is finitely generated over A as an A-algebra.*

PROOF. Writing $X =$ Spec B, $Y =$ Spec A, and $f = {}^a\varphi$, there exists a finite affine open cover $\{X_i \mid 1 \le i \le m\}$ of X such that $f|_{X_i}$: $X_i \to Y$ is of f.g. type for all i by Lemma 1.46. Each open set X_i is covered by open subsets of the form X_ψ $(= D(\psi) \subseteq$ Spec B, $\psi \in B)$, and $A(X_\psi) = B_\psi$ is finitely generated over A, since $A(X_\psi) = A(X_i)\ [1/\bar{\psi}]$, $\bar{\psi} = \psi|_{X_i}$, and since $A(X_i)$ is finitely generated over A. Define a subset Ψ of B by $\Psi = \{\psi \in B \mid X_\psi \subseteq X_i$ for some $i\}$. Then $\bigcup_{\psi \in \Psi} X_\psi = X$ and by Theorem 1.2, there exist $\psi_1, \ldots, \psi_l \in \Psi \subset B$ such that $\bigcup_{j=1}^l X_{\psi_j} = X$. Hence, one can find $b_{1j}, \ldots, b_{r_j} \in B$ and $N > 0$ such that $B_{\psi_j} = A[b_{1j}/\psi_j^N, \ldots, b_{rj}/\psi_j^N]$ for all j. Since $X = \bigcup_{j=1}^l X_{\psi_j}$ by the Corollary to Lemma 1.1 there exist $\gamma_1, \ldots, \gamma_l \in B$ such that $1 = \sum_{j=1}^l \gamma_j \psi_j$. We claim that $B = A[b_{11}, \ldots, b_{ij}, \ldots, b_{r_l}, \psi_1, \ldots, \psi_l, \gamma_1, \ldots, \gamma_l]$. For any $b \in B$, $b/1 \in B_{\psi_j}$; hence, $b\psi_j^{Nc} \in A[b_{1j}, \ldots, b_{rj}, \psi_j]$ for all j, with c independent of j. One has $1 = \sum_{j=1}^l \delta_j \psi_j^{Nc}$ for some $\delta_j \in A[\gamma_1, \ldots, \gamma_l, \psi_1, \ldots, \psi_l]$. Thus

$$b = \sum_{j=1}^{l} \delta_j b \psi_j^{Nc} \in A[b_{11}, \ldots, b_{rl}, \psi_1, \ldots, \psi_l, \gamma_1, \ldots, \gamma_l].$$ □

Proposition 1.47. *Let $f: X \to Y$ be a morphism of finite type. If V and U are affine open subsets of Y and X, respectively, such that $f(U) \subseteq V$, then $f|_U$: $U \to V$ is of f.g. type, i.e., $A(U)$ is finitely generated over $A(V)$.*

PROOF. $f^{-1}(V)$ is covered by affine open subsets $W_1, \ldots, W_n$ such that each $f|_{W_j}$: $W_j \to V$ is of f.g. type by Lemma 1.46. $U \cap W_j$ is covered by open subsets of the form $W_{j,\varphi}$ $(=D(\varphi),\ \varphi \in A(W_j))$. Thus U has an affine open cover $\{W_{j,\varphi} \mid 1 \leq j \leq n,\ W_{j,\varphi} \subseteq U\}$, in which all $A(W_{j,\varphi}) = A(W_j)[1/\varphi]$ are finitely generated over $A(V)$. Therefore, since U is quasi-compact by Theorem 1.2, $f|_U$: $U \to V$ is of finite type. By Lemma 1.47, $A(U)$ is finitely generated over $A(V)$. □

b. The following result is elementary.

Proposition 1.48

(i) *The composition of two morphisms of finite type is also a morphism of finite type.*
(ii) *The property of being a morphism of finite type is stable under base extension.*
(iii) *If j: $Y \to X$ is a closed immersion, then j is of finite type.*

PROOF. (i) Let f: $X \to Y$ and g: $Y \to Z$ be morphisms of finite type. By definition, there exists an affine open cover $\{Z_\lambda \mid \lambda \in \Lambda\}$ such that all $g^{-1}(Z_\lambda)$ are covered by affine open subsets $Y_{\lambda 1}, \ldots, Y_{\lambda l}$ with the property that all $g|_{Y_{\lambda j}}$: $Y_{\lambda j} \to Z_\lambda$ are of f.g. type. By Lemma 1.46, each $f^{-1}(Y_{\lambda j})$ is covered by affine open subsets $X_{\lambda j1}, \ldots, X_{\lambda jm}$. Thus by Lemma 1.47, all $f|_{X_{\lambda jt}}$: $X_{\lambda jt} \to Y_{\lambda j}$ are of f.g. type and hence $g \circ f|_{X_{\lambda jt}}$: $X_{\lambda jt} \to Z_\lambda$ are of f.g. type for all λ, j, t.

(ii) Let f: $X \to Y$ be a morphism of finite type and let g: $H \to Y$ be a morphism. For any affine open subset V of Y, one has an affine open subset W of H such that $g(W) \subseteq V$. Then by Lemma 1.46, $f^{-1}(V)$ is covered by affine open subsets $U_1, \ldots, U_n$ such that the $f|_{U_j}$: $U_j \to V$ are of f.g. type. Now, $f_H^{-1}(W) \cong X_H \times_H W \cong X \times_Y W \cong X \times_Y V \times_V W \cong f^{-1}(V) \times_V W = \bigcup_{j=1}^n (U_j \times_V W)$, and $A(U_j \times_V W) \cong A(U_j) \otimes_{A(V)} A(W)$, which is finitely generated over $A(W)$. Hence, f_H is of finite type. □

Proposition 1.49. *If j: $W \to X$ is an immersion and X is a Noetherian scheme, then j is of finite type.*

PROOF. By Proposition 1.48.(i) and (iii), one can assume W is an open subset and j is the inclusion map. Clearly, one can also assume that X is a Noetherian affine scheme Spec A. Any open subset W of X is of the form $D(\mathfrak{a})$ for some ideal $\mathfrak{a}$ of A. Since A is Noetherian, $\mathfrak{a}$ has a finite basis $f_1, \ldots, f_l$; thus $D(\mathfrak{a}) = D(f_1) \cup \cdots \cup D(f_l)$. Clearly, $A(D(f_j)) = A[1/f_j]$ is finitely generated over A. □

c. The condition of being of finite type is indispensable in order to connect local rings with rational maps.

Definition. If $X \to \operatorname{Spec} R$ is of finite type, then X is said to be *algebraic over* R or R-*algebraic*. In particular, when R is a field, we say that X is *algebraic*.

Proposition 1.50. *Let X and Y be R-schemes such that Y is algebraic over R.*

(i) *Let $f, g: X \to Y$ be R-morphisms. If $f(x) = g(x) = y$ and $\theta(f)_y^{\#} = \theta(g)_y^{\#}$: $\mathcal{O}_{Y,y} \to \mathcal{O}_{X,x}$ for some point $x \in X$, then there exists an open neighborhood U of x such that $f|_U = g|_U$.*

(ii) *Suppose that R is Noetherian. Given any local R-homomorphism Θ: $\mathcal{O}_{Y,y} \to \mathcal{O}_{X,x}$ for* some *points $x \in X$* and *$y \in Y$, there exists an open neighborhood U of x and an R-morphism $h: U \to Y$ such that $h(x) = y$ and $\theta(h)_y^{\#} = \Theta$.*

Proof. For both assertions, it suffices to consider the case where $X = \operatorname{Spec} B$, $Y = \operatorname{Spec} A$, and $A = R[\alpha_1, \ldots, \alpha_n]$ for certain $\alpha_1, \ldots, \alpha_n \in A$. The points x and y are identified with prime ideals $\mathfrak{p}$ and $\mathfrak{q}$, respectively.

(i) The morphisms f and g correspond to φ and $\psi: A \to B$, i.e., $\varphi = \Gamma(f)$ and $\psi = \Gamma(g)$. By hypothesis, φ and ψ induce the same local homomorphism $\theta(f)_y^{\#}: A_{\mathfrak{q}} \to B_{\mathfrak{p}}$. Hence $\varphi(\alpha_i)/1 = \psi(\alpha_i)/1 \in B_{\mathfrak{p}}$ for any i. Thus, $b(\varphi(\alpha_i) - \psi(\alpha_i)) = 0$, $1 \leq i \leq n$, for some $b \in B \backslash \mathfrak{p}$, i.e. $r_{b,1} \circ \varphi = r_{b,1} \circ \psi: A \to B_b$. This implies that $f|_{D(b)} = g|_{D(b)}$ and $\mathfrak{p} \in D(b)$.

(ii) One has $b \in B \backslash \mathfrak{p}$ such that $\Theta(\alpha_i/1) = b_i/b$, $b_i \in B$ for all i. Let ξ: $R[\mathsf{X}_1, \ldots, \mathsf{X}_n] \to B_b$ be the R-homomorphism defined by $\xi(\mathsf{X}_i) = b_i/b$ for all i, and let $\eta: R[\mathsf{X}_1, \ldots, \mathsf{X}_n] \to A$ be the R-homomorphism defined by $\eta(\mathsf{X}_i) = \alpha_i$ for all i, where $R[\mathsf{X}_1, \ldots, \mathsf{X}_n]$ is the polynomial ring over R.

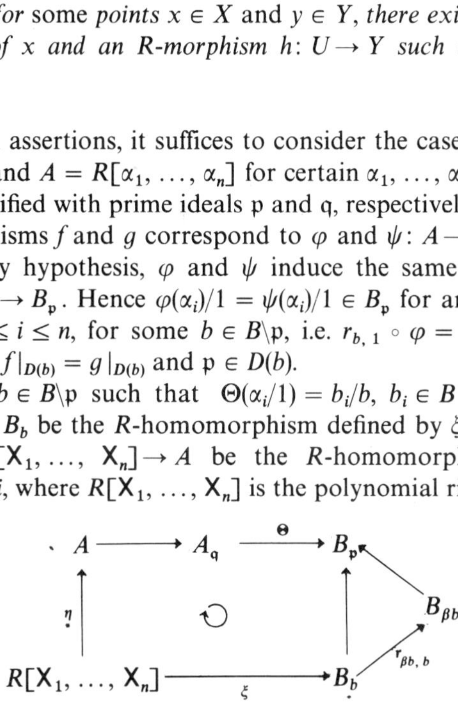

Since R is Noetherian, so is $R[\mathsf{X}_1, \ldots, \mathsf{X}_n]$ by the Hilbert basis theorem. Hence, the ideal Ker η has a finite basis $\gamma_1, \ldots, \gamma_s$. Then

$$\Theta(\eta(\mathsf{X}_i)/1) = \Theta(\alpha_i/1) = b_i/b = \xi(\mathsf{X}_i)/1 \quad \text{for all } i,$$

and the above diagram is commutative. Since $\gamma_j \in \operatorname{Ker} \eta$,

$$0 = \Theta(\eta(\gamma_j)/1) = \xi(\gamma_j)/1 \quad \text{in } B_{\mathfrak{p}};$$

hence there exists $\beta \in B \backslash \mathfrak{p}$ with $\beta\xi(\gamma_j) = 0$ for all j. Thus, $r_{\beta b, b}(\xi(\gamma_j)) = 0$, where $r_{\beta b, b}$ was defined in §1.11.**a**. ξ induces an R-homomorphism φ: $A \cong R[\mathsf{X}_1, \ldots, \mathsf{X}_n]/\ker \eta \to B_{\beta b}$, such that $f = {}^a\varphi: D(\beta b) \to \operatorname{Spec} A$, $f(x) = y$, and $\theta(f)_y^{\#} = \Theta$. □

§1.27 Affine Morphisms and Integral Morphisms

a. First we introduce the notion of direct image of an $\mathcal{O}_X$-module (cf. §1.14.**a**). Let f: $(X, \mathcal{O}_X) \to (Y, \mathcal{O}_Y)$ be a morphism of ringed spaces.

Definition. The direct image sheaf $f_* \mathcal{F}$ of an $\mathcal{O}_X$-module $\mathcal{F}$ has the structure of an $\mathcal{O}_Y$-module in the following way: for every open subset U of Y, $\alpha \in \mathcal{O}_Y(U)$, and $s \in (f_* \mathcal{F})(U)$, define $\alpha \cdot s$ to be $\theta(f)_U(\alpha) \cdot s$. The $\mathcal{O}_Y$-module $f_* \mathcal{F}$ is said to be the *direct image* ($\mathcal{O}_Y$-*module*) of $\mathcal{F}$.

Lemma 1.48. *Let* φ: $A \to B$ *be a ring homomorphism and let* M *be a* B-*module. Then* ${}^a\varphi_* \tilde{M} = \tilde{M}_{[\varphi]}$, *where* $M_{[\varphi]}$ *is* M *with the structure of* A-*module via* φ, *i.e.,* $a \cdot x = \varphi(a)x$ *for any* $a \in A$ *and* $x \in M$.

PROOF. For any $f \in A$, we have ${}^a\varphi_*(\tilde{M})(D(f)) = \tilde{M}(D(\varphi(f))) \cong M_{\varphi(f)} \doteq (M_{[\varphi]})_f$. □

Proposition 1.51. *Let* f: $X \to Y$ *be a separated morphism of schemes. Suppose that* Y *has an affine open cover* $\{Y_\lambda \mid \lambda \in \Lambda\}$ *such that all* $f^{-1}(Y_\lambda)$ *are quasi-compact. If* $\mathcal{F}$ *is a quasi-coherent* $\mathcal{O}_X$-*module, then* $f_* \mathcal{F}$ *is also quasi-coherent.*

PROOF. Clearly, we may assume that Y is affine, i.e., $Y = \operatorname{Spec} A$. X has a finite affine open cover $\{U_i \mid 1 \leq i \leq r\}$. We claim that $f_* \mathcal{F}$ satisfies condition (c) in Theorem 1.11. Thus, take $\alpha \in A$ and $s \in (f_* \mathcal{F})(Y) = \mathcal{F}(X)$ with $s|_{D(\alpha)} = 0$. Hence s, which is regarded as an element of $\mathcal{F}(X)$, satisfies $s|_{f^{-1}(D(\alpha))} = 0$, and so $s|_{U_i \cap f^{-1}(D(\alpha))} = 0$ for all i. But since $U_i \cap f^{-1}(D(\alpha)) = (U_i)_{\psi_i(\alpha)}$, where ψ_i is $\Gamma(f|_{U_i}) = r_{U_i, X} \circ \Gamma(f)$, one has $\psi_i(\alpha)^l \cdot s|_{U_i} = 0$ for some l independent of i, i.e., $\alpha^l \cdot s|_{U_i} = 0$ for all i; hence $\alpha^l \cdot s = 0$. Thus the condition (c.i) has been verified. By the similar argument, one can check the condition (c.ii). □

Therefore, if Y is affine, $f_* \mathcal{F} = \mathcal{F}(X)^{\sim}$. However, in the case when $\mathcal{F}$ is coherent, one cannot expect that $f_* \mathcal{F}$ is coherent. We shall give a sufficient condition for $f_* \mathcal{F}$ to be coherent in §7.6.

b. Definition. A morphism f: $X \to Y$ of schemes is said to be an *affine morphism*, if there exists an affine open cover $\{Y_\lambda \mid \lambda \in \Lambda\}$ such that the $X_\lambda = f^{-1}(Y_\lambda)$ are affine.

The next result is similar to the case of morphisms of finite type.

Lemma 1.49. *Let* f: $X \to Y$ *be an affine morphism.*

(*i*) *f is separated.*
(*ii*) *If Y is affine then so is X.*

Proof. (i) This follows from Proposition 1.35.(i) and Theorem 1.17.(v).

(ii) We use the notation in the definition of affine morphism. For any $\alpha \in A = A(Y)$ with $D(\alpha) \subseteq Y_\lambda, f^{-1}(D(\alpha)) = X_\lambda \times_{Y_\lambda} D(\alpha)$ is also affine. Hence, let Φ be $\{\alpha \in A \mid D(\alpha) \subseteq Y_\lambda \text{ for some } \lambda\}$. Since Y is quasi-compact by Theorem 1.2, there exist $\alpha_1, \ldots, \alpha_r \in \Phi$ such that $Y = \bigcup_{i=1}^r D(\alpha_i)$ and the $f^{-1}(D(\alpha_i))$ are affine. We set $Y_i = D(\alpha_i)$ and $X_i = f^{-1}(D(\alpha_i))$. By Theorem 1.15, one has the commutative diagram

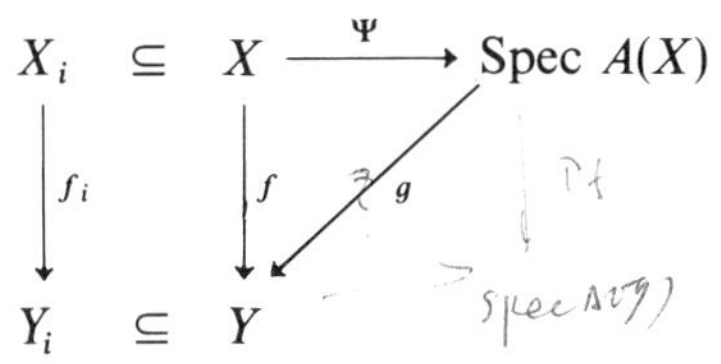

Thus $X_i = \Psi^{-1}(g^{-1}(Y_i))$ and so $\Psi|_{X_i}: X_i \to g^{-1}(Y_i)$. By Proposition 1.51, $f_*(\mathcal{O}_X) = \tilde{A}(X)$ as an $\tilde{A}$-module; hence $A(X_i) = \Gamma(Y_i, f_*(\mathcal{O}_X)) = \Gamma(Y_i, \tilde{A(X)}) = A(X)_{\alpha_i}$. Letting $Z = \text{Spec } A$, one has $A(g^{-1}(Y_i)) = \Gamma(Y_i, g_*(\mathcal{O}_Z)) \cong A(X)_{\alpha_i}$. Hence $\Psi|_{X_i}$ is an isomorphism for each i; thus Ψ is an isomorphism. □

Proposition 1.52. *Let $f: X \to Y$ be an affine morphism. If V is an affine open subset, then so is $f^{-1}(V)$.*

Proof. By definition, there exists an affine open cover $\{Y_\lambda \mid \lambda \in \Lambda\}$ such that all $X_\lambda = f^{-1}(Y_\lambda)$ are affine. $V \cap Y_\lambda$ has an open cover $\{(Y_\lambda)_\varphi \mid \varphi \in A(Y_\lambda)$ such that $(Y_\lambda)_\varphi \subseteq V\}$. Let Φ be $\{\varphi \in A(Y_\lambda) \mid (Y_\lambda)_\varphi \subseteq V\}$. Then all $f^{-1}((Y_\lambda)_\varphi) \cong X_\lambda \times_{Y_\lambda} (Y_\lambda)_\varphi \cong \text{Spec } A(X_\lambda)[1/(\theta(f)(\varphi))]$ are affine and $\{f^{-1}((Y_\lambda)_\varphi) \mid \varphi \in \Phi\}$ is an affine open cover; thus $f|_{f^{-1}(V)}: f^{-1}(V) \to V$ is affine. Hence by Lemma 1.49, $f^{-1}(V)$ is also affine. □

By an argument similar to the proof of Proposition 1.51, we can easily prove the next result.

Theorem 1.20.

(i) *Every closed immersion is affine.*
(ii) *The composition of two affine morphisms is affine.*
(iii) *The property of being an affine morphism is stable under base extension.*

Proof. (i) follows from the definition. (ii) and (iii) are left to the reader. □

c. **Definition.** A ring homomorphism $\varphi: A \to B$ is said to be *integral*, if B is integral over its subring $\varphi(A)$.

Lemma 1.50.

(i) *If* $\varphi\colon A \to B$ *is integral, then so is* $\varphi \otimes_A 1_C\colon C \to B \otimes_A C$ *for any A-algebra C.*
(ii) *If* $\varphi\colon A \to B$ *is integral, then* ${}^a\varphi$ *is a closed map.*

PROOF. (i) is clear. To prove (ii), let F be a closed subset $V(\mathfrak{a})$ of Spec B, where $\mathfrak{a}$ is an ideal of B. Then $B/\mathfrak{a}$ is integral over a subring isomorphic to $A/\varphi^{-1}(\mathfrak{a})$. By Theorem 1.7, $\operatorname{Spec}(B/\mathfrak{a}) \to \operatorname{Spec}(A/\varphi^{-1}(\mathfrak{a}))$ is onto; hence ${}^a\varphi(F) = {}^a\varphi(V(\mathfrak{a})) = V(\varphi^{-1}(\mathfrak{a}))$, which is a closed subset. □

Definition. A morphism $f\colon X \to Y$ is said to be *closed* if the associated continuous map between the base spaces is a closed map. Furthermore, f is said to be *universally closed* if $f_Z\colon X_Z \to Z$ is a closed morphism for every morphism $g\colon Z \to Y$.

The proof of the next result is left to the reader as an exercise.

Proposition 1.53

(i) *Every closed immersion is universally closed.*
(ii) *The composition of two universally closed morphisms is universally closed.*
(iii) *The property of being universally closed is stable under base extension.*
(iv) *Let* $f\colon X \to Y$ *be a morphism and let* $\{Y_\lambda \mid \lambda \in \Lambda\}$ *be an open cover of Y such that* $f|_{X_\lambda}\colon X_\lambda \to Y_\lambda$ *is universally closed for all* λ, *where* $X_\lambda = f^{-1}(Y_\lambda)$. *If the* $f|_{X_\lambda}$ *are universally closed, then so is f.*

d. Definition. A morphism f: Spec $B \to$ Spec A of affine schemes is provisionally said to be *strictly integral* if $\varphi = \Gamma(f)$ is integral.

Definition. Let $f\colon X \to Y$ be a morphism of schemes. The morphism f is said to be *integral* if there exists an affine open cover $\{Y_\lambda \mid \lambda \in \Lambda\}$ of Y such that $f_\lambda = f|_{X_\lambda}\colon X_\lambda \to Y_\lambda$ is strictly integral, i.e., $X_\lambda = f^{-1}(Y_\lambda)$ is affine and $\Gamma(f_\lambda)$ is integral, for all $\lambda \in \Lambda$.

Lemma 1.51. *Let* $\varphi\colon A \to B$ *be a ring homomorphism.*

(i) *For any* $a \in A$, $\varphi_a\colon A_a \to B_{\varphi(a)}$ *is integral if* φ *is integral.*
(ii) *Let* $a_1, \ldots, a_r \in A$ *satisfy* $A = \sum_{j=1}^r Aa_j$. *If the* $\varphi_{a_j}\colon A_{a_j} \to B_{\varphi(a_j)}$ *are integral, then* φ *is also integral.*

PROOF. (i) This is clear.

(ii) We can assume that A is a subring of B and φ is the inclusion map. Take $b \in B$. Then $(b/1)^m + (\alpha_1/a_j^l)(b/1)^{m-1} + \cdots + \alpha_m/a_j^l = 0$ for some $m > 0$, $l \geq 0$, and $\alpha_i \in A$. Hence $a_j^e b$ is integral over A for some e which can be chosen independently of j. Since $A = \sum_{j=1}^{r} Aa_j$, by Lemma 1.22 there exist $c_1, \ldots, c_r \in A$ such that $1 = \sum_{j=1}^{r} c_j a_j^e$; hence $b = \sum_{j=1}^{r} c_j a_j^e b$ is integral over A by Corollary 1 to Proposition 1.9. □

Proposition 1.54. *Let $f\colon X \to Y$ be an integral morphism. If V is an affine open subset of Y, then $f|_{f^{-1}(V)}\colon f^{-1}(V) \to V$ is strictly integral. In particular, if ${}^a\varphi$: Spec $B \to$ Spec A is integral, then ${}^a\varphi$ is strictly integral; in other words, $\varphi = \Gamma({}^a\varphi)$ is integral.*

Proof. Using the notation of the definition, we let Ψ be $\{\psi \in A(V) \mid V_\psi \subseteq Y_\lambda$ for some $\lambda \in \Lambda\}$. Then $\{V_\psi \mid \psi \in \Psi\}$ is an affine open cover and $f^{-1}(V_\psi) \cong X_\lambda \times_{Y_\lambda} V_\psi \to V_\psi$ is strictly integral for all $\psi \in \Psi$. Since $f^{-1}(V)$ is affine by Proposition 1.52 and there exist $\psi_1, \ldots, \psi_r \in \Psi$ such that $V = \bigcup_{j=1}^{r} V_{\psi_j}$, we can apply Lemma 1.51.(ii). □

Remark. As was proved in the last proposition, ${}^a\varphi$: Spec $B \to$ Spec A is integral if and only if ${}^a\varphi$ is strictly integral.

By the last lemma and Theorem 1.20, one can prove the next result.

Theorem 1.21

(i) *Every closed immersion is integral.*
(ii) *The composition of two integral morphisms is also integral.*
(iii) *The property of being an integral morphism is stable under base extension.*

Proof. (i) is clear. (ii) follows from Corollary 3 to Proposition 1.9. (iii) is clear by Lemma 1.50.(i). □

Proposition 1.55.

(i) *Any integral morphism is universally closed.*
(ii) *If $f\colon X \to Y$ is an integral and surjective morphism, then* dim X = dim Y.

Proof. (i) This follows from Lemma 1.50.(ii) and Theorem 1.21.(iii).

(ii) First, we assume that Y is affine, i.e., Y = Spec A. Then X is affine, i.e., X = Spec B. Since f is surjective, $\varphi = \Gamma(f)\colon A \to B$ has a nilpotent kernel by Proposition 1.4.(ii). Hence, dim Y = dim A = dim(A/Ker φ), and by the Corollary to Theorem 1.7, one has dim(A/Ker φ) = dim B, i.e., dim X = dim Y.

The general case follows from the affine case (cf. §1.9.**a**). □

e. Let X be a Noetherian scheme and let U be an open subscheme of X. The inclusion map $i\colon U \to X$ satisfies the hypothesis for f in Proposition 1.51. Thus, given a quasi-coherent $\mathcal{O}_U$-ideal $\mathcal{S}$, one has a quasi-coherent

$\mathcal{O}_X$-module $i_*(\mathcal{S})$, which is a subsheaf of $i_*(\mathcal{O}_U)$. Define $\sigma: \mathcal{O}_X \to i_*(\mathcal{O}_U)$ by $\sigma_W = r_{U \cap W, W}$ for any open subset W of X. $\mathcal{T} = \sigma^{-1}(i_*(\mathcal{S}))$ is a quasi-coherent $\mathcal{O}_X$-ideal such that $\mathcal{T}|_U = \mathcal{S}$. Hence $Z = \operatorname{supp}(\mathcal{O}_X/\mathcal{T})$ is a closed subset of X with $Z \cap U = Y$ and $Z \supseteq \bar{Y}$, where $\bar{Y}$ denotes the closure of Y in X. If $x \notin \bar{Y}$, then $X \backslash \bar{Y}$ is an open neighborhood of x and so $i_*(\mathcal{O}_U)_x = i_*(\mathcal{S})_x$. Thus, $\mathcal{T}_x = \mathcal{O}_{X,x}$, which implies $x \notin Z$; hence $Z = \bar{Y}$.

Definition. The closed subscheme $(\bar{Y}, (\mathcal{O}_X/\mathcal{T})|_{\bar{Y}})$ is said to be the *closure scheme* of Y in X, denoted $\bar{Y}$ for simplicity.

If Y is irreducible (respectively reduced), then so is $\bar{Y}$. Thus, in particular, if Y is an integral scheme, then $\bar{Y}$ is also an integral scheme.

§1.28 Proper Morphisms and Finite Morphisms

a. Definition. A *proper* morphism $f: X \to Y$ is a morphism of schemes that is separated, of finite type, and universally closed.

By Lemma 1.49.(i) and Proposition 1.55.(i), we know that any integral morphism of finite type is proper.

Theorem 1.22

(i) *Every closed immersion is proper.*
(ii) *The composition of two proper morphisms is proper.*
(iii) *The property of being a proper morphism is stable under base extension.*
(iv) *Let $\{Y_\lambda \mid \lambda \in \Lambda\}$ be an open cover, and let $X_\lambda = f^{-1}(Y_\lambda)$, $f_\lambda = f|_{X_\lambda}: X_\lambda \to Y_\lambda$. If every f_λ is proper, then so is f.*

Proof. (i) This is clear, since a closed immersion is integral and of finite type. (ii) and (iii) follow from Theorem 1.17, Proposition 1.48, and Proposition 1.53. (iv) is left to the reader. □

Definition. A scheme X over a field k is said to be *(k-)complete*, if the structure morphism $\varphi: X \to \operatorname{Spec} k$ is proper.

Thus, a k-complete scheme X is separated and algebraic over a field k.

b. Definition. A morphism $f: X \to Y$ is said to be a *finite morphism* if it is both integral and of finite type.

Lemma 1.52. *Let $\varphi: A \to B$ be a ring homomorphism. Then ${}^a\varphi$ is a finite morphism if and only if $B_{[\varphi]}$ is finitely generated as an A-module.*

PROOF. Suppose first that ${}^a\varphi$ is finite. By Proposition 1.47, B is finitely generated as an A-algebra via φ, i.e., $B = A[b_1, \ldots, b_n]$ for some $b_j \in B$. By Proposition 1.54, φ is integral. Recalling Proposition 1.9 and its Corollary 1, one concludes that $B_{[\varphi]}$ is a finitely generated A-module. The converse is obvious. □

Definition. If $B_{[\varphi]}$ is finitely generated as an A-module, φ is said to be a *finite homomorphism*. In this case, B is said to be *finite* over $\varphi(A)$.

Proposition 1.56

(i) *Every closed immersion is a finite morphism.*
(ii) *The composition of two finite morphisms is also finite.*
(iii) *The property of being a finite morphism is stable under base extension.*

PROOF. These follow from Proposition 1.48 and Theorem 1.21. □

Corollary

(i) *A finite morphism is a proper morphism.*
(ii) *If $f: X \to Y$ is a finite surjective morphism, then $f^{-1}(y)$ has dimension 0 for all $y \in Y$.*

PROOF. (i) This is clear.

(ii) By Proposition 1.56.(iii), $f^{-1}(y) = X \times_Y \operatorname{Spec} k(y) \to \operatorname{Spec} k(y)$ is finite. Hence, $f^{-1}(y)$ is an affine scheme Spec A, where A is finite over $k(y)$. Thus, A is Artinian and $\dim A = 0$. □

c. The method of the proof of Theorem 1.17.(iv) is quite useful. It can be used to prove a general theorem in the following way.

Theorem 1.23

(I) *Let $\mathbb{P}$ be a property of morphisms of schemes satisfying the following three conditions:*

(i) *All immersions have the property $\mathbb{P}$.*
(ii) *If both $f: X \to Y$ and $g: Y \to Z$ have the property $\mathbb{P}$, then so does $g \circ f$.*
(iii) *The property $\mathbb{P}$ is stable under base extension.*

Then $\mathbb{P}$ also satisfies

(iv) *Let $f: X \to Y$ and $g: Y \to Z$ be morphisms. If $g \circ f$ has the property $\mathbb{P}$, then so does f.*

(II) *Let $\mathbb{P}'$ be a property of morphisms satisfying* (ii) *and* (iii) *as well as*

(i′) *all closed immersions have the property $\mathbb{P}'$.*

Then $\mathbb{P}'$ *also satisfies*

(iv′) *Let* $f: X \to Y$ *and* $g: Y \to Z$ *be morphisms. If* g *is separated and* $g \circ f$ *has the property* $\mathbb{P}'$, *then* f *has the property* $\mathbb{P}'$.

PROOF. (i) We use the notation of Lemma 1.40. Then $q = (g \circ f)_Y$ has the property $\mathbb{P}$ by (iii). Γ_f has the property $\mathbb{P}$ by (i); hence $f = q \circ \Gamma_f$ has also the property $\mathbb{P}$ by (ii).

(ii) If g is separated, then Γ_f is a closed immersion by Proposition 1.41. Hence, $f = q \circ \Gamma_f$ has the property $\mathbb{P}'$ by (i′), (ii), and (iii). □

The properties of being a separated morphism and being an immersion satisfy (i), (ii), (iii) and hence (iv) by Theorems 1.17, and 1.23.

The properties of being an affine morphism, being a universally closed morphism, being an integral morphism, being a closed immersion, and of being a morphism of finite type satisfy (i′), (ii), (iii) and hence (iv′) by Theorems 1.23 and 1.21, and Propositions 1.38, 1.48, and 1.53. Thus, in particular, we obtain the first part of the next result.

Proposition 1.57. *Let* $f: X \to Y$ *and* $g: Y \to Z$ *be morphisms.*

(i) *If* $g \circ f$ *is proper and* g *is separated, then* f *is also proper.*
(ii) *In addition to the hypothesis of* (i), *assume that* f *is surjective and* g *is of finite type. Then* g *is also proper.*

PROOF OF (ii). First note that the property of being a surjective morphism is stable under base extension by the Corollary to Proposition 1.39. Thus, it suffices to prove that g is a closed morphism. For any closed subset F of Y, $g(F) = (g \circ f)(f^{-1}(F))$ is a closed subset, since $g \circ f$ is a closed morphism. □

d. k-complete schemes have similar properties to compact sets in a Hausdorff topological space as will be shown below.

Lemma 1.53. *If* Spec A *is k-complete, then* Spec A *is a discrete finite set.*

PROOF. Since A is finitely generated over k, Spec A is Noetherian. Thus, we may assume A is an integral domain. Letting $r = \dim A$, there exist $y_1, \ldots, y_r \in A$ such that $B = k[y_1, \ldots, y_r]$ is isomorphic to a polynomial ring in r variables, and A is finite over B. Hence, Spec $A \to$ Spec B is surjective by Theorem 1.7, and proper by Proposition 1.57.(i). Thus Spec B is k-complete by Proposition 1.57.(ii). $\mathbf{A}_k^r \cong$ Spec B is a dense subset of $\mathbf{P}_k^r$. The inclusion map i: Spec $B \to \mathbf{P}_k^r$ is a closed morphism by Proposition 1.57.(i) (cf. Example 1.30). Hence i(Spec B) = Spec B is closed; i.e., Spec $B = \mathbf{P}_k^r$. This implies that $r = 0$; thus $B = k$, and A is a field by Lemma 1.8. Hence, Spec A is a set consisting of a single point. □

Proposition 1.58.

(*i*) *Let X be a k-complete scheme and let F be a subscheme of X. F is k-complete if and only if F is a closed subscheme.*

(*ii*) *Let $f\colon X \to Y$ be a morphism, where X and Y are k-algebraic separated schemes.*

(*a*) *If f is proper and G is a subscheme of Y that is k-complete, then $f^{-1}(G)$ is also k-complete.*

(*b*) *If a subscheme F of X is k-complete, then $f(F)$ is the base space of a k-complete subscheme of Y.*

PROOF. (i) Let $\bar{F}$ be the closure of F. If F is k-complete, then $\bar{F} = F$ by Proposition 1.57.(i); hence F is a closed subscheme. The converse is obvious.

(ii) If the structure morphism $\xi\colon G \to \operatorname{Spec} k$ and f are proper, then so is $f^{-1}(G) = X \times_Y G \to \operatorname{Spec} k$, where $f^{-1}(G) \to \operatorname{Spec} k$ is the composition of ξ and $f_G\colon X \times_Y G \to G$. Proof of (*b*). Let F be a subscheme of X that is k-complete. Then $f|_F\colon F \to Y$ is proper by Proposition 1.57.(i). Hence $f(F)$ is a closed subset of Y. Then there exists a closed subscheme F_1 of Y whose base space is $f(F)$. Thus by Proposition 1.30, $f|_{F_{\mathrm{red}}}\colon F_{\mathrm{red}} \to F_1$ is a morphism. By Proposition 1.57.(ii), F_1 is k-complete. □

§1.29 Algebraic Varieties

a. Definition. A scheme V over a field k is said to be an *algebraic variety* over k if it is an integral separated scheme which is algebraic over k. In this case $\operatorname{Rat}_k(V)$ is a field finitely generated over k, called the *function field* of V.

An algebraic variety V is covered by a finite number of affine dense open subsets $V_1, \ldots, V_m$. If U is an affine dense open subset of V, then $\operatorname{Rat}_k(V) = \operatorname{Rat}_k(U) = Q(A(U))$. Furthermore, if W is another algebraic variety and $f\colon V \to W$ is a k-morphism, then f is of finite type and separated by Proposition 1.48 and Theorems 1.17 and 1.23.

If there is no danger of confusion, an algebraic variety over k is referred to simply as an (algebraic) variety. Subvarieties of an algebraic variety are subschemes which are varieties. If W is a subvariety of V, then the closure scheme of W is a closed subvariety (cf. §1.27.**e**).

In the following chapters, especially in Chapters 2, 5, 6, 8, 9, 10, and 11, we fix a base field k and consider varieties V that are algebraic over k. Thus, k-morphisms will be referred to as morphisms.

b. If k is an algebraically closed field, a variety V is covered by affine open dense subsets $V_1, \ldots, V_m$ and each $\operatorname{spm}(V) \cap V_i$ can be written as $\mathbf{V}(\mathfrak{p}_i)$, where $V_i = \text{Spec } k[\mathbf{X}_1, \ldots, \mathbf{X}_N]/\mathfrak{p}_i$ and $\mathbf{V}(\mathfrak{p}_i)$ denotes the set $\{p \in k^N \mid \varphi(p) = 0 \text{ for all } \varphi \in \mathfrak{p}_i\}$ (cf. §1.8.**d**). Any morphism $f\colon V \to W$, where W is a variety, induces a continuous map $\bar{f}\colon \text{spm}(V) \to \text{spm}(W)$ and we define

$$\text{Hom}(\text{spm}(V), \text{spm}(W)) = \{\bar{f} \mid f \in \text{Hom}_k(V, W)\}.$$

Then one has the bijection

$$\text{Hom}_k(V, W) \cong \text{Hom}(\text{spm}(V), \text{spm}(W)).$$

To verify this it suffices to see that if $\bar{f} = \bar{g}$, then $f = g$. For the proof, let $K = \text{Ker}(f, g)$, which is a closed subset because W is separated. Then by hypothesis, $\text{spm}(V)$ is contained in K. Since $\text{spm}(V)$ is dense by Theorem 1.5, one has $K = V$; hence, $f = g$ by Proposition 1.46.(i).

Remark. $\{\text{spm}(V)\}$ with $\{\text{Hom}(\text{spm}(V), \text{spm}(W))\}$ is the category of algebraic varieties in the classical sense.

c. Let V and W be algebric varieties. Let $\text{DRat}(V, W)$ be the set of dominating rational maps from V to W. Then letting v and w be generic points of V and W, respectively, one has, for any $f \in \text{DRat}(V, W)$, the homomorphism $f^* = \theta(f)_v^{\#}\colon \text{Rat}_k(W) \to \text{Rat}_k(V)$. Thus the map $\Psi\colon \text{DRat}(V, W) \to \text{Hom}(\text{Rat}_k(W), \text{Rat}_k(V))$ (= the set of ring homomorphisms) is given by $\Psi(f) = f^*$. By Proposition 1.50.(i), Ψ is injective. Furthermore, by (ii), Ψ is surjective.

Definition. A dominating rational map $f\colon V \to W$ is said to be a *birational* map if there is a rational map $g\colon W \to V$ such that $g \circ f = \text{id}$.

Then $\Psi(g \circ f) = \Psi(f) \circ \Psi(g) = \text{id}$, and so $\Psi(g) \circ \Psi(f) = \text{id}$. Hence, $f \circ g = \text{id}$ as a rational map; thus the set of all birational maps of V into itself becomes a group under composition.

Definition. The group $\text{Bir}(V) = \{f\colon V \to V \mid f \text{ is a birational map}\}$ is said to be the *birational automorphism group* of V.

Proposition 1.59

(*i*) $\Psi\colon \text{DRat}(V, W) \cong \text{Hom}(\text{Rat}_k(W), \text{Rat}_k(V))$.
(*ii*) $\text{Bir}(V) \cong \text{Aut}_k(\text{Rat}_k(V))$.

PROOF. Clear from the previous observation. □

Definition. If there exists a birational map $f\colon V \to W$, then V is said to be *birationally equivalent* to W. Furthermore, V (and also W) is said to be a *geometric model* of the field $\mathfrak{K}$ if $\mathfrak{K} \cong \text{Rat}_k(V)$.

In this case, $\text{Aut}_k(\mathfrak{K}) \cong \text{Bir}_k(V)$.

Remark. Let $\mathfrak{K}/k$ be a finitely generated extension of fields with tr. $\deg_k \mathfrak{K} = n$. If the characteristic of k is zero, then there exist $x_1, \ldots, x_{n+1} \in \mathfrak{K}$ such that $x_1, \ldots, x_n$ are algebraically independent over k and $\mathfrak{K} = k(x_1, \ldots, x_{n+1})$, where x_{n+1} is algebraic over $k(x_1, \ldots, x_n)$.

In this case, there exists an irreducible polynomial $F(\mathsf{X}_1, \ldots, \mathsf{X}_{n+1}) \in k[\mathsf{X}_1, \ldots, \mathsf{X}_{n+1}]$ such that $\mathfrak{K} = \mathrm{Rat}_k(V(F))$. Thus, $V(F)$ is a geometric model of $\mathfrak{K}$.

Proposition 1.60. *Let $f\colon V \to W$ be a dominating morphism, where both V and W are varieties. Then the following conditions are equivalent:*

(i) *f is birational.*
(ii) *There exists a nonempty open subset U of W such that $f|_{f^{-1}(U)}\colon f^{-1}(U) \to U$ is an isomorphism.*

PROOF. (ii) $\Rightarrow$ (i) This is clear by definition.

(i) $\Rightarrow$ (ii) There exist an open dense subset U of W and a morphism $\psi\colon U \to V$ such that $\psi \circ \varphi = \mathrm{id}$, where φ is $f|_{f^{-1}(U)}\colon f^{-1}(U) \to U$. By applying Proposition 1.59.(i), one has $\varphi \circ \psi = \mathrm{id}$; hence φ is an isomorphism. $\square$

d. Let V and W be varieties over k. The k-product $V \times_k W$ is not a variety in general (cf. Example 1.28). But, if k is algebraically closed, then $V \times_k W$ is a variety. To verify this, one has only to prove the next lemma, since $V \times_k W$ is covered by $\{V_i \times_k W_j \mid V_i$ is an affine open dense subvariety of V and W_j is an affine open dense subvariety of $W\}$.

Lemma 1.54. *Let A and B be k-algebras without zero divisors. If k is algebraically closed, then $A \otimes_k B$ has no zero divisors.*

PROOF. Since $A \otimes_k B \subseteq Q(A) \otimes_k Q(B)$, one can assume that A and B are fields. Let $\alpha = \sum_{i=1}^{r} a_i \otimes b_i (a_i \in A,\ b_i \in B)$ be a zero divisor of $A \otimes_k B$, i.e., there is $\beta = \sum_{j=1}^{s} c_j \otimes d_j$ such that $\alpha \cdot \beta = 0$. Replacing A by $k(\{a_i, c_j \mid 1 \leq i \leq r,\ 1 \leq j \leq s\})$, one also assumes that A is finitely generated over k. By a theorem of F. K. Schmidt [N1, p. 144], A is separable over k; i.e., A has a separable transcendence base $x_1, \ldots, x_n$ over k. Then there is an irreducible polynomial $F \in k[\mathsf{X}_1, \ldots, \mathsf{X}_{n+1}]$ such that $A \cong \mathrm{Rat}_k(V(F))$. By Proposition 1.11, F is also irreducible as a polynomial over B. Hence, $A \otimes_k B = B[\mathsf{X}_1, \ldots, \mathsf{X}_{n+1}]/(F)$ has no zero divisors. $\square$

Note that $\mathrm{Rat}(V \times_k W) = \mathrm{Rat}(V_i \times_k W_j) = Q(A(V_i) \otimes_k A(W_j)) = Q(\mathrm{Rat}(V) \otimes_k \mathrm{Rat}(W))$, and hence $\dim(V \times W) = \dim V + \dim W$.

By Lemma 1.43, $A(V) = \bigcap_{p \in V} \mathcal{O}_{V,p} = \bigcap_i \bigcap_{p \in V_i} \mathcal{O}_{V,p} = \bigcap_i A(V_i)$ as a subring of $\mathrm{Rat}(V)$. Thus one has

$$A(V \times_k W) = \bigcap_{i,j} A(V_i \times_k W_j) = \bigcap_{i,j} A(V_i) \otimes A(W_j)$$

$$= \bigcap_j \left(\bigcap_i A(V_i) \right) \bigotimes_k A(W_j) = A(V) \bigotimes_k A(W).$$

If V is a variety and V_i is an affine open dense subvariety of V, then

$$\dim V = \dim V_i = \dim A(V_i) = \mathrm{tr.\,deg}_k\, QA(V_i) = \mathrm{tr.\,deg}_k\, \mathrm{Rat}(V).$$

Proposition 1.61. *Let V be a variety and let U be an open subset of V. If $f \in \Gamma(U, \mathcal{O}_V)$, then $Z_U(f) = \{x \in U \mid \bar{f}(x) = 0\}$ is either $\varnothing$, U, or a union of $(n-1)$-dimensional closed irreducible subsets of U, where $n = \dim V$.*

Proof. One may assume $V = U$ and that V is an affine variety Spec A, where A is $A(V)$. By Theorem 1.3, there is a k-subalgebra B of A such that B is a polynomial ring $k[\mathsf{X}_1, \dots, \mathsf{X}_n]$ and A is integral over B.

Define L to be the separable closure of $Q(B)$ in $Q(A)$, i.e., $L = \{\alpha \in Q(A) \mid \alpha$ is separably algebraic over $Q(B)\}$. Then $Q(A)/L$ is purely inseparable and finite; hence there exists $e > 0$ such that $f_1 = f^e \in L$. Letting $\bar{L}$ be an algebraic closure of the field $Q(A)$, we define G to be the set of all $Q(B)$-homomorphisms from L to $\bar{L}$. Since G is finite, G can be written as $\{\sigma_1 = \mathrm{id}, \sigma_2, \dots, \sigma_r\}$. Define g to be $\sigma_1(f_1) \cdots \sigma_r(f_1)$, which belongs to $Q(B)$. As each $f_j = \sigma_j(f_1)$ is integral over B, g belongs to B, since B is a normal ring (For this, see Example 2.4 in §2.3). $C = A[f_1, \dots, f_r]$ is a subalgebra of $\bar{L}$, which is finite over A by Corollary 1 to Proposition 1.9. Associated with the inclusions $B \subseteq A \subseteq C$, one has finite morphisms

$$\psi\colon \mathrm{Spec}\, C \to \mathrm{Spec}\, A = V \qquad \text{and} \qquad \varphi\colon V \to \mathbf{A}_k^n.$$

Now, we use the symbol $V(gA)$ to denote the subset of Spec A defined by the ideal gA. If $g = 0$, then $V(gA) = V$. If g is a unit, then $V(gA) = \varnothing$. Thus, we assume g is neither zero nor a unit. Then $g \notin k$. Hence, $V(gB)$ consists of $(n-1)$-dimensional irreducible subsets $V(g_i B)$, where the g_i are irreducible polynomials dividing g. Since φ, ψ are finite morphisms, and $(\varphi \circ \psi)^{-1}(V(gB)) = V(gC) = V(f_1 C) \cup \cdots \cup V(f_r C)$, each $V(f_i C)$ consists of $(n-1)$-dimensional irreducible components. On the other hand, $V(f_1 C) = \psi^{-1}(V(f_1 A)) = \psi^{-1}(V(fA))$. Hence, each irreducible component of $V(fA)$ is $(n-1)$-dimensional. □

Corollary. *Let $f_1, \dots, f_r \in A(V)$. Then each irreducible component of $V(\sum_{j=1}^r f_j A)$ has dimension $\geq n - r$, whenever $V(\sum_{j=1}^r f_j A)$ is not empty.*

Proof. We show this by induction on r. The case of $r = 1$ follows from the last proposition. Assuming the case of $r - 1$, consider $W_r = V(\sum_{j=1}^r f_j A)$. Each irreducible component Y_j of W_r is an irreducible component of $V(f_r \bar{A})$, where $\bar{A} = A/\mathfrak{p}$ for some minimal prime ideal of $\sum_{j=1}^{r-1} f_j A$. Then by the induction hypothesis, dim $V(\mathfrak{p}) \geq n - r + 1$; hence dim $Y_j \geq$ dim $V(\mathfrak{p}) - 1 \geq n - r$ by the last proposition. □

Theorem 1.24. *Let V and W be closed subvarieties of $\mathbf{A}_k^n$ such that $V \cap W \neq \emptyset$. Suppose that k is algebraically closed and let* dim $V = r$, dim $W = s$. *Then* $\dim(V \cap W) \geq r + s - n$.

Proof. Since $V \cap W \cong \Delta_{\mathbf{A}^n}(\mathbf{A}_k^n) \cap (V \times_k W)$ by Proposition 1.40 and since $\Delta_{\mathbf{A}^n}(\mathbf{A}_k^n) = V(\mathrm{X}_1 \otimes 1 - 1 \otimes \mathrm{X}_1, \ldots, \mathrm{X}_n \otimes 1 - 1 \otimes \mathrm{X}_n)$, one has by the last corollary

$$\dim(V \cap W) \geq \dim(V \times_k W) - n = r + s - n,$$

if $V \cap W \neq \emptyset$. □

e. Complete varieties have beautiful properties and will be studied precisely in Chapters 7 and 10. Here, we note the next simple result.

Lemma 1.55. *Let V be a k-complete variety. Then $A(V)$ is a subfield of* $\mathrm{Rat}_k(V)$ *and is a finite extension of k.*

Proof. Take $\alpha \in A(V)$. Then there exists a regular function φ: $V \to \mathbf{A}_k^1 = \mathrm{Spec}\ k[\mathrm{X}]$ such that $\Gamma(\varphi)(\mathrm{X}) = \alpha$. Since V is complete, φ is proper by Proposition 1.57.(i). Hence $\varphi(V)$ is complete and so $\varphi(V)$ has dimension 0 by Proposition 1.58.(ii).(b) and Lemma 1.53. Thus $\varphi(V) = V(h)$, where h is an irreducible polynomial and $\mathrm{Ker}(\Gamma(\varphi)) = (h(\mathrm{X}))$. One has $h(\alpha) = h(\Gamma(\varphi)(\mathrm{X})) = \Gamma(\varphi)(h(\mathrm{X})) = 0$; hence α is algebraic over k. $A(V)$ is a subring of the field $\mathrm{Rat}_k(V)$ and hence $A(V)$ is an integral domain, finite over k. Then by Lemma 1.8, $A(V)$ is a field. □

The proof of the next result is far from being easy.

Theorem 1.25 (Nagata). *Given any variety, there exist a complete variety Z and an open immersion i: $V \to Z$.*

Proof. We refer to "Imbedding of an abstract variety in a complete variety" by M. Nagata in *J. Math. Kyoto Univ.* 2 (1962), 1–10. □

Definition. If (Z, i) satisfies the above condition, then (Z, i) or Z is said to be a *completion* of V. Furthermore, $Z \backslash i(V)$ is said to be the *algebraic boundary* of V in Z.

In general, there exist many completions of V.

Lemma 1.56. *A variety V is complete if and only if any open immersion i of V into a variety W is surjective; i.e.,* $i(V) = W$.

Proof. If V is complete and $i: V \to W$ is an open immersion, then i is proper by Proposition 1.57.(i); hence $i(V)$ is closed and so $i(V) = W$. Now, assume that V is not complete. By Theorem 1.25, one has a completion (Z, i) of V. Since V is not complete, $i(V) \neq Z$. This completes the proof. □

If V is an affine variety, then V is a closed subvariety of some affine space $\mathbf{A}_k^n$. Let $\mathbf{P}_k^n$ be such that $D_+(\mathsf{T}_0) = \mathbf{A}_k^n$. Thus V can be identified with a subvariety of $\mathbf{P}_k^n$. Let $\bar{V}$ be the closure variety of V. In Chapter 3, it will be proved that $\mathbf{P}_k^n$ is k-complete, and hence $\bar{V}$ is also k-complete. Thus, we have a completion $\bar{V}$ of V. In this way, any affine algebraic variety has a completion which is also projective. Theorem 1.25 asserts that any algebraic variety can be considered as an open subvariety of a complete variety.

A variety of dimension 1 is called a *curve*. Curves will be studied in Chapters 6 and 8.

A variety of dimension 2 is called a *surface*. Surfaces will be studied in Chapters 8, 9, and 10. A *curve C on a surface S* is a curve which is a closed subvariety of S. Curves on surfaces will be discussed in Chapter 8.

Remark. In some references, a variety is defined to be a reduced algebraic scheme. Then what we call a variety is referred to as an irreducible variety.

Exercises

1.1. Let f be an irreducible polynomial in two variables over a field k. Prove that the principal ideal (f) is a prime ideal. (This holds for any irreducible polynomial in several variables; see Example 2.4.)

1.2. Let $\varphi: A \to B$ be a homomorphism of rings and let $f \in A$. Then show the following equivalences:
(i) $\operatorname{Im}({}^a\varphi) \subseteq V(f) \Leftrightarrow \varphi(f)$ is nilpotent.
(ii) $\operatorname{Im}({}^a\varphi) \cap V(f) = \emptyset \Leftrightarrow \varphi(f)$ is a unit.

1.3. Let G_m^n denote Spec $k[\mathsf{X}_1, \ldots, \mathsf{X}_n, \mathsf{X}_1^{-1}, \ldots, \mathsf{X}_n^{-1}]$, where k is a field and $\mathsf{X}_1, \ldots, \mathsf{X}_n$ are algebraically independent over k.
(i) Describe $\operatorname{Hom}_k(G_m^n, G_m)$.
(ii) Show that $\operatorname{Aut}_k(G_m^n)$ is a group extension of $GL(n, \mathbb{Z})$ by $(k^*)^n$.

1.4. Let l_0, l_1, l_2 be three distinct lines on the projective plane $\mathbf{P}_k^2$ over a field k. What is the group $\operatorname{Aut}_k(\mathbf{P}_k^2 \backslash (l_0 \cup l_1 \cup l_2))$?

1.5. Let R be a polynomial ring in two variables X, Y over a field k. Find all P, $Q \in R$ satisfying $\varphi(P, Q) = \varphi(\mathsf{X}, \mathsf{Y})$ for a given $\varphi \in R$ in the following cases.
(i) $\varphi = \mathsf{Y}^2 - \mathsf{X}^m$, where $m = 1, 2, \ldots$.
(ii) $\varphi = A(\mathsf{X})\mathsf{Y} - B(\mathsf{X})$ for some polynomials A and B.
(iii) $\varphi = \mathsf{Y}^2 - \mathsf{X}(\mathsf{X} - 1)(\mathsf{X} - \lambda)$.

1.6. Let $\mathbf{P}_k^2$ be Proj $k[\mathsf{T}_0, \mathsf{T}_1, \mathsf{T}_2]$ and let $\mathsf{Y} = \mathsf{T}_2/\mathsf{T}_0$, $\mathsf{X} = \mathsf{T}_1/\mathsf{T}_0$. Let $C = V_+(\mathsf{T}_2\mathsf{T}_0^{m-1} - \mathsf{T}_1^m)$ and $S = \mathbf{P}_k^2 \backslash C$. For $\alpha, \beta \in k^*$, $\delta_1, \ldots, \delta_d \in k$, define an automorphism φ of $k(\mathsf{X}, \mathsf{Y})$ by

$$\varphi(\mathsf{X}) = \beta\mathsf{X} + \sum_{i=1}^{d} \delta_i/(\mathsf{Y} - \mathsf{X}^n)^i,$$

$$\varphi(\mathsf{Y}) = \alpha(\mathsf{Y} - \mathsf{X}^n) + (\beta\mathsf{X} + \sum_{i=1}^{d} \delta_i/(\mathsf{Y} - \mathsf{X}^n)^i)^n.$$

Show that φ gives rise to an automorphism of S.

1.7. Suppose that the field k has uncountably many elements. Let A be the ring generated by countably many elements over k. For any maximal ideal $\mathfrak{m}$ of A, show that $A/\mathfrak{m}$ is a field algebraic over k.

1.8. Let X be a ringed space and let $\{U_\lambda \mid \lambda \in \Lambda\}$ be an open cover of X. If morphisms f, g from X into a ringed space Y satisfy $f|_{U_\lambda} = g|_{U_\lambda}$ for all λ, then show that $f = g$.

1.9. Let $X = \operatorname{Spec} A$ for some ring A, and let U be an open subset of X. Suppose that U contains $x_1, \ldots, x_r$. Show that there exists $f \in A$ satisfying $\{x_1, \ldots, x_r\} \subseteq D(f) \subseteq U$.

1.10. Let V and W be varieties over a field k and let $f\colon V \to W$ be a dominating morphism.

(i) If both V and W are affine, then show that there exists a dominating morphism $\varphi\colon V \to \mathbf{A}_k^r \times W$, where $r = \dim V - \dim W$, such that $f = q \circ \varphi$, where $q\colon \mathbf{A}_k^r \times W \to W$ is the projection.

(ii) Suppose that there exists a dense subset S of W such that $f^{-1}(x)$ is discrete for all $x \in S$. Show that $\dim V = \dim W$.

Chapter 2

Normal Varieties

§2.1 Normal Rings

a. Definition. Let B be a subring of A. If every element of A which is integral over B belongs to B, then B is said to be *integrally closed* in A.

In §1.7.**a**, we introduced the notation $B'_A = \{\alpha \in A \mid \alpha \text{ is integral over } B\}$. Thus, B is integrally closed in A if and only if $B'_A = B$.

Definition. An integral domain that is integrally closed in its field of fractions is called a *normal ring*. If A is a subring of a field L, then by Corollary 4 to Proposition 1.9, A'_L is a normal ring, which is said to be the *integral closure* of A in L.

Clearly, $Q(A'_L)$ is the algebraic closure (the field of all elements of L that are algebraic over $Q(A)$) of $Q(A)$ in L.

EXAMPLE 2.1. Let A be the subring $k[\mathsf{X}^2, \mathsf{X}^3]$ of the polynomial ring $k[\mathsf{X}]$ over a field k. Then $Q(A) = k(\mathsf{X})$ and $\mathsf{X} \in Q(A)\backslash A$. But X is integral over A, since it is a root of a monic polynomial $F(\mathsf{T}) = \mathsf{T}^2 - \mathsf{X}^2 \in A[\mathsf{T}]$. Hence, the integral closure of A in $Q(A)$ is $k[\mathsf{X}]$.

b. Lemma 2.1. *Let S be a multiplicative subset of a normal ring A with $0 \notin S$. Then $S^{-1}A$ is also a normal ring.*

PROOF. Let a/b with $a, b \in A$, be integral over $S^{-1}A$, i.e., there exist $a_1, \ldots, a_m \in A$ and $s \in S$ such that

$$\left(\frac{a}{b}\right)^m + \frac{a_1}{s}\left(\frac{a}{b}\right)^{m-1} + \cdots + \frac{a_m}{s} = 0.$$

Then

$$\left(\frac{sa}{b}\right)^m + a_1\left(\frac{sa}{b}\right)^{m-1} + \cdots + a_m s^{m-1} = 0.$$

Since A is normal, one has $sa/b \in A$; hence $a/b \in S^{-1}A$. □

Proposition 2.1. *Let S be a multiplicative subset of an integral domain A and let L be a field containing A. If $0 \notin S$, then*

$$(S^{-1}A)'_L = S^{-1}(A'_L).$$

PROOF. By Lemma 2.1, $S^{-1}(A'_L)$ is normal and contains $S^{-1}A$. Clearly, $S^{-1}(A'_L)$ is integral over $S^{-1}A$ and $Q(S^{-1}(A'_L)) = Q(A'_L) = Q((S^{-1}A)'_L)$. Hence $S^{-1}(A'_L)$ is the integral closure of $S^{-1}A$ in L. □

Proposition 2.2. *Let $\{A_i \mid i \in J\}$ be a collection of subrings of an integral domain B. If each A_i is integrally closed in B, then so is $\bigcap_{i \in J} A_i$. In particular, if all the A_i are normal, then so is $\bigcap_{i \in J} A_i$.*

PROOF. If an element α of B is integral over $\bigcap_{i \in J} A_i$, then it is integral over each A_i; hence $\alpha \in A_i$ for all i, by assumption. □

Corollary. *Let A be an integral domain. A is normal if and only if $A_{\mathfrak{m}}$ is normal for all maximal ideals $\mathfrak{m}$ of A.*

PROOF. The "only if" part follows from Lemma 2.1. Suppose that $A_{\mathfrak{m}}$ is normal for all $\mathfrak{m}$. Then by Proposition 2.2, $\bigcap_{\mathfrak{m}} A_{\mathfrak{m}}$ is normal. But this ring coincides with A by Example 1.7. □

§2.2 Normal Points on Schemes

a. Definition. Let X be a quasi-compact scheme, and let $x \in X$. If $\mathcal{O}_{X,x}$ is a normal ring, then x is said to be a *normal point* of X. The set of all normal points on X is denoted Nor(X). If X is connected and Nor(X) = X, then X is said to be a *normal scheme*.

If an open subset U of an integral scheme X is contained in Nor(X), then $\Gamma(U, \mathcal{O}_X) = \bigcap_{x \in U} \mathcal{O}_{X,x}$ by Lemma 1.43, and is a normal ring by Corollary to Proposition 2.2.

By Proposition 2.2, a normal scheme X is an integral scheme. Hence, any $\Gamma(U, \mathcal{O}_X)$ is a normal ring. In particular, Spec A is a normal scheme if and only if A is normal.

Lemma 2.2. *Let A be an integral domain and let A' be its integral closure in $Q(A)$. Then for any prime ideal $\mathfrak{p}$ of A, $A_\mathfrak{p}$ is normal if and only if $(A'/A) \otimes_A A_\mathfrak{p} = 0$.*

PROOF. Putting $S = A\backslash\mathfrak{p}$, one has, by Lemma 1.2.(ii), the exact sequence

$$0 \to A_\mathfrak{p} \to S^{-1}A' \to S^{-1}(A'/A) \to 0.$$

By Proposition 2.1, $S^{-1}A'$ is $(A_\mathfrak{p})'_{Q(A)}$, and by Lemma 1.2.(i), $S^{-1}(A'/A) \cong (A'/A) \otimes_A A_\mathfrak{p}$. Thus,

$$(A_\mathfrak{p})'_{Q(A)}/A_\mathfrak{p} \cong (A'/A) \otimes_A A_\mathfrak{p}.$$

The result follows immediately from this isomorphism. □

Therefore, Nor(Spec A) is the complement of the support of $(A'/A)^\sim$ considered as an $\tilde{A}$-module (cf. §1.11.**e**). If A'/A is finitely generated, then supp(A'/A) is a closed subset by Proposition 1.22. Hence, Nor(Spec A) is an open subset. This observation motivates the next theorem.

b. Theorem 2.1. *Let A be an integral domain finitely generated over k as a k-algebra. If $L/Q(A)$ is a finite extension of fields, then A'_L is finitely generated as an A-module.*

PROOF. (1) We first prove this under the additional conditions that A is normal and $L/Q(A)$ is a finite separable extension, i.e., $L = Q(A)(\alpha)$ for some α separable over $Q(A)$. One can assume that $\alpha \in A'_L$. Let $f(\mathsf{X})$ be the monic minimal polynomial of α over $Q(A)$ and let the field $\mathfrak{K}$ be the Galois closure of L over $Q(A)$. Then $f(\mathsf{X})$ has distinct roots $\alpha_1 = \alpha, \alpha_2, \ldots, \alpha_m \in \mathfrak{K}$ and $\mathfrak{K} = Q(A)(\alpha_1, \ldots, \alpha_m)$, where $m = \deg f$. By Corollary 1 to Proposition 1.9, $B = A[\alpha_1, \ldots, \alpha_m]$ is a finitely generated A-module, since every α_j is integral over A. The Galois group $G = \{\sigma_1, \ldots, \sigma_s\}$ of $\mathfrak{K}/Q(A)$ acts on B. Any G-invariant element of B belongs to A, since A is normal. Letting $g_j(\mathsf{X}) = f(\mathsf{X})/(\mathsf{X} - \alpha_j)$ for $1 \leq j \leq m$, the derived polynomial f' can be written as

$$f'(\mathsf{X}) = \sum_{j=1}^{m} g_j(\mathsf{X}).$$

Since $f(\mathsf{X})$ is irreducible, for each α_j there exists $\sigma_j \in G$ that maps α to α_j. Furthermore, σ_1 can be chosen to be id. It follows that

$$g_j(\mathsf{X}) = \frac{f(\mathsf{X})}{\mathsf{X} - \alpha_j} = \sigma_j \frac{f(\mathsf{X})}{\mathsf{X} - \alpha} = (\sigma_j g_1)(\mathsf{X}).$$

Letting $g_1(\mathsf{X}) = \mathsf{X}^{m-1} + c_1 \mathsf{X}^{m-2} + \cdots + c_{m-1}$ where $c_i \in A[\alpha]$, one has

$$(\sigma_j g_1)(\alpha) = g_j(\alpha) = 0, \text{ if } \ j \neq 1.$$

For any $b \in A'_L$, one has

$$bf'(\alpha) = bg_1(\alpha) = bg_1(\alpha) + \sigma_2(b)\sigma_2(g_1)(\alpha) + \cdots + \sigma_m(b)\sigma_m(g_1)(\alpha)$$

$$= \sum_{i,j} \sigma_j(b)\sigma_j(c_i)\alpha^{m-i-1} = \sum_{i=0}^{m-1} a_i \alpha^{m-i-1},$$

where $c_0 = 1$ and $a_i = \sum_{j=1}^{m} \sigma_j(bc_i) \in \mathfrak{K}$ for all i. All the $bc_i \in A'_L[\alpha]$, since $b \in A'_L$ and $c_i \in A[\alpha]$. Hence, all the a_i are integral over A.

We claim that $a_i \in A$. In fact, since $bc_i \in A'_L[\alpha] \subseteq L = Q(A)[\alpha]$, there exists $\psi_i(\mathsf{X}) \in Q(A)[\mathsf{X}]$ such that $bc_i = \psi_i(\alpha)$. Then $a_i = \sum_{j=1}^{m} \sigma_j(\psi_i(\alpha)) = \sum_{j=1}^{m} \psi_i(\alpha_j)$ is G-invariant; hence $a_i \in Q(A)$. But, since A is normal and a_i is integral over A, this implies that $a_i \in A$ for all i.

Therefore, $bf'(\alpha) = \sum_{i=0}^{m-1} a_i \alpha^{m-i-1} \in A[\alpha]$. This implies that $b \in A[\alpha] \cdot (1/f'(\alpha))$, since $f'(\alpha) \neq 0$. Hence, $A'_L \subseteq A[\alpha] \cdot (1/f'(\alpha))$. Since A is Noetherian and A'_L is a submodule of a finitely generated A-module, A'_L is also finitely generated as an A-module.

(2) If A is not normal, we use Theorem 1.3 applied to A; there exist $\mathsf{Y}_1, \ldots, \mathsf{Y}_r \in A$ such that A is integral over $B = k[\mathsf{Y}_1, \ldots, \mathsf{Y}_r]$, which is isomorphic to a polynomial ring in r variables. We suppose that L is separable over $Q(B)$. Then by the first step, since B is normal (cf. Example 2.4), B'_L is finitely generated as a B-module. On the other hand, $Q(B'_L) = L$ and so $A'_L = B'_L$, since A is integral over B. Thus, A'_L is finitely generated as an A-module.

(3) The general case, in which $L/Q(A)$ may not be separable, is left to the reader (cf. [N1, pp. 131, 132]). □

EXAMPLE 2.2. Let L be an algebraic number field, i.e., L is a finite extension of $\mathbb{Q}$. Then $\mathbb{Z}'_L$ is the ring of algebraic integers of L. By the first part of the proof of Theorem 2.1, we see that $\mathbb{Z}'_L$ is a finitely generated $\mathbb{Z}$-module.

Proposition 2.3. *Let V be an algebraic variety. Then* Nor(V) *is a dense open subset of V.*

PROOF. Let $\{U_\lambda \mid \lambda \in \Lambda\}$ be an affine open cover of V. Then clearly $U_\lambda \cap \text{Nor}(V) = \text{Nor}(U_\lambda)$ and $\text{Nor}(U_\lambda)$ is open by the last theorem. Moreover, Nor(V) is not empty, since it contains the generic point of V. □

§2.3 Unique Factorization Domains

a. Let A be an integral domain.

Definition. A nonzero element a of A is said to be *reducible*, if a can be written as bc, where neither b nor c is a unit of A. A nonzero element which is neither a unit nor a reducible element is said to be an *irreducible element.*

Definition. If a principal ideal aA is a nonzero prime ideal of A, then a is said to be a *prime element*.

Clearly, any prime element is an irreducible element. But the converse is not true, as will be seen below.

EXAMPLE 2.3. Let $A = k[\mathsf{X}^2, \mathsf{X}^3]$ as in Example 2.1. Then $a = \mathsf{X}^2$ and $b = \mathsf{X}^3$ are irreducible elements. However, $b^2 \in aA$ and $b \notin aA$, which implies that a is not a prime element.

b. Definition. An integral domain A is said to be a *unique factorization domain*, abbreviated as UFD, or a *factorial ring* if any nonzero element a can be expressed as a finite product of prime elements, whenever a is not a unit. A prime element dividing $b \in A$ is called a *prime factor* of b.

Proposition 2.4. *If A is a* UFD, *then the factorization of any element of A into a finite number of prime elements is unique up to units.*

PROOF. Suppose that $a = p_1 \cdots p_m = q_1 \cdots q_n$ where the p_i and the q_j are prime elements. Since $p_1 A$ is a prime ideal, one of the q_j belongs to $p_1 A$, say $q_1 \in p_1 A$. Hence, $q_1 = p_1 \alpha_1$ for some $\alpha_1 \in A$. But now since $q_1 A$ is prime, $p_1 \in q_1 A$ or $\alpha_1 \in q_1 A$. If the second case occurs, then $\alpha_1 = q_1 \beta_1$ for some $\beta_1 \in A$; hence $q_1 = p_1 \alpha_1 = p_1 q_1 \beta_1$, i.e., $1 = p_1 \beta_1$. This implies p_1 is a unit, a contradiction. Thus $p_1 \in q_1 A$ and hence α_1 is a unit. One has

$$p_2 \cdots p_m = (\alpha_1 q_2) q_3 \cdots q_n,$$

and $q_2' = \alpha_1 q_2$ is also a prime element. The result now follows by induction on $\min\{n, m\}$. □

Remark. Every irreducible element of a UFD is obviously a prime element.

The uniqueness of such a factorization in the sense of the last proposition is a necessary and sufficient condition for A to be a UFD, as will be seen below.

Proposition 2.5. *Let A be an integral domain. If every nonzero element is expressible as a finite product of irreducible elements and the expression is unique up to units, then A is a* UFD. *The converse is also true.*

PROOF. First, we show that any irreducible element a is a prime element. If $bc \in aA$, i.e., $bc = a\alpha$ for some $\alpha \in A$, then by uniqueness, b or c is divisible by a. From this and Proposition 2.4, the result follows at once. □

Proposition 2.6. *Any* UFD *is a normal ring.*

Proof. Let A be a UFD and let $\alpha = a_1/a_2 \in Q(A)\backslash\{0\}$ be integral over A with $a_1, a_2 \in A$. We can assume that a_1 and a_2 have no common prime factors. Since α is integral, there exist $\alpha_1, \ldots, \alpha_m \in A$ such that

$$\left(\frac{a_1}{a_2}\right)^m + \alpha_1\left(\frac{a_1}{a_2}\right)^{m-1} + \cdots + \alpha_m = 0.$$

Then

$$a_1^m = -a_2(\alpha_1 a_1^{m-1} + \cdots + \alpha_m a_2^{m-1}).$$

By Proposition 2.4, any prime factor p of a_2 divides a_1, if a_2 is not a unit. This contradicts the hypothesis. Hence, a_2 is a unit and so

$$\alpha = a_1 \cdot a_2^{-1} \in A. \qquad \square$$

c. **Proposition 2.7.** *Let A be a* UFD.

(*i*) *For any multiplicative subset S of A with $0 \notin S$, $S^{-1}A$ becomes a* UFD.
(*ii*) *For any n, the polynomial ring $A[X_1, \ldots, X_n]$ over A is a* UFD.

Proof. (i) is clear. (ii) is a consequence of Gauss' lemma. For the proof, we refer the reader to [N1 p. 37]. $\square$

Example 2.4. The polynomial ring $k[X_1, \ldots, X_n]$ is a UFD; hence it is a normal ring by Proposition 2.6.

Example 2.5. Let A be a k-algebra which is a UFD. Suppose that the unit group $U(A)$ consists of $k^*(=k\backslash\{0\})$. If $\pi_1, \ldots, \pi_r$ are prime elements of A such that $\pi_i A \neq \pi_j A$ for all $i \neq j$, then $U(A[\pi_1^{-1}, \ldots, \pi_r^{-1}])/k^* \cong \mathbb{Z}^r$.

Example 2.6. Let A be the subring $\mathbb{C}[X_1^n, X_2^n, X_1 \cdot X_2]$ of the polynomial ring $\mathbb{C}[X_1, X_2]$ for some $n \geq 2$. The ring A is normal, since A can be written as $\{f \in \mathbb{C}[X_1, X_2] \mid f(\zeta X_1, \zeta^{-1}X_2) = f(X_1, X_2)\}$, where ζ is a primitive n-th root of 1. Furthermore one can prove that $\alpha = X_1^n$, $\beta = X_2^n$, and $\gamma = X_1 \cdot X_2$ are irreducible elements. However, $\alpha\beta = \gamma^n$, i.e., the factorization is not unique.

Proposition 2.8. *Let A be a* UFD. *If $\mathfrak{p}$ is a prime ideal of height* 1, *then $\mathfrak{p}$ is always principal.*

Proof. Suppose that A is not a field. Then one has $f \in \mathfrak{p}\backslash\{0\}$. Let $p_1 \cdots p_m$ be the factorization of f into prime elements. Since $f \in \mathfrak{p}$, there exists p_i such that $p_i \in \mathfrak{p}$; hence $p_i A \subseteq \mathfrak{p}$. Since $\mathfrak{p}$ is a prime ideal of height 1, $\mathfrak{p}$ coincides with $p_i A$. $\square$

§2.4 Primary Decomposition of Ideals

Throughout this section, we fix a Noetherian ring A and develop the theory of primary decomposition of ideals.

a. First, we recall some definitions.

Definition. A proper ideal $\mathfrak{a}$ is said to be *reducible* if it can be written as $I \cap J$, where I and J are ideals $\neq \mathfrak{a}$. A proper ideal which is not reducible is said to be *an irreducible ideal.*

Definition. A proper ideal $\mathfrak{q}$ is said to be *primary*, if $xy \in \mathfrak{q}$ for any x, y implies $x \in \mathfrak{q}$ or $y^n \in \mathfrak{q}$ for some $n > 0$.

In this case, $\mathfrak{p} = \sqrt{\mathfrak{q}}$ is a prime ideal and $\mathfrak{q}$ is said to be $\mathfrak{p}$-*primary.*

Lemma 2.3. *Any irreducible ideal* $\mathfrak{q}$ *is a primary ideal.*

Proof. Suppose that $\mathfrak{q}$ is not primary, i.e., there exist a, b such that $ab \in \mathfrak{q}$, $a \notin \mathfrak{q}$, and $b \notin \sqrt{\mathfrak{q}}$. Letting $\mathfrak{q}_n = (\mathfrak{q}: b^n)$ $(= \{\alpha \in A \mid \alpha b^n \in \mathfrak{q}\}$, by definition) for $n = 1, 2, \ldots$, one has $\mathfrak{q}_{n+1} = (\mathfrak{q}_n: b) \supseteq \mathfrak{q}_n$ for any $n \geq 1$. Since $b \notin \sqrt{\mathfrak{q}}$, the $\mathfrak{q}_n$ are proper ideals. Since A is Noetherian, there exists N such that $\mathfrak{q}_n = \mathfrak{q}_{n+1} = \cdots$ for any $n \geq N$.

We claim that $(\mathfrak{q} + aA) \cap (\mathfrak{q} + b^N A) = \mathfrak{q}$. First, $(\mathfrak{q} + aA) \cap (\mathfrak{q} + b^N A) \supseteq \mathfrak{q}$ is obvious. Thus, take $\alpha \in (\mathfrak{q} + aA) \cap (\mathfrak{q} + b^N A)$, i.e., $\alpha = q_1 + ax_1 = q_2 + b^N x_2$ where q_1, $q_2 \in \mathfrak{q}$ and x_1, $x_2 \in A$. From $ab \in \mathfrak{q}$, it follows that $\alpha b = q_1 b + abx_1 \in \mathfrak{q}$. Furthermore, $\alpha b = q_2 b + b^{N+1} x_2$ implies that $b^{N+1} x_2 = \alpha b - q_2 b \in \mathfrak{q}$; hence $x_2 \in (\mathfrak{q}: b^{N+1})$. But since $(\mathfrak{q}: b^{N+1}) = (\mathfrak{q}: b^N)$, one has $b^N x_2 \in \mathfrak{q}$, i.e., $\alpha = q_2 + b^N x_2 \in \mathfrak{q}$. Hence, $\mathfrak{q} = (\mathfrak{q} + aA) \cap (\mathfrak{q} + b^N A)$ is reducible. □

Lemma 2.4. *Any proper ideal* $\mathfrak{a}$ *is an intersection of a finite number of primary ideals.*

Proof. Let $\mathbb{F}$ be the set of ideals for which the above assertion fails. If $\mathbb{F} \neq \varnothing$, then there exists a maximal element $\mathfrak{a}$ of $\mathbb{F}$, since A is Noetherian.

By Lemma 2.3, $\mathfrak{a}$ is reducible, i.e., $\mathfrak{a} = I \cap J$ where I and J are ideals such that $I \supset \mathfrak{a}$ and $J \supset \mathfrak{a}$. Since $I \notin F$ and $J \notin F$, one has the decompositions into primary ideals $I = \mathfrak{q}_1 \cap \cdots \cap \mathfrak{q}_r$, and $J = \mathfrak{q}'_1 \cap \cdots \cap \mathfrak{q}'_s$; hence $\mathfrak{a} = \mathfrak{q}_1 \cap \cdots \cap \mathfrak{q}_r \cap \mathfrak{q}'_1 \cap \cdots \cap \cdots \cap \mathfrak{q}'_s$. This implies $\mathfrak{a} \notin \mathbb{F}$, which contradicts the choice of $\mathfrak{a}$. □

Remark. If $\mathfrak{q}$ is a $\mathfrak{p}$-primary ideal, then $\sqrt{\mathfrak{q}} = \mathfrak{p}$ and so there exists $N > 0$ such that $\mathfrak{q} \supseteq \mathfrak{p}^N$, since $\mathfrak{p}$ has a finite base.

Lemma 2.5. *Let* $\mathfrak{q}_1$ *and* $\mathfrak{q}_2$ *be primary ideals with* $\mathfrak{p} = \sqrt{\mathfrak{q}_1} = \sqrt{\mathfrak{q}_2}$. *Then* $\mathfrak{q}_1 \cap \mathfrak{q}_2$ *is primary and* $\mathfrak{p} = \sqrt{\mathfrak{q}_1 \cap \mathfrak{q}_2}$.

PROOF. If $xy \in \mathfrak{q}_1 \cap \mathfrak{q}_2$ with $x \notin \mathfrak{p}$, then $y \in \mathfrak{q}_1$ and $y \in \mathfrak{q}_2$; hence $y \in \mathfrak{q}_1 \cap \mathfrak{q}_2$. Since $\sqrt{\mathfrak{q}_1 \cap \mathfrak{q}_2} = \sqrt{\mathfrak{q}_1} \cap \sqrt{\mathfrak{q}_2}$, one obtains the assertion. □

Thus, from Lemmas 2.3, 2.4, and 2.5, one obtains the next result.

Theorem 2.2 (Lasker's Theorem). *Any proper ideal* $\mathfrak{a}$ *of a Noetherian ring* A *can be expressed irredundantly in the form* $\mathfrak{q}_1 \cap \cdots \cap \mathfrak{q}_r$, *where the* $\mathfrak{q}_i$ *are primary ideals and* $\mathfrak{p}_1 = \sqrt{\mathfrak{q}_1}, \ldots, \mathfrak{p}_r = \sqrt{\mathfrak{q}_r}$ *are distinct prime ideals.* (This expression is called the *shortest primary decomposition of* $\mathfrak{a}$).

b. Definition. A prime ideal $\mathfrak{p}$ containing $\mathfrak{a}$ is said to be *a minimal prime ideal* of $\mathfrak{a}$, if there is no prime ideal between $\mathfrak{p}$ and $\mathfrak{a}$, except $\mathfrak{p}$.

If $\mathfrak{a} = \mathfrak{q}_1 \cap \cdots \cap \mathfrak{q}_r$ is the shortest primary decomposition and $\mathfrak{p}_i = \sqrt{\mathfrak{q}_i}$ for all i, then

$$\sqrt{\mathfrak{a}} = \mathfrak{p}_1 \cap \cdots \cap \mathfrak{p}_r.$$

Let $\mathfrak{p}_1, \ldots, \mathfrak{p}_s$ be minimal prime ideals of $\mathfrak{a}$, after a permutation of the indices, if necessary. Then every $\mathfrak{p}_j$ for $s < j \le r$ contains some $\mathfrak{p}_i$, for $1 \le i \le s$. One has the decomposition of $V(\mathfrak{a})$ into irreducible components $V(\mathfrak{p}_i)$ for $1 \le i \le s$. Thus, each $V(\mathfrak{p}_j)$ for $s < j \le r$ is embedded in $V(\mathfrak{p}_i)$ for some $1 \le i \le s$. The $\mathfrak{p}_j$ for $s < j \le r$ are called the *embedded prime ideals associated with* $\mathfrak{a}$.

Lemma 2.6. *Let* $\mathfrak{a} = \mathfrak{q}_1 \cap \cdots \cap \mathfrak{q}_r$ *be the shortest primary decomposition. Then each* $\mathfrak{p}_l = \sqrt{\mathfrak{q}_l}$ *for* $1 \le l \le r$ *is of the form* $(\mathfrak{a}: x)$ *for some* x.

PROOF. If $\mathfrak{q}_l = \mathfrak{p}_l$, then taking $x \in \bigcap_{j \neq l} \mathfrak{q}_j \backslash \mathfrak{q}_l$, one has $(\mathfrak{a}: x) = \bigcap_{j \neq l} (\mathfrak{q}_j: x) \cap (\mathfrak{q}_l: x) = \mathfrak{q}_l$. If $\mathfrak{p}_l \neq \mathfrak{q}_l$, then $\mathfrak{q}_l \supseteq \mathfrak{p}_l^{m+1}$ and $\mathfrak{q}_l \not\supseteq \mathfrak{p}_l^m$ for some m; hence, $\mathfrak{a} \not\supseteq \bigcap_{j \neq l} \mathfrak{q}_j \cap \mathfrak{p}_l^m$. Taking $x \in \bigcap_{j \neq l} \mathfrak{q}_j \cap \mathfrak{p}_l^m \backslash \mathfrak{a}$, one has $(\mathfrak{a}: x) = (\mathfrak{q}_l: x) = \mathfrak{p}_l$. □

Remark. The converse of the above assertion holds, i.e., if a prime ideal $\mathfrak{p}$ can be written as $(\mathfrak{a}: y)$ for some $y \in A$, then $\mathfrak{p}$ coincides with some $\mathfrak{p}_i$. In fact, since $\mathfrak{p} = (\mathfrak{a}: y) = \bigcap_{j=1}^r (\mathfrak{q}_j: y)$, one has $\mathfrak{p} = (\mathfrak{q}_i: y)$ for some i. Then $\mathfrak{p} = \sqrt{\mathfrak{q}_i} = \mathfrak{p}_i$.

Furthermore, the proof of the next remark is left to the reader as an exercise (see [N1, Theorem 4.7.1]).

Remark. If S is a multiplicative subset of A and $\mathfrak{a} = \mathfrak{q}_1 \cap \cdots \cap \mathfrak{q}_r$ is the shortest primary decomposition, then $\mathfrak{q}_j S^{-1}A$ is either $S^{-1}A$ or a primary ideal with $\sqrt{\mathfrak{q}_j S^{-1}A} = \sqrt{\mathfrak{q}_j}S^{-1}A$. Moreover,

$$\mathfrak{a}S^{-1}A = \bigcap_{j=1}^{r} \mathfrak{q}_j S^{-1}A.$$

In particular, take $\mathfrak{p} = \mathfrak{p}_i$ and let $S = A \backslash \mathfrak{p}$. Then

$$\mathfrak{a}A_{\mathfrak{p}} = \bigcap_{j=1}^{r} \mathfrak{q}_j A_{\mathfrak{p}} = \bigcap_{\mathfrak{q}_j \subseteq \mathfrak{p}} \mathfrak{q}_j A_{\mathfrak{p}}.$$

Since $\mathfrak{q}_j \subseteq \mathfrak{p}$ if and only if $\mathfrak{p}_j \subseteq \mathfrak{p}$, for a minimal prime ideal $\mathfrak{p}_i$ of $\mathfrak{a}$, one has $\mathfrak{a}A_{\mathfrak{p}_i} = \mathfrak{q}_i A_{\mathfrak{p}_i}$. Thus $\mathfrak{q}_i = \psi_S^{-1}(\mathfrak{q}_i A_{\mathfrak{p}}) = \psi_S^{-1}(\mathfrak{a}A_{\mathfrak{p}_i})$. Therefore, we obtain the next uniqueness theorem.

Theorem 2.3. *Let $\mathfrak{a} = \bigcap_{j=1}^{r} \mathfrak{q}_j$ be the shortest primary decomposition. If $\mathfrak{p}_i$ is minimal, then $\mathfrak{q}_i = \psi_i^{-1}(\mathfrak{a}A_{\mathfrak{p}_i})$, where $\psi_i \colon A \to A_{\mathfrak{p}_i}$ is the canonical homomorphism. In particular, the primary ideals corresponding to minimal prime ideals of $\mathfrak{a}$ depend only on $\mathfrak{a}$ (not on the particular choice of the decomposition).*

Remarks (1). $\psi_i^{-1}(\mathfrak{p}_i A_{\mathfrak{p}_i})$ coincides with $\mathfrak{p}_i$ for all i (cf. the proof of Lemma 1.3).

(2) If $\mathfrak{m}$ is a maximal ideal, then $\mathfrak{m}^n$ is $\mathfrak{m}$-primary for any $n \geq 1$. In fact, if $xy \in \mathfrak{m}^n$ and $y \notin \mathfrak{m}$, then $\mathfrak{m}^n + yA = A$, i.e., $1 = \alpha y + \beta$ for $\alpha \in A$ and $\beta \in \mathfrak{m}^n$; hence $x = \alpha xy + \beta x \in \mathfrak{m}^n$.

§2.5 Intersection Theorem and Complete Local Rings

a. Lemma 2.7. *Let $\mathfrak{a}$ and $\mathfrak{b}$ be proper ideals of a Noetherian ring A. Then there exist an ideal $\mathfrak{a}'$ and $n > 0$ such that*

$$\mathfrak{a} \cdot \mathfrak{b} = \mathfrak{a}' \cap \mathfrak{b} \quad \textit{and} \quad \mathfrak{a}' \supseteq \mathfrak{a}^n.$$

Proof. Let $\mathfrak{a} \cdot \mathfrak{b} = \mathfrak{q}_1 \cap \cdots \cap \mathfrak{q}_r$ be the shortest primary decomposition and assume that $\mathfrak{a} \subseteq \mathfrak{p}_1, \ldots, \mathfrak{a} \subseteq \mathfrak{p}_t$, $\mathfrak{a} \nsubseteq \mathfrak{p}_{t+1}, \ldots, \mathfrak{a} \nsubseteq \mathfrak{p}_r$ where $\mathfrak{p}_i = \sqrt{\mathfrak{q}_i}$ for $1 \leq i \leq r$. Letting $\mathfrak{a}' = \mathfrak{q}_1 \cap \cdots \cap \mathfrak{q}_t$ and $\mathfrak{b}' = \mathfrak{q}_{t+1} \cap \cdots \cap \mathfrak{q}_r$, one has $\mathfrak{a} \cdot \mathfrak{b} = \mathfrak{a}' \cap \mathfrak{b}' \subseteq \mathfrak{a}'$.

But since there exists $n > 0$ such that $\mathfrak{q}_i \supseteq \mathfrak{p}_i^n \supseteq \mathfrak{a}^n$ for all $1 \leq i \leq t$, it follows that $\mathfrak{a}' \supseteq \mathfrak{a}^n$. If $t + 1 \leq j \leq r$, then $\mathfrak{a} \nsubseteq \mathfrak{p}_j$ and one may take $a_j \in \mathfrak{a} \backslash \mathfrak{p}_j$. We claim that $\mathfrak{b} \subseteq \mathfrak{b}'$.

For any $x \in \mathfrak{b}$, $x \cdot a_j \in \mathfrak{a} \cdot \mathfrak{b} = \mathfrak{a}' \cap \mathfrak{b}' \subseteq \mathfrak{b}' = \bigcap_{i=t+1}^{r} \mathfrak{q}_i$. Hence, by the definition of primary ideals, $x \in \bigcap_{i=t+1}^{r} \mathfrak{q}_i = \mathfrak{b}'$; therefore, $\mathfrak{b} \subseteq \mathfrak{b}'$. Thus, $\mathfrak{a} \cdot \mathfrak{b} = \mathfrak{a}' \cap \mathfrak{b}' \supseteq \mathfrak{a}' \cap \mathfrak{b} \supseteq (\mathfrak{a} \cdot \mathfrak{b}) \cap \mathfrak{b} = \mathfrak{a} \cdot \mathfrak{b}$; hence $\mathfrak{a} \cdot \mathfrak{b} = \mathfrak{a}' \cap \mathfrak{b}' = \mathfrak{a}' \cap \mathfrak{b}$. □

Corollary.
(i) *For any n, there exists N such that*

$$\mathfrak{a}^n \cdot \mathfrak{b} \supseteq \mathfrak{a}^N \cap \mathfrak{b}.$$

(ii) *If* $I = \bigcap_{n=1}^{\infty} \mathfrak{a}^n$, *then for any n,*

$$\mathfrak{a}^n \cdot I = I.$$

PROOF. (i) follows immediately from the last lemma, and (ii) is a direct consequence of (i). □

b. To prove the intersection theorem for ideals, we need the next simple but very important result.

Lemma 2.8 (Nakayama's Lemma). *Let M be a finitely generated A-module and let* $\mathfrak{a}$ *be an ideal contained in the Jacobson radical* $\mathfrak{R}(0_A)$ *of A, i.e.,* $\mathfrak{a}$ *is contained in every maximal ideal of A. If N is an A-submodule of M with* $N + \mathfrak{a}M = M$, *then* $N = M$.

PROOF. Replacing M by M/N, we can assume that $N = 0$. There exist generators $x_1, \ldots, x_n$ of M, i.e., $M = \sum_{j=1}^{n} Ax_j$. By assumption, $M = \mathfrak{a}M = \sum_{j=1}^{n} \mathfrak{a}x_j$, and so each x_i can be written as $\sum_{j=1}^{n} r_{ij}x_j$ with $r_{ij} \in \mathfrak{a}$. By making use of matrix notation, one has $[\delta_{ij} - r_{ij}][x_j] = 0$, where δ_{ij} is the Kronecker δ. Hence, let $\eta = \det[\delta_{ij} - r_{ij}] - 1$, which belongs to $\mathfrak{a}$ and satisfies $(1 + \eta)x_j = \det[\delta_{ij} - r_{ij}]\, x_j = 0$ for all j.

If $1 + \eta$ were not a unit, then $1 + \eta$ would be contained in a maximal ideal $\mathfrak{m}$ of A by Lemma 1.1; hence $-1 = \eta - (1 + \eta) \in \mathfrak{m}$, which would imply $A = 0$. Thus $1 + \eta$ is a unit and so $x_1 = \cdots = x_n = 0$. □

Corollary. *If* $\mathfrak{a}$ *is a proper ideal of a Noetherian local ring such that* $\mathfrak{a} = \mathfrak{a}^2$, *then* $\mathfrak{a} = 0$.

Theorem 2.4 (Intersection theorem of Krull). *Let* $\mathfrak{a}$ *be an ideal contained in the Jacobson radical of a Noetherian ring A. Then* $\bigcap_{n=1}^{\infty} \mathfrak{a}^n = 0$. *Furthermore, for any ideal* $\mathfrak{b}$, $\bigcap_{n=1}^{\infty} (\mathfrak{a}^n + \mathfrak{b}) = \mathfrak{b}$.

PROOF. Letting $I = \bigcap_{n=1}^{\infty} \mathfrak{a}^n$, one has $\mathfrak{a} \cdot I = I$ by Corollary (ii) to Lemma 2.7. By the last lemma, one obtains $I = 0$.

The general case can be reduced to the first case by considering $A/\mathfrak{b}$. □

EXAMPLE 2.7. Let $\bar{A}$ be a ring of germs of C^{∞} functions in x. Then $\bar{A}$ is not Noetherian and the maximal ideal $\mathfrak{m}$ of $\bar{A}$ is $x\bar{A}$. Since there exist many nonzero elements $\varphi \in \mathfrak{m}$ such that the j-th derivatives of φ vanish at 0 for all $j \geq 0$, one sees $\bigcap_{n=1}^{\infty} \mathfrak{m}^n \neq 0$.

c. In this subsection, let R denote a Noetherian local ring and let $\mathfrak{m}$ be its maximal ideal. Let $R_m = R/\mathfrak{m}^{m+1}$ and let $p_m\colon R_m \to R_{m-1}$ be the natural homomorphism for any $m \geq 1$. Then one has the inverse limit (or, the projective limit) projlim R_m, denoted $\hat{R}$. By definition, any element ξ of $\hat{R}$ corresponds to $\{\xi_m\}_{m \geq 0}$ with $\xi_m \in R_m$ and $p_m(\xi_m) = \xi_{m-1}$ for all $m \geq 0$. For any $a \in R$, one has $i(a) = \{a \bmod \mathfrak{m}^m\}_{m \geq 0} \in \hat{R}$. Clearly, $\hat{R}$ is a ring and $i\colon R \to \hat{R}$ is a ring homomorphism. If $a \in \operatorname{Ker} i$, then $a \in \bigcap_{m=1}^{\infty} \mathfrak{m}^m$, which is zero by Theorem 2.4. Thus i is one-to-one.

The proof of the following lemma is left to the reader.

Lemma 2.9. *The formal power series ring* $R[[\mathsf{X}_1, \ldots, \mathsf{X}_n]]$ *is a Noetherian local ring, whose maximal ideal* $\mathfrak{M}$ *is generated by* $\mathfrak{m}, \mathsf{X}_1, \ldots, \mathsf{X}_n$. *Hence* $R[[\mathsf{X}_1, \ldots, \mathsf{X}_n]]/\mathfrak{M} = R/\mathfrak{m}$.

PROOF. We refer to [$N2$, Theorem 15.3]. □

Corollary. *R is a Noetherian local ring whose maximal ideal is* $\mathfrak{m}\hat{R}$.

PROOF. Letting $x_1, \ldots, x_n$ be a basis of $\mathfrak{m}$, one has a ring homomorphism $\varphi\colon R[[\mathsf{X}_1, \ldots, \mathsf{X}_n]] \to \hat{R}$ by setting $\varphi(f(\mathsf{X}_1, \ldots, \mathsf{X}_n)) = f(x_1, \ldots, x_n)$, which is defined as follows: let $f_{(m)}$ be the m-th homogeneous part of f with respect to $\mathsf{X}_1, \ldots, \mathsf{X}_n$. Then setting $\xi_m = f_{(0)}(x_1, \ldots, x_n) + \cdots + f_{(m)}(x_1, \ldots, x_n) \bmod \mathfrak{m}^{m+1}$, one has $p_m(\xi_m) = \xi_{m-1}$ for all $m \geq 1$. Thus, we can define $f(x_1, \ldots, x_n)$ to be $\{\xi_m\}_{m \geq 0} \in \hat{R}$. We claim that φ is surjective. Given any $\eta = \{\eta_m\}_{m \geq 0}$, we choose $r_m \in R$ such that $r_m \bmod \mathfrak{m}^{m+1} = \eta_m$ for all m. Since $r_m - r_{m-1} \in \mathfrak{m}^m$ for all $m \geq 1$, there is a homogeneous polynomial $\psi_m(\mathsf{X}_1, \ldots, \mathsf{X}_n)$ of degree m such that $r_m - r_{m-1} = \psi_m(x_1, \ldots, x_n)$. Putting $\psi_0 = r_1$, define h to be $\sum_{m=0}^{\infty} \psi_m(\mathsf{X}_1, \ldots, \mathsf{X}_n) \in R[[\mathsf{X}_1, \ldots, \mathsf{X}_n]]$. Then clearly, $\varphi(h) = \eta$.

Since $\hat{R}$ is a homomorphic image of the Noetherian local ring $R[[\mathsf{X}_1, \ldots, \mathsf{X}_n]]$ by Lemma 2.9, $\hat{R}$ is a Noetherian local ring by the Hilbert basis theorem, whose maximal ideal is $\varphi(\mathfrak{M})\hat{R} = \mathfrak{m}\hat{R}$. □

Lemma 2.10. *The induced homomorphism* $i_m\colon R_m \to \hat{R}_m = \hat{R}/\mathfrak{m}^{m+1}\hat{R}$ *is an isomorphism for all m.*

PROOF. The surjectivity is clear. In fact, given any $\eta \in \hat{R}$, one has $\eta \equiv \sum_{j=0}^{m} \psi_j(x_1, \ldots, x_n) \bmod \mathfrak{m}^{m+1}$ by using the above notation. To show the injectivity, we use induction on m. Letting $k = R/\mathfrak{m}$, one has $k = \hat{R}/\mathfrak{m}\hat{R}$ by Lemma 2.9. Assuming i_{m-1} is an isomorphism, let I_m be $\operatorname{Ker} i_m$. Then $I_m \subseteq \mathfrak{m}^m/\mathfrak{m}^{m+1}$, and hence I_m is the kernel of the induced homomorphism $i_m^*\colon \mathfrak{m}^m/\mathfrak{m}^{m+1} \to \mathfrak{m}^m\hat{R}/\mathfrak{m}^{m+1}\hat{R}$. But if $\mathfrak{m}^m/\mathfrak{m}^{m+1} \cong (R/\mathfrak{m})^u$ (i.e., a vector space of dimension u), then $\mathfrak{m}^m\hat{R}/\mathfrak{m}^{m+1}\hat{R} \cong \mathfrak{m}^m/\mathfrak{m}^{m+1} \otimes_R \hat{R} \cong (\hat{R}/\mathfrak{m}\hat{R})^u = (R/\mathfrak{m})^u$. Since the vector spaces have the same dimensions, i_m^* is an isomorphism; hence $I_m = 0$. □

Definition. If $i: R \to \hat{R}$ is an isomorphism, R is said to be a *complete local ring*. In general, $\hat{R}$ is called the *completion* of R.

By the Corollary to Lemma 2.9 and Lemma 2.10, $\hat{R}$ is a Noetherian complete local ring.

EXAMPLE 2.8. Let $k[X_1, \ldots, X_n]$ be a polynomial ring and $\mathfrak{m}$ be a maximal ideal $(X_1 - a_1, \ldots, X_n - a_n)$. Then the completion of $k[X_1, \ldots, X_n]_{\mathfrak{m}}$ is canonically isomorphic to $k[[X_1 - a_1, \ldots, X_n - a_n]]$.

EXAMPLE 2.9. Let $R = \mathbb{C}[X_1, X_2]/(X_1 X_2 + X_1^3 + X_2^3)$. Then R is an integral domain. Letting $\mathfrak{m} = X_1 R + X_2 R$ and $x_i = X_i \bmod(X_1 X_2 + X_1^3 + X_2^3)$ for $i = 1, 2$, one has the completion $\hat{R}_{\mathfrak{m}} = \mathbb{C}[[x_1, x_2]]$ with $x_1 x_2 + x_1^3 + x_2^3 = 0$, which is not an integral domain. In fact, $(x_1 + x_2^2 + \frac{1}{2}x_1^2 x_2 + \cdots)(x_2 + x_1^2 - \frac{3}{2}x_1 x_2^2 + \cdots) = 0$.

Thus, the completion of an integral domain may have zero divisors.

d. To state an important result about $\hat{R}$, we make the following definitions. Let R be a Noetherian local ring with maximal ideal $\mathfrak{m}$. $\mathfrak{m}^j/\mathfrak{m}^{j+1}$ is an $R/\mathfrak{m}$-vector space, whose dimension is denoted by $\sigma_j(R)$ for $j \geq 0$.

Definition. Let $H_R(m) = \sum_{j=0}^{m} \sigma_j(R)$. $H_R(m)$ is said to be the *Samuel function* of R.

Theorem 2.5. *There exists an m_1 such that $H_R(m)$ is a polynomial of degree* dim R *for all $m \geq m_1$.*

PROOF. We refer the reader to [A–K, III, Theorem 1.4]. A geometric proof will be given in §7.6.**d**. □

Definition. The polynomial defined in Theorem 2.5 is said to be the *characteristic polynomial* of R, and is denoted by $\sigma_R(m)$.

Then we have $\sigma_R(m) = (a_0/n!)m^n$ + terms of lower degree, with a_0 a positive integer, where $n = \dim R$.

Definition. a_0 is said to be the *multiplicity* of R, denoted by $e(R)$.

By Lemma 2.10, one obtains the next result.

Proposition 2.9. $H_{\hat{R}}(m) = H_R(m)$, $\sigma_{\hat{R}}(m) = \sigma_R(m)$, $\dim \hat{R} = \dim R$, *and* $e(\hat{R}) = e(R)$.

Corollary. *The formal power series ring $k[[X_1, \ldots, X_n]]$ has dimension n.*

We give a direct consequence of Lemma 2.8 (Nakayama's Lemma).

Lemma 2.11. *Let R be a Noetherian local ring with maximal ideal $\mathfrak{m}$. If $x_1, \ldots, x_n \in \mathfrak{m}$ are such that $x_1 \bmod \mathfrak{m}^2, \ldots, x_n \bmod \mathfrak{m}^2$ span $\mathfrak{m}/\mathfrak{m}^2$ as an $R/\mathfrak{m}$-vector space, then $\mathfrak{m}$ is generated by $x_1, \ldots, x_n$.*

PROOF. By hypothesis, $\sum_{i=1}^n Rx_i + \mathfrak{m}^2 = \mathfrak{m}$. Since $\mathfrak{m}$ has a finite basis and since $\mathfrak{R}(0_{\mathfrak{R}}) = \mathfrak{m}$, by Lemma 2.8 one obtains $\mathfrak{m} = \sum_{i=1}^n Rx_i$. □

Theorem 2.6 (Quasi-finiteness theorem). *Let A be a subring of B. Suppose that A and B are Noetherian local rings with maximal ideals $\mathfrak{m}$ and $\mathfrak{n}$, respectively, such that $\mathfrak{m}B \subseteq \mathfrak{n}$. If there exist $x_1, \ldots, x_n \in B$ such that $B/\mathfrak{m}B \cong B \otimes_A (A/\mathfrak{m}) = \sum_{i=1}^n A/\mathfrak{m}(x_i \otimes_A 1)$, and if A is complete, then $B = \sum_{i=1}^n Ax_i$.*

PROOF. By hypothesis, $B = \mathfrak{m}B + \sum_{i=1}^n Ax_i$. If $y_1, \ldots, y_s$ form a basis of $\mathfrak{m}$, then any $\beta \in B$ can be written in the form $\beta = \sum_{i=1}^n a_i x_i + \sum_{j=1}^s \beta_j y_j$, for $a_i \in A$, $\beta_j \in B$. Furthermore, we have

$$\beta_j = \sum_{i=1}^n a_{ji} x_i + \sum_{l=1}^s \beta_{jl} y_l,$$

where $a_{ji} \in A$, $\beta_{jl} \in B$.

Hence letting $a_i^{(0)} = a_i$, $a_i^{(1)} = \sum_{j=1}^s a_{ji} y_j \in \mathfrak{m}$, one has

$$\beta = \sum_{i=1}^n (a_i^{(0)} + a_i^{(1)})x_i + \sum_{j,l} y_j y_l \beta_{jl}.$$

Repeating this argument, we have $a_i^{(p)} \in \mathfrak{m}^p$ for any $p \geq 0$ such that $\beta \equiv \sum_{i=1}^n (\sum_{p=0}^{t-1} a_i^{(p)})x_i \bmod \mathfrak{m}^t B$ for any $t \geq 1$. Since A is complete, $\alpha_i = \sum_{p=0}^{\infty} a_i^{(p)}$ exist in A for all i which satisfy

$$\beta - \sum_{i=1}^n \alpha_i x_i \in \bigcap_{t=1}^{\infty} \mathfrak{m}^t B \subseteq \bigcap_{t=1}^{\infty} \mathfrak{n}^t.$$

By Theorem 2.4, $\bigcap_{t=1}^{\infty} \mathfrak{n}^t = 0$, and hence one obtains the assertion. □

Remark. The result asserts that B is finitely generated as an A-module. Thus, Theorem 2.6 can be restated as follows: If $({}^a i)^{-1}(\mathfrak{m})$ is a finite scheme over $A/\mathfrak{m}$ and A is complete, then ${}^a i$ is a finite morphism, where i is an inclusion $A \subset B$.

Corollary 1. *Let R be a Noetherian complete local ring with $n = \dim R$. Suppose that the maximal ideal $\mathfrak{m}$ is generated by n elements $x_1, \ldots, x_n$ and that R contains a field k isomorphic to $R/\mathfrak{m}$. Then R is isomorphic to the formal power series ring $k[[\mathsf{X}_1, \ldots, \mathsf{X}_n]]$.*

PROOF. The subring $R_1 = k[[x_1, \ldots, x_n]]$ of R is clearly complete and $\mathfrak{m}_1 = \sum_{i=1}^n x_i R_1$ is the maximal ideal; hence $R \otimes_{R_1} (R_1/\mathfrak{m}_1) \cong R/\mathfrak{m} = k$. By the last theorem, $R_1 = R$. Define a k-homomorphism $\Psi\colon k[[\mathsf{X}_1, \dots, \mathsf{X}_n]] \to R_1$ by $\Psi(\mathsf{X}_i) = x_i$ for all i. Then $k[[\mathsf{X}_1, \dots, \mathsf{X}_n]]/\operatorname{Ker}\Psi \cong R_1 = R$. Noting that $\dim k[[\mathsf{X}_1, \dots, \mathsf{X}_n]] = n$ by the Corollary to Proposition 2.9 and $\dim R = n$, one has $\operatorname{Ker}\Psi = \{0\}$, i.e., Ψ is an isomorphism. □

Corollary 2. *Let B be a formal power series ring $k[[\mathsf{X}_1, \dots, \mathsf{X}_n]]$. If $Y_1, \dots, Y_n \in B$, $Y_j(0) = 0$ for all j, and*

$$\det\left[\frac{\partial Y_i}{\partial \mathsf{X}_j}(0)\right]_{0 \le i,\, j \le n} \neq 0,$$

then

$$k[[Y_1, \dots, Y_n]] = k[[\mathsf{X}_1, \dots, \mathsf{X}_n]],$$

i.e., there exist $G_1, \dots, G_n \in k[[Y_1, \dots, Y_n]]$ such that $G_j(0) = 0$ and $G_j(Y_1(\mathsf{X}_1, \dots, \mathsf{X}_n), \dots, Y_n(\mathsf{X}_1, \dots, \mathsf{X}_n)) = \mathsf{X}_j$ for all j.

Proof. Let $A = k[[Y_1, \dots, Y_n]]$, which is a subring of B, and let $\mathfrak{m} = \sum_{j=1}^n Y_j A$ and $\mathfrak{n} = \sum_{j=1}^n \mathsf{X}_j B$. Then $(\mathfrak{m}B + \mathfrak{n}^2)/\mathfrak{n}^2 = \mathfrak{n}/\mathfrak{n}^2$ by assumption. Hence, $\mathfrak{m}B + \mathfrak{n}^2 = \mathfrak{n}$ and so $\mathfrak{n} = \mathfrak{m}B$ by Lemma 2.8. Noting that $B \otimes_A (A/\mathfrak{m}) = B/\mathfrak{n} \cong k$, we have $B = A$ by Theorem 2.6. □

Corollary 3. *Let $f \in B = k[[\mathsf{X}_1, \dots, \mathsf{X}_n]]$ with $f(0, \dots, 0, \mathsf{X}_n) = a_1 \mathsf{X}_n^s + a_2 \mathsf{X}_n^{s+1} + \cdots$ such that $a_1 \neq 0$, $s > 0$. Then any $g \in B$ can be written in the form $\sum_{j=0}^{s-1} F_j(\mathsf{X}_1, \dots, \mathsf{X}_{n-1})\mathsf{X}_n^j + G(\mathsf{X}_1, \dots, \mathsf{X}_n) \cdot f(\mathsf{X}_1, \dots, \mathsf{X}_n)$, where the $F_j \in k[[\mathsf{X}_1, \dots, \mathsf{X}_{n-1}]]$ and $G \in B$.*

Proof. Let $A = k[[\mathsf{X}_1, \dots, \mathsf{X}_{n-1}]] \subseteq B$ and $\mathfrak{m} = \sum_{j=1}^{n-1} \mathsf{X}_j A$. Then, A is a subring of $\bar{B} = B/fB$ and

$$\bar{B} \otimes_A (A/\mathfrak{m}) \cong k[[\mathsf{X}_n]]/(f(0, \dots, 0, \mathsf{X}_n)).$$

Hence by Theorem 2.6, one has

$$\bar{B} = \sum_{j=0}^{s-1} A \cdot \mathsf{X}_n^j,$$

which implies the assertion. □

Corollary 4 (Weierstrass' preparation theorem). *Under the same assumptions as in Corollary 3, there exists a unit $\varepsilon \in B$ such that*

$$\varepsilon \cdot f = \mathsf{X}_n^s + h_1(\mathsf{X}_1, \dots, \mathsf{X}_{n-1})\mathsf{X}_n^{s-1} + \cdots + h_s(\mathsf{X}_1, \dots, \mathsf{X}_{n-1}),$$

where the $h_j \in B$ and $h_j(0, \dots, 0) = 0$ for $1 \le j \le s$.

Proof. Applying Corollary 3 for $g = \mathsf{X}_n^s$, one obtains

$$\mathsf{X}_n^s = \sum_{j=0}^{s-1} F_j(\mathsf{X}_1, \dots, \mathsf{X}_{n-1})\mathsf{X}_n^j + G(\mathsf{X}_1, \dots, \mathsf{X}_n) f(\mathsf{X}_1, \dots, \mathsf{X}_n).$$

From this,

$$\mathsf{X}_n^s = \sum_{j=0}^{s-1} F_j(0, \ldots, 0)\mathsf{X}_n^j + G(0, \ldots, 0, \mathsf{X}_n)(a_1 \mathsf{X}_n^s + a_2 \mathsf{X}_n^{s+1} + \cdots).$$

Thus $F_j(0, \ldots, 0) = 0$ and $G(0, \ldots, 0, \mathsf{X}_n) = b_1 + b_2 \mathsf{X}_n + \cdots$ with $b_1 \neq 0$, i.e., G is a unit. Letting $\varepsilon = G$ and $h_j = -F_j$, we obtain the desired form. □

§2.6 Regular Local Rings

a. We start by proving the next result due to Krull.

Theorem 2.7. *Let R be a Noetherian integral domain and f be a nonzero non-unit element. Then every minimal prime ideal of fR is of height* 1; *hence,* $\mathrm{ht}(fR) = 1$.

PROOF. It suffices to show that $\dim R_{\mathfrak{m}} = 1$ for any minimal prime ideal $\mathfrak{m}$ of fR. Replacing $R_{\mathfrak{m}}$ by R, we can assume that R is local and $\mathfrak{m}$ is its maximal ideal. We shall prove that any prime ideal $\mathfrak{p}$ is (0), if $\mathfrak{p} \subset \mathfrak{m}$. By Theorem 1.10, R/fR is an Artinian local ring. For any $n > 0$, $\mathfrak{p}^n R_{\mathfrak{p}}$ is $\mathfrak{p}R_{\mathfrak{p}}$-primary (by the remark in §2.4.**b**.) and so $\mathfrak{p}^{(n)} = R \cap \mathfrak{p}^n R_{\mathfrak{p}}$ is $\mathfrak{p}$-primary. One has a sequence of ideals

$$\cdots \supseteq \mathfrak{p}^{(n)} + fR \supseteq \mathfrak{p}^{(n+1)} + fR \supseteq \cdots.$$

Since R/fR is Artinian, there exists m such that $\mathfrak{p}^{(n+1)} + fR = \mathfrak{p}^{(n)} + fR$ for all $n \geq m$. Then, one has $f \cdot (\mathfrak{p}^{(n)} : fR) + \mathfrak{p}^{(n+1)} = \mathfrak{p}^{(n)}$ for all $n \geq m$. On the other hand, $\mathfrak{p}^{(n)}$ is $\mathfrak{p}$-primary and $f \notin \mathfrak{p}$; hence $(\mathfrak{p}^{(n)} : fR) = \mathfrak{p}^{(n)}$. Thus $f \cdot \mathfrak{p}^{(n)} + \mathfrak{p}^{(n+1)} = \mathfrak{p}^{(n)}$. By Lemma 2.8, one has $\mathfrak{p}^{(n+1)} = \mathfrak{p}^{(n)}$ for all $n \geq m$. Applying Theorem 2.4, $\bigcap_{n \geq m} \mathfrak{p}^{(n)} = R \cap (\bigcap_{n \geq m} \mathfrak{p}^n R_{\mathfrak{p}}) = 0$ is obtained, and so $\mathfrak{p}^{(m)} = \bigcap_{n \geq m} \mathfrak{p}^{(n)} = 0$. Any $x \in \mathfrak{p}$ satisfies $x^m \in \mathfrak{p}^{(m)} = 0$; hence $x = 0$. Thus $\mathfrak{p} = 0$. □

Theorem 2.8. *Let R be a Noetherian ring and let $f_1, \ldots, f_s \in R$. If $\mathfrak{a} = f_1 R + \cdots + f_s R$ is a proper ideal, then* $\mathrm{ht}(\mathfrak{a}) \leq s$.

PROOF. Taking a minimal prime ideal $\mathfrak{m}$ of $\mathfrak{a}$, and replacing $R_{\mathfrak{m}}$ by R, we may assume that R is a local ring, where the maximal ideal $\mathfrak{m}$ is a minimal prime ideal of $\mathfrak{a}$. We shall prove that $\mathrm{ht}(\mathfrak{m}) = \dim R \leq s$ by induction on s. We suppose that $\mathrm{ht}(\mathfrak{a}) > s$ and $s \geq 1$, i.e., there exists a prime ideal $\mathfrak{p} \subset \mathfrak{m}$ such that there are no prime ideals between $\mathfrak{p}$ and $\mathfrak{m}$. Since $\mathfrak{a} \not\subseteq \mathfrak{p}$, there exists some f_i that is not contained in $\mathfrak{p}$, say $f_s \notin \mathfrak{p}$. The ideal $\mathfrak{p} + f_s R$ has the property that $V(\mathfrak{p} + f_s R) = \{\mathfrak{m}\}$; hence $\sqrt{\mathfrak{p} + f_s R} = \mathfrak{m}$. Thus, each $f_i \in \sqrt{\mathfrak{p} + f_s R}$, i.e., $f_i^m = c_i + f_s x_i$ for some $m > 0$, $c_i \in \mathfrak{p}$, $x_i \in R$, where m

can be chosen independently of i. We claim that $\mathfrak{p}$ is a minimal prime ideal of $\mathfrak{a}' = c_1R + \cdots + c_{s-1}R$. In fact, if $\mathfrak{q}$ is a prime ideal such that $\mathfrak{a}' \subseteq \mathfrak{q} \subseteq \mathfrak{p}$, then applying Theorem 2.7 to $R/\mathfrak{q}$ and f_s mod $\mathfrak{q}$, one has $\mathfrak{q} = \mathfrak{p}$.

Therefore, by the induction hypothesis, $\mathrm{ht}(\mathfrak{p}) \leq s - 1$ and so $\mathrm{ht}(\mathfrak{m}) \leq s$ is obtained. □

b. Using this result, we can prove the next proposition.

Proposition 2.10. *Let R be a Noetherian local ring and $x_1, \ldots, x_n$ be elements of the maximal ideal $\mathfrak{m}$. Then* $\mathrm{ht}(\sum_{i=1}^{n} Rx_i) \leq n$. *In particular, if $\mathfrak{m}$ is generated by n elements, then* $\dim R \leq n$.

Letting $K = R/\mathfrak{m}$, we consider $\mathfrak{m}/\mathfrak{m}^2$ as a vector space over K. Recalling Lemma 2.11, one has $\dim_K(\mathfrak{m}/\mathfrak{m}^2)$ elements which generate $\mathfrak{m}$. Thus by the last proposition, $\dim R \leq \dim_K(\mathfrak{m}/\mathfrak{m}^2)$.

Definition. The dual vector space $\mathrm{Hom}_K(\mathfrak{m}/\mathfrak{m}^2, K)$ is said to be the *Zariski tangent space* of R, and is denoted by $T_{\mathfrak{m}}(R)$.

c. Definition. If a Noetherian local ring R satisfies $\dim R = \dim_K T_{\mathfrak{m}}(R)$, then R is called a *regular local ring*.

Then, there exist n generators of $\mathfrak{m}$, n being $\dim R$.

Definition. A system of generators $(x_1, \ldots, x_n)$ of $\mathfrak{m}$ is said to be a *regular system of parameters* of the regular local ring R of dimension n.

EXAMPLE 2.10. Let $\mathfrak{p}$ be a prime ideal of $k[\mathsf{X}_1, \ldots, \mathsf{X}_N]$ and let V be $V(\mathfrak{p}) \cong \mathrm{Spec}\, k[\mathsf{X}_1, \ldots, \mathsf{X}_N]/\mathfrak{p}$. Suppose that $n = \dim V$ and $\mathfrak{p} \subseteq (\mathsf{X}_1, \ldots, \mathsf{X}_N)$, denoted 0, i.e., $0 \in V$. If $\mathcal{O}_{V,0}$ is a regular local ring, then there exist $f_1, \ldots, f_{N-n} \in \mathfrak{p}$ and $\mathsf{X}_1, \ldots, \mathsf{X}_{N-n}$, after renumbering the X_i, such that $[(\partial f_i/\partial \mathsf{X}_j)(0)]_{1 \leq i, j \leq N-n}$ is a nonsingular matrix, i.e., $\det[(\partial f_i/\partial \mathsf{X}_j)(0)] \neq 0$. Then $z_1 = \mathsf{X}_{N-n+1}$ mod $\mathfrak{p}, \ldots, z_n = \mathsf{X}_N$ mod $\mathfrak{p}$ form a regular system of parameters.

In order to verify this, letting M be the maximal ideal of $\mathcal{O}_{A^N,0}$ and $\mathfrak{m}$ be the maximal ideal of $\mathcal{O}_{V,0}$, one has

$$\mathfrak{m}/\mathfrak{m}^2 = M/(M^2 + \mathfrak{p}_1) \cong \left(\sum_{i=1}^{N} k\xi_i\right)\Big/ I,$$

where $\mathfrak{p}_1 = \mathfrak{p}\mathcal{O}_{A^N,0}$, $\xi_i = \mathsf{X}_i$ mod M^2, and I is the subspace spanned by $\sum_{j=1}^{N} (\partial f/\partial \mathsf{X}_j)(0)\xi_j$, for all $f \in \mathfrak{p}$.

Thus, $\dim_k T_{\mathfrak{m}}(\mathcal{O}_{V,0}) = \dim_k((\sum_{i=1}^{N} k\xi_i)/I) = N - \dim_k I$. Hence, $n = \dim_k T_{\mathfrak{m}}(\mathcal{O}_{V,0})$ if and only if $\dim_k I = N - n$, i.e., if and only if there exist

$f_1, \ldots, f_{N-n}$ such that $\det[(\partial f_i/\partial X_j)(0)]_{1 \le i, j \le N-n} \neq 0$, after a permutation of the indices j.

d. Let R be a Noetherian local ring with maximal ideal $\mathfrak{m}$. For any $f \in R \setminus \{0\}$, there exists ν such that $f \in \mathfrak{m}^\nu$ but $f \notin \mathfrak{m}^{\nu+1}$, since $\bigcap_{m>0} \mathfrak{m}^m = 0$ by Theorem 2.4.

Definition. The ν defined above is denoted by $\nu_R(f)$. In this case, f mod $\mathfrak{m}^{\nu+1}$ is a nonzero element of $\mathfrak{m}^\nu/\mathfrak{m}^{\nu+1}$, denoted by $\mathrm{in}_\mathfrak{m}(f)$. One says that $\mathrm{in}_\mathfrak{m}(f)$ is the *initial form* of f with respect to $\mathfrak{m}$.

If $x \in \mathfrak{m}^n$ and $y \in \mathfrak{m}^m$, then define $(x \bmod \mathfrak{m}^{n+1}) \cdot (y \bmod \mathfrak{m}^{m+1})$ to be xy mod $\mathfrak{m}^{n+m+1}$. Under this multiplication, $\bigoplus_{n=0}^\infty \mathfrak{m}^n/\mathfrak{m}^{n+1}$ becomes a graded ring, which is an $R/\mathfrak{m}$-algebra.

Definition. $\bigoplus_{n=0}^\infty \mathfrak{m}^n/\mathfrak{m}^{n+1}$ is said to be the *graded ring associated with* $(R, \mathfrak{m})$, denoted by $\mathrm{gr}_\mathfrak{m}^{\cdot}(R) = \bigoplus_{n=0}^\infty \mathrm{gr}_\mathfrak{m}^n(R)$, where $\mathrm{gr}_\mathfrak{m}^n(R) = \mathfrak{m}^n/\mathfrak{m}^{n+1}$.

Lemma 2.12. $\mathrm{gr}_\mathfrak{m}^{\cdot}(R)$ *is a finitely generated* $R/\mathfrak{m}$*-algebra, and hence is Noetherian.*

PROOF. Since R is Noetherian, $\mathfrak{m}$ is generated by $x_1, \ldots, x_n$ satisfying $x_j \notin \mathfrak{m}^2$ for all j. Then $\mathrm{gr}_\mathfrak{m}^n(R)$ is an $R/\mathfrak{m}$-vector space spanned by $\prod_{i=1}^m \mathrm{in}_\mathfrak{m}(x_i)^{\alpha_i}$ where the $\alpha_i \ge 0$ with $\sum_{i=1}^m \alpha_i = n$. In fact, any $\xi \in \mathfrak{m}^n \setminus \mathfrak{m}^{n+1}$ can be written as $\sum_\alpha a_\alpha x_1^{\alpha_1} \cdots x_m^{\alpha_m}$, where $\alpha = (\alpha_1, \ldots, \alpha_m)$, $\sum_{i=1}^m \alpha_i = n$, and $a_\alpha \in R$. One may assume that $\nu_R(a_\alpha x_1^{\alpha_1} \cdots x_m^{\alpha_m}) = n$ and then

$$\mathrm{in}_\mathfrak{m}(\xi) = \sum_\alpha \mathrm{in}_\mathfrak{m}(a_\alpha) \prod_{i=1}^m \mathrm{in}_\mathfrak{m}(x_i)^{\alpha_i}.$$

Therefore, $\mathrm{gr}_\mathfrak{m}^{\cdot}(R) = R/\mathfrak{m}\,[\mathrm{in}_\mathfrak{m}(x_1), \ldots, \mathrm{in}_\mathfrak{m}(x_m)]$. □

Suppose that $\mathrm{gr}_\mathfrak{m}^{\cdot}(R)$ is an integral domain. For $x \neq 0$, $y \neq 0$ of R, one has $\mathrm{in}_\mathfrak{m}(x) \cdot \mathrm{in}_\mathfrak{m}(y) \neq 0$; hence $\nu_R(xy) = \nu_R(x) + \nu_R(y)$. Thus $\mathrm{in}_\mathfrak{m}(xy) = \mathrm{in}_\mathfrak{m}(x) \cdot \mathrm{in}_\mathfrak{m}(y)$. In particular, $xy \neq 0$. Hence, one obtains the next result.

Lemma 2.13. *If* $\mathrm{gr}_\mathfrak{m}^{\cdot}(R)$ *is an integral domain, then so is* R. *In this case,* $\mathrm{in}_\mathfrak{m}(xy) = \mathrm{in}_\mathfrak{m}(x) \cdot \mathrm{in}_\mathfrak{m}(y)$, *if* $xy \neq 0$.

e. Lemma 2.14. *Furthermore, if* $\mathrm{gr}_\mathfrak{m}^{\cdot}(R)$ *is a normal ring, then so is* R.

PROOF. By the last lemma, R is an integral domain. Take a nonzero α of $Q(R)$ which is integral over R; $R[\alpha]$ is a finitely generated R-module by Proposition 1.9. Thus, there exists a nonzero c of R such that $cR[\alpha] \subseteq R$, i.e., $c\alpha^n \in R$ for any $n \ge 1$. Letting $b = c\alpha \in R$, we have $b^n/c^{n-1} = c\alpha^n \in R$. Letting $\bar{b} = \mathrm{in}_\mathfrak{m}(b)$ and $\bar{c} = \mathrm{in}_\mathfrak{m}(c)$, we have $(\bar{b}/\bar{c})^n = (\bar{b}^n/\bar{c}^{n-1}) \cdot 1/\bar{c} =$

$\mathrm{in}_{\mathfrak{m}}(b^n/c^{n-1}) \cdot 1/\bar{c} = \mathrm{in}_{\mathfrak{m}}(c\alpha^n) \cdot 1/\bar{c} \in \mathrm{gr}^{\cdot}_{\mathfrak{m}}(R) \cdot 1/\bar{c} \subseteq Q(\mathrm{gr}^{\cdot}_{\mathfrak{m}}(R))$ for any $n \geq 0$. Hence, $\mathrm{gr}^{\cdot}_{\mathfrak{m}}(R)[\bar{b}/\bar{c}] \subseteq \mathrm{gr}^{\cdot}_{\mathfrak{m}}(R) \cdot 1/\bar{c}$, which is finitely generated as a $\mathrm{gr}^{\cdot}_{\mathfrak{m}}(R)$-module, since $\mathrm{gr}^{\cdot}_{\mathfrak{m}}(R)$ is a Noetherian ring by Lemma 2.12. Hence, $\bar{b}/\bar{c}$ is integral over $\mathrm{gr}^{\cdot}_{\mathfrak{m}}(R)$ by Proposition 1.9. Then by assumption, $\bar{b}/\bar{c} \in \mathrm{gr}^{\cdot}_{\mathfrak{m}}(R)$, i.e., there exists $a \in R$ such that $\bar{b}/\bar{c} = \mathrm{in}_{\mathfrak{m}}(a)$. From this, it follows that $v_R(b - ca) > v_R(b)$.

Furthermore, $(b - ca)/c = \alpha - a$ is integral over R. Repeating the above procedure, there exist $a', a'', \ldots, a^{(s)} \in R$ for any s such that

$$v_R(b-ca-ca' - \cdots - ca^{(s)}) > v_R(b-ca - \cdots - ca^{(s-1)}) > \cdots > v_R(b).$$

Hence, $b \in cR + \mathfrak{m}^N$ for any $N \geq 1$, i.e.,

$$b \in \bigcap_{N \geq 1} (cR + \mathfrak{m}^N)$$

But by Theorem 2.4, the right-hand side is cR; hence $\alpha = b/c \in R$. □

f. Theorem 2.9. *Let R be a regular local ring with maximal ideal $\mathfrak{m}$. If $n = \dim R$ and $K = R/\mathfrak{m}$, then $\mathrm{gr}^{\cdot}_{\mathfrak{m}}(R) \cong K[\mathsf{X}_1, \ldots, \mathsf{X}_n]$. Hence, $H_R(m) = (m + n) \cdots (m + 1)/n! = m^n/n! + \cdots$, so $e(R) = 1$.*

Proof. Suppose that $(x_1, \ldots, x_n)$ is a regular system of parameters of R. Letting $\xi_j = \mathrm{in}_{\mathfrak{m}}(x_j)$ for all j, one has $\mathrm{gr}^{\cdot}_{\mathfrak{m}}(R) = K[\xi_1, \ldots, \xi_n]$. Define Ψ: $K[\mathsf{X}_1, \ldots, \mathsf{X}_n] \to \mathrm{gr}^{\cdot}_{\mathfrak{m}}(R)$ by $\Psi(\mathsf{X}_j) = \xi_j$ for all j. We claim that $\mathrm{Ker}\, \Psi = 0$. If $\mathrm{Ker}\, \Psi$ is not zero, it is a homogeneous ideal (see §3.1) and we choose a nonzero homogeneous polynomial F from $\mathrm{Ker}\, \Psi$. Thus,

$$\dim_K \mathrm{gr}^d_{\mathfrak{m}}(R) \leq \dim_K(K[\mathsf{X}_1, \ldots, \mathsf{X}_n]/(F))_d \qquad \text{for all} \qquad d \geq 0. \qquad (*)$$

Here, $(K[\mathsf{X}_1, \ldots, \mathsf{X}_n]/(F))_d$ denotes the space of all homogeneous elements of degree d. It is easy to check that $\psi_F(d) = \dim_K(K[\mathsf{X}_1, \ldots, \mathsf{X}_n]/(F))_d$ is a polynomial in d of degree $n - 2$ for all $d \gg 0$. Hence $H_R(d) - H_R(d - 1) = \dim_K \mathrm{gr}^d_{\mathfrak{m}}(R) \leq \psi_F(d)$ and this means that $\sigma_R(d) \leq cd^{n-1}$ for all $d \gg 0$, where $c > 0$. This is contradictory to the result in Theorem 2.5.

Therefore, $\mathrm{Ker}\, \Psi = 0$ and thus the desired isomorphism Ψ is obtained. □

Combining this with Corollary 1 to Theorem 2.6 and Lemma 2.14, we obtain the following results on regular local rings.

Theorem 2.10. *Let R be a regular local ring of dimension n. Then*

(i) *R is a normal ring with $e(R) = 1$.*
(ii) *$\hat{R}$ is also a regular local ring.*
(iii) *If $\hat{R}$ contains a field isomorphic to $K = \hat{R}/\mathfrak{m}\hat{R} = R/\mathfrak{m}$, then $\hat{R} = K[[\mathsf{X}_1, \ldots, \mathsf{X}_n]]$.*

PROOF. (i) This follows from Theorem 2.9 and Lemma 2.14.
(ii) This is obvious, since $\mathfrak{m}\hat{R}$ is the maximal ideal of $\hat{R}$.
(iii) This is clear from Corollary 1 to Theorem 2.6. □

The following result is hard to prove.

Theorem 2.11 (Auslander-Buchsbaum). *Any regular local ring is a* UFD.

PROOF. We refer to [N2, Theorem 28.7]. □

The previous theorem and also the next one are consequences of the homological theory of local rings.

Theorem 2.12 (Serre). *If R is a regular local ring and $\mathfrak{p}$ is a prime ideal, then $R_{\mathfrak{p}}$ is also a regular local ring.*

PROOF. We refer to [N2, Theorem 28.3]. □

g. Let X be a Noetherian scheme.

Definition. Let $x \in X$. If $\mathcal{O}_{X,x}$ is a regular local ring, then x is said to be a *regular point*. Define Reg X to be the set of all regular points on X, and Sing X to be $X \setminus \text{Reg } X$.

If X is an integral scheme, Reg X contains the generic point of X. But Reg X is not an open subset, in general. By Theorem 2.12, if $x \in \text{Reg } X$, then any generalization of x is contained in Reg X.

Suppose that k is algebraically closed and V is an affine variety which is a closed subvariety $V(\mathfrak{p})$ of $\mathbf{A}^N$. Then if $p \in \text{spm}(V) \cap \text{Reg } V$ and $\dim V = n$, there exist $f_1, \ldots, f_{N-n} \in \mathfrak{p}$ and $i_1, \ldots, i_{N-n}$ such that

$$\frac{\partial(f_1, \ldots, f_{N-n})}{\partial(\mathsf{X}_{i_1}, \ldots, \mathsf{X}_{i_{N-n}})}(p) \neq 0,$$

as was shown in Example 2.10.

Here, when k is algebraically closed and p is a closed regular *point* of an algebraic variety, we say that p is a *nonsingular point*. If $V = \text{Reg } V$, V is said to be a *nonsingular variety*.

Definition. Let x be a point of X. The *multiplicity* of X at x is defined to be the multiplicity of $\mathcal{O}_{X,x}$ and is denoted by $e(x, X)$ (see §2.5.**d**), i.e., $e(x, X) = e(\mathcal{O}_{X,x})$.

In particular, when x is a regular point on a Noetherian scheme X, $e(x, X) = 1$ by Theorem 2.9.

Proposition 2.11. *Let X be an affine Noetherian scheme* Spec A *and let x be a regular point of X. If $f \in A$ satisfies $f(x) = 0$ and $V =$* $\operatorname{Spec}(A/(f))$, *then $e(x, V) = \nu_R(f)$ where $R = \mathcal{O}_{X,x}$ and ν_R is the order function defined in* §2.5.**d**.

PROOF. Let $\bar{R} = \mathcal{O}_{V,x} \cong R/fR$ and $\bar{\mathfrak{m}} = \mathfrak{m}_{V,x} \cong \mathfrak{m}/fR$, where $\mathfrak{m}$ is the maximal ideal of R. Then since $\operatorname{gr}_{\mathfrak{m}}^{\cdot} R \cong R/\mathfrak{m}[\mathsf{X}_1, \ldots, \mathsf{X}_n]$, where $n = \dim R$, by Theorem 2.9, one has

$$\operatorname{gr}_{\bar{\mathfrak{m}}}^{\cdot}(\bar{R}) \cong R/\mathfrak{m}[\mathsf{X}_1, \ldots, \mathsf{X}_n]/(\operatorname{in}_{\mathfrak{m}}(f)).$$

Hence $\sigma_{\bar{R}}(d) = (\nu/(n-1)!)d^{n-1} +$ lower terms for $d \gg 0$, where $\nu = \deg(\operatorname{in}_{\mathfrak{m}}(f)) = \nu_R(f)$. Thus $e(\bar{R}) = \nu$. □

§2.7 Normal Points on Algebraic Curves and Extension Theorem

a. As was seen in the last section, a regular local ring is a UFD, and a UFD is normal. Moreover, we shall prove that if R is a Noetherian normal local ring of dimension 1, then R is regular.

We start by proving the next simple lemma.

Lemma 2.15. *Let A be a Noetherian integral domain and let $\mathfrak{a}$ be a nonzero ideal of A. If $a \in Q(A)$ satisfies $a\mathfrak{a} \subseteq \mathfrak{a}$, then a is integral over A.*

PROOF. Taking $b \in \mathfrak{a}\backslash\{0\}$, one has $a^m b \in \mathfrak{a}$ for any $m \geq 1$. Hence $a^m \in \mathfrak{a} \cdot 1/b \subseteq A \cdot 1/b$, i.e., $A[a] \subseteq A \cdot 1/b$. Since A is Noetherian, $A[a]$ is finitely generated as an A-module; thus a is integral over A by Proposition 1.9. □

Theorem 2.13. *Let R be a Noetherian normal local ring of dimension* 1. *Then R is a regular local ring.*

PROOF. Letting $\mathfrak{m}$ be the maximal ideal of R, one defines $\mathfrak{m}^*$ to be $\{x \in Q(R) \mid x\mathfrak{m} \subseteq R\}$; i.e., $\mathfrak{m}\mathfrak{m}^* \subseteq R$. If $a \in \mathfrak{m}\backslash\{0\}$, then $\dim R/aR = 0$; hence R/aR is Artinian by Theorem 1.10. Thus $\mathfrak{m}^s \subseteq aR$ for some $s > 0$ by Lemma 2.8.

We let $r = \min\{s \mid \mathfrak{m}^s \subseteq aR\}$. Then $r \geq 1$ and $\mathfrak{m}^{r-1} \not\subseteq aR$, i.e., there exists $b \in \mathfrak{m}^{r-1}\backslash aR$. Hence, $b/a \notin R$ and $b\mathfrak{m} \subseteq \mathfrak{m}^r \subseteq aR$, which implies that $b/a \in \mathfrak{m}^*$; thus $R \subset \mathfrak{m}^*$.

Now, since $R \supseteq \mathfrak{m}\mathfrak{m}^* \supseteq \mathfrak{m}$, we have two cases: $\mathfrak{m}\mathfrak{m}^* = \mathfrak{m}$ or $\mathfrak{m}\mathfrak{m}^* = R$. If $\mathfrak{m}\mathfrak{m}^* = \mathfrak{m}$, then any $y \in \mathfrak{m}^*$ satisfies $y\mathfrak{m} \subseteq \mathfrak{m}$. Hence by the last lemma, y is integral over R. Since R is normal, $y \in R$; thus $\mathfrak{m}^* = R$. This contradicts $R \subset \mathfrak{m}^*$.

Thus $\mathfrak{m}\mathfrak{m}^*$ must be R. By the Corollary to Lemma 2.8, $\mathfrak{m} \neq \mathfrak{m}^2$. Hence, taking an element π of $\mathfrak{m}\backslash\mathfrak{m}^2$, one has $\pi\mathfrak{m}^* \subseteq \mathfrak{m}\mathfrak{m}^* = R$; thus $\pi\mathfrak{m}^* \subseteq \mathfrak{m}$ or

$\pi\mathfrak{m}^* = R$. If $\pi\mathfrak{m}^* \subseteq \mathfrak{m}$, then $\pi R = \pi\mathfrak{m}^*\mathfrak{m} \subseteq \mathfrak{m}^2$, and hence $\pi \in \mathfrak{m}^2$, which is inconsistent with the choice of π. Therefore, $\pi\mathfrak{m}^* = R$ and so $\pi R = \pi\mathfrak{m}^*\mathfrak{m} = \mathfrak{m}$, i.e., R is a regular local ring of dimension 1. □

Lemma 2.16. *Let R be a regular local ring of dimension* 1 *and π be a generator of the maximal ideal. Then any nonzero proper ideal $\mathfrak{a}$ can be written in the form $\pi^r R$. Hence R is a principal ideal domain, and so in particular, a* UFD.

PROOF. For any $x \in R\backslash\{0\}$, define $\mathrm{ord}_R(x)$ to be $\max\{r \mid x/\pi^r \in R\}$. If $r = \mathrm{ord}_R(x)$, then $x/\pi^r \notin \mathfrak{m}$; hence it is a unit u, i.e., $x = \pi^r u$. Now, for any ideal $\mathfrak{a}$, define $\mathrm{ord}_R(\mathfrak{a})$ to be $\min\{\mathrm{ord}_R(x) \mid x \in \mathfrak{a}\backslash\{0\}\}$. Then, if $s = \mathrm{ord}_R(x)$, one has $\mathfrak{a} = \pi^s \cdot (\mathfrak{a} : \pi^s) = \pi^s R$. □

b. Definition. A regular local ring R of dimension 1 is often said to be a *discrete valuation ring*, abbreviated as DVR.

The discrete valuation of $\mathfrak{K} = Q(R)$ with respect to R, is introduced as follows: any nonzero $x = a/b$ with $a, b \in R$ can be written as $\pi^{r-s}uv^{-1}$, where $r = \mathrm{ord}_R(a)$, $s = \mathrm{ord}_R(b)$, and u, v, uv^{-1} are units of R. Since $r - s$ does not depend on the choice of a and b, we can define $\mathrm{ord}_R(x) = r - s$.

Proposition 2.12. *The discrete valuation* ord_R *satisfies the following properties*:

(*i*) $\mathrm{ord}_R(xy) = \mathrm{ord}_R(x) + \mathrm{ord}_R(y)$, *if* $xy \neq 0$.
(*ii*) $\mathrm{ord}_R(x + y) \geq \min\{\mathrm{ord}_R(x), \mathrm{ord}_R(y)\}$, *if* $xy \neq 0$ *and* $x + y \neq 0$.
(*iii*) $\mathrm{ord}_R(x + y) = \mathrm{ord}_R(y)$, *if* $\mathrm{ord}_R(x) > \mathrm{ord}_R(y)$ *and* $xy \neq 0$.
(*iv*) $R = \{\alpha \in \mathfrak{K}\backslash\{0\} \mid \mathrm{ord}_R(\alpha) \geq 0\} \cup \{0\}$,
$\mathfrak{m} = \{\alpha \in \mathfrak{K}\backslash\{0\} \mid \mathrm{ord}_R(\alpha) > 0\} \cup \{0\}$.

PROOF. (i) is clear.

(ii) and (iii). Letting $r = \mathrm{ord}_R(x) \geq s = \mathrm{ord}_R(y)$, one has $x = \pi^r u$ and $y = \pi^s v$, where u and v are units. Then $x + y = \pi^s(\pi^{r-s} u + v)$ and so $\mathrm{ord}_R(x + y) \geq s$. If $r > s$, then $\pi^{r-s} u + v$ is a unit and so $\mathrm{ord}_R(x + y) = s$.

(iv). This is clear. □

c. Let C be an affine normal curve Spec A, i.e., A is a normal ring finitely generated over a field k with $\dim A = 1$. Then any local ring at a closed point p is a DVR. The discrete valuation with respect to $\mathcal{O}_{C,p}$ is denoted by ord_p, or ord_t, where t denotes a regular parameter of $\mathcal{O}_{C,p}$.

EXAMPLE 2.11. Let $R = k[\mathsf{X}]_{(\mathsf{X})} = \{\varphi(\mathsf{X})/\psi(\mathsf{X}) \mid \varphi, \psi$ are polynomials and $\psi(0) \neq 0\}$. X is a regular parameter of R and for a rational function $F(\mathsf{X})$, one has $\mathrm{ord}_{\mathsf{X}}(F(\mathsf{X})) = r$, where $F(\mathsf{X}) = \mathsf{X}^r G(\mathsf{X})$, $G(0) \neq 0$.

EXAMPLE 2.12. $R = k[[X]]$ is a DVR, and so $\mathrm{ord}_R(\varphi(X)) = v_R(\varphi(X))$ for $\varphi \in R$. Then

$$\dim_k(k[[X]]/(\varphi(X))) = \mathrm{ord}_R(\varphi(X)).$$

Meanwhile, if φ is a polynomial, then

$$\dim_k(k[X]/(\varphi(X))) = \deg \varphi(X).$$

d. The following result is a higher-dimensional analogue of Theorem 2.13.

Theorem 2.14 (Krull). *Let A be a Noetherian integral domain. A is a normal ring if and only if the following two conditions are satisfied:*

(1) *For any $\mathfrak{p} \in \operatorname{Spec} A$ with $\mathrm{ht}(\mathfrak{p}) = 1$, $A_\mathfrak{p}$ is a DVR.*
(2) *Any principal ideal fA has no embedded prime ideals.*

PROOF. If A is normal and $\mathrm{ht}(\mathfrak{p}) = 1$, then $A_\mathfrak{p}$ is normal by Lemma 2.1 and $\dim A_\mathfrak{p} = 1$; hence $A_\mathfrak{p}$ is a DVR by Theorem 2.13. To check the condition (2), suppose that $\mathfrak{p}$ is an embedded prime ideal of fA, where f is not a unit. Then by Lemma 2.6, $\mathfrak{p}$ can be written as $(fA: b)$ for some $b \in A$. $R = A_\mathfrak{p}$ is a normal local ring with $\dim R \geq 2$, since $\mathfrak{p}$ is an embedded prime ideal of fR. We let $\mathfrak{m} = \mathfrak{p}A_\mathfrak{p}$.

As in the proof of Theorem 2.13, we let $\mathfrak{m}^* = \{x \in Q(R) \mid x\mathfrak{m} \subseteq R\}$, which contains b/f. Then $\mathfrak{m}\mathfrak{m}^* \subseteq R$ and $\mathfrak{m} \subseteq \mathfrak{m}\mathfrak{m}^*$. If $\mathfrak{m}\mathfrak{m}^* = R$, then taking $\alpha \in \mathfrak{m}\backslash\mathfrak{m}^2$, one has $\alpha R = \mathfrak{m}$ by the argument in the proof of Theorem 2.13. Then $\dim R \leq 1$ by Proposition 2.10.

Therefore, $\mathfrak{m}\mathfrak{m}^* = \mathfrak{m}$. By Lemma 2.15, any element of $\mathfrak{m}^*$ is integral over R. R being normal, one has $\mathfrak{m}^* = R$; thus $b/f \in R = A_\mathfrak{p}$ and so $b/f = c/a$ for some $c \in A$ and $a \in A\backslash\mathfrak{p}$. Then $\mathfrak{p} = (fA: b) = (faA: ba) = (faA: cf) = (aA: c)$ and hence $a \in \mathfrak{p}$, which contradicts the choice of a.

Conversely, suppose that A satisfies the conditions (1) and (2). If b/f ($b, f \in A$) is integral over A, then let

$$fA = \mathfrak{q}_1 \cap \cdots \cap \mathfrak{q}_r$$

be the shortest primary decomposition. By (2), every prime ideal $\mathfrak{p}_i = \sqrt{\mathfrak{q}_i}$ associated with fA is not embedded; hence $\mathfrak{q}_j = A \cap fA_{\mathfrak{p}_j}$ for all j by Theorem 2.3. Since b/f is integral over $A_{\mathfrak{p}_j}$, it belongs to $A_{\mathfrak{p}_j}$ by condition (1); $b \in fA_{\mathfrak{p}_j}$ and so $b \in \mathfrak{q}_j = A \cap fA_{\mathfrak{p}_j}$. Therefore, $b \in \bigcap_{j=1}^r \mathfrak{q}_j = fA$, i.e., $b/f \in A$. □

Corollary. *If A is a Noetherian normal ring, then $A = \bigcap_\mathfrak{p} A_\mathfrak{p}$, where the intersection is taken over all $\mathfrak{p} \in \operatorname{Spec} A$ with $\mathrm{ht}(\mathfrak{p}) = 1$.*

PROOF. Since clearly $A \subseteq \bigcap_\mathfrak{p} A_\mathfrak{p}$, take an arbitrary element b/f of $\bigcap_\mathfrak{p} A_\mathfrak{p}$ for $b, f \in A$. Let $\mathfrak{q}_1 \cap \cdots \cap \mathfrak{q}_r$ be the shortest primary decomposition of fA and let $\mathfrak{p}_j = \sqrt{\mathfrak{q}_j}$. Then $\mathrm{ht}(\mathfrak{p}_j) = 1$ by Theorem 2.7, and so $b/f \in A_{\mathfrak{p}_j}$ i.e., $b \in fA_{\mathfrak{p}_j}$.

But since $\mathfrak{q}_j = A \cap fA_{\mathfrak{p}_j}$ by Theorem 2.3, one has $b \in \mathfrak{q}_j$ for all j. Hence, $b \in \bigcap_j \mathfrak{q}_j = fA$, i.e., $b/f \in A$. □

By using the corollary, one can prove an extension theorem which is similar to the extension theorem of Hartogs in the theory of functions of several complex variables.

Theorem 2.15 (Extension theorem). *Let X be a Noetherian normal scheme and let F be a closed subset of X with* $\operatorname{codim}(F) \geq 2$. *Then* $\Gamma(X \backslash F, \mathcal{O}_X) = \Gamma(X, \mathcal{O}_X)$.

PROOF. If X is affine, i.e., if $X = \operatorname{Spec} A$, then by Lemma 1.43,

$$\Gamma(X \backslash F, \mathcal{O}_X) = \bigcap_{\mathfrak{p} \in X \backslash F} A_{\mathfrak{p}} \supseteq \Gamma(X, \mathcal{O}_X) = \bigcap_{\mathfrak{p} \in X} A_{\mathfrak{p}}.$$

But by the last corollary, $\Gamma(X, \mathcal{O}_X) = \bigcap_{\mathfrak{p} \in X, \operatorname{ht}(\mathfrak{p}) = 1} A_{\mathfrak{p}}$. Then $\operatorname{codim}(F) \geq 2$ implies that $\mathfrak{p} \notin F$ if $\operatorname{ht}(\mathfrak{p}) = 1$. Hence,

$$\bigcap_{\mathfrak{p} \in X \backslash F} A_{\mathfrak{p}} \subseteq \Gamma(X, \mathcal{O}_X), \text{ i.e., } \Gamma(X \backslash F, \mathcal{O}_X) = \Gamma(X, \mathcal{O}_X).$$

In general, assuming that $\{V_\lambda \mid \lambda \in \Lambda\}$ is an affine open cover, one has by Lemma 1.43

$$\Gamma(X \backslash F, \mathcal{O}_X) = \bigcap_{\lambda \in \Lambda} \Gamma(V_\lambda \backslash F, \mathcal{O}_X) = \bigcap_{\lambda \in \Lambda} \Gamma(V_\lambda, \mathcal{O}_X) = \Gamma(X, \mathcal{O}_X). \qquad \square$$

Corollary. *Let V be a normal variety and let $\mathbf{A}_k^1$ be the affine line over k. If F is a closed subset of V with* $\operatorname{codim}(F) \geq 2$, *then* $\operatorname{Hom}(V, \mathbf{A}_k^1) \cong \operatorname{Hom}(V \backslash F, \mathbf{A}_k^1)$, *i.e., any regular function on $V \backslash F$ is obtained as the restriction of a regular function on V.*

§2.8 Divisors on a Normal Variety

a. Let V be a normal variety.

Definition. A *prime divisor* Γ on V is defined to be a closed subvariety of V with $\operatorname{codim}(\Gamma) = 1$.

Since Γ is irreducible, it has a generic point w and $\mathcal{O}_{V,w}$ is a normal ring of dimension 1, which is a DVR by Theorem 2.13.

Definition. The *valuation* of $\operatorname{Rat}(V)$ defined by the DVR $\mathcal{O}_{V,w}$, which is also called the *valuation* on V at Γ, is denoted by $\operatorname{ord}_\Gamma$.

Rewriting the Corollary to Theorem 2.14, we obtain the next result.

Lemma 2.17. *Let* $\varphi \in \mathrm{Rat}(V)\backslash\{0\}$. *If* $\mathrm{ord}_\Gamma(\varphi) \geq 0$ *for any prime divisor* Γ *on* V, *then* $\varphi \in A(V)$.

Proposition 2.13. *Given any* $\varphi \neq 0$, *there exist a finite number of prime divisors* $\Gamma_1, \ldots, \Gamma_r$ *such that* $\mathrm{ord}_\Gamma(\varphi) = 0$ *if* $\Gamma \neq \Gamma_j$, *for* $1 \leq j \leq r$.

PROOF. Since $\varphi \in \mathrm{Rat}(V) = \mathrm{injlim}_U\, A(U)$, where the U are nonempty affine open subsets of V, φ is regular on some affine open subset U. Hence, $\varphi \in A(U)$ and $U_\varphi = \{x \in U \mid \bar{\varphi}(x) \neq 0\}$ is an open subset of V. Let $\Gamma_1, \ldots, \Gamma_r$ be the prime divisors which are irreducible components of $V\backslash U_\varphi$. Then if $\Gamma \neq \Gamma_j$ for all j, then $\Gamma \cap U_\varphi \neq \varnothing$; thus $\mathrm{ord}_\Gamma(\varphi) = 0$. □

b. Definition. The free Abelian group generated by the prime divisors on V is said to be the *divisor group* of V, denoted by $\mathbb{G}(V)$. Any element D of $\mathbb{G}(V)$ is said to be a *divisor* on V.

Any divisor D can be written as a finite formal sum of prime divisors Γ_i, i.e., $D = \sum_{i=1}^r m_i\Gamma_i$, $m_i \in \mathbb{Z}$.

Definition. If each $m_i \geq 0$, then $D = \sum_{i=1}^r m_i\Gamma_i$ is said to be *effective* and $\bigcup_{m_i>0} \Gamma_i$ is called the *support* of D, denoted by supp D.

When there is no danger of confusion, D is also used to denote supp D. We use the notation $D_1 \geq D_2$ to indicate that $D_1 - D_2$ is effective.

Definition. Given any divisor $D = \sum_{i=1}^r m_i\Gamma_i$, we set $D_+ = \sum_{m_i\geq 0} m_i\Gamma_i$, and $D_- = -\sum_{m_i\leq 0} m_i\Gamma_i$, which are called the *positive and negative parts* of D, respectively.

c. By Proposition 2.13, $\sum_\Gamma \mathrm{ord}_\Gamma(\varphi)\Gamma$ becomes a divisor for any nonzero rational function φ on V.

Definition. $\sum_\Gamma \mathrm{ord}_\Gamma(\varphi)\Gamma$ is said to be the *divisor defined by* φ, denoted by $\mathrm{div}(\varphi)$. Any divisor written in the form $\mathrm{div}(\varphi)$ for some φ is called a *principal divisor*. $(\varphi)_\infty$ is defined to be $\mathrm{div}(\varphi)_-$ and $(\varphi)_\lambda = \mathrm{div}(\varphi - \lambda)_+$ for any $\lambda \in k$.

Definition. Two divisors D_1 and D_2 are said to be *linearly equivalent*, if $D_1 - D_2$ is a principal divisor. In this case, we write $D_1 \sim D_2$.

Thus $D \sim 0$ if and only if D is a principal divisor.

d. Let V be an affine space $\mathbf{A}_k^n = \mathrm{Spec}\, k[\mathsf{X}_1, \ldots, \mathsf{X}_n]$. Since the polynomial ring is a UFD, any irreducible polynomial f determines a prime divisor $V(f)$ and conversely any prime divisor Γ can be written $V(f)$ for some irreducible polynomial f. Thus, any polynomial determines an effective divisor on V.

Now, let V be a projective space $\mathbf{P}_k^n = \bigcup_{i=0}^n D_+(\mathrm{T}_i)$ with $(\mathrm{T}_0, \ldots, \mathrm{T}_n)$ the homogeneous coordinate system (cf. §1.18). Clearly, $\mathbf{P}_k^n$ is an integral scheme algebraic over k. Furthermore, in §3.5 it will be proved that $\mathbf{P}_k^n$ is separated. $\mathbf{P}_k^n$ is a normal variety, since $\Gamma(D_+(\mathrm{T}_i), \mathcal{O}_{\mathbf{P}_k^n})$ is a polynomial ring for all i.

Given any homogeneous polynomial $F(\mathrm{T}_0, \ldots, \mathrm{T}_n)$ of degree r, letting $f_i = F(\mathrm{T}_0, \ldots, \mathrm{T}_n)/\mathrm{T}_i^r \in k[\mathrm{X}_i^{(0)}, \ldots, \mathrm{X}_i^{(n)}]$, one has the closed subsets $V(f_i)$ of $D_+(\mathrm{T}_i)$. Then $V(f_i) \cap D_+(\mathrm{T}_i\,\mathrm{T}_j) = V(f_j) \cap D_+(\mathrm{T}_i\,\mathrm{T}_j)$ for all i, j. Hence, $\bigcup_{i=0}^n V(f_i)$ is a closed subset, denoted by $V_+(F)$, such that $V_+(F) \cap D_+(\mathrm{T}_i) = V(f_i)$ for all i. Clearly, if F is an irreducible homogeneous polynomial, then $V_+(F)$ is also irreducible; hence it is a closed subvariety of $\mathbf{P}_k^n$ of codimension 1 by Corollary 2 to Theorem 1.8, i.e., $V_+(F)$ is a prime divisor on $\mathbf{P}_k^n$.

Conversely, let Γ be a prime divisor on $\mathbf{P}_k^n$. If $\Gamma \cap D_+(\mathrm{T}_0) = \varnothing$, then $\Gamma \subseteq V_+(\mathrm{T}_0)$, and so $\Gamma = V_+(\mathrm{T}_0)$. Now, if $\Gamma \cap D_+(\mathrm{T}_0) \neq \varnothing$, then it is a prime divisor on $D_+(\mathrm{T}_0)$, and hence $\Gamma \cap D_+(\mathrm{T}_0)$ can be written as $V(f_0)$ in $D_+(\mathrm{T}_0)$, for some irreducible polynomial f_0 in $\mathrm{T}_1/\mathrm{T}_0, \ldots, \mathrm{T}_n/\mathrm{T}_0$ of degree r. Letting $F = f_0\,\mathrm{T}_0^r$, which is an irreducible homogeneous polynomial, one has $V_+(F)$ such that $V_+(F) \cap D_+(\mathrm{T}_0) = V(f_0) = \Gamma \cap D_+(\mathrm{T}_0)$. Since Γ and $V_+(F)$ are prime divisors, one concludes that $\Gamma = V_+(F)$.

Regarding f_0 as a rational function on $\mathbf{P}_k^n$, one has $\operatorname{div}(f_0) = V_+(F) - rV_+(\mathrm{T}_0)$, i.e., $V_+(F) \sim rV_+(\mathrm{T}_0)$. Therefore, any divisor on $\mathbf{P}_k^n$ is linearly equivalent to $\delta V_+(\mathrm{T}_0)$, for some $\delta \in \mathbb{Z}$.

e. Let V be a complete normal variety such that k is algebraically closed in $\mathrm{Rat}_k(V)$ (for example, if k is an algebraically closed field). Then $A(V) = k$ by Lemma 1.53.

Definition. For any divisor D on V, we define

$$\mathbf{L}_V(D) = \{\varphi \in \mathrm{Rat}(V) \mid \varphi = 0 \quad \text{or} \quad D + \operatorname{div}(\varphi) \geq 0\}.$$

Then $\mathbf{L}_V(0) = A(V) = k$ by Lemma 2.17. By making use of Proposition 2.12, one can show that $\mathbf{L}_V(D)$ is a vector space over k.

The following result is one of the corollaries to the finiteness theorem (Theorem 7.13) which will be proved in Chapter 7.

Theorem 2.16. *$\mathbf{L}_V(D)$ is a finite-dimensional vector space over k, if V is complete and normal.*

Assuming this, we continue the discussion about divisors. $\dim_k \mathbf{L}_V(D)$ is denoted $l_V(D)$ or $l(D)$. If $D_1 \sim D_2$, then, clearly, $\mathbf{L}_V(D_1) \cong \mathbf{L}_V(D_2)$; hence $l(D_1) = l(D_2)$.

EXAMPLE 2.13. Let D be a divisor on $\mathbf{P}_k^n$. Then $D \sim \delta H$, where H is $V_+(\mathsf{T}_0)$; thus $l(D) = l(\delta H)$. $\varphi \in \mathbf{L}_{\mathbf{P}^n}(\delta H)$ if and only if $\operatorname{div}(\varphi) + \delta V_+(\mathsf{T}_0) \geq 0$, which is equivalent to $\mathsf{T}_0^\delta\, \varphi$ being a polynomial, which will therefore be a homogeneous polynomial of degree δ. Hence, $\mathbf{L}(\delta H)$ is spanned by monomials of the form $\mathsf{T}_0^{\alpha_0} \cdots \mathsf{T}_n^{\alpha_n}$, where $\alpha_i \geq 0$ and $\sum_{i=0}^n \alpha_i = \delta$, and so

$$l(\delta H) = \binom{n+\delta}{\delta} = \frac{(n+\delta)\cdots(n+1)}{\delta!},$$

if $\delta \geq 0$. Clearly, if $\delta < 0$ then $l(\delta H) = 0$.

§2.9 Linear Systems

a. Let V be a complete normal variety over algebraically closed field k.

Definition. The *complete linear system* associated with a divisor D on V is defined by

$$|D| = \{D + \operatorname{div}(\varphi) \mid \varphi \in \mathbf{L}_V(D)\backslash\{0\}\}.$$

Then $|D| = \{D' \mid D' \text{ is an effective divisor} \sim D\}$. One has the canonical surjective map $D + \operatorname{div}\colon \mathbf{L}_V(D)\backslash\{0\} \to |D|$. If $\operatorname{div} \varphi = \operatorname{div} \psi$, then $\operatorname{div}(\varphi\psi^{-1}) = 0$, and so $\varphi = \lambda\psi$ for some $\lambda \in k$. Hence, $(\mathbf{L}_V(D)\backslash\{0\})/\sim \;\cong |D|$, where $\varphi \sim \psi$ if $\varphi = \lambda\psi$ for some $\lambda \in k^*$, where $k^* = k\backslash\{0\}$.

b. Let $\mathbf{E}$ be a vector space over k with $\dim_k \mathbf{E} = n < \infty$. Define $\mathbf{P}(\mathbf{E})'$ to be the set of all linear subspaces $\mathbf{F}$ of $\mathbf{E}$ with $\dim_k \mathbf{F} = n - 1$.

Choosing a basis $\mathbf{e}_1, \ldots, \mathbf{e}_n$ of $\mathbf{E}$, we identify $\mathbf{E}$ with k^n, i.e.,

$$\sum_{i=1}^n \lambda_i \mathbf{e}_i \in \mathbf{E} \leftrightarrow (\lambda_1, \ldots, \lambda_n) \in k^n.$$

Then $\mathbf{F}$ is described as

$$\mathbf{L}_c = \{\lambda = (\lambda_1, \ldots, \lambda_n) \in k^n \mid \sum_{i=1}^n \lambda_i c_i = 0\}$$

for some $(c_1, \ldots, c_n) = c \in k^n\backslash\{0\}$. Clearly, there is a bijection between $\mathbf{P}(\mathbf{E})'$ and $\mathbf{P}_k^{n-1}(k)$ via the correspondence: $\mathbf{L}_c \leftrightarrow (c_1 : c_2 : \cdots : c_n) \in \mathbf{P}_k^{n-1}(k)$, i.e., $\mathbf{P}(\mathbf{E})' \cong \mathbf{P}_k^{n-1}(k) = \operatorname{spm}(\mathbf{P}_k^{n-1})$. Thus, denoting $\mathbf{P}_k^{n-1}$ by $\mathbf{P}(\mathbf{E})$, we identify $\mathbf{P}(\mathbf{E})'$ with $\operatorname{spm}(\mathbf{P}(\mathbf{E}))$.

Furthermore, let $\mathbf{E}^\vee$ be the dual vector space of $\mathbf{E}$, i.e., $\mathbf{E}^\vee = \operatorname{Hom}_k(\mathbf{E}, k)$. Then $\mathbf{P}(\mathbf{E}^\vee)'$ is the set of all linear subspaces of dimension 1 of $\mathbf{E}$, in other words, $\mathbf{P}(\mathbf{E}^\vee)' = \{ky \mid y \in \mathbf{E}\backslash\{0\}\}$. Hence,

$$\operatorname{spm}(\mathbf{P}(\mathbf{L}_V(D)^\vee)) = \{k\varphi \mid \varphi \in \mathbf{L}_V(D)\backslash\{0\}\} \cong |D|.$$

Definition. Define the projective spaces $\mathbf{P}|D|$ to be $\mathbf{P}(\mathbf{L}_V(D))$, and $\mathbf{P}^*|D|$ to be $\mathbf{P}(\mathbf{L}_V(D)^\vee)$.

Then, the above isomorphism is rewritten as $\operatorname{spm}(\mathbf{P}^*|D|) \cong |D|$.

Definition. $\dim|D|$ is defined to be $\dim(\mathbf{P}|D|)$.

Thus, if $l(D) \geq 1$, then $\dim|D| = l(D) - 1$. By convention, we put $\dim|D| = -1$, if $|D| = \varnothing$.

c. Given a linear subspace $\mathbf{M}$ of $\mathbf{L}_V(D)$, define $\Lambda(\mathbf{M}, D)$, the *linear system defined by* $\mathbf{M}$, to be $\{D + \operatorname{div}\varphi \mid \varphi \in \mathbf{M}\setminus\{0\}\}$. $\Lambda(\mathbf{M}, D)$ is a linear subspace of $|D|$, if $|D|$ is identified with the projective space $\mathbf{P}^*|D|(k)$ in the classical sense. The following result is easily shown.

Lemma 2.18. $\Lambda(\mathbf{M}, D) = \Lambda(\mathbf{M}_1, D_1)$ *if and only if* $D \sim D_1$ *and* $\mathbf{M} = \mathbf{M}_1 \cdot \varphi$, *where* $D = D_1 - \operatorname{div}(\varphi)$.

If $\Lambda = \Lambda(\mathbf{M}, D)$ is a linear system, then one has the bijection $\operatorname{spm}(\mathbf{P}(\mathbf{M}^\vee)) \cong \Lambda$ which is compatible with $\operatorname{spm}(\mathbf{P}^*|D|) \cong |D|$; $\dim \Lambda$ is defined to be $\dim \mathbf{M} - 1$. Since $\mathbf{P}(\mathbf{M})$ depends only on Λ, $\mathbf{P}(\mathbf{M})$ (respectively $\mathbf{P}(\mathbf{M}^\vee)$) is often denoted $\mathbf{P}\Lambda$ (respectively $\mathbf{P}^*\Lambda$). Thus $\operatorname{spm}(\mathbf{P}^*\Lambda) \cong \Lambda$.

d. Let Λ be a nonempty linear system defined by $\mathbf{M}$. Take a basis $\varphi_1, \ldots, \varphi_l$ of $\mathbf{M}$ and let $D_j = D + \operatorname{div}(\varphi_j)$. Then, letting $\mathbf{M}_j$ be $\mathbf{M} \cdot \varphi_j^{-1}$, we have $\Lambda = \Lambda(\mathbf{M}_j, D_j)$ for all j.

For any closed point $p \in V\setminus D_j$, define $\Psi_j(p)$ to be $\{\psi \in \mathbf{M}_j \mid \psi(p) = 0\}$, which is a linear subspace of dimension $l - 1$. Hence, one has a map Ψ_j: $\operatorname{spm}(V\setminus D_j) \to \operatorname{spm} \mathbf{P}(\mathbf{M}_j) \cong \operatorname{spm}(\mathbf{P}\Lambda)$. Since $\operatorname{spm} \mathbf{P}(\mathbf{M}_j) = \operatorname{spm} \mathbf{P}(\mathbf{M}_i)$ for all i, j by definition, Ψ_j is obtained as the restriction of the morphism Φ_j: $V\setminus D_j \to \mathbf{P}(\Lambda)$ defined by $(\varphi_1/\varphi_j, \ldots, \varphi_l/\varphi_j)_k$, and Φ_j coincides with Φ_i on $V\setminus(D_i \cup D_j)$ for all i, j. Thus, $(V\setminus D_j, \Phi_j) \sim (V\setminus D_i, \Phi_i)$ where $\sim$ was defined when we introduced rational maps in §1.25.**a**.

Definition. The *rational map* Φ_Λ: $V \to \mathbf{P}(\Lambda)$ *associated with* Λ is defined to be $\{(V\setminus D_j, \Phi_j)\}_{1 \leq j \leq l}$. Furthermore, Φ_D denotes $\Phi_{|D|}$. If $\varphi_1, \ldots, \varphi_l$ form a basis of $\mathbf{M}$, we write as $(\varphi_1 : \cdots : \varphi_l)$ instead of Φ_Λ.

Definition. The *set of base points* of Λ is defined to be the closed subset $\bigcap_{j=1}^l D_j$, denoted by $\operatorname{Bs}\Lambda$. $\operatorname{Bs}\Lambda$ is said to be the *base locus* of Λ.

It is easy to see that $\operatorname{Bs}\Lambda = \bigcap_{D \in \Lambda} D$. Furthermore, $\bigcup_{j=1}^l (V\setminus D_j) = V\setminus \operatorname{Bs}\Lambda$, on which Φ_Λ becomes a morphism.

e. Example 2.14. Let $V = \mathbf{P}_k^n$ and let $(\mathsf{T}_0, \ldots, \mathsf{T}_n)$ be a homogeneous coordinate system. If $H = V_+(\mathsf{T}_0)$, then

$$|H| = \{H + \operatorname{div}\left(\sum_{i=0}^{n} \lambda_i \mathsf{T}_i/\mathsf{T}_0\right) | (\lambda_0, \ldots, \lambda_n) \in k^{n+1}\backslash\{0\}\}.$$

For any homogeneous polynomial $F(\mathsf{T}_0, \ldots, \mathsf{T}_n)$, we define div F to be $\sum_{i=1}^r s_i V_+(G_i)$, where $F = \prod_{i=1}^r G_i^{s_i}$ is the product of irreducible homogeneous polynomials G_i. Thus, div $F = \delta H + \operatorname{div}(F/\mathsf{T}_0^\delta)$, where δ is deg F. Therefore, $\mathrm{Bs}|H| = \varnothing$ and $\Phi_H: \mathbf{P}_k^n \to \mathbf{P}_k^n$ is an isomorphism. If $\Lambda = \{\operatorname{div}(\sum_{i=0}^m \lambda_i \mathsf{T}_i) | (\lambda_0, \ldots, \lambda_m) \in k^{m+1}\backslash\{0\}\}$, then $\mathrm{Bs}\,\Lambda = V_+(\mathsf{T}_0) \cap \cdots \cap V_+(\mathsf{T}_m)$ and $\Phi_\Lambda = (\mathsf{T}_0 : \cdots : \mathsf{T}_m)$ is said to be the *projection* of the space $\mathbf{P}_k^n$ to $\mathbf{P}_k^m$ with the center $V_+(\mathsf{T}_0) \cap \cdots \cap V_+(\mathsf{T}_m)$.

§2.10 Domain of a Rational Map

a. Theorem 2.17. *Let $f: V \to Z$ be a rational map. If V is a normal variety and Z is a closed subvariety of $\mathbf{P}_k^n$, then* $\operatorname{codim}(V\backslash\operatorname{dom}(f)) \geq 2$.

Proof. Suppose that $\operatorname{codim}(V\backslash\operatorname{dom}(f)) = 1$, i.e., there is a prime divisor Γ on V contained in $V\backslash\operatorname{dom}(f)$. Since $f_0 = f|_{\operatorname{dom}(f)}: \operatorname{dom}(f) \to Z \subseteq \mathbf{P}_k^n$ is a morphism by Proposition 1.46.(iii), we can assume that $\operatorname{Im}(f_0) \not\subseteq V_+(\mathsf{T}_0)$, i.e., $V_0 = f_0^{-1}(D_+(\mathsf{T}_0))$ is not empty. $f_1 = f_0|_{V_0}$ is a morphism into $D_+(\mathsf{T}_0) \cong \mathbf{A}_k^n$. Letting $p_i: \mathbf{A}_k^n \to \mathbf{A}_k^1$ be the i-th projection for $1 \leq i \leq n$, and letting $\varphi_i = p_i \cdot f_1$, one has $f_1 = (\varphi_1, \ldots, \varphi_n)_k$. Furthermore, putting $\varphi_0 = 1$, define α to be $\min\{\operatorname{ord}_\Gamma(\varphi_j) | 0 \leq j \leq n\}$. If $\alpha = \operatorname{ord}_\Gamma(\varphi_r)$, then $\varphi_i/\varphi_r \in \mathcal{O}_{V,w}$, where w is the generic point of Γ; therefore all the φ_i/φ_r are morphisms on some affine open subset U containing w, by Proposition 1.50. $(\varphi_0/\varphi_r, \ldots, \varphi_n/\varphi_r)_k$ is a morphism from U into $\mathbf{P}_k^n$ that defines the rational map $f: V \to \mathbf{P}_k^n$. Thus $w \in \operatorname{dom}(f)$, by the next lemma. □

Lemma 2.19. *Let Z be a closed subscheme of a scheme W and let V_0 be a dense open subset of a variety V. If a rational map (respectively morphism) $f: V \to W$ is a morphism φ on V_0 such that $\varphi(V_0) \subseteq Z$, then f is a rational map (respectively morphism) into Z.*

Proof. One may assume that f is a morphism. Then $f(V) = f(\overline{V_0}) \subseteq \overline{\varphi(V_0)} \subseteq Z$, and apply Proposition 1.30. □

Corollary to Theorem 2.17.

(i) *If C is a normal curve and $f: C \to \mathbf{P}_k^n$ is a rational map, then f becomes a morphism.*
(ii) *If C is a normal curve which is a closed subvariety of a projective space, then* $\operatorname{Bir}_k(C) = \operatorname{Aut}_k(C)$.

b. Now, we consider a rational map into an affine space.

Theorem 2.18. *Let $f\colon V \to Z$ be a rational map, where V is normal and Z is an affine variety. If* $\operatorname{codim}(V \backslash \operatorname{dom}(f)) \geq 2$, *then f is a morphism.*

PROOF. By assumption, Z is a closed subvariety of some affine space $\mathbf{A}_k^n$. By Lemma 2.19, one can assume $Z = \mathbf{A}_k^n$. Then $f|_{\operatorname{dom}(f)} = (f_1, \ldots, f_n)_k$ where the f_j are regular functions on $\operatorname{dom}(f)$. By Theorem 2.15, all f_j are regular functions on V and so f is a morphism. □

c. Now, assuming V is a complete normal variety over the algebraically closed field k, we study Φ_Λ for a linear system Λ on V. By Theorem 2.17, one has $\operatorname{codim}(V \backslash \operatorname{dom} \Phi_\Lambda) \geq 2$, even when $\operatorname{Bs} \Lambda$ has codimension 1. We have the next result.

Proposition 2.14. *If* $\operatorname{codim}(\operatorname{Bs} \Lambda) \geq 2$, *then* $\operatorname{Bs} \Lambda = V \backslash \operatorname{dom} \Phi_\Lambda$.

PROOF. Left to the reader as an exercise. □

Definition. An effective divisor Δ on V is said to be a *fixed component* of Λ, if $\Delta \leq D$ for every member D of Λ.

If $\Delta = \sum_i m_i \Gamma_i$ and $\Delta' = \sum_i n_i \Gamma_i$ are fixed components of Λ, then so is $\Delta'' = \sum_i \max(m_i, n_i) \Gamma_i$, where the Γ_i are prime divisors. Hence, there exists a maximal fixed component F of Λ, called the *fixed part* of Λ, and denoted by Λ_{fix}.

Definition. The linear system Λ_{red} defined to be $\{D - F \mid D \in \Lambda\}$, where F is the fixed part of Λ, is said to be the *reduced linear system* obtained from Λ.

Clearly, one has $\Lambda_{\text{red}} = \Lambda(\mathbf{M}, D - F)$, if $\Lambda = \Lambda(\mathbf{M}, D)$. In particular, $\mathbf{P}(\Lambda_{\text{red}}) = \mathbf{P}(\Lambda)$ and $\Phi_\Lambda = \Phi_{\Lambda_{\text{red}}}$. Furthermore, by Proposition 2.14, $V \backslash \operatorname{dom} \Phi_\Lambda = V \backslash \operatorname{dom} \Phi_{\Lambda_{\text{red}}} = \operatorname{Bs}(\Lambda_{\text{red}})$.

We have already constructed the rational map from a given linear system. Conversely, we can start from a rational map $f\colon V \to \mathbf{P}_k^n$ and construct a linear system. To do this, we have to introduce the notion of pullback of a divisor.

§2.11 Pullback of a Divisor

a. Let V, Z denote normal varieties. For a morphism $f\colon V \to Z$ and a divisor $D = \sum_{i=1}^r m_i \Gamma_i$, one wishes to define the pullback of D by f. First consider the case where D is a prime divisor Γ. The pullback $f^*\Gamma$ should be an effective divisor whose support is the inverse image $f^{-1}(\Gamma)$ as a set. But, in the following cases, one fails to define $f^*(\Gamma)$.

(1) If $f(V) \subseteq \Gamma$, then $f^{-1}(\Gamma) = V$.
(2) If $f(V) \cap \Gamma = \varnothing$, then $f^{-1}(\Gamma) = \varnothing$.
(3) Even if $f(V) \nsubseteq \Gamma$ and $f(V) \cap \Gamma \neq \varnothing$, there exist f and Γ such that $f^{-1}(\Gamma)$ is not equidimensional.

EXAMPLE 2.15. Let B be a polynomial ring $k[\mathsf{X}_1, \mathsf{X}_2, \mathsf{X}_3, \mathsf{X}_4]$ and put $\alpha = \mathsf{X}_1\mathsf{X}_3$, $\beta = \mathsf{X}_1\mathsf{X}_4$, $\gamma = \mathsf{X}_2\mathsf{X}_3$, $\delta = \mathsf{X}_2\mathsf{X}_4$. Then $\alpha\delta = \beta\gamma$ and $A = k[\alpha, \beta, \gamma, \delta]$ is a subring of B, and A is isomorphic to $k[\mathsf{T}_1, \mathsf{T}_2, \mathsf{T}_3, \mathsf{T}_4]/(\mathsf{T}_1\mathsf{T}_4 - \mathsf{T}_2\mathsf{T}_3)$, which is normal. It is easy to see that $\mathfrak{p} = A\alpha + A\beta$ is a prime ideal with $\mathrm{ht}(\mathfrak{p}) = 1$ and $\mathfrak{p}B = (B\mathsf{X}_1) \cap (B\mathsf{X}_3 + B\mathsf{X}_4)$. Thus, letting $i: A \to B$ be the inclusion map, one has $f = {}^a i: V = \operatorname{Spec} B \to Z = \operatorname{Spec} A$, and $f^{-1}(V(\mathfrak{p})) = V(\mathfrak{p}B)$ is a union of a prime divisor $V(\mathsf{X}_1)$ and a surface $V(\mathsf{X}_3) \cap V(\mathsf{X}_4)$.

b. Thus, in order to define a pullback appropriately, we had better assume that Z is a locally factorial variety as defined below.

Definition. A variety Z is said to be *locally factorial* if every stalk $\mathcal{O}_{Z,x}$ is a UFD.

By Proposition 2.6, any locally factorial variety is normal and any prime divisor Γ on a locally factorial variety Z is locally principal, i.e., there exists an affine open cover $\{U_\lambda \mid \lambda \in \Lambda\}$ of Z such that, for all λ, $\Gamma \cap U_\lambda$ is written as $V(\varphi_\lambda)$ for some φ_λ. To show this, we have to prove the next result.

Lemma 2.20. *Let A be a Noetherian integral domain and let $\mathfrak{m}$ be a maximal ideal of A such that $A_\mathfrak{m}$ is a* UFD. *If $\mathfrak{p}$ is a prime ideal of A with* $\mathrm{ht}(\mathfrak{p}) = 1$, *then there exist f and $a \in A$ such that $\mathfrak{m} \in D(f)$ and $\mathfrak{p}A_f = aA_f$.*

PROOF. Since $A_\mathfrak{m}$ is a UFD, $\mathfrak{p}A_\mathfrak{m}$ is principal by Proposition 2.8, i.e., $\mathfrak{p}A_\mathfrak{m} = aA_\mathfrak{m}$ for some $a \in A$. $a/1 \in \mathfrak{p}A_\mathfrak{m}$ and so $a/1 = p/s$, for $p \in \mathfrak{p}$ and $s \notin \mathfrak{m}$. Replacing as by a, we can assume $a \in \mathfrak{p}$. Thus $aA \subseteq \mathfrak{p}$ and $aA_\mathfrak{m} = \mathfrak{p}A_\mathfrak{m}$. Letting $f_1, \ldots, f_r$ be a basis of $\mathfrak{p}$, one has $f_i/1 \in aA_\mathfrak{m}$ i.e., $f_i/1 = ab_i/s_i$ for $b_i \in A$ and $s_i \notin \mathfrak{m}$. Hence, $f = s_1 \cdots s_r$ satisfies $\mathfrak{p}A_f = aA_f$. □

c. Now, let V be a normal variety and let Z be a locally factorial variety. The pullback $f^*\Gamma$ of a prime divisor Γ by a morphism $f: V \to Z$ is defined as follows, if $f(V) \cap \Gamma \neq \varnothing$ and $f(V) \nsubseteq \Gamma$. By Theorem 2.7, $f^{-1}(\Gamma)$ is a union of prime divisors $W_1, \ldots, W_r$. Letting w_i be the generic point of W_i, we choose an affine open neighborhood U_λ of $p_i = f(w_i)$ such that $\Gamma \cap U_\lambda = V(\varphi_\lambda)$ for some prime element φ_λ of $A(U_\lambda)$ by Lemma 2.20. Since $\theta(f)^\#_{w_i}: \mathcal{O}_{Z,p_i} \to \mathcal{O}_{V,w_i}$ is a local homomorphism, one has $\theta(f)^\#_{w_i}(\varphi_\lambda)$, which is denoted by $f^*(\varphi_\lambda)$.

Definition. $m_i = \text{ord}_{W_i}(f^*(\varphi_\lambda))$ depends only on Γ and f, the divisor $\sum_{i=1}^r m_i W_i$ is well-defined. This divisor is denoted by $f^*(\Gamma)$, and is called the *pullback of* Γ *by* f.

By the definition of $f^*(\Gamma)$, $\text{supp}(f^*(\Gamma))$ coincides with $f^{-1}(\Gamma)$ as a set.

Remark. In §1.22.**c**, we defined the scheme-theoretic inverse image $f^{-1}(\Gamma)$ of Γ. If $X = f^{-1}(\Gamma)$, then $\mathcal{O}_{X, W_i}$ is a local Artinian ring with m_i equal to the length of $\mathcal{O}_{X, W_i}$. The scheme X will not be used in what follows.

Furthermore, if $f(V) \cap \Gamma = \emptyset$, then we define $f^*\Gamma = 0$. In general, for a divisor $D = \sum_{i=1}^s m_i \Gamma_i$, if $f(V) \not\subseteq \Gamma_j$ for all j, then f^*D is defined to be $\sum_{i=1}^s m_i f^*(\Gamma_i)$.

d. Proposition 2.15. *As in the previous subsection, we assume Z is a locally factorial variety. If $D_1 - D_2 = \text{div}(\varphi)$ for some $\varphi \in \text{Rat}(Z)^*$ and if both $f^*(D_i)$, $i = 1, 2$, can be defined, then $f^*D_1 - f^*D_2 = \text{div}(f^*\varphi)$.*

PROOF. We can suppose that V and Z are affine varieties. At any closed point on Z which is the maximal ideal $\mathfrak{m}$ of $A = A(Z)$, one has $\varphi = a/b \in Q(A_\mathfrak{m})$. Since $A_\mathfrak{m}$ is a UFD, one can assume that a and b are elements of $A_\mathfrak{m}$ without common prime factors. We have $a = p_1 \cdots p_e$, $b = q_1 \cdots q_n$ where the p_i and the q_j are prime elements of $A_\mathfrak{m}$. Then $p_i \in A_\alpha$ and $q_j \in A_\alpha$ for all i, j, if one chooses a suitable $\alpha \in A \backslash \mathfrak{m}$. Replacing A_α by A, we may assume that the p_i and the q_j belong to A. Furthermore, after a similar replacement, all the (p_i) and (q_j) may be assumed to be prime. If $f^*(p_i) = 0$, then $f(V) \subseteq V(p_i)$, which contradicts the hypothesis. Thus $f^*(p_i) \neq 0$ and $f^*(q_j) \neq 0$. Now, assume $f^*(p_i)$ is a unit of $B = A(V)$. Then $\text{div}(f^*(p_i)) = 0$ and $f^*(\text{div}(p_i)) = 0$, since $f(V) \cap V(p_i) = \emptyset$. If a nonzero element $f^*(p_i)$ is not a unit of B, we let $\mathfrak{q}_1 \cap \cdots \cap \mathfrak{q}_r$ be the shortest primary decomposition of $p_i B$. All the $\mathfrak{p}_j = \sqrt{\mathfrak{q}_j}$ are prime ideals of height 1 by Theorem 2.7 and if one lets $B_j = B_{\mathfrak{p}_j}$, one has $p_i B_j = \mathfrak{q}_j B_j = \mathfrak{p}_j^{e_j} B_j$ by Theorem 2.3, where $e_j = \text{ord}_j(p_i B_j)$. Here ord_j denotes the valuation defined by $V(\mathfrak{p}_j)$. Then $\text{div}(f^*p_i) = \sum_{j=1}^r e_j \Gamma_j$. On the other hand, if $\mathfrak{p}$ is a minimal prime ideal containing $f^*(p_i)$, then $\mathfrak{p} = \mathfrak{p}_j$ for some j. Hence, $f^*(\text{div}(p_i)) = \sum_{j=1}^r e_j \Gamma_j$, and so $f^*(\text{div}(p_i)) = \text{div}(f^*(p_i))$.

Since $\varphi = \prod_{i=1}^e p_i / \prod_{j=1}^n q_j$, one has

$$\begin{aligned} f^* \text{div}(\varphi) &= \sum_{i=1}^{e} f^* \text{div}(p_i) - \sum_{j=1}^{n} f^* \text{div}(q_j) \\ &= \sum_{i=1}^{e} \text{div}(f^*(p_i)) - \sum_{j=1}^{n} \text{div}(f^*(q_j)) = \text{div}(f^*(\varphi)). \end{aligned}$$ □

Remark. On a normal variety Z, the formula $f^* \text{div}(\varphi) = \text{div} f^*(\varphi)$ does not always hold (See Exercise 2.1).

e. Let V_0 be a dense open subset of a normal variety V. If a prime divisor Γ on V_0 is given, then the closure $\bar{\Gamma}$ becomes a divisor on V. Further, for a divisor $D = \sum_{i=1}^{r} m_i \Gamma_i$ on V_0, where the Γ_i are prime divisors, one defines the closure divisor $\bar{D}$ to be $\sum_{i=1}^{r} m_i \bar{\Gamma}_i$.

We claim that if $\operatorname{codim}(V \backslash V_0) \geq 2$ and $\varphi \in \operatorname{Rat}(V_0) \backslash \{0\}$, then $\operatorname{div}(\varphi)$ on V is the closure in V of $\operatorname{div}(\varphi)$ on V_0.

To show this, let Γ be a prime divisor on V. Then $\Gamma \not\subseteq V \backslash V_0$ and so $\Gamma \cap V_0$ is an open dense subset of Γ. Hence $\operatorname{ord}_\Gamma(\varphi) = \operatorname{ord}_{\Gamma \cap V_0}(\varphi)$. From this, the claim follows.

Therefore, if the divisors D_1 and D_2 on V_0 are linearly equivalent, then so are the closures $\bar{D}_1$ and $\bar{D}_2$.

f. Now, let $f: V \to \mathbf{P}_k^n$ be a rational map where V is a complete normal variety over an algebraically closed field k. If V_0 is $\operatorname{dom}(f)$, then $\operatorname{codim}(V \backslash V_0) \geq 2$ and $f_0 = f|_{V_0}: V_0 \to \mathbf{P}_k^n$ is a morphism. If the image $f_0(V_0)$ is contained in a hyperplane L, i.e., $V_+(\sum_{i=0}^{n} \lambda_i \mathrm{T}_i)$ for some $(\lambda_0 : \cdots : \lambda_n) \in \mathbf{P}_k^n(k)$, then one can replace $\mathbf{P}_k^n$ by L. After such replacements, one can assume that $f_0(V_0)$ is not contained in any hyperplane. For $\lambda = (\lambda_0 : \cdots : \lambda_n) \in \mathbf{P}_k^n(k)$, define D_λ to be the closure of the pullback $f_0^*(V_+(\sum_{i=0}^{n} \lambda_i \mathrm{T}_i))$. Then by Proposition 2.15 and Lemma 2.20, all the D_λ are linearly equivalent. Thus, we have the linear system $\Lambda = \{D_\lambda \mid \lambda \in \mathbf{P}_k^n(k)\}$, and then $\Phi_\Lambda = f$.

§2.12 Strictly Rational Maps

a. Let $f: V \to W$ be a rational map of varieties. Then $\operatorname{dom}(f)$ is an open subvariety and so $\operatorname{dom}(f) \times W$ is an open subset of the product k-scheme $V \times W$. Letting φ be $f|_{\operatorname{dom}(f)}$, $\Gamma_\varphi: \operatorname{dom}(f) \to \Gamma_\varphi(\operatorname{dom}(f))$ is an isomorphism and the closure scheme of $\Gamma_\varphi(\operatorname{dom}(f))$ in $V \times W$ is also a variety (see §1.27.**e**).

Definition. The closure scheme of $\Gamma_\varphi(\operatorname{dom}(f))$ in $V \times W$ is said to be the *graph of the rational map f*, denoted by $\Gamma_f(V)$ or Γ_f. In particular, Γ_φ denotes $\Gamma_\varphi(\operatorname{dom}(f))$.

Letting $p: V \times W \to V$, $q: V \times W \to W$ be projections and $i: \Gamma_f \to V \times_k W$ be the closed immersion, one has morphisms $\pi = p \circ i: \Gamma_f \to V$ and $g = q \circ i: \Gamma_f \to W$. Since $\pi|_{\Gamma_\varphi}: \Gamma_\varphi \to \operatorname{dom}(f)$ is an isomorphism, π is the birational morphism such that $f \circ \pi = g$.

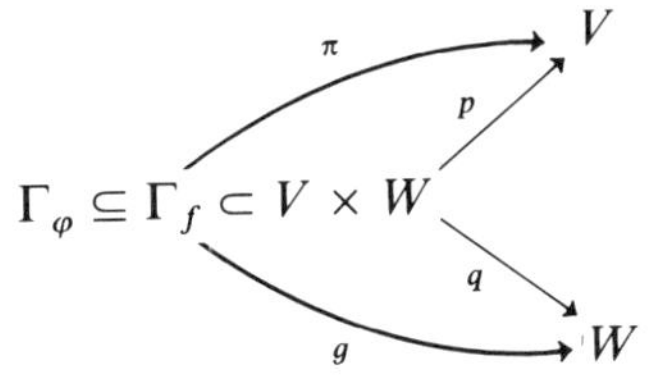

Definition. For $x \in V$, the value $f(x)$ is defined to be $g(\pi^{-1}(x))$. $f(x)$ is a set, which may be empty.

The pair (Γ_f, π) is one of the reduction models of f, which will be defined below.

b. Definition. A pair (Z, μ) consisting of a variety Z and a birational morphism $\mu: Z \to V$ is said to be a *reduction model* of f if $f \circ \mu$ is a morphism from Z into W.

Lemma 2.21. *If (Z, μ) is a reduction model of f, then $f \circ \mu(\mu^{-1}(x)) \subseteq f(x)$ for any $x \in V$. Furthermore, let $\alpha = (\mu, f \circ \mu)_k: Z \to V \times_k W$. Then $f \circ \mu(\mu^{-1}(x)) = f(x)$, whenever $\alpha(Z) = \Gamma_f$.*

Proof. There exists an open dense subset U of V such that $\varphi = f|_U: U \to W$ is a morphism. Since $\alpha|_{\mu^{-1}(U)} = (\mu', \varphi \circ \mu')_k = \Gamma_\varphi \circ \mu'$, where μ' denotes the restriction of μ to $\mu^{-1}(U)$, α is a morphism into Γ_f. Thus, we obtain the next diagram.

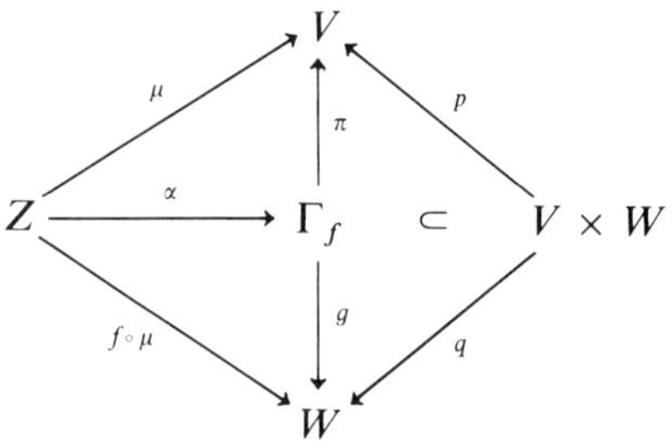

(∗)

Then $(f \circ \mu)(\mu^{-1}(x)) = (g \circ \alpha)(\alpha^{-1}(\pi^{-1}(x))) = g(\pi^{-1}(x) \cap \alpha(Z)) \subseteq g(\pi^{-1}(x)) = f(x)$. Furthermore, if $\alpha(Z) = \Gamma_f$, one has $f \circ \mu(\mu^{-1}(x)) = f(x)$. □

c. Definition. If the morphism $\pi: \Gamma_f \to V$ is proper, then f is said to be a *strictly rational map*.

Remarks. (1) The notion of strictly rational map is different from that of "application rationelle stricte" defined by Grothendieck and Dieudonné in [EGA, IV. 20.2.1].

(2) If $f: V \to W$ is a birational morphism, then f^{-1} can be regarded as a rational map and the value $f^{-1}(y)$ coincides with the set-theoretic inverse image $f^{-1}(y)$ for any $y \in W$.

(3) If $f: V \to W$ is a strictly rational map and x is a closed point of V, then $f(x)$ is a k-complete reduced subscheme of W in view of Proposition 1.58.

Lemma 2.22. *If there exists a reduction model (Z, μ) of f such that μ is proper, then f is a strictly rational map.*

PROOF. We observe the diagram in Lemma 2.21. Note that α is proper, since $\mu = \pi \circ \alpha$ is proper, by Proposition 1.57.(i). Since $\alpha\colon Z \to \Gamma_f$ is birational, $\alpha(Z)$ coincides with Γ_f. Hence by Proposition 1.57.(ii), π is a proper morphism. □

Lemma 2.23. *Let $f\colon V \to W$ be a rational map and let $g\colon W \to Y$ be a morphism such that $h = g \circ f$ is a morphism. If g is proper, then f is a strictly rational map.*

PROOF. Instead of the (Spec k)-graph, we use the Y-graph. Namely, let $U = \operatorname{dom}(f)$ and $\varphi = f|_U$. Since $g \circ \varphi = h|_U$, one has $\Gamma_\varphi(U) \subseteq U \times_Y W \subseteq V \times_Y W$. But since $V \times_Y W$ is a closed subscheme of $V \times W$ (to check this, use §§1.22.**d**, 1.23.**a**), it follows that $\Gamma_f \subseteq V \times_Y W$. The projection $p'\colon V \times_Y W \to V$ is proper, because g is proper, by Theorem 1.22. Hence, $\pi = p'|_{\Gamma_f}$ is proper; thus f is a strictly rational map.

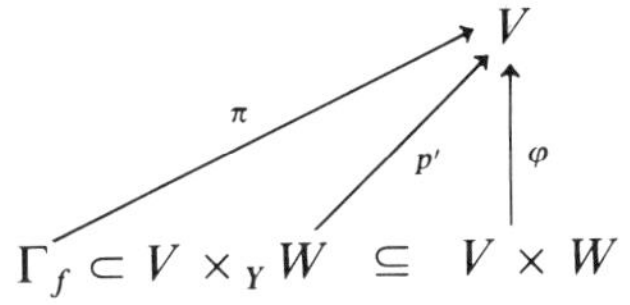

□

Corollary. *If $f\colon V \to W$ is a rational map and W is complete, then f is a strictly rational map.*

PROOF. This follows immediately from the last lemma. □

If $f\colon V \to W$ is a strictly rational map, then $f(x)$ is not empty for any $x \in V$, since $\pi\colon \Gamma_f \to V$ is surjective. For example, let $j\colon W \to V$ be an open immersion and let f be j^{-1}. If f is strictly rational, $f(x) \neq \varnothing$ for all $x \in V$; hence $W = V$. It is quite reasonable to exclude such a rational map j^{-1} with $W \neq V$.

d. Proposition 2.16. *Let $f_1\colon V_1 \to V_2$ and $f_2\colon V_2 \to V_3$ be strictly rational maps such that $f_2 \circ f_1$ is defined. Then $f_2 \circ f_1$ is a strictly rational map.*

PROOF. Let (Z_i, μ_i) be a reduction model of f_i such that μ_i is proper and $F_i = f_i \circ \mu_i$ is a morphism for $i = 1, 2$. Let (Y, p, q) be the V_2-product of Z_1 and Z_2.

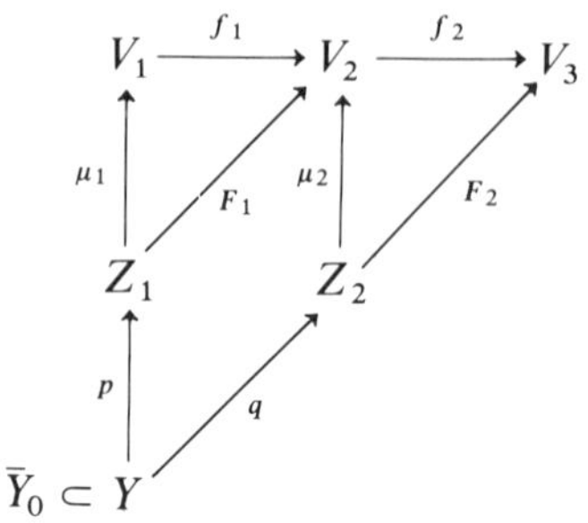

Since μ_2 is proper, the projection p is proper and so $\mu_1 \circ p$ is proper. Since $f_2 \circ f_1$ can be defined, $F_1(Z_1) \cap \mathrm{dom}(\mu_2^{-1}) \neq \emptyset$. Hence, letting $U = \mathrm{dom}(\mu_2^{-1})$, one has $F_1^{-1}(U) \neq \emptyset$ and $\mu_2^{-1}(U) \cong U$; thus Y contains an open subscheme $Y_0 = F_1^{-1}(U) \times_U \mu_2^{-1}(U) \cong F_1^{-1}(U)$. The closure scheme $\bar{Y}_0$ is a variety and contains $Y_0 \cong F_1^{-1}(U)$, i.e., $p|_{\bar{Y}_0}: \bar{Y}_0 \to Z_1$ is proper birational. Therefore, one obtains the reduction model $(\bar{Y}_0, \mu_1 \circ (p|_{\bar{Y}_0}))$ of $f_2 \circ f_1$, where $\mu_1 \circ (p|_{\bar{Y}_0})$ is proper. □

e. Definition. A strictly rational map $f: V \to W$ is said to be a *proper rational map* if the morphism $g = f \circ \pi: \Gamma_f \to W$ is proper.

Lemma 2.24. *Let (Z, μ) be a reduction model of a strictly rational map f such that μ is proper. Then $f \circ \mu$ is proper if and only if $f \circ \pi$ is proper.*

Proof. By using the diagram in the proof of Lemma 2.21, one can prove this easily. □

Proposition 2.17

(i) *Let $f: V \to W$ be a birational map. Then f is proper if and only if both f and f^{-1} are strictly rational maps.*
(ii) *Let $f_1: V_1 \to V_2$ and $f_2: V_2 \to V_3$ be proper rational maps such that the composition $f_2 \circ f_1$ is defined. Then $f_2 \circ f_1$ is also proper.*
(iii) *The set* $\mathrm{PBir}(V) = \{f: V \to V \mid f$ *is a proper birational map*$\}$ *becomes a group under composition of rational maps.*
(iv) *Let $f: V \to W$ be a strictly rational map and let $g: W \to Y$ be a morphism such that $g \circ f$ is a proper morphism. Then f is proper.*

Proof. (i) We prove that if both f and f^{-1} are strictly rational, then f is a proper rational map. By hypothesis, there exist dense open subsets $U_1 \subseteq V$ and $U_2 \subseteq W$ such that $\varphi = f|_{U_1}: U_1 \to U_2$ is an isomorphism. Then $\Gamma_\varphi(U_1) = \tau(\Gamma_{\varphi^{-1}}(U_2))$, where $\tau: W \times V \cong V \times W$ was defined in Proposition 1.32. (ii). Hence, $\Gamma_f = \tau(\Gamma_{f^{-1}})$. Letting p, q and p', q' be the projections of $V \times W$ and $W \times V$, respectively, $p' = q \circ \tau$ and $q' = p \circ \tau$. Since $p'|_{\Gamma_{f^{-1}}}$ is proper, so is $q \circ \tau|_{\Gamma_{f^{-1}}} = q|_{\Gamma_f}$; hence f is a proper birational map.

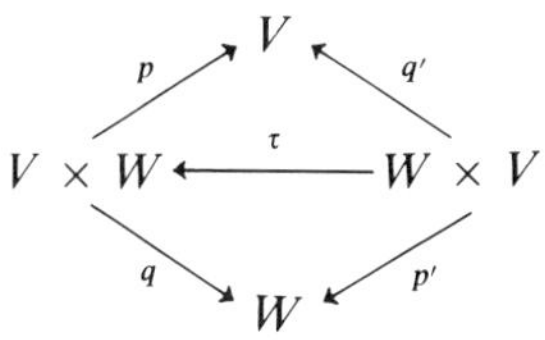

(ii) This is proved similarly to the proof of Proposition 2.16.

(iii) This follows from (i) and (ii) immediately.

(iv) This is left to the reader. □

Definition. A variety V is said to be *proper birationally equivalent* to a variety W, if there exists a proper birational map $f: V \to W$.

Lemma 2.25. *Let $f: V \to W$ be a strictly rational map and let $g: W \to Y$ be a morphism. Then $(g \circ f)(x) = g(f(x))$ for any $x \in V$. In particular, if $(\bar{W}, i)$ is a completion of W, then $(i \circ f)(x) = f(x)$ for any $x \in V$ and so* $\operatorname{dom}(f) = \operatorname{dom}(i \circ f)$.

PROOF. Take a reduction model (Z, μ) of f such that μ is proper. $F = f \circ \mu: Z \to W$ is a morphism and $(g \circ F)(\mu^{-1}(x)) = g(f(x))$; therefore $(g \circ f)(x) = g(f(x))$, by Lemma 2.21.

Clearly, $\operatorname{dom}(f) \subseteq \operatorname{dom}(i \circ f)$. Hence, it suffices to prove that $\varphi = i \circ f|_{\operatorname{dom}(i \circ f)}$ is a morphism into W. But this is obvious, since $\varphi(V) \subseteq W$. □

Now, we want to state a main result about strictly rational map which is an extension of Theorem 2.17. But, for the proof, we have to use two results which will be proved in Chapter 3.

(1) *Any projective space $\mathbf{P}_k^n$ is complete* (see §3.5).

Definition. A variety which is a closed subvariety of some projective space is called a *projective* variety.

Thus by Theorem 1.23.(i), any projective variety is complete.

(2) *For any complete variety W, there exist a projective variety Z and a proper birational morphism $\mu: Z \to W$* (see §3.6). This is *Chow's Lemma.*

Theorem 2.19. *Let $f: V \to W$ be a strictly rational map. If V is normal, then* $\operatorname{codim}(V \backslash \operatorname{dom}(f)) \geq 2$.

PROOF. Replacing $\operatorname{Im} f$ by W, we can assume that f is dominating. By Lemma 2.25, W can be replaced by $\bar{W}$, i.e., we can assume that W is complete. Then by Chow's Lemma (see (2) above) one has a birational morphism $\mu: Z \to W$, and Theorem 2.17 applied to $\mu^{-1} \circ f$ asserts that $\operatorname{codim}(V \setminus \operatorname{dom}(\mu^{-1} \circ f)) \geq 2$. But since $\operatorname{dom}(f) \supseteq \operatorname{dom}(\mu^{-1} \circ f)$, one obtains the assertion. □

Corollary. *If V is normal and W is affine, then every strictly rational map $f: V \to W$ is a morphism.*

PROOF. This follows immediately from Theorems 2.18 and 2.19. □

§2.13 Connectedness Theorem

a. The following result seems very natural from a geometrical viewpoint.

Theorem 2.20 (Connectedness Theorem of Zariski, or Degeneracy Principle of Enriques). *Let $f: V \to W$ be a proper surjective morphism such that the field extension* $\operatorname{Rat}(V)/\operatorname{Rat}(W)$ *is algebraically closed. If W is normal, then every fiber $f^{-1}(y)$ is a connected set.*

PROOF. We refer the reader to [EGA, III.4.3.12 or H, III Corollary 11.4]. □

Note that these proofs require the magnificient theory of formal schemes due to Zariski and Grothendieck.

EXAMPLE 2.16. Let $A = k[X_1, X_2]$ and $B = k[T]$ be polynomial rings. Define $\varphi: B \to A$ by $\varphi(T) = X_1X_2$, and let $V = \operatorname{Spec} A \setminus \{(0, 0)\}$ and $W = \operatorname{Spec} B$. Then $f = {}^a\varphi|_V: V \to W$ is not proper but is surjective. Further, $\operatorname{Rat}(V)/\operatorname{Rat}(W)$ is algebraically closed. $f^{-1}(0) = V(X_1X_2) \setminus \{(0, 0)\}$ in $\mathbf{A}_k^2$, which is disconnected. Thus, the assumption of being proper is indispensable in the connectedness theorem.

b. Definition. Let F be a closed set of a topological space. If a connected component of F is a set $\{p\}$ containing a single point, one says that F has an *isolated point component* $\{p\}$.

Theorem 2.21. *Let $f: V \to W$ be a strictly rational map. If V is a normal variety, then $f(x)$ is connected for all $x \in V$. In particular, if $f(x)$ has an isolated point component $\{y\}$, then $f(x) = y$.*

PROOF. Since $\pi: \Gamma_f \to V$ is a proper birational morphism, $\pi^{-1}(x)$ is connected by Theorem 2.20; hence $f(x) = (f \circ \pi)(\pi^{-1}(x))$ is also connected. □

By this and Theorem 2.19, we prove the next result, which is fundamental in the theory of strictly rational maps.

Theorem 2.22. *Let $f\colon V \to W$ be a strictly rational map, where V is a normal variety. If for a closed point x, $f(x)$ has an isolated point component, then $x \in \mathrm{dom}(f)$.*

PROOF. By the last theorem, $f(x) = y$. As in §2.12.**a**, we consider the graph Γ_f of f and $\pi\colon \Gamma_f \to V$. Define $g = f \circ \pi\colon \Gamma_f \to W$. Then since $f(x) = g(\pi^{-1}(x)) = y$, one has $\pi^{-1}(x) \subseteq g^{-1}(y)$. Taking an affine open neighborhood U of y, $Y = g^{-1}(U) \supseteq \pi^{-1}(x)$ and $\pi(\Gamma_f \backslash Y)$ is a closed set, since π is proper. Defining V_0 to be $V \backslash \pi(\Gamma_f \backslash Y)$, one can easily check that

$$x_1 \in \pi^{-1}(V_0) \Leftrightarrow \pi(x_1) \in V_0 \Leftrightarrow \pi(x_1) \notin \pi(\Gamma_f \backslash Y) \Rightarrow x_1 \in Y.$$

Hence, it follows that $\pi^{-1}(V_0) \subseteq Y = g^{-1}(U)$; thus $f|_{V_0}$ is a strictly rational map into the affine variety U. By Theorem 2.18, $V_0 = \mathrm{dom}(f|_{V_0})$ or $\mathrm{codim}(V_0 \backslash \mathrm{dom}(f|_{V_0})) = 1$. However, the latter case does not occur, since $\mathrm{codim}(V \backslash \mathrm{dom}(f)) \geq 2$ by Theorem 2.19. □

Corollary. *Let V be a normal variety and let W be an affine variety. Then every strictly rational map: $V \to W$ is a morphism. In particular, if both V and W are normal affine varieties, then every proper birational map: $V \to W$ is an isomorphism.*

PROOF. This follows from the last theorem, since $f(x)$ is a complete subscheme for any closed point by Remark (3) in §2.12.**c**. □

Note that this has been proved as a corollary to Theorem 2.19.

Thus proper birational equivalence for normal affine varieties is reduced to isomorphism equivalence. Further, birational equivalence for complete varieties becomes proper birational equivalence. Therefore, *proper birational geometry*, which means the study of proper birational equivalence classes, includes birational geometry for complete varieties as well as the theory of normal rings finitely generated over fields.

§2.14 Normalization of Varieties

a. The integral closure A'_L of an integral domain A in a given field L was defined in §2.1.**a**. Here, we introduce the integral closure of an integral scheme X in a given field L containing $\mathrm{Rat}(X)$ by glueing $\{\mathrm{Spec}(A_\lambda)'_L \mid \lambda \in \Lambda\}$, where $\{U_\lambda \cong \mathrm{Spec}\, A_\lambda \mid \lambda \in \Lambda\}$ is an affine open cover of X. Precisely speaking, let Ω^* be the set of all affine open subsets of X and define $\mathscr{A}(U)$ to be the integral closure of $A(U)$ in L, i.e., $\mathscr{A}(U) = A(U)'_L$ for any $U \in \Omega^*$. If

$U \supseteq V$ for $V \in \Omega^*$, one has $A(U) \subseteq A(V)$; hence $\mathscr{A}(U) \subseteq \mathscr{A}(V)$. Taking these inclusions as restriction maps, $\mathscr{A}$ satisfies conditions (1), (2), (3) in the definition of sheaf in §1.10.**a**, where Ω is replaced by Ω^*. Now, we want to check conditions (4) and (5) for $\mathscr{A}$. First note that if $U \in \Omega^*$, $A = A(U)$, and $a \in A$, then $\mathscr{A}(D(a)) = (A(U)_a)'_L = (A_a)'_L = A'_L \otimes_A A_a = \mathscr{A}(U) \otimes_A A_a$ by Proposition 2.1. Let $\{U_i \mid i \in J\}$ be an affine open cover of U. Then clearly $\mathscr{A}(U) \subseteq \bigcap_{i \in J} \mathscr{A}(U_i)$. Each U_i is covered by affine open subsets of the form $D(a_{ij})$ where $a_{ij} \in A(U)$. Hence $\mathscr{A}(U_i) \subseteq \bigcap_j \mathscr{A}(D(a_{ij}))$. Since U is quasi-compact by Theorem 1.2, there exists a finite subset $\{a_1, \ldots, a_r\}$ of $\{a_{ij}\}$ such that $U = \bigcup_{i=1}^r D(a_i)$ and hence $\bigcap_{i,j} \mathscr{A}(D(a_{ij})) = \bigcap_{i=1}^r \mathscr{A}(D(a_i)) = \bigcap_{i=1}^r (\mathscr{A}(U) \otimes_A A_{a_i}) = \mathscr{A}(U)$; $\mathscr{A}(U) = \bigcap_{i \in J} \mathscr{A}(U_i)$.

Thus, conditions (4) and (5) have been checked. By Lemma 1.21, one has a sheaf $\mathscr{A}$ of rings on X such that $\mathscr{A}(U) = A(U)'_L$ for any $U \in \Omega^*$.

Definition. The sheaf $\mathscr{A}$ is said to be the *integral closure* of $\mathscr{O}_X$ in L, denoted by $(\mathscr{O}_X)'_L$.

b. For any affine open subset U of X, define Y_U to be Spec $A(U)'_L$ and μ_U to be ${}^a i_U$, where i_U is the inclusion $A(U) \to A(U)'_L$. If $a \in A(U)$, then $Y_{D(a)} \cong \mu_U^{-1}(D(a))$. Thus, it is easy to verify that these Y_U and μ_U can be glued, and so we obtain a scheme Y and a morphism $\mu: Y \to X$ in such a way that $\mu^{-1}(U) \cong Y_U$ for all $U \in \Omega'$.

Definition. The above pair (Y, μ), or simply Y, is said to be the *integral closure* of an integral scheme X in L or the *L-normalization* of X, denoted by X'_L. If $L = \text{Rat}(X)$, then the Rat(X)-normalization of X is said to be the *normalization* of X.

X'_L is a normal scheme and $\text{Rat}(X'_L)$ is the algebraic closure of Rat(X) in L. Thus, in particular, if L is algebraic over Rat(X), then $\text{Rat}(X'_L) = L$. Furthermore, $\mu: X'_L \to X$ is an integral morphism. If V is a variety over a field k and $L/\text{Rat}(V)$ is finitely generated, then $\mu: V'_L \to V$ is a finite morphism by Theorem 2.1. Hence, V'_L is also a variety (over k). Thus, we obtain the next result.

Theorem 2.23. *For a variety V and a field L which is finitely generated over* Rat(V), *there exist a normal variety Y and a finite surjective morphism $\mu: Y \to V$ such that* Rat(Y) *is the algebraic closure of* Rat(V) *in L.*

Corollary. *If $\mathfrak{K}$ is a field finitely generated over k such that* $\text{tr.deg}_k\, \mathfrak{K} = 1$, *then there exists a complete normal curve C with* $\text{Rat}(C) = \mathfrak{K}$.

Proof. There exists $t \in \mathfrak{K}$ that is transcendental over k. Then $\mathfrak{K}$ is algebraic over $k(t)$. Since $k(t)$ is $\text{Rat}_k(\mathbf{P}_k^1)$, define C to be the $\mathfrak{K}$-normalization of $\mathbf{P}_k^1$. Then $\text{Rat}(C) = \mathfrak{K}$ and C is a complete normal curve (cf. §3.5.**a**). □

If k is an algebraically closed field, the normal curve is nonsingular. Thus, the corollary asserts that for such a $\mathfrak{K}$, there exists a complete nonsingular geometric model of $\mathfrak{K}$ (see §1.29.**c**). This proof is purely algebraic, depending on Theorem 2.13. In Chapter 9, we will give another proof of the corollary (in the case of plane curves), using a geometric method due to Max Noether. The latter method is rather complicated and laborious, but it is indispensable in resolving the singularities of a given higher-dimensional variety.

c. We want to show the uniqueness of the L-normalization of a given V.

Theorem 2.24. *If Y is a normal variety and $\lambda\colon Y \to V$ is a finite surjective morphism, then Y is isomorphic to the* Rat(Y)*-normalization of V.*

Proof. Let (V'_L, μ) be the $L\,(= \text{Rat}(Y))$-normalization of V. Since $\text{Rat}(V'_L)$ is isomorphic to $\text{Rat}(Y)$ over $\text{Rat}(V)$, one has a birational map $h\colon V'_L \to Y$ such that $\lambda \circ h = \mu$. Since λ and μ are proper, h is a proper birational map by Proposition 2.17.(iv). For any closed point $x \in V'_L$, $(\lambda \circ h)(x) = \mu(x)$; thus $h(x) \subseteq \lambda^{-1}(\mu(x))$, which is a finite closed subset. Hence by Theorem 2.22, h is a morphism. Similarly, one can prove that h^{-1} is a morphism; hence h is an isomorphism. Thus, (Y, λ) is the $\text{Rat}(Y)$-normalization of V. □

Therefore, when $\text{tr. deg}_k\, \mathfrak{K} = 1$, the nonsingular complete geometric model of $\mathfrak{K}$ is unique.

Theorem 2.25. *Let $f\colon V \to W$ be a morphism of varieties. Letting (V', μ) and (W', λ) be the normalizations of V and W, respectively, there exists a morphism $f'\colon V' \to W'$ such that $\lambda \circ f' = f \circ \mu$.*

Proof. Define a rational map f' to be $\lambda^{-1} \circ f \circ \mu$. Since $\lambda \circ f' = f \circ \mu$, f' is a strictly rational map by Lemma 2.23. Furthermore, $f'(x) \subseteq \lambda^{-1}(f \circ \mu)(x)$ is a finite closed set for any closed point $x \in V$. Hence, by Theorem 2.22, f' is a morphism. □

Definition. $f'\colon V' \to W'$ is said to be the *normalization* of $f\colon V \to W$.

d. Another application of Theorem 2.22 is the Stein factorization theorem.

Theorem 2.26. *Let $f\colon V \to W$ be a dominating morphism. If V is normal, there exist a normal variety Y, a finite morphism $\mu\colon Y \to W$, and a dominating morphism $g\colon V \to Y$ such that $f = \mu \circ g$ and the field extension* $\text{Rat}(V)/\text{Rat}(Y)$ *is algebraically closed.*

Proof. Let (Y, μ) be the $\text{Rat}(V)$-normalization of W. Then one has a rational map $g\colon V \to Y$ corresponding to the inclusion $\text{Rat}(Y) \subseteq \text{Rat}(V)$.

Since $\mu \circ g = f$, g is strictly rational by Lemma 2.23 and $g(x) \subseteq \mu^{-1}(f(x))$ is a finite closed set for any closed point $x \in V$. Hence by Theorem 2.22, g is a morphism. □

Remark. If f is proper, then so is g by Proposition 1.57.(i). In this case, $g^{-1}(y)$ is a connected component of $f^{-1}(\mu(y))$.

Definition. The factorization $\mu \circ g$ of f is said to be the *Stein factorization* of f.

But Theorem 2.26 can be proved without assuming V to be normal (see [H, III. Corollary 11.5]), where Y may not be normal.

Theorem 2.27. *Let $f\colon V \to W$ be a proper morphism such that $f^{-1}(x)$ is empty or a finite set for any closed point $x \in W$. If V is normal, then f is a finite morphism.*

Proof. Replacing $\operatorname{Im} f$ by W, we can assume that f is surjective. We let $\mu \circ g$ be the Stein factorization of f as in Theorem 2.26. Since g is birational (cf. Exercise 1.10.) and $\mu = f \circ g^{-1}$, g^{-1} is strictly rational by Lemma 2.23. Since $g^{-1}(y)$ is a single point set for all $y \in Y$, g^{-1} is a morphism by Theorem 2.22. Hence g is an isomorphism; f is a finite morphism. □

Remark. This theorem still holds without the assumption that V is normal. This can be proved using the factorization theorem of Stein for not necessarily normal varieties.

Corollary. *Let C and Y be complete normal curves, and let $f\colon C \to Y$ be a nonconstant morphism. Then f is finite.*

Proof. If $f(C) \neq Y$, then $f(C)$ is a closed point, since f is proper. Hence, by hypothesis, $f(C) = Y$ and one can apply Theorem 2.27. □

§2.15 Degree of a Morphism and a Rational Map

a. Let $f\colon V \to W$ be a finite surjective morphism between normal varieties. For a prime divisor Γ on W, $f^{-1}(\Gamma)$ is purely 1-codimensional, i.e., $f^{-1}(\Gamma) = D_1 \cup \cdots \cup D_r$, where the D_i are prime divisors. Letting y_i be the generic point of D_i for $1 \le i \le r$, $f(y_i)$ is the generic point η of Γ. Take a generator t of the maximal ideal of $\mathcal{O}_{W,\eta}$. For each i, $f^*(t) \in \mathcal{O}_{V,y_i}$ is assigned. Define $e_i = \operatorname{ord}_{D_i}(f^*(t))$ and $f^*\Gamma = \sum_{i=1}^{r} e_i D_i$ (cf. §2.11.**c**).

Lemma 2.26. *With the notation and assumptions above, one has*

$$\sum_{i=1}^{r} e_i[\mathrm{Rat}(D_i)\colon \mathrm{Rat}(\Gamma)] = [\mathrm{Rat}(V)\colon \mathrm{Rat}(W)].$$

Here, for a field extension $\mathfrak{K}/k$, $[\mathfrak{K}\colon k]$ denotes the dimension of $\mathfrak{K}$ as a k-vector space. Rat(W) *is identified with* $f^*\mathrm{Rat}(W)$, *and* Rat(Γ) *with* $f_i^*\mathrm{Rat}(\Gamma)$, *where* f_i *is* $f|_{D_i}\colon D_i \to \Gamma$.

PROOF. Let R be $\mathcal{O}_{W,\eta}$. Then $V \times_W \operatorname{Spec} R$ is an affine integral scheme $\operatorname{Spec} B$ finite over $\operatorname{Spec} R$. Since B has no zero divisors and R is a DVR, B is a free R-module, i.e., $B \cong R^n$, and $B_0 = B \otimes_R Q(R) \cong Q(R)^n$. B_0 is an integral domain. Hence, by Lemma 1.8, B_0 is a field, which is Rat(V); thus $n = [\mathrm{Rat}(V)\colon \mathrm{Rat}(W)]$. Let $\mathfrak{p}_1, \ldots, \mathfrak{p}_r$ be minimal prime ideals of the ideal tB, which correspond to $D_1, \ldots, D_r$, respectively. Then $\mathfrak{p}_i + \bigcap_{j\neq i} \mathfrak{p}_j = B$ for any $1 \leq i \leq r$, since $\dim B/\mathfrak{p}_i B = 0$. Let $\mathfrak{q}_1 \cap \cdots \cap \mathfrak{q}_r$ be the shortest primary decomposition of tB such that $\sqrt{\mathfrak{q}_i} = \mathfrak{p}_i$ for all i. Then, since $\mathfrak{q}_i + \bigcap_{j\neq i} \mathfrak{q}_j = B$ for $1 \leq i \leq r$, one has by the Chinese remainder theorem

$$B \otimes_R (R/tR) \cong B/tB \cong B/\mathfrak{q}_1 \oplus \cdots \oplus B/\mathfrak{q}_r.$$

Since $B/\mathfrak{q}_i$ has only one prime ideal, $B/\mathfrak{q}_i = B_i/\mathfrak{q}_i B_i$, where B_i is $B_{\mathfrak{p}_i}$. Let π_i be a generator of $\mathfrak{p}_i B_i$. Then $\mathfrak{q}_i B_i = \pi_i^{e_i} B_i$ and $\mathrm{Rat}(D_i) = B_i/\pi_i B_i$. Thus, if $L = R/tR = \mathrm{Rat}(\Gamma)$, one has

$$\dim_L(B/tB) = \sum_{i=1}^{r} \dim_L(B/\mathfrak{q}_i) = \sum_{i=1}^{r} e_i[\mathrm{Rat}(D_i)\colon \mathrm{Rat}(\Gamma)].$$

Since $B \cong R^n$, it follows that $n = \dim_L(B \otimes_R L)$. □

Definition. Let $f\colon V \to W$ be a dominating rational map. If $\dim V = \dim W$, the field extension Rat(V)/Rat(W) is finite and one defines the *degree* of f to be $[\mathrm{Rat}(V)\colon \mathrm{Rat}(W)]$, denoted by $\deg f$. Occasionally, if $\dim V > \dim W$ or f is not dominating, we put $\deg f = 0$.

Then the above formula in the last lemma can be stated as

$$\sum_{i=1}^{r} e_i \deg f|_{D_i} = \deg f.$$

b. Let $g\colon V \to Y$ be a proper birational morphism between normal varieties. If $Y_0 = \mathrm{dom}(g^{-1})$, then $\mathrm{codim}(Y \setminus Y_0) \geq 2$ by Theorem 2.19 and $g|_{g^{-1}(Y_0)}\colon g^{-1}(Y_0) \to Y_0$ is an isomorphism. For any prime divisor Γ on Y with η the generic point, one has $\eta \in Y_0$, and $\xi = g^{-1}(\eta)$ is the generic point of a prime divisor Γ' such that $\mathcal{O}_{\Gamma',\xi} \cong \mathcal{O}_{\Gamma,\eta}$. Γ' is said to be the *proper transform* of Γ by g^{-1}.

The set $g^{-1}(\Gamma)$ is a union of Γ' and a closed subset E such that $E \cap g^{-1}(Y_0) = \varnothing$, since $g^{-1}(\Gamma)|_{g^{-1}(Y_0)} \cong \Gamma|_{Y_0}$.

c. Let $f\colon V \to W$ be a proper dominating rational map between normal varieties V and W with $\dim V = \dim W$. Let $V_0 = \mathrm{dom}(f)$ and define $\varphi = f|_{V_0}\colon V_0 \to W$. For any prime divisor Γ on W, $\varphi^{-1}(\Gamma)$ can be written as $D_1^\circ \cup \cdots \cup D_r^\circ$, where the D_i° are the irreducible components. Denote by D_j the closure of D_j°, and define $J = \{j \mid D_j$ is a prime divisor and $\varphi(D_j^\circ)$ is dense in $\Gamma\}$. Then letting η and ξ_j be generic points of Γ and D_j°, respectively, one has $f(\xi_j) = \eta$ and thus $e_j = \mathrm{ord}_{D_j}(f^*(t))$ is defined, where t is a generator of the maximal ideal of $\mathcal{O}_{W,\eta}$.

Theorem 2.28. *With the above notation, it follows that*

$$\sum_{j \in J} e_j \deg(f|_{D_j}) = \deg f,$$

where $f\colon V \to W$ is a proper dominating rational map such that $\dim V = \dim W$.

PROOF. Clearly, one can assume that f is a proper morphism. Let $f = \mu \circ g$ be the Stein factorization, where $g\colon V \to Y$ is a proper birational morphism and $\mu\colon Y \to W$ is a finite surjective morphism. Applying Lemma 2.26 to μ and recalling the result on birational morphisms in §2.15.**b**, we obtain the formula. □

Remark. Let φ be a non-constant rational function on a complete normal variety, i.e., φ is a rational map: $V \to \mathbf{P}_k^1$. Then for any $\lambda \in k$, $(\varphi)_\lambda$ agrees with $\varphi_0^*(1:\lambda)$, where φ_0 is $\varphi|_{\mathrm{dom}(\varphi)}$, since $\mathrm{codim}(V \backslash \mathrm{dom}(\varphi)) \geq 2$. Assume that V is a complete normal curve. Then $V = V_0$, and $\varphi^*(\infty) = (\varphi)_\infty$. If one writes $(\varphi)_\lambda$ as $\sum_{i=1}^r m_i p_i$, then $\sum_{i=1}^r m_i[k(p_i): k] = \deg \varphi$ by Lemma 2.26, i.e., $\deg(\varphi)_\infty = \deg(\varphi)_\lambda = \deg \varphi$ for any $\lambda \in k$.

§2.16 Inverse Image Sheaves

a. Let $\psi\colon X \to Y$ be a continuous map between topological spaces. For a sheaf $\mathcal{G}$ on Y, we shall define the inverse image (sheaf) $\psi^*\mathcal{G}$ of $\mathcal{G}$ by ψ; the étale space $(\mathbf{F}, \pi, X)$ associated with $\psi^*\mathcal{G}$ is explicitly constructed as follows.

The total space $\mathbf{F}$ is $\coprod_{x \in X} \mathcal{G}_{\psi(x)}$ and $\pi\colon \mathbf{F} \to X$ is the projection, i.e., $\pi(a) = x$ if $a \in \mathcal{G}_{\psi(x)}$. We take as an open base of the topology on $\mathbf{F}$ the family of sets of the form $\mathbf{W}^*(U, \alpha) = \{\alpha_{\psi(x)} \mid x \in U\}$ where U is any open subset of X, V is an open subset of Y such that $\psi(U) \subseteq V$, and $\alpha \in \mathcal{G}(V)$. It is easy to see that $\mathbf{F} = (\mathbf{F}, \pi, X)$ is an étale space (see §1.10.**d**).

Definition. The sheaf associated with $\mathbf{F}$ is said to be the *inverse image sheaf* of $\mathcal{G}$, denoted $\psi^*\mathcal{G}$.

For any $x \in X$, $(\psi^*\mathscr{G})_x \cong \pi^{-1}(x) = \mathscr{G}_{\psi(x)}$. If V is an open subset of Y, then one has the map $\sigma_{\mathscr{G},V}: \mathscr{G}(V) \to \Gamma(V, \psi_*\psi^*\mathscr{G}) = \Gamma(\psi^{-1}(V), \psi^*\mathscr{G}) = \Gamma(\psi^{-1}(V), \mathbf{F})$ defined by $\sigma_{\mathscr{G},V}(\alpha)(x) = \alpha_{\psi(x)}$ for any $x \in \psi^{-1}(V)$. Clearly, $\{\sigma_{\mathscr{G},V}\}$ commute with restriction maps of the sheaves and so it gives a sheaf homomorphism $\sigma_{\mathscr{G}}: \mathscr{G} \to \psi_*\psi^*\mathscr{G}$, called the canonical homomorphism.

Given any sheaf $\mathscr{F}$ on X and a homomorphism $\theta: \psi^*\mathscr{G} \to \mathscr{F}$, we define $\psi_*(\theta): \psi_*\psi^*\mathscr{G} \to \psi_*\mathscr{F}$ by $\psi_*(\theta)_V = \theta_{\psi^{-1}(V)}$ for any open subset V of Y, and thus $\psi_*(\theta) \circ \sigma_{\mathscr{G}}: \mathscr{G} \to \psi_*(\mathscr{F})$ is obtained.

Lemma 2.27. *The homomorphism h: $\mathrm{Hom}(\psi^*\mathscr{G}, \mathscr{F}) \to \mathrm{Hom}(\mathscr{G}, \psi_*\mathscr{F})$ defined by $h(\theta) = \psi_*(\theta) \circ \sigma_{\mathscr{G}}$ is an isomorphism.*

Proof. First, we define a homomorphism $\rho_{\mathscr{F}}: \psi^*\psi_*\mathscr{F} \to \mathscr{F}$. For an open subset U of X, $\sigma \in \Gamma(U, \psi^*\psi_*\mathscr{F})$ is represented by $\{\alpha_\lambda \in (\psi_*\mathscr{F})(V_\lambda) = \mathscr{F}(\psi^{-1}(V_\lambda)) \mid \lambda \in \Lambda\}$, where $\{U_\lambda \mid \lambda \in \Lambda\}$ is an open cover of U and the V_λ are open subsets of Y with $\psi(U_\lambda) \subseteq V_\lambda$, and the $\alpha_\lambda \in (\psi_*\mathscr{F})(V_\lambda)$ satisfy $(\alpha_\lambda)_{\psi(x)} = (\alpha_\mu)_{\psi(x)}$ if $x \in U_\lambda \cap U_\mu$. Then $\bar{\alpha}_\lambda = \alpha_\lambda|_{U_\lambda}$ satisfy $\bar{\alpha}_\lambda|_{U_\lambda \cap U_\mu} = \bar{\alpha}_\mu|_{U_\lambda \cap U_\mu}$ and so $\alpha \in \mathscr{F}(U)$ is determined such that $\alpha|_{U_\lambda} = \bar{\alpha}_\lambda$ for all λ. Thus, by $\rho_{\mathscr{F},U}(\sigma) = \alpha$, the homomorphism $\rho_{\mathscr{F}}$ is defined.

It is easy to check that $h(\rho_{\mathscr{F}}) = \mathrm{id}: \psi_*\mathscr{F} \to \psi_*\mathscr{F}$. By making use of this, we can verify that h is bijective. The rest is left to the reader. □

Example 2.17. Let X be an open subset of Y and let $i: X \to Y$ the inclusion map. For any sheaf $\mathscr{G}$ on Y, $i^*\mathscr{G}$ coincides with the restriction of $\mathscr{G}$ to X, i.e., $i^*\mathscr{G} = \mathscr{G}|_X$. $\sigma_{\mathscr{G},V}: \mathscr{G}(V) \to \mathscr{G}(V \cap X) = i_*i^*\mathscr{G}(V)$ is the restriction for any open subset V of Y.

If $\mathscr{K}$ is a subsheaf of $\mathscr{G}|_X$, then the sheaf $\mathscr{F} = \sigma_{\mathscr{G}}^{-1}(i_*(\mathscr{K}))$ is a subsheaf of $\mathscr{G}$ such that $\mathscr{F}|_X = \mathscr{K}$. $\mathscr{F}$ is said to be the *extension of $\mathscr{K}$ as a subsheaf* of $\mathscr{G}$.

Note that, for any sheaf $\mathscr{F}$ on X, $\rho_{\mathscr{F}}$ is the isomorphism: $i^*i_*\mathscr{F} \to \mathscr{F}$.

b. Next, let $(X, \mathscr{O}_X)$ and $(Y, \mathscr{O}_Y)$ be ringed spaces, and let $f = (\psi, \theta)$: $(X, \mathscr{O}_X) \to (Y, \mathscr{O}_Y)$ be a morphism. $\psi^*(\theta): \psi^*\mathscr{O}_Y \to \psi^*\psi_*\mathscr{O}_X$ is naturally induced by θ.

Definition. For any $\mathscr{O}_Y$-module $\mathscr{G}$, the *inverse image $\mathscr{O}_X$-module* of $\mathscr{G}$ by f, denoted $f^*\mathscr{G}$, is defined to be $\psi^*\mathscr{G} \otimes_{\psi^*\mathscr{O}_Y} \mathscr{O}_X$ where $\psi^*\mathscr{G}$ is regarded as a $\psi^*\mathscr{O}_Y$-module naturally and $\mathscr{O}_X$ as a $\psi^*\mathscr{O}_Y$-module via $\rho_{\mathscr{O}_X} \circ \psi^*(\theta)$.

Note that, if $x \in X$ and $y = \psi(x)$, then $(f^*\mathscr{G})_x = \mathscr{G}_y \otimes_{\mathscr{O}_{Y,y}} \mathscr{O}_{X,x}$. By using the canonical homomorphisms: $\mathscr{O}_Y \to \psi_*\psi^*\mathscr{O}_Y$, $\mathscr{G} \to \psi_*\psi^*\mathscr{G}$ and θ: $\mathscr{O}_Y \to \psi_*\mathscr{O}_X$, we have a homomorphism $\mathscr{G} \cong \mathscr{G} \otimes_{\mathscr{O}_Y} \mathscr{O}_Y \to \psi_*\psi^*\mathscr{G} \otimes_{\psi_*\psi^*\mathscr{O}_Y} \psi_*\mathscr{O}_X$. From the last term to $\psi_*f^*\mathscr{G}$, we have the natural homomorphism.

Composing these homomorphisms, the homomorphism: $\mathscr{G} \to \psi_* f^* \mathscr{G} = f_* f^* \mathscr{G}$ is obtained, which is also denoted by $\sigma_{\mathscr{G}}$. Similarly, we can define $\rho_{\mathscr{F}}: f^* f_* \mathscr{F} \to \mathscr{F}$ for any $\mathscr{O}_X$-module $\mathscr{F}$. As in the case of sheaves, we have the natural homomorphism h: $\mathrm{Hom}_{\mathscr{O}_X}(f^*\mathscr{G}, \mathscr{F}) \to \mathrm{Hom}_{\mathscr{O}_Y}(\mathscr{G}, f_*\mathscr{F})$ defined by $h(\eta) = f_*(\eta) \circ \sigma_{\mathscr{G}}$. In the case $\mathscr{G} = f_*\mathscr{F}$, we have $h(\rho_{\mathscr{F}}) = \mathrm{id}$. By this, we can verify that h is a bijection.

Lemma 2.28. *If $\mathscr{G}$ is quasi-coherent, or of finite type, then so is $f^*\mathscr{G}$.*

PROOF. This follows immediately from Proposition 1.16. □

Definition. An $\mathscr{O}_X$-module $\mathscr{L}$ is said to be *locally free*, if there exists an open cover $\{U_\lambda \mid \lambda \in \Lambda\}$ such that each $\mathscr{L} \mid_{U_\lambda}$ is a free $\mathscr{O}_{U_\lambda}$-module.

If a locally free $\mathscr{L}$ is of finite type, then each $\mathscr{L}|_{U_\lambda}$ is isomorphic to $\mathscr{O}_{U_\lambda}^m$ for some m.

c. The following elementary lemma is called the *projection formula*.

Lemma 2.29. *With the previous notation, let $\mathscr{F}$ be an $\mathscr{O}_X$-module and let $\mathscr{L}$ be an $\mathscr{O}_Y$-module which is locally free of finite type. Then*

$$(f_*\mathscr{F}) \otimes_{\mathscr{O}_Y} \mathscr{L} \cong f_*(\mathscr{F} \otimes_{\mathscr{O}_X} f^*\mathscr{L}).$$

In particular, $(f_* \mathscr{O}_X) \otimes_{\mathscr{O}_Y} \mathscr{L} \cong f_* f^* \mathscr{L}$.

PROOF. From the tensor product of $\mathrm{id}: f_*\mathscr{F} \to f_*\mathscr{F}$ and $\sigma_{\mathscr{L}}: \mathscr{L} \to f_* f^*\mathscr{L}$, one obtains $(f_*\mathscr{F}) \otimes_{\mathscr{O}_Y} \mathscr{L} \to (f_*\mathscr{F}) \otimes_{\mathscr{O}_Y} (f_* f^*\mathscr{L})$. Then, composing this with the natural homomorphism $(f_*\mathscr{F}) \otimes_{\mathscr{O}_Y} (f_* f^*\mathscr{L}) \to f_*(\mathscr{F} \otimes_{\mathscr{O}_X} f^*\mathscr{L})$, one has an $\mathscr{O}_Y$-homomorphism

$$\tau: (f_*\mathscr{F}) \otimes_{\mathscr{O}_Y} \mathscr{L} \to f_*(\mathscr{F} \otimes_{\mathscr{O}_X} f^*\mathscr{L}).$$

In order to show that τ is an isomorphism, it suffices to prove it locally. For every point $y \in Y$, choose an open neighborhood U of y such that $\mathscr{L}|_U \cong \mathscr{O}_U^m$. Then

$$(f_*\mathscr{F}) \otimes_{\mathscr{O}_Y} \mathscr{L}|_U \cong (f_*\mathscr{F})|_U \otimes_{\mathscr{O}_U} \mathscr{L}|_U \cong (f_*\mathscr{F}|_U)^m.$$

Letting $f_1 = f|_{\psi^{-1}(U)}$, one has

$$\begin{aligned} f_*(\mathscr{F} \otimes f^*\mathscr{L})|_U &\cong f_{1*}(\mathscr{F}|_{\psi^{-1}(U)} \otimes f_1^*(\mathscr{L}|_U)) \\ &\cong f_{1*}(\mathscr{F}|_{\psi^{-1}(U)})^m = (f_*\mathscr{F})^m|_U. \end{aligned}$$ □

d. Now, we consider inverse image $\mathscr{O}$-modules in the category of schemes.

Lemma 2.30. *Let $\varphi: A \to B$ be a ring homomorphism and let M be an A-module. Then, letting $f = ({}^a\varphi, \tilde{\varphi})$, one has*

$$f^*\tilde{M} \cong (M \otimes_A B)^\sim.$$

PROOF. For any B-module N, we have by Lemma 2.27

$$\mathrm{Hom}_{\tilde{B}}(f^*\tilde{M}, \tilde{N}) \cong \mathrm{Hom}_{\tilde{A}}(\tilde{M}, f_*\tilde{N}) \cong \mathrm{Hom}_{\tilde{A}}(\tilde{M}, \tilde{N}_{[\varphi]}).$$

Furthermore, since $f^*\tilde{M}$ is quasi-coherent by Lemma 2.28, there is a B-module L such that $f^*\tilde{M} \cong \tilde{L}$. Thus by Lemma 1.24, we have

$$\mathrm{Hom}_{\tilde{B}}(\tilde{L}, \tilde{N}) \cong \mathrm{Hom}_B(L, N) \quad \text{and} \quad \mathrm{Hom}_{\tilde{A}}(\tilde{M}, \tilde{N}_{[\varphi]}) \cong \mathrm{Hom}_A(M, N_{[\varphi]}).$$

Hence,

$$\mathrm{Hom}_B(L, N) \cong \mathrm{Hom}_A(M, N_{[\varphi]}).$$

On the other hand, we have an isomorphism $\mu\colon \mathrm{Hom}_A(M, N_{[\varphi]}) \cong \mathrm{Hom}_B(M \otimes_A B, N)$ defined by $\mu(\eta)(x \otimes b) = \eta(x)b$ for $x \in M$ and $b \in B$. Hence,

$$\mathrm{Hom}_B(L, N) \cong \mathrm{Hom}_B(M \otimes_A B, N).$$

From this, it follows that $L \cong M \otimes_A B$. □

e. Let $f\colon X \to Y$ be a morphism of schemes and $\mathscr{T}$ be an $\mathscr{O}_Y$-ideal which is quasi-coherent. Then $f^*\mathscr{T}$ is quasi-coherent and a homomorphism $f^*\mathscr{T} \to f^*\mathscr{O}_Y = \mathscr{O}_X$ is derived from the inclusion $\mathscr{T} \to \mathscr{O}_Y$.

Definition. The $\mathscr{O}_X$-ideal generated by the image of $f^*\mathscr{T} \to \mathscr{O}_X$ is said to be the *inverse image $\mathscr{O}_X$-ideal*, denoted by $f^{-1}\mathscr{T}$.

Proposition 2.18. *The ideal $f^{-1}\mathscr{T}$ is quasi-coherent and the subscheme defined by $f^{-1}\mathscr{T}$ is the inverse image scheme of the closed subscheme defined by $\mathscr{T}$* (§1.22.**c**).

PROOF. If X and Y are affine, then $X = \mathrm{Spec}\, B$, $Y = \mathrm{Spec}\, A$ and $\mathscr{T} = \tilde{I}$, where I is an ideal of A. Clearly, $f^{-1}\mathscr{T} = (IB)^\sim$. The inverse image scheme of $V(I)$ is associated with a ring $B \otimes_A (A/I)$. The result follows immediately from $B \otimes_A A/I \cong B/IB$. □

§2.17 The Pullback Theorem

a. Theorem 2.29. *Let $f\colon V \to W$ be a proper surjective morphism of varieties. If W is normal and the field extension* $\mathrm{Rat}(V)/\mathrm{Rat}(W)$ *is algebraically closed, then $f_*\mathscr{O}_V = \mathscr{O}_W$. In particular, $A(V) = A(W)$.*

PROOF. By Theorem 1.19, $\sigma_{\mathscr{O}_W}\colon \mathscr{O}_W \to f_*f^*\mathscr{O}_W = f_*\mathscr{O}_V$ is injective. Thus, we can assume W is an affine scheme $\mathrm{Spec}\, R$, where the R is a normal ring finitely generated over k. If $*$ is the generic point of W, we have by Lemma 1.44

$$\Gamma(f^{-1}(*), \mathscr{O}) \cong \Gamma(V, \mathscr{O}_V) \otimes_R Q(R).$$

Since $f^{-1}(*)$ is a $k(*)$-complete variety, we have

$$\Gamma(f^{-1}(*), \mathcal{O}) \cong k(*) = \mathrm{Rat}(W) = Q(R)$$

by Lemma 1.54, where $\mathrm{Rat}(W)$ is algebraically closed in $\mathrm{Rat}(f^{-1}(*)) = \mathrm{Rat}(V)$. Hence, $R \subseteq A(V) = \Gamma(V, \mathcal{O}_V) \subseteq Q(R)$.

Taking a nonzero b in $A(V)$, we define a subring $R[b]$ of $A(V)$. Corresponding to the inclusions $R \subseteq R[b] \subseteq A(V)$, one has the next commutative triangle by Proposition 1.31.

$$\begin{array}{ccc} V & \xrightarrow{\ f\ } & W = \mathrm{Spec}\, R \\ {\scriptstyle g}\big\downarrow & \nearrow_{h} & \\ \mathrm{Spec}\, R[b] & & \end{array}$$

g is proper, since f is proper, by Proposition 1.57.(i). Hence, $g(V)$ is a closed subset. Since $b \in Q(R)$, h is a birational morphism. Hence, g is dominating and so g is surjective, since g is proper. Since $f = h \circ g$ is proper, h is also proper by Proposition 1.57.(ii). If $b \notin R$, then by the Corollary to Theorem 2.14, there exists a prime ideal $\mathfrak{p}$ of height 1 such that $b \notin R_{\mathfrak{p}}$; thus $b^{-1} \in \mathfrak{p}R_{\mathfrak{p}}$. Extending the base scheme of Spec $R[b]$ from Spec R to Spec $R_{\mathfrak{p}}$, one has q: Spec $R[b] \otimes_R R_{\mathfrak{p}} \to \mathrm{Spec}\ \mathrm{R}_{\mathfrak{p}}$ which is also proper by Theorem 1.22.(iii). Hence Im q must be a closed subset. However, Im $q = D(b^{-1})$ which is not closed. Therefore, $b \in R$ and so $A(V) = R$. □

b. Theorem 2.30. *Let $f: V \to W$ be a proper surjective morphism. If both V and W are normal, then $f_*\mathcal{O}_V = (\mathcal{O}_W)'_L$ where $L = \mathrm{Rat}(V)$. Furthermore, $A(V)$ is integral over $A(W)$.*

Proof. By the Stein factorization of f (see Theorem 2.26) and the last theorem, we can assume f is finite. If W is affine, i.e., $W = \mathrm{Spec}\, R$, then $A(V) = \Gamma(V, \mathcal{O}_V)$ is finite over R and $A(V)$ is normal by Proposition 2.2, i.e., $A(V) = R'_L$. Hence, $f_*\mathcal{O}_V = (\mathcal{O}_W)'_L$.

Take an affine open cover $\{W_i \mid i \in J\}$ of W. Then f is finite, and $V_i = f^{-1}(W_i)$ is an affine variety such that $A(V_i)$ is finite over $A(W_i)$ for all $i \in J$. Take a nonzero α of $A(V)$. Then α is algebraic over $\mathrm{Rat}(W)$ and so there exists a monic minimal polynomial $\Phi(\mathsf{X}) \in \mathrm{Rat}(W)[\mathsf{X}]$ with $\Phi(\alpha) = 0$. We claim that $\Phi(\mathsf{X}) \in A(W)[\mathsf{X}]$, i.e., α is integral over $A(W)$. Since $A(V) = \bigcap_{i \in J} A(V_i)$ and $A(W) = \bigcap_{i \in J} A(W_i)$, we use the following lemma.

Lemma 2.31. *Let R be a Noetherian normal ring and let α be an integral element over R. If $\Phi(\mathsf{X}) \in R[\mathsf{X}]$ is a monic minimal polynomial such that $\Phi(\alpha) = 0$, then $\Phi(\mathsf{X})$ is irreducible as a polynomial over $Q(R)$.*

Proof. By the Corollary to Theorem 2.14, we can assume R is a DVR. Thus, one has the valuation ord_R corresponding to R. For any nonzero

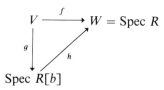

$f(\mathsf{X}) = \sum_{i=0}^{n} a_i \mathsf{X}^i \in Q(R)[\mathsf{X}]$, define $\operatorname{ord}_R f = \min\{\operatorname{ord}_R(a_i) \mid a_i \neq 0\}$. One can prove immediately that $\operatorname{ord}_R(fg) = \operatorname{ord}_R f + \operatorname{ord}_R g$, if f and g are nonzero polynomials with coefficients in $Q(R)$.

Suppose $\Phi(\mathsf{X})$ is reducible as a polynomial over $Q(R)$. Then there exist monic polynomials $\Phi_1, \Phi_2 \in Q(R)[\mathsf{X}]$ such that $\Phi = \Phi_1 \cdot \Phi_2$. Since Φ is a monic polynomial with coefficients in R, it follows that $\operatorname{ord}_R f = 0$. Hence $0 = \operatorname{ord}_R \Phi_1 + \operatorname{ord}_R \Phi_2$. Thus if $\Phi_1 \notin R[\mathsf{X}]$, then $\operatorname{ord}_R \Phi_1 < 0$, and so $\operatorname{ord}_R \Phi_2 > 0$. This implies $\Phi_2 \in R[\mathsf{X}]$. But since Φ_2 is a monic polynomial, one has $\operatorname{ord}_R \Phi_2 \leq 0$, which is absurd. □

By this lemma, $\alpha \in A(V_i)$ implies that $\Phi(\mathsf{X}) \in A(W_i)[\mathsf{X}]$ for all i; hence $\Phi(\mathsf{X}) \in A(W)[\mathsf{X}]$. Thus from $\Phi(\alpha) = 0$, it follows that α is integral over $A(W)$. □

Corollary. *Let f: $V \to W$ be a proper surjective morphism between normal varieties. Then $\gamma_k(V) = \gamma_k(W)$.*

The invariant $\gamma_k(V)$ was introduced in §1.24.**d**.

Remark. The above corollary was communicated to the author by Tsunoda.

c. Theorem 2.31. *Let $\mathscr{L}$ be a locally free $\mathscr{O}_W$-module of finite type and let f: $V \to W$ be a proper surjective morphism. If W is normal and the extension* $\operatorname{Rat}(V)/\operatorname{Rat}(W)$ *is algebraically closed, then $\Gamma(W, \mathscr{L}) \cong \Gamma(V, f^*\mathscr{L})$.*

Proof. By Lemma 2.29, one has $f_*f^*\mathscr{L} \cong (f_*\mathscr{O}_V) \otimes \mathscr{L}$. Since W is normal, it follows from Theorem 2.29 that $f_*\mathscr{O}_V = \mathscr{O}_W$. Hence, $\Gamma(V, f^*\mathscr{L}) = \Gamma(W, f_*f^*\mathscr{L}) \cong \Gamma(W, \mathscr{L})$. □

This result is one of the pullback theorems which will play essential roles in the birational geometry developed in Chapters 10 and 11.

If W is complete, then by the finiteness theorem (see Theorem 7.14), the k-vector space $\Gamma(W, \mathscr{L})$ is finite-dimensional.

Definition. Let V be a complete variety and let (V', μ) be the normalization of V. For any locally free $\mathscr{O}_V$-module $\mathscr{L}$ of finite type, define $l_V(\mathscr{L})$ to be $\dim_k \Gamma(V', \mu^*\mathscr{L})$.

Note that since μ: $V' \to V$ is finite, V' is also a complete variety by Theorem 1.22.(ii), and hence $l_V(\mathscr{L}) < \infty$.

Theorem 2.32. *Let f: $V \to W$ be a proper surjective morphism and let $\mathscr{L}$ be a locally free $\mathscr{O}_W$-module of finite type. If W is complete and the field extension* $\operatorname{Rat}(V)/\operatorname{Rat}(W)$ *is algebraically closed, then $l_V(f^*\mathscr{L}) = l_W(\mathscr{L})$.*

PROOF. Letting (V', μ) and (W', λ) be the normalizations of V and W, respectively, one has the normalization f': $V' \to W'$ of f. By the pullback theorem (Theorem 2.31),

$$\dim_k \Gamma(V', f'^*\lambda^*\mathscr{L}) = \dim_k \Gamma(W', \lambda^*\mathscr{L}).$$

But since $f \circ \mu = \lambda \circ f'$, $\dim_k \Gamma(V', f'^*\lambda^*\mathscr{L}) = \dim_k \Gamma(V', \mu^* f^*\mathscr{L})$. Hence, $l_V(f^*\mathscr{L}) = \dim_k \Gamma(V', \mu^* f^*\mathscr{L}) = \dim_k \Gamma(W', \lambda^*\mathscr{L}) = l_W(\mathscr{L})$. □

d. Now, let F be a closed proper subset of a normal variety V and let j: $V_0 = V \backslash F \to V$ be the natural inclusion map. For any locally free $\mathcal{O}_V$-module $\mathscr{L}$ of finite type, one has the injective homomorphism

$$\sigma_{\mathscr{L}}: \mathscr{L} \to j_*(\mathscr{L}|_{V_0}).$$

Lemma 2.32. *If* $\operatorname{codim}(F) \geq 2$, *then* $\sigma_{\mathscr{L}}: \mathscr{L} \to j_*(\mathscr{L}|_{V_0})$ *is an isomorphism. In particular*, $\Gamma(V, \mathscr{L}) \cong \Gamma(V_0, \mathscr{L})$.

PROOF. Take an open cover $\{U_\lambda | \lambda \in \Lambda\}$ of V such that $\mathscr{L}|_{U_\lambda} \cong \mathcal{O}_{U_\lambda}^m$. It suffices to prove that $\sigma_{\mathscr{L}}|_{U_\lambda}$ is an isomorphism for each λ. Thus, one may assume $\mathscr{L} = \mathcal{O}_V^m$, and then $\Gamma(U_\lambda, j_*(\mathscr{L}|_{V_0})) = \Gamma(U_\lambda \backslash F, \mathcal{O}_V)^m = \Gamma(U_\lambda, \mathcal{O}_V)^m = \Gamma(U_\lambda, \mathscr{L})^m$ by Theorem 2.15. □

Theorem 2.33. *Let* f: $V \to W$ *be a proper surjective morphism such that the field extension* $\mathrm{Rat}(V)/\mathrm{Rat}(W)$ *is algebraically closed. If* W *is a complete normal variety and* F *is a closed subset of* W *with* $\operatorname{codim}(F) \geq 2$, *then for any locally free* $\mathcal{O}_W$*-module* $\mathscr{L}$ *of finite type, one has an isomorphism*

$$f^*: \Gamma(W, \mathscr{L}) \cong \Gamma(V \backslash f^{-1}(F), f^*\mathscr{L}); \text{ thus } l_W(\mathscr{L}) = \dim_k \Gamma(V \backslash f^{-1}(F), f^*\mathscr{L}).$$

PROOF. Letting $W_0 = W \backslash F$ and $V_0 = V \backslash f^{-1}(F) = f^{-1}(W_0)$, one has $\Gamma(W_0, \mathscr{L}) \cong \Gamma(W, \mathscr{L})$, and $\Gamma(W, \mathscr{L}) \cong \Gamma(V, f^*\mathscr{L})$, $\Gamma(W_0, \mathscr{L}) \cong \Gamma(V_0, f^*\mathscr{L})$ by Theorem 2.31. Hence,

$$\Gamma(W, \mathscr{L}) \cong \Gamma(V_0, f^*\mathscr{L}).$$ □

§2.18 Invertible Sheaves

a. Let $(X, \mathcal{O}_X)$ be a ringed space.

Definition. An $\mathcal{O}_X$-module $\mathscr{L}$ is said to be *invertible* if it is locally isomorphic to $\mathcal{O}_X$, i.e., there exists an open cover $\{U_\lambda | \lambda \in \Lambda\}$ of X such that $\mathscr{L}|_{U_\lambda} \cong \mathcal{O}_{U_\lambda}$ for all λ. Any invertible $\mathcal{O}_X$-module is often referred to as *an invertible sheaf* on X.

Proposition 2.19

(i) *If* $\mathscr{L}$ *and* $\mathscr{M}$ *are invertible sheaves, then so is* $\mathscr{L} \otimes_{\mathscr{O}_X} \mathscr{M}$.
(ii) *If* $\mathscr{L}$ *is invertible, then* $\mathscr{H}om_{\mathscr{O}_X}(\mathscr{L}, \mathscr{O}_X)$ *is also invertible,* $\mathscr{L} \otimes_{\mathscr{O}_X} \mathscr{H}om_{\mathscr{O}_X}(\mathscr{L}, \mathscr{O}_X) \cong \mathscr{O}_X$, *and* $\mathscr{H}om_{\mathscr{O}_X}(\mathscr{L}, \mathscr{L}) \cong \mathscr{O}_X$.

PROOF. (i) Choosing a suitable open cover $\{U_\lambda \mid \lambda \in \Lambda\}$ of X, we can assume that $\mathscr{L}|_{U_\lambda} \cong \mathscr{O}_{U_\lambda}$ and $\mathscr{M}|_{U_\lambda} \cong \mathscr{O}_{U_\lambda}$ for all λ. Then $(\mathscr{L} \otimes_{\mathscr{O}_X} \mathscr{M})|_{U_\lambda} \cong \mathscr{L}|_{U_\lambda} \otimes_{\mathscr{O}_{U_\lambda}} \mathscr{M}|_{U_\lambda} \cong \mathscr{O}_{U_\lambda}$ by Proposition 1.16.(iv).

(ii) For any open subset U of X, define $\xi_U: \mathscr{L}(U) \otimes \operatorname{Hom}_{\mathscr{O}_U}(\mathscr{L}|_U, \mathscr{O}_U) \to \operatorname{Hom}_{\mathscr{O}_U}(\mathscr{L}|_U, \mathscr{L}|_U)$ and $\eta_U: \mathscr{O}_X(U) \to \operatorname{Hom}_{\mathscr{O}_U}(\mathscr{L}|_U, \mathscr{L}|_U)$ by $\xi_U(s \otimes \varphi) = s\varphi$ and $\eta_U(\alpha)(\beta) = \alpha|_W \cdot \beta$ for $\alpha \in \mathscr{O}_X(U)$, $\beta \in \mathscr{L}(W)$, where W is any open subset of U. From these $\{\xi_U\}$ and $\{\eta_U\}$, we form homomorphisms of sheaves $\xi: \mathscr{L} \otimes_{\mathscr{O}_X} \mathscr{H}om_{\mathscr{O}_X}(\mathscr{L}, \mathscr{O}_X) \to \mathscr{H}om_{\mathscr{O}}(\mathscr{L}, \mathscr{L})$ and $\eta: \mathscr{O}_X \to \mathscr{H}om_{\mathscr{O}_X}(\mathscr{L}, \mathscr{L})$. Then ξ and η are isomorphisms, since $\mathscr{L}$ is locally isomorphic to $\mathscr{O}_X$. □

Defining $\mathscr{L}^{-1} = \mathscr{H}om_{\mathscr{O}_X}(\mathscr{L}, \mathscr{O}_X)$, $\mathscr{L}^{\otimes -n} = (\mathscr{L}^{\otimes n})^{-1}$ for any $n \geq 1$, one has $(\mathscr{L}^{\otimes n}) \otimes (\mathscr{L}^{\otimes m}) \cong \mathscr{L}^{\otimes (n+m)}$ for all $n, m \in \mathbb{Z}$. Thus we have an Abelian group structure on $\operatorname{Pic}(X)$, that is defined to be the set of all isomorphism classes of invertible sheaves on X.

Definition. $\operatorname{Pic}(X)$ is said to be the *Picard group* of X.

b. Let $\mathscr{L}$ be an invertible sheaf on X, i.e., there exist an open cover $\{U_\lambda \mid \lambda \in \Lambda\}$ of X and $\mathscr{O}_{U_\lambda}$-isomorphisms $\theta_\lambda: \mathscr{O}_{U_\lambda} \xrightarrow{\sim} \mathscr{L}|_{U_\lambda}$ for all $\lambda \in \Lambda$. One has $\beta_\lambda = \theta_\lambda(1) \in \mathscr{L}(U_\lambda)$ and defines an $\mathscr{O}_{U_\lambda}$-module $\mathscr{O}_{U_\lambda} \cdot \beta_\lambda$ by $(\mathscr{O}_{U_\lambda} \cdot \beta_\lambda)(W) = \mathscr{O}_{U_\lambda}(W) \cdot \beta_\lambda|_W$, where W is any open subset of U_λ. Then $\mathscr{L}|_{U_\lambda} = \mathscr{O}_{U_\lambda} \cdot \beta_\lambda$ and $\mathscr{L}|_{U_\lambda \cap U_\mu} = \mathscr{O}_{U_\lambda \cap U_\mu} \cdot \beta'_\lambda = \mathscr{O}_{U_\lambda \cap U_\mu} \cdot \beta'_\mu$, in which the β'_λ denote the restrictions to $U_\lambda \cap U_\mu$. Since $\mathscr{O}_{U_\lambda \cap U_\mu} \cdot \beta'_\lambda = \mathscr{O}_{U_\lambda \cap U_\mu} \cdot \beta'_\mu$, there exist $s_{\lambda\mu} \in \Gamma(U_\lambda \cap U_\mu, \mathscr{O}_X)$ such that $\beta'_\lambda = s_{\mu\lambda} \cdot \beta'_\mu$ for all λ, μ. These $s_{\lambda\mu}$ satisfy the conditions: $s_{\lambda\lambda} = 1$, $s_{\lambda\mu} \cdot s_{\mu\lambda} = 1$, and $s''_{\lambda\mu} = s''_{\lambda\nu} \cdot s''_{\nu\mu}$, where the $s''_{\lambda\mu}$ denote the restrictions to $U_\lambda \cap U_\mu \cap U_\nu$ for all λ, μ, ν. Thus $s_{\lambda\mu}$ is a unit of the ring $\Gamma(U_\lambda \cap U_\mu, \mathscr{O}_X)$.

Remark. These β_λ are called *bases* of $\mathscr{L}$.

Definition. For a ringed space $(X, \mathscr{O}_X)$, define a sheaf $\mathscr{F}$ of multiplicative groups by letting $\mathscr{F}(U)$ be the group of units of $\Gamma(U, \mathscr{O}_X)$ for any open subset U of X. $\mathscr{F}$ is said to be the *sheaf of units* of X, denoted by $\mathscr{O}_X^*$.

Then, $s_{\lambda\mu} \in \Gamma(U_\lambda \cap U_\mu, \mathscr{O}_X^*)$. Furthermore, any $\alpha \in \Gamma(X, \mathscr{L})$ gives rise to $\{a_\lambda\}$ such that $\alpha|_{U_\lambda} = a_\lambda \cdot \beta_\lambda$ where $a_\lambda \in \Gamma(U_\lambda, \mathscr{O}_X)$ for all λ. Since $\alpha|_{U_\lambda \cap U_\mu} = a'_\lambda \cdot \beta'_\lambda = a'_\lambda \cdot s_{\mu\lambda} \cdot \beta'_\mu = a'_\mu \cdot \beta'_\mu$, one has $a'_\mu = s_{\mu\lambda} \cdot a'_\lambda$ for all λ, μ, where a'_λ, β'_λ and a'_μ, β'_μ are restrictions to $U_\lambda \cap U_\mu$. Conversely, given $\{a_\lambda\} = \{a_\lambda \in \Gamma(U_\lambda, \mathscr{O}_X) \mid a'_\lambda = s_{\lambda\mu} a'_\mu$ for all $\lambda, \mu \in \Lambda\}$, one can construct a section α of $\mathscr{L}$ in such a way that $\alpha|_{U_\lambda} = a_\lambda \cdot \beta_\lambda$ for all λ.

EXAMPLE 2.18. On $\mathbf{P}_k^n$ with homogeneous coordinate system $(\mathsf{T}_0, \ldots, \mathsf{T}_n)$, one can define an invertible sheaf $\mathcal{O}(1)$ by $\mathcal{O}(1)|_{D_+(\mathsf{T}_i)} = \mathcal{O}_{D_+(\mathsf{T}_i)} \cdot \mathsf{T}_i$. Then $\mathcal{O}(1)^{\otimes m}|_{D_+(\mathsf{T}_i)} = \mathcal{O}_{D_+(\mathsf{T}_i)} \cdot \mathsf{T}_i^m$, and $\mathcal{O}(1)^{\otimes m}$ is denoted $\mathcal{O}(m)$.

Any $\varphi \in \Gamma(\mathbf{P}_k^n, \mathcal{O}(m))$ corresponds to $\{\varphi_j(\mathsf{T}_0/\mathsf{T}_j, \ldots, \mathsf{T}_n/\mathsf{T}_j)\}_{0 \le j \le n}$ such that $\varphi|_{D_+(\mathsf{T}_j)} = \varphi_j(\mathsf{T}_0/\mathsf{T}_j, \ldots, \mathsf{T}_n/\mathsf{T}_j)\mathsf{T}_j^m$, for all j. Since $\varphi_j(\mathsf{T}_0/\mathsf{T}_j, \ldots, \mathsf{T}_n/\mathsf{T}_j) = \varphi_0(\mathsf{T}_1/\mathsf{T}_0, \ldots, \mathsf{T}_n/\mathsf{T}_0)(\mathsf{T}_0/\mathsf{T}_j)^m$ are polynomials for all j, $\varphi_0(\mathsf{T}_1/\mathsf{T}_0, \ldots, \mathsf{T}_n/\mathsf{T}_0)$ is a polynomial in $\mathsf{T}_1/\mathsf{T}_0, \ldots, \mathsf{T}_n/\mathsf{T}_0$ of degree m; thus $\varphi_j(\mathsf{T}_0/\mathsf{T}_j, \ldots, \mathsf{T}_n/\mathsf{T}_j)\mathsf{T}_j^m = \varphi_0(\mathsf{T}_1/\mathsf{T}_0, \ldots, \mathsf{T}_n/\mathsf{T}_0)\mathsf{T}_0^m$ is a homogeneous polynomial of degree m. Conversely, given a homogeneous polynomial F of degree m, one has a section φ of $\mathcal{O}(m)$ defined by $\varphi|_{D_+(\mathsf{T}_j)} = F(\mathsf{T}_0/\mathsf{T}_j, \ldots, 1, \ldots, \mathsf{T}_n/\mathsf{T}_j) \cdot \mathsf{T}_j^m$ for all j. Hence, $\Gamma(\mathbf{P}_k^n, \mathcal{O}(m))$ can be identified with the space of all homogeneous polynomials of degree m.

c. If α is a section of an invertible sheaf $\mathcal{L}$ over a scheme X and $x \in X$, then $\bar{\alpha}(x) = \alpha_x \bmod \mathfrak{m}_x \mathcal{L}_x \in \mathcal{L}_x/\mathfrak{m}_x \mathcal{L}_x \cong k(x)$ is defined, since $\mathcal{L}_x \cong \mathcal{O}_{X,x}$.

Definition. Let $X_\alpha = \{x \in X \mid \bar{\alpha}(x) \neq 0\}$. $X \backslash X_\alpha$ is said to be the *set of zeros* of α.

If X is covered by affine open subsets U_λ for $\lambda \in \Lambda$ such that $\mathcal{L}|_{U_\lambda} \cong \mathcal{O}_{U_\lambda}$, then $X_\alpha \cap U_\lambda = D(a_\lambda)$, where $a_\lambda \cdot \beta_\lambda = \alpha|_{U_\lambda} \in \mathcal{L}(U_\lambda)$ for all λ. Thus X_α is an open subset of X.

Remark. If φ is a section of $\mathcal{L}$ over X, then $\varphi|_{X_\alpha}$ is written as $\psi\alpha|_{X_\alpha}$ for some $\psi \in A(X_\alpha)$; i.e., φ/α is a regular function on X_α.

§2.19 Rational Sections of an Invertible Sheaf

a. Definition. For an integral scheme V, define the sheaf $\mathcal{R}at_V$ of *rational functions* on V by $\mathcal{R}at_V(U) = \mathrm{Rat}(V)$ if U is a nonempty open subset, and $\mathcal{R}at_V(\emptyset) = 0$, whose restriction maps $r_{W,U}$ are id, if $W \neq \emptyset$; and $r_{\emptyset,U}$ is the zero map.

$\mathcal{R}at_V$ is a constant sheaf on V which is also an $\mathcal{O}_V$-module.

Definition. For an $\mathcal{O}_V$-module $\mathcal{F}$, any section of $\mathcal{R}at_V \otimes_{\mathcal{O}_V} \mathcal{F}$ is said to be a *rational section* of $\mathcal{F}$ over V. The space of all rational sections of $\mathcal{F}$ over V is denoted by $\Gamma_{\mathrm{rat}}(V, \mathcal{F})$.

Thus, $\Gamma_{\mathrm{rat}}(V, \mathcal{F}) = \Gamma(V, \mathcal{R}at_V \otimes_{\mathcal{O}_V} \mathcal{F})$. When one wants to avoid confusion, any section of $\mathcal{F}$ over V is said to be a *regular section* of $\mathcal{F}$.

b. Given an invertible sheaf $\mathcal{L}$ on V, choose an affine open cover $\{U_\lambda \mid \lambda \in \Lambda\}$ of V with bases $\{\beta_\lambda\}$ such that $\mathcal{L}|_{U_\lambda} = \mathcal{O}_{U_\lambda} \cdot \beta_\lambda$ for all λ (see

§2.18.**b**). Then any rational section α of $\mathscr{L}$ can be written as $\alpha|_{U_\lambda} = a_\lambda \cdot \beta_\lambda$ for some $a_\lambda \in \operatorname{Rat}(V)$. These $\{a_\lambda\}$ satisfy $a_\lambda = s_{\lambda\mu} \cdot a_\mu$ as rational functions, and so one has a bijection $\Gamma_{\text{rat}}(V, \mathscr{L}) \cong \{\{a_\lambda\} \mid a_\lambda \in \operatorname{Rat}(V)$ and $a_\lambda = s_{\lambda\mu} \cdot a_\mu$ for all $\lambda, \mu\}$. Then, we say that a *rational section* α *of* $\mathscr{L}$ *corresponds to* $\{a_\lambda\}$ with respect to bases $\{\beta_\lambda\}$ of $\mathscr{L}$.

Lemma 2.33. *Given any invertible sheaf* $\mathscr{L}$ *on* V, *there exists a nonzero rational section* γ *of* $\mathscr{L}$ *over* V. *Furthermore, if* γ *is a nonzero rational section, then* $\Gamma_{\text{rat}}(V, \mathscr{L}) = \operatorname{Rat}(V)\gamma$.

Proof. Specifying an index 1 of Λ, define $c_\lambda = s_{\lambda, 1}$, which is clearly a nonzero rational function of V. Since $c_\lambda = s_{\lambda\mu} c_\mu$ for all λ, μ, one obtains a rational section γ corresponding to $\{c_\lambda\}$.

If $\{a_\lambda\}$ corresponds to a given rational section α, then $a_\lambda/c_\lambda = (s_{\lambda\mu} \cdot a_\mu)/(s_{\lambda\mu} \cdot c_\mu) = a_\mu/c_\mu$ for all λ, μ, which defines a rational function φ; hence $\alpha = \varphi\gamma$. □

c. Given a nonzero $\alpha \in \Gamma_{\text{rat}}(V, \mathscr{L})$ corresponding to $\{a_\lambda\}$ with respect to $\{\beta_\lambda\}$, we define an $\mathscr{O}_V$-module $\mathscr{F}$ by $\mathscr{F}|_{U_\lambda} = \mathscr{O}_{U_\lambda} \cdot a_\lambda^{-1}$ for all λ. Then $\mathscr{F}$ is an invertible sheaf which is a subsheaf of $\mathscr{R}at_V$. We have an $\mathscr{O}_V$-isomorphism $\theta\colon \mathscr{L} \to \mathscr{F}$ such that for all λ, $\theta_{U_\lambda}(a\beta_\lambda) = a \cdot a_\lambda^{-1}$, where $a \in \mathscr{O}(U_\lambda)$.

θ induces an isomorphism $\bar{\theta}_V\colon \Gamma_{\text{rat}}(V, \mathscr{L}) \to \Gamma_{\text{rat}}(V, \mathscr{F})$ which sends α to 1, where 1 is defined by $1_{U_\lambda} = a_\lambda \cdot a_\lambda^{-1}$. It is easy to verify that $\mathscr{F}$ is uniquely determined by α, independently of the choice of $\{\beta_\lambda\}$.

Definition. Any invertible sheaf $\mathscr{F}$ which is a sub $\mathscr{O}_V$-module of $\mathscr{R}at_V$ is said to be a *Cartier sheaf* on V.

Thus, corresponding to a nonzero section α of $\mathscr{L}$, one has the Cartier sheaf $\mathscr{F}$, denoted by $\operatorname{Ca}(\alpha)$.

Definition. The set of all Cartier sheaves on V becomes an Abelian group, denoted by $\operatorname{Ca}(V)$, under the operation of tensor product.

Let $RS(V)$ denote the set of all nonzero rational sections of invertible sheaves on V. If $\alpha \in \Gamma_{\text{rat}}(V, \mathscr{L})$ and $\alpha' \in \Gamma_{\text{rat}}(V, \mathscr{L}')$, then $\alpha \cdot \alpha'$ is defined to be $\alpha \otimes \alpha' \in \Gamma_{\text{rat}}(V, \mathscr{L} \otimes \mathscr{L}')$, where $\alpha \otimes \alpha'$ is the tensor product over $\mathscr{R}at_V$. Then, $\operatorname{Ca}(\alpha \cdot \alpha') = \operatorname{Ca}(\alpha) \otimes \operatorname{Ca}(\alpha')$.

Proposition 2.20. *Let* V *be a complete variety over a field* k *such that* $\operatorname{Rat}(V)/k$ *is algebraically closed. For any nonzero rational sections* α, β *of* $\mathscr{L}$, *one has* $\alpha = \lambda\beta$ *for some* $\lambda \in k$ *if and only if* $\operatorname{Ca}(\alpha) = \operatorname{Ca}(\beta)$.

PROOF. If $\mathrm{Ca}(\alpha) = \mathrm{Ca}(\beta)$, then $\varphi = \alpha \otimes \beta^{-1} \in \mathrm{Rat}(V)$ satisfies $\mathrm{Ca}(\varphi) = \mathcal{O}_V$, i.e., $\varphi \in \Gamma(V, \mathcal{O}_V^*) \subset A(V)$. By Lemma 1.54, $A(V) = k$; hence $\varphi \in k$. The converse is obvious. □

Remark. If $\alpha \in \Gamma_{\mathrm{rat}}(V, \mathcal{L}) \setminus \{0\}$, then $\mathcal{L} \cong \mathrm{Ca}(\alpha)$. Therefore, for any nonzero rational sections α, β of $\mathcal{L}$, one has $\mathrm{Ca}(\alpha) \cong \mathrm{Ca}(\beta)$. Furthermore, $\mathrm{Ca}(\varphi\alpha) = \varphi^{-1} \cdot \mathrm{Ca}(\alpha)$, where $\varphi \in \mathrm{Rat}(V)^*$.

§2.20 Divisors and Invertible Sheaves

a. Let V be a normal variety and let α be a nonzero rational section of some invertible sheaf $\mathcal{L}$ on V, which corresponds to $\{a_\lambda\}$ with respect to a fixed open cover $\{U_\lambda \mid \lambda \in \Lambda\}$ of V and bases $\{\beta_\lambda\}$ of $\mathcal{L}$. Since $\mathrm{Rat}(U_\lambda) = \mathrm{Rat}(V)$, a_λ is regarded as a rational function on U_λ. $D_\lambda = \mathrm{div}(a_\lambda)$ is a divisor on U_λ and $D_\lambda|_{U_\lambda \cap U_\mu} = D_\mu|_{D_\lambda \cap U_\mu}$ for all λ, μ, since $a_\lambda = s_{\lambda\mu} a_\mu$ for $s_{\lambda\mu} \in \Gamma(U_\lambda \cap U_\mu, \mathcal{O}_V^*)$. Thus one has a divisor D on V such that $D|_{U_\lambda} = D_\lambda$, which does not depend on the choices of $\{U_\lambda \mid \lambda \in \Lambda\}$ and $\{\beta_\lambda\}$.

Definition. The divisor D is denoted by $\mathrm{div}(\alpha)$, and called the *divisor defined by the section* α of $\mathcal{L}$.

If α and $\beta \in \Gamma_{\mathrm{rat}}(V, \mathcal{L}) \setminus \{0\}$, then $\alpha = \varphi\beta$ for some $\varphi \in \mathrm{Rat}(V)$; hence $\mathrm{div}\,\alpha = \mathrm{div}\,\varphi + \mathrm{div}\,\beta$, where $\mathrm{div}\,\varphi$ is the divisor of the rational function φ (cf. §2.8.**c**). Hence, $\mathrm{div}(\alpha)$ is linearly equivalent to $\mathrm{div}(\beta)$.

Definition. Given an invertible sheaf $\mathcal{L}$ and a divisor D on a normal complete variety V such that $\mathrm{Rat}(V)/k$ is algebraically closed, define a k-vector space by

$$\mathbf{L}_V(\mathcal{L} + D) = \{\alpha \in \Gamma_{\mathrm{rat}}(V, \mathcal{L}) \mid \alpha = 0 \quad \text{or} \quad \mathrm{div}(\alpha) + D \geq 0\}.$$

Note that $\mathbf{L}_V(\mathcal{O}_V + D) = \mathbf{L}_V(D)$ and $\mathbf{L}_V(\mathcal{L} + 0) = \Gamma(V, \mathcal{L})$.

Lemma 2.34. *If γ is a nonzero rational section of $\mathcal{L}$, then*

$$\eta\colon \mathbf{L}_V(\mathcal{L} + D) \cong \mathbf{L}_V(\mathrm{div}(\gamma) + D)$$

defined by $\alpha = \eta(\alpha) \cdot \gamma$ is an isomorphism.

PROOF. For any nonzero $\alpha \in \Gamma_{\mathrm{rat}}(V, \mathcal{L})$, one has φ with $\alpha = \varphi\gamma$ and then

$$\mathrm{div}(\alpha) + D = \mathrm{div}(\varphi) + \mathrm{div}(\gamma) + D.$$

Thus, $\alpha \in \mathbf{L}_V(\mathcal{L} + D)$ if and only if $\eta(\alpha) = \varphi \in \mathbf{L}_V(\mathrm{div}(\gamma) + D)$. □

By Lemma 2.33, there exists a nonzero rational section γ of $\mathcal{L}$ and so applying Theorem 2.15, one sees that $\mathbf{L}_V(\mathscr{L} + D)$ is a finite-dimensional vector space over k.

By using a basis of the vector space $\mathbf{L}_V(\mathscr{L} + D)$, we can define a rational map $\Phi_{\mathscr{L}+D}: V \to \mathbf{P}(\mathbf{L}_V(\mathscr{L} + D))$ (denoted by $\mathbf{P}(\mathscr{L} + D)$), which coincides with Φ_M, where $M = \operatorname{div}(\gamma) + D$. If $D = 0$, one can define $\Phi_{\mathscr{L}}(= \Phi_{\mathscr{L}+0})$ as follows. First, take a basis $\alpha_1, \ldots, \alpha_l$ of the vector space $\Gamma(V, \mathscr{L})$, and put $V_j = V_{\alpha_j}$ for $1 \leq j \leq l$. For each j, one has $\varphi_j^1, \ldots, \varphi_j^{l-1} \in \operatorname{Rat}(V)$ such that $\alpha_1 = \varphi_j^1 \alpha_j, \ldots, \alpha_{j-1} = \varphi_j^{j-1}\alpha_j, \alpha_{j+1} = \varphi_j^j \alpha_j, \ldots, \alpha_l = \varphi_j^{l-1}\alpha_j$. Then the $\varphi_j^i|_{V_j}$ are regular functions (see Remark in §2.18.**c**) and thus one has a morphism $\Phi_j = (\varphi_j^1, \ldots, \varphi_j^{l-1})_k: V_j \to \mathbf{A}_k^{l-1} \cong D_+(\mathsf{T}_j) \subseteq \mathbf{P}_k^{l-1}$, where $\mathbf{P}_k^{l-1}$ is a projective space with homogeneous coordinate system $(\mathsf{T}_1, \ldots, \mathsf{T}_l)$. Regarding the Φ_j as morphisms from V_j to $\mathbf{P}_k^{l-1}$, one has $\Phi_i|_{V_i \cap V_j} = \Phi_j|_{V_i \cap V_j}$ for all i, j. Thus $\{\Phi_i\}$ determines a rational map: $V \to \mathbf{P}_k^{l-1}$, which is denoted by $\Phi_{\mathscr{L}}$. If V_0 denotes $\bigcup_{j=1}^l V_j$, then $f = \Phi_{\mathscr{L}}|_{V_0}: V_0 \to \mathbf{P}_k^{l-1}$ is a morphism, which is obtained by glueing $\{\Phi_i\}$ by Lemma 1.30.

We write $\mathbf{P}(\mathscr{L})$ instead of $\mathbf{P}_k^{l-1}$ in this case. There exists an invertible sheaf $\mathscr{O}(1)$ on $\mathbf{P}(\mathscr{L})$ with $\mathscr{O}(1)|_{D+(\mathsf{T}_i)} = \mathscr{O}_{D+(\mathsf{T}_i)} \cdot \mathsf{T}_i$ for all i (see §2.18.**b**). Then $f^*(\mathscr{O}(1))|_{V_i} = \mathscr{O}_{V_i} \cdot f^*(\mathsf{T}_i)$ and $f^*(\mathsf{T}_j)|_{V_i \cap V_j} = \mathsf{T}_j \otimes 1 = (\mathsf{T}_i \cdot \mathsf{T}_j/\mathsf{T}_i) \otimes 1 = \mathsf{T}_i \otimes (\alpha_j/\alpha_i) = \alpha_j/\alpha_i \cdot f^*(\mathsf{T}_i)|_{V_i \cap V_j}$ for all i, j.

Now, define θ_i: $\mathscr{L}|_{V_i} \to f^*\mathscr{O}(1)|_{V_i} = \mathscr{O}_{V_i} \cdot f^*(\mathsf{T}_i)$ by $\theta_{i,W}(\alpha) = \alpha/\alpha_i \cdot f^*(\mathsf{T}_i)|_W$ for $\alpha \in \Gamma(W, \mathscr{L})$, where W is any open subset of U_i. Then the θ_i are isomorphisms and satisfy $\theta_i|_{V_i \cap V_j} = \theta_j|_{V_i \cap V_j}$, since $f^*(\mathsf{T}_j)|_{V_i \cap V_j} = (\alpha_j/\alpha_i) \cdot f^*(\mathsf{T}_i)|_{V_i \cap V_j}$. Hence $\{\theta_i\}$ determines an isomorphism θ: $\mathscr{L}|_{V_0} \cong f^*(\mathscr{O}(1))$. Thus one obtains the next result.

Proposition 2.21. *Let* $V_0 = \bigcup_{j=1}^l V_j$ *and* $f = \Phi_{\mathscr{L}}|_{V_0}$. *Then* f *is a morphism and* $f^*(\mathscr{O}(1)) \cong \mathscr{L}|_{V_0}$.

Remark. $\Phi_{\mathscr{L}}$ is also denoted $(\alpha_1 : \alpha_2 : \cdots : \alpha_l)$.

b. In general, let $\{U_\lambda | \lambda \in \Lambda\}$ be an open cover of a normal variety V. If $\{a_\lambda\}$ denotes a collection of $a_\lambda \in \operatorname{Rat}(V)^*$, $\lambda \in \Lambda$ such that $a_\lambda \cdot a_\mu^{-1} \in \Gamma(U_\lambda \cap U_\mu, \mathscr{O}_V^*)$ for all λ, μ, then one has a Cartier sheaf $\mathscr{F}$ defined by $\mathscr{F}|_{U_\lambda} = \mathscr{O}_{U_\lambda} \cdot a_\lambda^{-1}$ and the section 1 of $\mathscr{F}$ corresponds to $\{a_\lambda\}$. $D = \operatorname{div}(1)$ satisfies $D|_{U_\lambda} = \operatorname{div}(a_\lambda)$ for all λ. We use the notation $\operatorname{div}\{a_\lambda\}$ to denote the divisor D.

Definition. If a divisor D on V is written as $\operatorname{div}\{a_\lambda\}$ for some $\{a_\lambda\}$, then D is said to be a *locally principal divisor*.

If $D = \operatorname{div}\{a_\lambda\} = \operatorname{div}\{b_u\}$, where $\{b_u\} = \{b_u \in \operatorname{Rat}(V)^* | b_u b_v^{-1} \in \Gamma(U_u \cap U_v, \mathscr{O}_V^*)$ for $u, v \in J\}$, and $\{U_u | u \in J\}$ is an open cover of V, then the Cartier sheaf defined by $\{a_\lambda\}$ coincides with that defined by $\{b_u\}$. Thus, one has the Cartier sheaf $\mathscr{O}(D)$ uniquely determined by a given locally principal divisor D. Note that $\mathscr{O}(D)$ has the section 1 and $\operatorname{div}(1) = D$.

EXAMPLE 2.19. On $\mathbf{P}_k^{l-1}$ one has the hyperplane $H = V_+(\mathsf{T}_1)$. Then

$$\mathcal{O}(H)|_{D_+(\mathsf{T}_i)} = \mathcal{O}_{D_+(\mathsf{T}_i)} \cdot (\mathsf{T}_i/\mathsf{T}_1) \quad \text{for all} \quad i;$$

hence

$$\mathcal{O}(H) \cdot \mathsf{T}_1 \cong \mathcal{O}(1).$$

If $\alpha \in \Gamma(\mathbf{P}_k^{l-1}, \mathcal{O}(m))\backslash\{0\}$, then one has a homogeneous polynomial $F(\mathsf{T}_1, \ldots, \mathsf{T}_l)$ with $\alpha = (F/\mathsf{T}_i^m)\mathsf{T}_i^m$ for all i (Example 2.18 in §2.18.**b**.). If $\prod_{i=1}^s F_i^{n_i}$ is the decomposition of F into irreducible homogeneous polynomials, then $\operatorname{div}(\alpha) = \sum_{i=1}^s n_i V_+(F_i)$.

EXAMPLE 2.20. Put $l = n + 2$ and let F be an irreducible homogeneous polynomial. $V = V_+(F)$ is said to be a *hypersurface* of $\mathbf{P}_k^{n+1}$. If $t_i = \mathsf{T}_i|_V$, one has $\mathcal{O}_V(1) = \mathcal{O}_{\mathbf{P}_k^{n+1}}(1)|_V$ and $\mathcal{O}_V(m)|_{V_i} = \mathcal{O}_{V_i} \cdot t_i^m$, where $V_i = V \cap D_+(\mathsf{T}_i)$ for all i. $\alpha \in \Gamma(V, \mathcal{O}_V(m))$ corresponds to a homogeneous $a(t_0, \ldots, t_{n+1}) \in k[t_0, \ldots, t_{n+1}] = k[\mathsf{T}_0, \ldots, \mathsf{T}_{n+1}]/(F)$.

Definition. Any invertible sheaf $\mathcal{L}$ on a complete variety V is said to be *very ample*, if $\Gamma(V, \mathcal{L})$ has a base $\alpha_0, \ldots, \alpha_l$ such that $V = \{x \in V \mid \alpha_j(x) \neq 0$ for some $j\}$ and $\Phi_{\mathcal{L}} = (\alpha_0 : \cdots : \alpha_l)\colon V \to \mathbf{P}(\mathcal{L})$ is a closed immersion.

Definition. If D is a locally principal divisor on a complete variety such that $\mathcal{O}(D)$ is very ample, then D is said to be a *very ample divisor* or a *hyperplane section*. Furthermore, D is said to be *ample*, whenever some positive multiple mD is very ample.

D is very ample if and only if $\operatorname{Bs}|D| = \varnothing$ and $\Phi_D\colon V \to \mathbf{P}|D|$ is a closed immersion. In this case $\mathcal{O}(D) \cong \Phi_D^* \mathcal{O}_{\mathbf{P}|D|}(1)$. For example, any hyperplane on $\mathbf{P}_k^{l-1}$ is very ample, and $\mathcal{O}(m)$ on $\mathbf{P}_k^{l-1}$ is very ample for all $m > 0$.

By Proposition 2.8 and Lemma 2.20, every prime divisor on V is locally principal if and only if every stalk $\mathcal{O}_{V,x}$ is a UFD. Thus, one obtains the next result.

Proposition 2.22. *Every divisor on V is locally principal if and only if V is a locally factorial variety.*

c. Let $f\colon V \to Z$ be a morphism of varieties and let $\mathcal{L}$ be an invertible sheaf on Z. The inverse image $\mathcal{O}_V$-module $f^*\mathcal{L}$ is always defined, and is an invertible sheaf on V. But, if $\mathcal{L}$ is a Cartier sheaf, $f^*\mathcal{L}$ need not be a Cartier sheaf. Furthermore, suppose that both V and Z are normal, and let D be a locally principal divisor on Z such that f^*D is defined. Then, in some cases, $f^*\mathcal{O}(D)$ is not isomorphic to $\mathcal{O}(f^*D)$.

However, when Z is locally factorial, one has the next simple result.

Proposition 2.23. *Let α be a nonzero rational section of an invertible sheaf $\mathscr{L}$ on a locally factorial variety Z such that f^* $\operatorname{div}(\alpha)$ can be defined. Then $f^*\alpha$ is defined and $\operatorname{div}(f^*\alpha) = f^* \operatorname{div}(\alpha)$.*

Further, $\mathcal{O}(f^* \operatorname{div}(\alpha)) \cong f^*\mathcal{O}(\operatorname{div}(\alpha))$.

Proof. One can assume that V and Z are affine, and $\mathscr{L} = \mathcal{O}_V$. Then since f^* $\operatorname{div}(\alpha)$ is defined, one has a rational function $f^*\alpha$. Applying the formula of Proposition 2.15, one obtains the assertion. □

d. Now, let $\mathscr{F}$ be a coherent $\mathcal{O}_Z$-module. Then $\sigma_{\mathscr{F}}: \mathscr{F} \to f_* f^* \mathscr{F}$ was defined in §2.16.**b**. Thus the homomorphism

$$\sigma_{\mathscr{F}, Z}: \Gamma(Z, \mathscr{F}) \to \Gamma(Z, f_* f^* \mathscr{F}) = \Gamma(V, f^* \mathscr{F})$$

is obtained, which is denoted by $f^\#$, for simplicity. For example, if $V = \operatorname{Spec} B$, $Z = \operatorname{Spec} A$, and $\mathscr{F} = \tilde{M}$, then $f^\#\alpha = \alpha \otimes 1 \in M \otimes_A B$.

Lemma 2.35. *If f is dominating and $\mathscr{F}$ is locally free of finite type, then $f^\#: \Gamma(Z, \mathscr{F}) \to \Gamma(V, f^*\mathscr{F})$ is one-to-one.*

Proof. By Lemma 2.29, $f_*(f^*\mathscr{F}) = f_*\mathcal{O}_V \otimes \mathscr{F}$. Since f is dominating, $\mathcal{O}_Z$ is contained in $f_*\mathcal{O}_V$ by Theorem 1.19. Therefore, $f_*(\mathcal{O}_V) \otimes \mathscr{F} \supseteq \mathscr{F}$. Hence, $\Gamma(Z, \mathscr{F}) \subseteq \Gamma(Z, f_*(\mathcal{O}_V) \otimes \mathscr{F}) \cong \Gamma(V, f^*\mathscr{F})$. □

Let $\mathscr{L}$ be an invertible sheaf on a normal variety V with $\Gamma(V, \mathscr{L}) \neq 0$. Take a basis $\alpha_1, \ldots, \alpha_l$ of $\Gamma(V, \mathscr{L})$ and define $V_0 = \bigcup_{j=1}^{l} V_{\alpha_j}$. Suppose that $\operatorname{codim}(V \setminus V_0) \geq 2$. If $f = \Phi_{\mathscr{L}}|_{V_0}$, then $f^*\mathcal{O}(1) \cong \mathscr{L}|_{V_0}$ by Proposition 2.21. Letting Y be the closed image $\Phi_{\mathscr{L}}(V)$ and $j: Y \to \mathbf{P}(\mathscr{L})$ be the inclusion map, one has a dominating morphism $g: V_0 \to Y$ such that $f = j \circ g$. Let $\mathscr{M} = j^*\mathcal{O}_{\mathbf{P}}(1)$. Then $g^*\mathscr{M} \cong f^*\mathcal{O}_{\mathbf{P}}(1) \cong \mathscr{L}|_{V_0}$. Since g is dominating, $g^\#: \Gamma(Y, \mathscr{M}) \to \Gamma(V_0, g^*\mathscr{M}) \cong \Gamma(V_0, \mathscr{L})$ is one-to-one. But $f^\# = g^\# \circ j^\#: \Gamma(\mathbf{P}(\mathscr{L}), \mathcal{O}_{\mathbf{P}}(1)) \to \Gamma(V_0, \mathscr{L})$ is an isomorphism, since $\Gamma(V_0, \mathscr{L}) = \Gamma(V, \mathscr{L})$ by Lemma 2.32. Thus $g^\#$ is surjective; hence $g^\#: \Gamma(Y, \mathscr{M}) \cong \Gamma(V_0, g^*\mathscr{M})$ and $j^\#: \Gamma(\mathbf{P}(\mathscr{L}), \mathcal{O}_{\mathbf{P}}(1)) \cong \Gamma(Y, \mathscr{M})$.

Proposition 2.24. *Let D be a divisor on a locally factorial variety V with $|D| \neq \varnothing$ and let Y be the closed image $\Phi_D(V)$. Then $\Gamma(\mathbf{P}(D), \mathcal{O}_{\mathbf{P}}(1)) \cong \Gamma(Y, j^*\mathcal{O}_{\mathbf{P}}(1)) \cong \Gamma(V, \mathcal{O}(D))$, where $j: Y \to \mathbf{P}(D)$ is the inclusion map.*

Proof. If $\dim |D| = 0$, then the assertion is trivial. If $\dim |D| \geq 1$, we can suppose that $|D|$ has no fixed components. Then, $\operatorname{codim} \operatorname{Bs}|D| \geq 2$, and one can apply the preceding argument. □

e. Let $\mathscr{L}$ be an invertible sheaf on a normal complete variety V and let $\mathbf{L}$ be a subspace of $\Gamma(V, \mathscr{L})$.

Definition. If $\mathbf{L} \neq 0$, define $\Lambda(\mathbf{L})$ to be $\{\operatorname{div}(\alpha) \mid \alpha \in \mathbf{L}\backslash\{0\}\}$, which is said to be the *linear system defined by* $\mathbf{L}$.

Taking a nonzero γ of $\Gamma(V, \mathscr{L})$, one has an effective divisor $D = \operatorname{div}(\gamma)$ and an isomorphism $\eta: \Gamma(V, \mathscr{L}) \cong \mathbf{L}_V(D)$ defined by $\eta(\alpha) = \alpha/\gamma$. Then $\operatorname{div}(\alpha) = \operatorname{div}(\eta(\alpha)) + D$ and so $\Lambda(\mathbf{L})$ coincides with $\Lambda(\eta(\mathbf{L}), D)$ defined in §2.9.**c**.

Now, let $f: V \to Z$ be a morphism and let $\mathbf{L}$ be a nonzero subspace of $\Gamma(Z, \mathscr{L})$, where $\mathscr{L}$ is an invertible sheaf on Z. By using the linear map $f^{\#}: \Gamma(Z, \mathscr{L}) \to \Gamma(V, f^*\mathscr{L})$, one has the subspace $f^{\#}\mathbf{L}$ of $\Gamma(V, f^*\mathscr{L})$. If $f^{\#}\mathbf{L} \neq \{0\}$, then the *pullback of the linear system* $\Lambda(\mathbf{L})$ is defined to be $\Lambda(f^{\#}\mathbf{L})$, and denoted by $f^{\#}(\Lambda(\mathbf{L}))$.

One can define the rational map $\Phi_\Lambda: Z \to \mathbf{P}(\mathbf{L})$ by using a basis of $\mathbf{L}$, i.e., $\Phi_\Lambda = (\alpha_1 : \cdots : \alpha_l)$, where $(\alpha_1, \ldots, \alpha_l)$ is a basis of $\mathbf{L}$. Then, if $f^{\#}\mathbf{L} \neq 0$, $\Phi_{f^{\#}\Lambda} = \Phi_\Lambda \circ f$.

Note that when V is not locally factorial, f^*D need not be a member of $f^{\#}\Lambda$, where $D \in \Lambda$. In particular, regarding D as a linear system consisting of a single member, one has the pullback $f^{\#}D$ as a linear system, whenever $f^*(\alpha) \neq 0$, where α is a section of some invertible sheaf $\mathscr{L}$ such that $\operatorname{div}(\alpha) = D$. In other words, the pullback $f^{\#}D$ as a linear system is defined to be $\operatorname{div}(f^*(\alpha))$, which may be different from $f^*(\operatorname{div}(\alpha)) = f^*D$.

f. Let $f: V \to Z$ be a proper surjective morphism between normal varieties.

Definition. An effective divisor E on V is said to be *exceptional with respect to* f if $\operatorname{codim}(f(E)) \geq 2$.

Lemma 2.36. *With V, Z, and f as above, assume further that* Rat(V)/Rat(Z) *is algebraically closed. If E is an effective locally principal divisor on V which is exceptional with respect to f, then $f_*\mathscr{O}(E) = \mathscr{O}_Z$.*

Proof. Since $\mathscr{O}_V \subseteq \mathscr{O}(E)$, one has $f_*\mathscr{O}_V \subseteq f_*\mathscr{O}(E)$. By Theorem 2.29, $f_*\mathscr{O}_V = \mathscr{O}_Z$. Now, for any open subset W of Z, $\Gamma(W, f_*\mathscr{O}(E)) \subseteq \Gamma(W\backslash f(E), f_*(\mathscr{O}_V(E)))$. But, since $f_*(\mathscr{O}(E))|_{W\backslash f(E)} = f_*(\mathscr{O}_V|_{V\backslash f^{-1}(f(E))}) = (f_*\mathscr{O}_V)|_{W\backslash f(E)}$, it follows that $\Gamma(W\backslash f(E), f_*(\mathscr{O}_V(E))) = \Gamma(W\backslash f(E), f_*\mathscr{O}_V) = \Gamma(W\backslash f(E), \mathscr{O}_Z) = \Gamma(W, \mathscr{O}_Z)$ by Theorem 2.15. Hence $\mathscr{O}_Z \cong f_*(\mathscr{O}_V(E))$. □

Proposition 2.25. *With the notation as in Lemma 2.36, let $\mathscr{L}$ be a locally free $\mathscr{O}_Z$-module of finite type. Then $f_*((f^*\mathscr{L}) \otimes \mathscr{O}(E)) \cong \mathscr{L}$. In particular, $\Gamma(V, f^*\mathscr{L} \otimes \mathscr{O}(E)) \cong \Gamma(Z, \mathscr{L})$.*

Proof. By Lemma 2.29 and the previous lemma, one has $f_*((f^*\mathscr{L}) \otimes \mathscr{O}(E)) \cong \mathscr{L} \otimes f_*(\mathscr{O}(E)) = \mathscr{L}$. □

Exercises

2.1. Let A be a subring $k[X, XY^2, XY]$ of the polynomial ring $B = k[X, Y]$. Then $\varphi = XY$ defines effective divisors $\operatorname{div}_A(\varphi)$ on $V = \operatorname{Spec} A$ and $\operatorname{div}_B(\varphi)$ on $W = \operatorname{Spec} B$. Let $f: W \to V$ be a morphism associated with the inclusion. Show that $f^* \operatorname{div}_A(\varphi) - \operatorname{div}_B(\varphi) \neq 0$.

2.2. Let $B \supseteq A$ be an extension of integral domains such that $Q(A)$ is algebraically closed in $Q(B)$. Furthermore, let $A_1 \supseteq A$ be a finite extension of integral domains such that $Q(A_1)/Q(A)$ is a separable extension of fields. Show that $B \otimes_A A_1 \subseteq Q(B) \otimes_A A_1$ and that $B \otimes_A A_1$ is an integral domain.

2.3. Let $f_1, \ldots, f_r \in k[X_1, \ldots, X_n]$ such that (1) the f_j are nonconstant irreducible polynomials (2) $\operatorname{tr.deg}_k k(f_1, \ldots, f_r) = 1$. Show that $k[f_1] = \cdots = k[f_r]$.

2.4. Let f, g be polynomials in $X_1, \ldots, X_n$ such that (1) g is irreducible and (2) f, g are algebraically dependent. Show that f is a polynomial of g.

2.5. Let V be a variety and F be a closed subset of V. Suppose that $V \backslash F$ is an affine variety. Show that either $\operatorname{codim}(F) = 1$ or $F = \varnothing$.

2.6. Let A be an integral domain finitely generated over k. Let $U(A)$ denote the group of units of A. The rank of $U(A)/k^*$ is denoted by $\rho(A)$. If $f \in A$ satisfies $\rho(A_f) = \rho(A)$, then show that f is a unit.

2.7. Let R be a Noetherian local ring with maximal ideal $\mathfrak{m}$. If M is a finitely generated R-module, show that $\hat{M} = \operatorname{projlim}_m M/\mathfrak{m}^m M$ is isomorphic to $M \otimes_R \hat{R}$ as an $\hat{R}$-module. Furthermore, show that $M = 0$ if and only if $\hat{M} = 0$.

2.8. Let A be a Noetherian ring. An ideal $\mathfrak{a}$ is said to be secondary if $\sqrt{\mathfrak{a}}$ is a maximal ideal. Show that:
   (i) A secondary ideal is primary.
   (ii) Given an ideal J, define $S(J) = \{\mathfrak{a} \mid \mathfrak{a}$ is a secondary ideal and $J \subseteq \mathfrak{a}\}$. Then $J = \bigcap_{\mathfrak{a} \in S(J)} \mathfrak{a}$.
   (iii) If $A = \mathbb{Z}[x_1, \ldots, x_n]$, then $\mathfrak{a}$ is secondary if and only if $A/\mathfrak{a}$ is a finite ring.

2.9. Let $S_{p,q,r}$ be $V(X^p + Y^q - Z^r) \subset \operatorname{Spec} k[X, Y, Z]$. Determine all dominating morphisms $\varphi: \mathbf{A}_k^2 \to S_{p,q,r}$ in the following cases:
   (i) $p = q = r \geqq 2$.
   (ii) $p \geqq q = r = 2$.
   (iii) $p = 3$, $q = 3$, $r = 2$.
   (iv) $p = 4$, $q = 3$, $r = 2$.
   (v) $p = 5$, $q = 3$, $r = 2$.

2.10. Let V and W be normal varieties, and let $f: V \to W$ be a strictly rational map. Then one has the induced homomorphism $f^*: A(W) \to A(V)$. Show that if f is proper birational, then f^* is an isomorphism.

2.11. Let $A = \mathbb{C}[X_1, X_2, \ldots, X_n, X_1^{-1}, \ldots, X_n^{-1}, (X_1 + \cdots + X_n + 1)^{-1}]$. Show that $\operatorname{Aut}_{\mathbb{C}} A$ is the symmetric group on n letters.

2.12. Let $R = \mathbb{C}[X, X^{-1}, (X - 1)^{-1}]$ and $A = R \otimes_{\mathbb{C}} \cdots \otimes_{\mathbb{C}} R$. Compute the order of $\operatorname{Aut}_{\mathbb{C}} A$.

2.13. Let V be an affine open subset of an affine space $\mathbf{A}_k^n$. Show that V is of the form $D(f)$ for some polynomial f.

Chapter 3

Projective Schemes

§3.1 Graded Rings

a. Throughout this chapter, R denotes a ring.

Definition. If an R-algebra A satisfies the following conditions, it is said to be a *graded R-algebra* or a *graded ring* (over R).

(i) $A = \bigoplus_{d=0}^{\infty} A_d$, where the A_d are R-modules.
(ii) $A_d \cdot A_e \subseteq A_{d+e}$ for any $d, e \geq 0$.

Definition. Let A be a graded R-algebra and let M be an A-module. If M has the following properties it is said to be a *graded A-module.*

(i) $M = \bigoplus_{n=-\infty}^{\infty} M_n$, where the M_n are R-modules.
(ii) $A_d \cdot M_n \subseteq M_{d+n}$ for any $d \geq 0$ and $n \in \mathbb{Z}$.

Definition. $x \in M$ is said to be a *homogeneous element* if $x \in M_n$ for some n.

Given any graded A-module M and an integer l, define a graded A-module $M(l)$ as follows: $M(l) = M$ as an A-module and $M(l)_n = M_{n+l}$ for any n.

If a graded A-module I is a graded submodule of A, then I is said to be a *homogeneous ideal* of A. Clearly, an ideal I of A is homogeneous if and only if $pr_d(I) \subseteq I_d$ for any $d \geq 0$, where $pr_d(\sum_{n=0} x_n) = x_d$.

Thus, a homogeneous ideal is an ideal which has a basis consisting of homogeneous elements.

Given a graded ring A, the direct sum $A_+ = \bigoplus_{d=1}^{\infty} A_d$ is a homogeneous ideal.

b. Let $S = R[T_0, \ldots, T_n]$ be a polynomial ring over R in $T_0, \ldots, T_n$. Then, defining S_d to be the set of homogeneous polynomials over R of degree d, one can regard S as a graded R-algebra. In all that follows, a polynomial ring is regarded as a graded ring in this way.

§3.2 Homogeneous Spectra

a. From the viewpoint of geometry, the homogeneous ideals of a graded ring are more important than the usual ideals.

Definition. Let A be a graded R-algebra. The set $\operatorname{Proj} A = \{\mathfrak{p} \in \operatorname{Spec} A \mid \mathfrak{p}$ is homogeneous and $\mathfrak{p} \not\supseteq A_+\}$ is said to be the *homogeneous spectrum associated with A*.

EXAMPLE 3.1. Let $A = k[T_0, T_1]$ be a polynomial ring over an algebraically closed field k. Then any $\mathfrak{p} \in \operatorname{Proj} A$ can be written in one of the forms (0), $T_0 A$, or $(T_1 - \lambda T_0)A$ for some $\lambda \in k$. Thus identifying $\operatorname{Spec} k[T_1/T_0]$ with $\operatorname{Proj} A \setminus \{T_0 A\}$ by $(0) \leftrightarrow (0)$ and $(T_1/T_0 - \lambda) \leftrightarrow (T_1 - \lambda T_0)$, one has

$$\operatorname{Proj} A = \operatorname{Spec} k[T_1/T_0] \cup \{T_0 A\}.$$

EXAMPLE 3.2. If a graded ring A satisfies $A_m = 0$ for all $m \geq m_0$, for a fixed integer m_0, then $\operatorname{Proj} A = \varnothing$. In fact, for any $m > 0$ and $x \in A_m$, there exists n with $nm \geq m_0$; hence $x^n = 0$. If $\mathfrak{p}$ is a prime ideal of A, then $x \in \mathfrak{p}$, since $x^n = 0 \in \mathfrak{p}$. Therefore $\mathfrak{p} \supseteq A_+$, i.e., $\operatorname{Proj} A = \varnothing$.

b. Let A be a graded ring. Since $\operatorname{Proj} A$ is a subset of $\operatorname{Spec} A$, we regard $\operatorname{Proj} A$ as a subspace of the spectrum $\operatorname{Spec} A$. Then a set $\{D(f) \cap \operatorname{Proj} A \mid f \in A\}$ is an open base of the topology. For any $f \in A$, we denote by $f_{(d)}$ the d-th homogeneous part of f. If $\mathfrak{p}$ is a homogeneous prime ideal of A, then

$$f \notin \mathfrak{p} \Leftrightarrow f_{(d)} \notin \mathfrak{p} \quad \text{for some } d.$$

In other words, $D(f) \cap \operatorname{Proj} A = \bigcup_{d \geq 0} (D(f_{(d)}) \cap \operatorname{Proj} A)$.

Definition. For any homogeneous element f, define $V_+(f)$ to be $V(f) \cap \operatorname{Proj} A$, and $D_+(f)$ to be $D(f) \cap \operatorname{Proj} A$. Furthermore, for any homogeneous ideal $\mathfrak{a}$ of A, define $V_+(\mathfrak{a})$ to be $V(\mathfrak{a}) \cap \operatorname{Proj} A$ and $D_+(\mathfrak{a})$ to be $D(\mathfrak{a}) \cap \operatorname{Proj} A$.

$\{D_+(f) \mid f$ is homogeneous$\}$ is an open base of the topology, and $\{V_+(\mathfrak{a}) \mid \mathfrak{a}$ is a homogeneous ideal$\}$ is the set of all closed subsets of $\operatorname{Proj} A$.

EXAMPLE 3.3. Let A be a polynomial ring $k[\mathsf{T}_0, \ldots, \mathsf{T}_n]$ over a field k and let $X = \operatorname{Proj} A$. If $\mathfrak{p} \in D_+(\mathsf{T}_0)$ and $\mathsf{X}_i = \mathsf{T}_i/\mathsf{T}_0$ for $1 \leq i \leq n$, then define $\psi_0(\mathfrak{p})$ to be $\{\varphi \in k[\mathsf{X}_1, \ldots, \mathsf{X}_n] \mid \varphi \mathsf{T}_0^s \in \mathfrak{p}$ for some $s \geq 0\}$. It is easy to check that $\psi_0 : D_+(\mathsf{T}_0) \to \operatorname{Spec} k[\mathsf{X}_1, \ldots, \mathsf{X}_n]$ is a homeomorphism.

c. Let A and B be graded R-algebras, and let $\varphi: A \to B$ be an R-algebra homomorphism preserving degree, i.e., $\varphi(A_d) \subseteq B_d$ for any $d \geq 0$. ${}^a\varphi$: Spec $B \to$ Spec A has the property that ${}^a\varphi$ (homogeneous prime ideal) = homogeneous prime ideal. But ${}^a\varphi(\operatorname{Proj} B)$ is not always contained in Proj A. Clearly, ${}^a\varphi^{-1}(\operatorname{Proj} A) = D_+(\varphi(A_+)B)$.

Definition. Define $G(\varphi)$ to be $D_+(\varphi(A_+)B)$ and Proj φ to be ${}^a\varphi|_{G(\varphi)}: G(\varphi) \to$ Proj A. Proj φ is said to be the *continuous map associated* with φ.

EXAMPLE 3.4. Suppose that the ideal B_+ is generated by B_1. If there exist $\alpha_0, \ldots, \alpha_n \in A_1$ such that $B_1 = \sum_{j=0}^{n} R\varphi(\alpha_j)$, then $G(\varphi) = \operatorname{Proj} B$.

EXAMPLE 3.5. Let $\varphi: k[\mathsf{T}_0, \ldots, \mathsf{T}_n] \to k[\mathsf{Y}_0, \ldots, \mathsf{Y}_m]$ be a nonzero k-homomorphism of polynomial rings defined by $\varphi(\mathsf{T}_i) = \sum_{j=0}^{m} \lambda_{ij} \mathsf{Y}_j$ for $0 \leq i \leq n$ and $\lambda_{ij} \in k$. Then $G(\varphi) = D_+(\varphi(\mathsf{T}_0)) \cup \cdots \cup D_+(\varphi(\mathsf{T}_n))$. One has a rational map f: $\mathbf{P}_k^m \to \mathbf{P}_k^n$ defined by ${}^a\varphi_i$: $\mathbf{A}_k^m \to \mathbf{A}_k^n$ with $\varphi_i(\mathsf{T}_j/\mathsf{T}_i) = \varphi(\mathsf{T}_j)/\varphi(\mathsf{T}_i)$ for all j. Then $\operatorname{dom}(f) = G(\varphi)$, and f is a projection of $\mathbf{P}_k^m$ to $\mathbf{P}_k^n$ (see § 2.9.**e**).

§3.3 Finitely Generated Graded Rings

a. Definition. Suppose that a graded R-algebra A satisfies the following conditions:

(i) $A_0 = R$ (or (i)′ A_0 is finitely generated as an R-module).
(ii) A_1 is finitely generated as an R-module.
(iii) $A_d \cdot A_1 = A_{d+1}$ for all $d \geq 1$.

Then A is said to be a *finitely generated graded R-algebra* or a *graded ring finitely generated over R* (*by degree one elements*).

By (i) and (ii), there exist $\alpha_0, \ldots, \alpha_r \in A_1$ such that $A_1 = \sum_{i=0}^{r} R\alpha_i$. By (iii), any element f of A_d is written in the form $F(\alpha_0, \ldots, \alpha_r)$, where $F(\mathsf{T}_0, \ldots, \mathsf{T}_r)$ is a homogeneous polynomial of degree d over R.

Letting S be the polynomial ring $R[\mathsf{T}_0, \ldots, \mathsf{T}_r]$, we have a surjective R-algebra homomorphism $\Psi: S \to A$ defined by $\Psi(\mathsf{T}_i) = \alpha_i$ for all i, which preserves degree and has $G(\Psi) = \operatorname{Proj} A$. $G(\Psi) \approx V_+(\operatorname{Ker} \Psi)$ is a closed subset of $\mathbf{P}_R^r = \operatorname{Proj} S$ and Proj Ψ is the inclusion map.

In what follows, we exclusively consider finitely generated R-algebras, and use the above notation.

b. If $\mathfrak{p} \in \text{Proj } A$, then $\mathfrak{p} \not\supseteq A_+$, and so $\alpha_i \notin \mathfrak{p}$ for some i. Hence, $\text{Proj } A = D_+(\alpha_0) \cup \cdots \cup D_+(\alpha_r)$.

For $\alpha \in A_d$, we write $A[1/\alpha]$ instead of $\{\alpha^j \mid j = 0, 1, \ldots\}^{-1}A$. Then, $A[1/\alpha]$ is a graded ring with positive and negative gradation, i.e., letting

$$(A[1/\alpha])_p = \sum_m A_{p+dm} \cdot 1/\alpha^m \quad \text{for any } p \in \mathbb{Z},$$

one has

$$A[1/\alpha] = \bigoplus_{p=-\infty}^{\infty} (A[1/\alpha])_p.$$

$(A[1/\alpha])_0$ is a ring, and is denoted by $A_{[\alpha]}$.

If $\mathfrak{p} \in D_+(\alpha) \subseteq \text{Proj } A$, then $\mathfrak{p}A[1/\alpha]$ is also a (homogeneous) prime ideal and therefore so is $A_{[\alpha]} \cap \mathfrak{p}A[1/\alpha]$. Thus, we have a map $\psi_{(\alpha)}: D_+(\alpha) \to \text{Spec } A_{[\alpha]}$ defined by $\psi_{(\alpha)}(\mathfrak{p}) = A_{[\alpha]} \cap \mathfrak{p}A[1/\alpha]$.

If M is a graded A-module, then define $M[1/\alpha]$ to be $M \otimes_A A[1/\alpha]$ and $M_{[\alpha]}$ to be $(M[1/\alpha])_0$. $M_{[\alpha]}$ is an $A_{[\alpha]}$-module in a natural way.

c. Lemma 3.1. $\psi_{(\alpha)}: D_+(\alpha) \to \text{Spec } A_{[\alpha]}$ *is a homeomorphism.*

PROOF. (1) $\psi_{(\alpha)}$ is continuous. In fact, for any $b \in A_{md}$, $b/\alpha^m \in \psi_{(\alpha)}(\mathfrak{p})$ if and only if $b \in \mathfrak{p}$. In other words, $\psi_{(\alpha)}^{-1}(D(b/\alpha^m)) = D_+(b) \cap D_+(\alpha)$. Hence, $\psi_{(\alpha)}$ is continuous.

(2) $\psi_{(\alpha)}$ is one-to-one. In fact, let $\mathfrak{p}_1 \neq \mathfrak{p}_2 \in D_+(\alpha)$, say $\mathfrak{p}_1 \not\subseteq \mathfrak{p}_2$. Choosing a homogeneous element $b \in (\mathfrak{p}_1 \cap A_m) \backslash \mathfrak{p}_2$ for some m, we have

$$b^d/\alpha^m \in \psi_{(\alpha)}(\mathfrak{p}_1) \backslash \psi_{(\alpha)}(\mathfrak{p}_2).$$

Hence, $\psi_{(\alpha)}(\mathfrak{p}_1) \neq \psi_{(\alpha)}(\mathfrak{p}_2)$.

(3) $\psi_{(\alpha)}$ is surjective. Given $\mathfrak{p}' \in \text{Spec } A_{[\alpha]}$, we define $\mathfrak{a}_m$ to be $\{b \in A_m \mid b^d/\alpha^m \in \mathfrak{p}'\}$ and a homogeneous ideal $\mathfrak{a}$ to be $\bigoplus_{m=0}^{\infty} \mathfrak{a}_m$. We claim that $\mathfrak{a}$ is prime. Leaving the general case to the reader, we consider only homogeneous elements. If $bc \in \mathfrak{a}_{n+m}$ for $b \in A_n$, $c \in A_m$, then $(bc)^d/\alpha^{n+m} \in \mathfrak{p}'$. But then b^d/α^n or $c^d/\alpha^m \in \mathfrak{p}'$ since $\mathfrak{p}'$ is prime. Furthermore, if $\alpha \in \mathfrak{a}$, then $1 = \alpha/\alpha \in \mathfrak{p}'$, which contradicts the hypothesis that $\mathfrak{p}'$ is prime. Thus $\mathfrak{a} \in D_+(\alpha)$ and $\psi_{(\alpha)}(\mathfrak{a}) = \mathfrak{p}'$.

(4) $\psi_{(\alpha)}^{-1}$ is continuous. By (1) and (3), one has

$$\text{Spec } A_{[\alpha]} \supseteq D(b/\alpha^m) = \psi_{(\alpha)}(D_+(b) \cap D_+(\alpha)).$$

Since $\{D_+(b) \cap D_+(\alpha) \mid b \text{ is homogeneous}\}$ is an open base of $D_+(\alpha)$, one sees that $\psi_{(\alpha)}^{-1}$ is continuous. □

d. In §1.1.c, we showed that $A = \{0\}$ if and only if $\text{Spec } A = \varnothing$. We shall consider the analogue for a graded ring A and $\text{Proj } A$.

Proposition 3.1. *Let A be a graded ring finitely generated over R. Then $\text{Proj } A = \varnothing$ if and only if $A_m = 0$ for all m greater than or equal to some fixed integer m_0.*

PROOF. The "if" part was proved in Example 3.2. Now, assume Proj $A = \emptyset$. Then since Proj $A = D_+(\alpha_0) \cup \cdots \cup D_+(\alpha_r)$ with the notation in §3.3.**b**, one has $D_+(\alpha_0) = \cdots = D_+(\alpha_r) = \emptyset$. Hence $A_{[\alpha_0]} = \cdots = A_{[\alpha_r]} = \{0\}$. $A_{[\alpha_j]}$ is written as $A_0[\alpha_0/\alpha_j, \ldots, \alpha_r/\alpha_j]$. Thus, $A_{[\alpha_j]} = 0$ implies $\alpha_i \alpha_j^N = 0$ for every $0 \leq i \leq r$ and some N which is chosen independently of i and j. Consequently, letting $m_0 = (N+1)(r+1)$, we see that $A_m = 0$ for all $m \geq m_0$. □

Proposition 3.2. *Let A be a polynomial ring $R[\mathsf{T}]$. Then* Proj S = Spec R.

PROOF. This follows immediately from $R[\mathsf{T}]_{[\mathsf{T}]} \cong R$. □

e. Definition. For a graded R-algebra A and a positive integer $e > 0$, define a graded R-algebra $A^{(e)}$ by setting

$$(A^{(e)})_d = A_{ed} \quad \text{for all } d \geq 0.$$

Clearly, if A is finitely generated, so is $A^{(e)}$.

Proposition 3.3. *Let A be a finitely generated graded R-algebra. Then* Proj $A^{(e)}$ *is naturally homeomorphic to* Proj A.

PROOF. The natural inclusion map $i\colon A^{(e)} \to A$ does not preserve degree. But, ${}^a i$ maps Proj A into Proj $A^{(e)}$. Furthermore, for any $\alpha \in A_d$, ${}^a i(D_+(\alpha)) \subseteq D_+(\alpha^e)$. It is easy to check that

$$A^{(e)}_{[\alpha^e]} \cong A_{[\alpha]}.$$

Hence, ${}^a i|_{D_+(\alpha)}\colon D_+(\alpha) \to D_+(\alpha^e)$ is a homeomorphism for all α. Now, $A^{(e)}$ is generated by $\alpha_0^e, \alpha_0^{e-1}\alpha_1, \ldots, \alpha_r^e$. But since $D_+(\alpha_0 \cdots \alpha_r) = D_+(\alpha_0) \cap \cdots \cap D_+(\alpha_r)$, Proj$(A^{(e)})$ is covered by $D_+(\alpha_0^e), \ldots, D_+(\alpha_r^e)$. Therefore, ${}^a i$ is a homeomorphism from Proj A onto Proj $A^{(e)}$. □

§3.4 Construction of Projective Schemes

a. Let A be a (finitely generated) graded ring over R and let $\alpha \in A_d$, $\beta \in A_e$. By Lemma 1.2.(iii), we have the isomorphisms

$$A\left[\frac{1}{\alpha\beta}\right] \cong A\left[\frac{1}{\alpha}\right]\left[\frac{1}{\beta}\right] \hookleftarrow A\left[\frac{1}{\alpha}\right]\left[\frac{1}{\beta^d}\right],$$

$$A_{[\alpha\beta]} \cong \left(A\left[\frac{1}{\alpha}\right]\left[\frac{1}{\beta^d}\right]\right)_0.$$

Furthermore,

$$\left(A\left[\frac{1}{\alpha}\right]\left[\frac{1}{\beta^d}\right]\right)_0 \cong \sum_{m=0}^{\infty} A\left[\frac{1}{\alpha}\right]_{mde} \cdot \frac{1}{\beta^{dm}} = \sum_{m,\,j} A_{(me+j)d} \cdot \frac{1}{\alpha^j} \cdot \frac{1}{\beta^{dm}}$$

$$\cong \sum_{m,\,j} A_{(me+j)d} \cdot \frac{1}{\alpha^{j+me}} \cdot \left(\frac{\alpha^e}{\beta^d}\right)^m = A_{[\alpha]}\left[\frac{\alpha^e}{\beta^d}\right].$$

Thus we obtain the following result.

Lemma 3.2. *There exists an A_0-isomorphism*

$$\varphi_{\alpha,\,\beta}\colon A_{[\alpha\beta]} \cong A_{[\alpha]}\left[\frac{\alpha^e}{\beta^d}\right].$$

Furthermore, $\psi_{(\alpha\beta)}^{-1} \circ {}^a\varphi_{\alpha,\,\beta}\colon D(\beta^d/\alpha^e) \to D_+(\alpha\beta)$ is the restriction of $\psi_{(\alpha)}^{-1}$: Spec $A_{[\alpha]} \to D_+(\alpha)$ to $D(\beta^d/\alpha^e)$, i.e., the following diagram is commutative ($\psi_{(\alpha)}$ was defined in §3.3).

$$\begin{array}{ccc} D(\beta^d/\alpha^e) & \cong & \text{Spec } A_{[\alpha\beta]} \\ & & \wr\| \\ & & D_+(\alpha\beta) \\ & & \cap| \\ \text{Spec } A_{[\alpha]} & \cong & D_+(\alpha) \end{array}$$

Proof. Left to the reader. □

b. $X = \operatorname{Proj} A$ is covered by the open subsets $\{D_+(\alpha) \mid \alpha \in A_d \text{ with } d > 0\}$. By the homeomorphism $\psi_{(\alpha)}\colon D_+(\alpha) \to \text{Spec } A_{[\alpha]}$, we have $\mathcal{O}_\alpha = \psi_{(\alpha)}^* \tilde{A}_{[\alpha]}$, which is a sheaf of rings. By Lemma 3.2, $\mathcal{O}_\alpha|_{D_+(\alpha\beta)} = \mathcal{O}_\beta|_{D_+(\alpha\beta)}$ for any homogeneous α, β. Hence $\{\mathcal{O}_\alpha\}$ defines the sheaf of rings $\mathcal{O}$ on X such that $\mathcal{O}|_{D_+(\alpha)} \cong \mathcal{O}_\alpha$. Since $(D_+(\alpha), \mathcal{O}_\alpha) \cong \text{Spec } A_{[\alpha]}$, $(X, \mathcal{O})$ is a scheme.

Definition. The scheme $(X, \mathcal{O})$ is said to be the *projective scheme* associated with the graded ring A.

Note that $\mathcal{O}$ is denoted by $\mathcal{O}_X$ and called the *structure sheaf of the projective scheme* Proj A, which is of finite type over Spec R.

Example 3.6. If S is a polynomial ring over R in $n + 1$ variables, then Proj S is said to be the *projective n-space* over R, denoted $\mathbf{P}_R^n$. Clearly, this agrees with the definition of the projective space in §1.18.

c. Now, let M be a graded A-module. For $\alpha \in A_d$, the open subscheme $D_+(\alpha)$ of $X = \operatorname{Proj} A$ is isomorphic to Spec $A_{[\alpha]}$ via $\psi_{(\alpha)} = (\psi_{(\alpha)}, \text{id})$ and one defines $\mathscr{F}_\alpha$ on $D_+(\alpha)$ to be $\psi_\alpha^* \widetilde{M_{[\alpha]}}$. Then by Lemma 3.2, $\mathscr{F}_\alpha|_{D_+(\alpha\beta)} = \mathscr{F}_\beta|_{D_+(\alpha\beta)}$ for any homogeneous α and β. Thus, by glueing these $\mathscr{F}_\alpha$, one obtains an $\mathscr{O}_X$-module $\mathscr{F}$ such that $\mathscr{F}|_{D_+(\alpha)} \cong \mathscr{F}_\alpha$.

Definition. $\mathscr{F}$ is said to be the $\mathscr{O}_X$-module associated with the graded A-module M, denoted $\tilde{M}$.

Remark. In §1.11.**e**, the $\tilde{A}$-module $\tilde{M}$ associated with an A-module M was defined, which is obviously different from the $\mathscr{O}_X$-module $\tilde{M}$ defined above.

Since $A(1)_m = (\sum_{i=0}^r A\alpha_i)_m$ for $m \geq 0$, one has

$$A(1)_{[\alpha]} = \sum_{i=0}^{r} A_{[\alpha]}\alpha_i$$

for any $\alpha \in A_1$. Hence

$$A(1)^\sim|_{D_+(\alpha)} = \sum_{i=0}^{r} \mathscr{O}_X|_{D_+(\alpha)} \cdot \alpha_i.$$

But since

$$\alpha_i \cdot A_{[\alpha]} = (\alpha_i/\alpha) \cdot \alpha A_{[\alpha]} \subseteq \alpha \cdot A_{[\alpha]} \cong A_{[\alpha]},$$

one has

$$\sum_{i=0}^{r} \mathscr{O}_X|_{D_+(\alpha)} \cdot \alpha_i = \mathscr{O}_X|_{D_+(\alpha)} \cdot \alpha \cong \mathscr{O}_X|_{D_+(\alpha)}.$$

Thus, if $X = \mathbf{P}_R^r$, $A(1)^\sim$ coincides with $\mathscr{O}_X(1)$, which was defined in §2.18.**b**. Hence, in general, we define $\mathscr{O}_X(1)$ to be $A(1)^\sim$, which is said to be the *fundamental sheaf* (or *the tautological sheaf*) *on* $X = \operatorname{Proj} A$.

d. The next lemma will be used to study closed subschemes of Proj A.

Lemma 3.3.

(i) *If $\mathfrak{A}$ is a homogeneous ideal of A, then all $\mathfrak{A}_{[\alpha]} = A_{[\alpha]} \cap \mathfrak{A}A[1/\alpha]$ for α homogeneous satisfy*

$$\varphi_{\alpha,\beta}(\mathfrak{A}_{[\alpha\beta]}) = \mathfrak{A}_{[\alpha]} A_{[\alpha]}\left[\frac{\alpha^e}{\beta^d}\right]$$

for all $\alpha \in A_d$ and $\beta \in A_e$.

(ii) *If ideals $\mathfrak{a}\langle\alpha\rangle$ of $A_{[\alpha]}$ are given for all $\alpha \in A_d$ such that $\varphi_{\alpha,\beta}(\mathfrak{a}\langle\alpha\beta\rangle) = \mathfrak{a}\langle\alpha\rangle A_{[\alpha]}[\alpha^e/\beta^d]$ for all $\alpha \in A_d$, $\beta \in A_e$, then there exists a homogeneous ideal $\mathfrak{A}$ of A with $\mathfrak{A}_{[\alpha]} = \mathfrak{a}\langle\alpha\rangle$ for all $\alpha \in A_d$.*

Proof. (i) follows from the definition.

(ii) For each $d > 0$, put

$$\mathfrak{A}_d = \{\alpha \in A_d \mid \alpha/\alpha_i^d \in \mathfrak{a}\langle\alpha_i\rangle \quad \text{for all} \quad 0 \leq i \leq r\},$$

which becomes an additive group. Also, $0 \oplus \mathfrak{A}_1 \oplus \cdots \oplus \mathfrak{A}_d \oplus \cdots$ is a homogeneous ideal, denoted by $\mathfrak{A}$. Then $\mathfrak{A}_{[\alpha_i]} = \mathfrak{a}\langle \alpha_i \rangle$ for any i by definition. Since for $\beta \in A_e$, any element of $\mathfrak{A}_{[\alpha_i]} A_{[\alpha_i]} [\alpha_i^e/\beta]$ is written as

$$\frac{x}{\alpha_i^m}\left(\frac{\alpha_i^e}{\beta}\right)^m = \frac{x\alpha_i^{em}}{(\alpha_i \beta)^m} \quad \text{for } x \in \mathfrak{A}_m,$$

one has

$$\varphi_{\alpha_i, \beta}^{-1}\left(\mathfrak{A}_{[\alpha_i]} \cdot A_{[\alpha_i]}\left[\frac{\alpha_i^e}{\beta}\right]\right) = \mathfrak{A}_{[\beta\alpha_i]},$$

$$\varphi_{\beta, \alpha_i}^{-1}\left(\mathfrak{A}_{[\beta]} \cdot A_{[\beta]}\left[\frac{\beta}{\alpha_i^e}\right]\right) = \mathfrak{A}_{[\beta\alpha_i]}.$$

Recalling condition (ii) and $\mathfrak{A}_{[\alpha_i]} = \mathfrak{a}\langle \alpha_i \rangle$, one has

$$\mathfrak{A}_{[\beta]}\left[\frac{\beta}{\alpha_i^e}\right] = \mathfrak{a}\langle \beta \rangle \left[\frac{\beta}{\alpha_i^e}\right] \quad \text{as ideals of} \quad A_{[\beta]}\left[\frac{\beta}{\alpha_i^e}\right].$$

But since $D_+(\beta) = \bigcup_{i=0}^r D_+(\beta\alpha_i)$, it follows that

$$\mathfrak{A}_{[\beta]} = \mathfrak{a}\langle \beta \rangle. \qquad \square$$

e. Let $\varphi: A \to B$ be an R-homomorphism between graded R-algebras preserving degree. For any $\alpha \in A_d$ with $d > 0$, φ induces an R-homomorphism $\varphi_{[\alpha]}: A_{[\alpha]} \to B_{[\varphi(\alpha)]}$ defined by $\varphi_{[\alpha]}(x/\alpha^m) = \varphi(x)/\varphi(\alpha)^m$ for any $x \in A_{dm}$. ${}^a\varphi_{[\alpha]}$: Spec $B_{[\varphi(\alpha)]} \to$ Spec $A_{[\alpha]}$ defines a morphism $f_\alpha: D_+(\varphi(\alpha)) \to D_+(\alpha)$. Clearly, $f_\alpha|_{D_+(\alpha\beta)} = f_{\alpha\beta}$ for homogeneous α, β; hence $\{f_\alpha\}$ determines a morphism $f: G(\varphi) \to$ Proj A with $f|_{D_+(\alpha)} = f_\alpha$, whose continuous map between base spaces is Proj φ, i.e., the following square is commutative.

$$\begin{array}{ccc} D_+(\varphi(\alpha)) & \xrightarrow{\text{Proj } \varphi|_{D_+(\varphi(\alpha))}} & D_+(\alpha) \\ {\scriptstyle \psi_{(\varphi\alpha)}} \| \wr & & \| \wr \\ \text{Spec } A_{[\varphi\alpha]} & \xrightarrow[{}^a\varphi_{[\alpha]}]{} & \text{Spec } A_{[\alpha]} \end{array}$$

For the same of convenience, f is also denoted by Proj φ.

§3.5 Some Properties of Projective Schemes

a. Proposition 3.4. *Let A be a finitely generated graded R-algebra. Then*

(*i*) Proj A *is separated over* Spec R.
(*ii*) *Let* $\mathfrak{a} = A_+ \cap \sqrt{(0_A)}$. *Then* $\bar{A} = A/\mathfrak{a}$ *is finitely generated over* R *and* Proj $\bar{A} = (\text{Proj } A)_{\text{red}}$.
(*iii*) *Any closed subscheme* Y *of* Proj A *is also a projective scheme.*

(iv) *For any ring homomorphism* $\varphi\colon R \to R'$, $A' = A \otimes_R R'$ *is naturally a finitely generated graded* R'*-algebra and* Proj $A' \cong (\text{Proj } A) \otimes_R R'$ *as* R'*-schemes.*

PROOF. (i) Let $X = \text{Proj } A$ and let $\Delta_{X/R}\colon X \to X \times_R X$ be the diagonal morphism. Since the natural homomorphisms $A_{[\alpha]} \otimes_R A_{[\beta]} \to A_{[\alpha\beta]}$ are surjective for all homogeneous α and β (cf. Lemma 3.2), all $D_+(\alpha\beta)$ are closed subschemes of $D_+(\alpha) \times_R D_+(\beta)$. Since $(D_+(\alpha) \times_R D_+(\beta)) \cap \Delta_{X/R}(X) \cong D_+(\alpha) \cap D_+(\beta) = D_+(\alpha\beta)$ by Proposition 1.40 and $X \times_R X$ is covered by $(D_+(\alpha) \times_R D_+(\beta))$, $\Delta_{X/R}(X)$ is a closed subscheme of $X \times_R X$.

(ii) Noting that $(\bar{A})_0 = A_0 = R$, it is easily checked that $\bar{A}$ is finitely generated and the natural map $\pi\colon A \to \bar{A}$ preserves degree. Then Proj π: Proj $\bar{A} \to$ Proj A is obtained. The rest of the proof is left to the reader.

(iii) On each $D_+(\alpha)$ with $\alpha \in A_d$, $Y \cap D_+(\alpha)$ is a closed subscheme of $D_+(\alpha)$, which can be written as $V(\mathfrak{a}\langle\alpha\rangle) \cong \text{Spec}(A_{[\alpha]}/\mathfrak{a}\langle\alpha\rangle)$ for some ideal $\mathfrak{a}\langle\alpha\rangle$. Since as schemes $(Y \cap D_+(\alpha)) \cap D_+(\beta) = Y \cap (D_+(\alpha\beta))$ for any $\beta \in A_e$, we have $\varphi_{\alpha,\beta}(\mathfrak{a}\langle\alpha\beta\rangle) = \mathfrak{a}\langle\alpha\rangle A_{[\alpha]}[\alpha^e/\beta^d]$. Hence, by Lemma 3.3.(ii), one obtains a homogeneous ideal $\mathfrak{A}$ of A such that $\mathfrak{A}_{[\alpha]} = \mathfrak{a}\langle\alpha\rangle$ for all homogeneous α, thus, $Y \cong \text{Proj}(A/\mathfrak{A})$.

(iv) The homomorphism $\psi\colon A \to A'$ defined by $\psi(\alpha) = \alpha \otimes 1$, preserves degree and $\psi(A_+)A' = A'_+$. Hence, ${}^a\psi$ maps Proj A' into Proj A, and for any $\alpha \in A_d$, $(D_+(\alpha \otimes 1)) = {}^a\psi^{-1}(D_+(\alpha))$. We have the morphism Φ: Proj $A' \to$ Proj A associated with ψ, and $\Phi' = (\Phi, f')_R$: Proj $A' \to (\text{Proj } A) \otimes_R R'$ is obtained. Φ' is an isomorphism, since $A'_{[\alpha\otimes 1]} \cong A_{[\alpha]} \otimes_R R'$. □

Proposition 3.5. *If* A *is a finitely generated graded* R*-algebra, then the structure morphism* f: Proj $A \to$ Spec R *is a closed morphism.*

PROOF. By Proposition 3.4.(iii), we have only to show that $f(\text{Proj } A)$ is a closed subset. $f(\text{Proj } A)$ can be written as $\{\mathfrak{p} \in \text{Spec } R \mid f^{-1}(\mathfrak{p}) \neq \varnothing\}$. Since $f^{-1}(\mathfrak{p}) \cong (\text{Proj } A) \otimes_R R_\mathfrak{p}/\mathfrak{p}R_\mathfrak{p} \cong \text{Proj}(A \otimes_R R_\mathfrak{p}/(\mathfrak{p} \cdot A \otimes_R R_\mathfrak{p}))$ by Propositions 1.36 and 3.4.(iv), $f^{-1}(\mathfrak{p}) = \varnothing$ implies that $(A \otimes_R R_\mathfrak{p}/(\mathfrak{p} \cdot A \otimes_R R_\mathfrak{p}))_m = 0$ for all m greater than or equal to a fixed integer m_0 by Proposition 3.1. Since $(A \otimes_R R_\mathfrak{p})_{m_0}$ is finitely generated as an $R_\mathfrak{p}$-module, one has $A_{m_0} \otimes_R R_\mathfrak{p} = (A \otimes_R R_\mathfrak{p})_{m_0} = 0$ by Lemma 2.8. Thus we can find $u \notin \mathfrak{p}$ such that $u \cdot A_{m_0} = 0$; hence $u \cdot A_m = u \cdot A_{m_0} \cdot A_{m-m_0} = 0$ for all $m \geq m_0$. Putting $I^{(m)} = (0\colon A_m)$ in R, one has an ideal $I = \bigcup_{m=m_0}^{\infty} I^{(m)}$, which contains u. Hence, $\mathfrak{p} \in D(u) \subseteq D(I)$. Conversely, if $\mathfrak{p} \in D(I)$, then there exists $v \in I\backslash\mathfrak{p}$, i.e., $v \in I^{(n)}$ for some $n > 0$, i.e., $vA_n = 0$. Thus $(A \otimes_R R_\mathfrak{p})_n = 0$ and so $(A \otimes_R R_\mathfrak{p})_N = 0$ for all $N \geq n$. Hence $f^{-1}(\mathfrak{p}) = \varnothing$. Therefore, $f(\text{Proj } A) = V(I)$. □

b. Definition. Let $f\colon X \to Y$ be a morphism of schemes. If there exists an affine open cover $\{U_\lambda \mid \lambda \in \Lambda\}$ of Y such that each $X_\lambda = f^{-1}(U_\lambda)$ is isomorphic to some projective scheme Proj A_λ over U_λ, where A_λ is a finitely generated graded $A(U_\lambda)$-algebra, then f is said to be *a locally projective morphism.*

By Proposition 3.5, a locally projective morphism is a closed morphism. Furthermore, by Proposition 3.4.(iv), the property of being a locally projective morphism is stable under base extension. Hence, we obtain the following result.

Theorem 3.1. *Any locally projective morphism is a proper morphism.*

Corollary. $\mathbf{P}_k^n$ *over a field k is complete.*

Theorem 3.1. can be considered as a fundamental theorem in elimination theory.

Proposition 3.6. *Let* $k[\mathsf{T}_0, \ldots, \mathsf{T}_n, \mathsf{Y}_1, \ldots, \mathsf{Y}_m]$ *be a polynomial ring over an algebraically closed field k. Given a system of polynomials* $F_1(\mathsf{T}_0, \ldots, \mathsf{T}_n, \mathsf{Y}_1, \ldots, \mathsf{Y}_m), \ldots, F_r(\mathsf{T}_0, \ldots, \mathsf{T}_n, \mathsf{Y}_1, \ldots, \mathsf{Y}_m)$ *such that all* $F_i(\mathsf{T}_0, \ldots, \mathsf{T}_n, \mathsf{Y}_1, \ldots, \mathsf{Y}_m)$ *are homogeneous with respect to* $\mathsf{T}_0, \ldots, \mathsf{T}_n$*, there exist polynomials* $G_1(\mathsf{Y}_1, \ldots, \mathsf{Y}_m), \ldots, G_s(\mathsf{Y}_1, \ldots, \mathsf{Y}_m)$ *such that any* $(x, y) \in k^{n+1+m}$ *with* $x \neq 0$ *and* $F_1(x, y) = \cdots = F_r(x, y) = 0$ *satisfies* $G_1(y) = \cdots = G_s(y) = 0$; *and conversely, for any* $y \in k^m$ *with* $G_1(y) = \cdots = G_s(y) = 0$*, there exists* $x \in k^{n+1} \backslash \{0\}$ *with* $F_1(x, y) = \cdots = F_r(x, y) = 0$.

($G_1(\mathsf{Y}) = \cdots = G_s(\mathsf{Y}) = 0$ are said to be the result of eliminating $\mathsf{T}_0, \ldots, \mathsf{T}_n$ from $F_1(\mathsf{T}, \mathsf{Y}) = \cdots = F_r(\mathsf{T}, \mathsf{Y}) = 0$).

PROOF. Let $R = k[\mathsf{Y}_1, \ldots, \mathsf{Y}_m]$ and $S = R[\mathsf{T}_0, \ldots, \mathsf{T}_n]$. Define $\mathfrak{A} = \sum_{i=1}^r SF_i$, which is a homogeneous ideal of the R-algebra S. Letting f: $\mathbf{P}_R^n \to \operatorname{Spec} R = \mathbf{A}^m$ be the structure morphism, $f(V_+(\mathfrak{A}))$ is a closed subset $V(\mathfrak{a})$, since f is a closed morphism. The ideal $\mathfrak{a}$ has a finite basis $G_1(\mathsf{Y}), \ldots, G_s(\mathsf{Y})$. These $G_1, \ldots, G_s$ have the required property, by Theorem 1.4 and §1.8. □

EXAMPLE 3.7. Let $F_1 = \mathsf{T}_0 - \mathsf{Y}_1\mathsf{Y}_2$ and $F_2 = \mathsf{T}_1 - \mathsf{Y}_2$. For $a \neq 0$, there does not exist $(x, y) \in k^4$ with $F_1(x, y) = F_2(x, y) = 0$ and $y = (a, 0), x \neq (0, 0)$. Thus, by eliminating $\mathsf{T}_0, \mathsf{T}_1$ from $F_1 = F_2 = 0$, one cannot find a system of the form $G_1 = \cdots = G_s = 0$. Note that these F_1 and F_2 are not homogeneous with respect to T_0 and T_1. □

c. **Proposition 3.7.** *If* f: $X \to Y$ *and* g: $X' \to Y$ *are locally projective morphisms, then so is* $f \times_Y g$: $X \times_Y X' \to Y$.

PROOF. One can assume that Y is affine, i.e., $Y = \operatorname{Spec} R$ for some ring R, and that $X = \operatorname{Proj} A$, $X' = \operatorname{Proj} B$ for finitely generated graded R-algebras

A and B. Put $C_0 = R$, $C_1 = A_1 \otimes_R B_1, \ldots, C_d = A_d \otimes_R B_d, \ldots$ and define C to be $\bigoplus_{j=0}^{\infty} C_j$, which is also a graded R-algebra. Then if $\alpha_0, \ldots, \alpha_r$ and $\beta_0, \ldots, \beta_s$ are generators of the R-modules A and B, respectively, $\alpha_0 \otimes \beta_0, \ldots, \alpha_r \otimes \beta_s$ become the generators of C_1 and $C = R[\alpha_0 \otimes \beta_0, \ldots, \alpha_r \otimes \beta_s]$. Furthermore, one can easily check that $C_{[\alpha_i \otimes \beta_j]} \cong A_{[\alpha_i]} \otimes_R B_{[\beta_j]}$ for all $0 \le i \le r$, $0 \le j \le s$. Thus $D_+(\alpha_i \otimes \beta_j) \cong \operatorname{Spec} C_{[\alpha_i \otimes \beta_j]} \cong \operatorname{Spec} A_{[\alpha_i]} \times_R \operatorname{Spec} B_{[\beta_j]} \cong D_+(\alpha_i) \times D_+(\beta_j)$. This implies that $\operatorname{Proj} C \cong X \times_Y X'$. □

Definition. The graded ring C defined above is denoted $A \# B$, called the *Segre product* A and B over R.

EXAMPLE 3.8. Let $A = k[\mathsf{T}_0, \ldots, \mathsf{T}_r]$ and $B = k[\mathsf{Y}_0, \ldots, \mathsf{Y}_s]$ be polynomial rings. Then $A \# B = k[\mathsf{T}_0 \otimes \mathsf{Y}_0, \ldots, \mathsf{T}_r \otimes \mathsf{Y}_s]$, which is isomorphic to $k[\mathsf{T}_0\mathsf{Y}_0, \ldots, \mathsf{T}_r\mathsf{Y}_s]$, a subring of the polynomial ring $k[\mathsf{T}_0, \ldots, \mathsf{T}_r, \mathsf{Y}_0, \ldots, \mathsf{Y}_s]$ in $r + 1 + s + 1$ variables.

Let S be a polynomial ring $k[\mathsf{Z}_0, \mathsf{Z}_1, \ldots, \mathsf{Z}_N]$, where N is $rs + r + s$, and define $\psi_{r,s}\colon S \to k[\mathsf{T}_0\mathsf{Y}_0, \ldots, \mathsf{T}_r\mathsf{Y}_s]$ by $\psi_{r,s}(\mathsf{Z}_0) = \mathsf{T}_0\mathsf{Y}_0, \ldots, \psi_{r,s}(\mathsf{Z}_s) = \mathsf{T}_0\mathsf{Y}_s$, $\psi_{r,s}(\mathsf{Z}_{s+1}) = \mathsf{T}_1\mathsf{Y}_0, \ldots, \psi_{r,s}(\mathsf{Z}_N) = \mathsf{T}_r\mathsf{Y}_s$. Since $\psi_{r,s}$ is surjective, $k[\mathsf{T}_0\mathsf{Y}_0, \ldots, \mathsf{T}_r\mathsf{Y}_s] \cong S/\operatorname{Ker} \psi_{r,s}$; hence $\mathbf{P}_k^r \times \mathbf{P}_k^s$ is isomorphic to $V_+(\operatorname{Ker} \psi_{r,s})$ which is a subvariety of $\mathbf{P}_k^{rs+r+s}$. The closed immersion $\mathbf{P}_k^r \times \mathbf{P}_k^s \to \mathbf{P}_k^{rs+r+s}$ is said to be the *Segre morphism*, denoted by $\zeta_{r,s}$.

In the case $r = s = 1$, $\operatorname{Ker} \psi_{1,1}$ is generated by $\mathsf{Z}_0\mathsf{Z}_3 - \mathsf{Z}_1\mathsf{Z}_2$. Hence $\mathbf{P}_k^1 \times \mathbf{P}_k^1$ is isomorphic to a quadric surface.

§3.6 Chow's Lemma

a. The next result is fundamental, because it connects complete varieties with projective varieties.

Theorem 3.2 (Chow's lemma). *Let X be an irreducible k-complete scheme. Then there exist an irreducible k-projective scheme X' and a proper morphism $f\colon X' \to X$ with an open dense subset U of X such that $f|_{f^{-1}(U)}\colon f^{-1}(U) \to U$ is an isomorphism.*

PROOF. Let $\{U_i \mid 1 \le i \le m\}$ be an affine open cover of X with $U_i \ne \varnothing$ for all i. Since each U_i is affine, one has a projective scheme P_i which contains U_i as an open subset. $U = U_1 \cap \cdots \cap U_m$ is also open and dense, and the open immersions $j_i\colon U \to U_i$ determine the morphism $\sigma = (j_1, \ldots, j_m)\colon U \to U_1 \times_k \cdots \times_k U_m$, which is an immersion. Furthermore, letting the $w_i\colon U_i \to P_i$ be the open immersions, one has the open immersion $\rho = w_1 \times \cdots \times w_m\colon U_1 \times_k \cdots \times_k U_m \to P = P_1 \times_k \cdots \times_k P_m$, which is a projective scheme by the last proposition. $\varphi = \rho \circ \sigma\colon U \to P$ is an immersion and the graph $\Gamma_\varphi(U)$ is a closed subscheme of $U \times P$, since P is separated by Proposition 3.4.(i). The closure scheme of $\Gamma_\varphi(U)$ in $X \times P$ (cf. §1.27.**e**), which is denoted by X', is a closed subscheme of $X \times P$. Composing the closed immersion $i\colon X' \to X \times P$ with the projection morphisms, we obtain $f\colon X' \to X$ and $g\colon X' \to P$. Since P is projective, the projection $X \times P \to X$ is proper (by Theorems 3.1 and 1.22.(iii)); hence so is f. Furthermore, $f|_{\Gamma_\varphi(U)}\colon \Gamma_\varphi(U) \to U$ is an isomorphism.

$$\begin{array}{ccccc} & & U \times P & & \\ & \cup & & \subset & \\ \Gamma_\varphi(U) & & & & X \times P \\ & \subset & & \cup & \\ & & X' & & \end{array} \qquad \begin{array}{c} X' \subset X \times P \xrightarrow{g} P \\ X' \subset X \times P \xrightarrow{f} X \end{array}$$

We claim that g is a closed immersion. Replacing P_i by U_i in P, one has W_i, which isomorphic to $U_i \times P^{(i)}$, where $P^{(i)}$ is the product of $P_1, \ldots, P_{i-1}, P_{i+1}, \ldots, P_m$, for any $1 \le i \le m$. To prove that $f^{-1}(U_i) \subseteq g^{-1}(W_i)$, we let $p_i\colon U_i \times P \to U_i \times P_i$ be the natural projection. $p_i(f^{-1}(U_i)) = p_i(X' \cap (U_i \times P))$ is contained in the closure of Δ_U in $U_i \times P_i$; hence $p_i(f^{-1}(U_i)) \subseteq \Delta_{U_i} \subset U_i \times U_i$, by the next lemma. Then $f^{-1}(U_i) \subseteq p_i^{-1}(U_i \times U_i) \subseteq g^{-1}(W_i)$.

Lemma 3.4. *Let U be an open dense subscheme of a separated scheme X. Then the diagonal $\Delta_U(U)$ is closed in $U \times X$.*

Proof. If $j\colon U \to X$ is the open immersion, then Γ_j is closed in $U \times X$ by Proposition 1.41. But then since $\Gamma_j = \Delta_U$, Δ_U is also closed in $U \times X$. □

Since $X' = \bigcup_{i=1}^m f^{-1}(U_i)$, it follows that $\{g^{-1}(W_i) \mid 1 \le i \le m\}$ is an open cover of X'. Furthermore, $f^{-1}(U_i) = X' \cap (U_i \times P) \subseteq g^{-1}(W_i)$ implies that $f^{-1}(U_i) \subset U_i \times W_i$. Hence $f^{-1}(U_i)$ is the closure of $\Gamma_\varphi(U)$ in $U_i \times W_i$. One has the isomorphisms $h_i\colon W_i \simeq U_i \times P^{(i)}$ and $H_i = \mathrm{id} \times h_i\colon U_i \times W_i \simeq U_i \times U_i \times P^{(i)}$. Then $H_i(\Gamma_\varphi(U)) \subset \Delta_i \times P^{(i)}$, where Δ_i is the diagonal of $U_i \times U_i$, and so $H_i(f^{-1}(U_i))$ (= the closure of $H_i(\Gamma_\varphi(U))$ in $U_i \times U_i \times P^{(i)}$), is contained in $\Delta_i \times P^{(i)}$. Letting π_i be the projection of $U_i \times W_i$ onto W_i, $\pi_i \circ H_i^{-1}|_{\Delta_i \times P^{(i)}}\colon \Delta_i \times P^{(i)} \to W_i$ is an isomorphism. Therefore $\pi_i|_{f^{-1}(U_i)}\colon f^{-1}(U_i) \to W_i$ is a closed immersion. Since $f^{-1}(U_i) \subseteq g^{-1}(W_i)$ and $g|_{f^{-1}(U_i)} = \pi_i|_{f^{-1}(U_i)}$, this implies that g is the closed immersion.

$$\begin{array}{ccccc} \Gamma_\varphi(U) \subset f^{-1}(U_i) & \subset & U_i \times W_i & \overset{H_i}{\cong} & U_i \times U_i \times P^{(i)} \\ & \searrow^{g} & \downarrow^{\pi_i} & & \cup \\ & & W_i & \longleftarrow & \Delta_i \times P^{(i)} \end{array}$$

□

Remark. If X is reduced, then so is X'. Thus one obtains the next result.

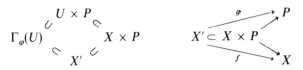

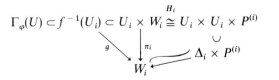

Corollary. *Let X be a complete variety. Then there exist a projective variety X' and a proper birational morphism $X' \to X$.*

Remarks (1). Any complete algebraic curve is projective (see §6.4.**b**).
(2) Any complete nonsingular surface is projective (see §8.5.**c**).
(3) There exists a complete normal surface which is not projective.

b. To generalize Theorem 3.2, we need a concept of direct sum of schemes.

Definition. The *direct sum* of the ringed spaces $(X_i, \mathcal{O}_{X_i})$, $1 \leq i \leq n$, is the ringed space $(X, \mathcal{D})$, where the topological space X is the disjoint union $X_1 \coprod X_2 \coprod \cdots \coprod X_n$ and $\mathcal{D}(U_1 \coprod U_2 \coprod \cdots \coprod U_n)$ is the direct sum of rings $\mathcal{O}_{X_1}(U_1) \oplus \mathcal{O}_{X_2}(U) \oplus \cdots \oplus \mathcal{O}_{X_n}(U_n)$, where U_i is an open subset of X_i.

It is easy to prove that if the X_i are schemes (respectively algebraic schemes; projective schemes), then so is the direct sum of the X_i.

Theorem 3.3. *Let X be a k-complete scheme. Then there exist a k-projective scheme X', a proper morphism f: $X' \to X$, and an open dense subset U of X such that $f|_{U'}$: $U' \xrightarrow{\sim} U$ is an isomorphism, where U' is $f^{-1}(U)$.*

Proof. Let $x_1, \ldots, x_m$ be the generic points of the irreducible components of the base space X. There exists an open subset V_i such that $x_i \in V_i$ and $x_j \notin V_i$ for all $j \neq i$. Now, define X_i to be the closure scheme of V_i. By the last theorem, there exist a k-projective scheme X_i' and a proper morphism f_i: $X_i' \to X_i$ such that $f_i^{-1}(V_i^0) \cong V_i^0$ for some open dense $V_i^0 \subseteq V_i$. Define X' to be the direct sum $\coprod_{i=1}^m X_i'$ and f: $X' \to X$ to be the direct sum of the f_i, $1 \leq i \leq m$. If one lets $U_i = V_i^0 \setminus \bigcup_{j \neq i} X_j$ and $U = U_1 \cup \cdots \cup U_m$, then $U' = f^{-1}(U) \cong U$. □

In the above proofs, one can replace k by a Noetherian ring R. Thus, one obtains the following result.

Theorem 3.4. *Let R be a Noetherian ring and let φ: $X \to$ Spec R be a proper morphism. Then, there exists a proper morphism f: $X' \to X$ such that*

(*i*) *$\psi = \varphi \circ f$: $X' \to$ Spec R is projective, i.e., $X' =$ Proj A', where A' is a finitely generated graded R-algebra, and ψ is the structure morphism of the R-scheme X', and*
(*ii*) *There exists an open dense subset U of X such that $f|_{f^{-1}(U)}$: $f^{-1}(U) \to U$ is an isomorphism.*

The last theorem will be used for the proof of the finiteness theorem for coherent scheaves (see §7.6.**c**).

EXERCISES

3.1. Let $f\colon V \to W$ be a dominating morphism between varieties. Then show that $f(V)$ contains an open dense subset W_0 of W.

3.2. Exhibit a basis of Ker $\psi_{r,s}$ in §3.5.**c**.

3.3. Let A be a graded R-algebra. Let $A[\mathsf{T}]$ be a polynomial ring over A and define the set B by $B = \{\sum_{i=0}^{m} a_i \mathsf{T}^i \mid a_i \in A_i,\ m \geq 0\}$. Show that B is a graded R-algebra and that the mapping $\psi\colon A \to B$ defined by $\psi(\sum_{i=0}^{m} a_i) = \sum_{i=0}^{m} a_i \mathsf{T}^i$ is an R-isomorphism preserving degree.

Let M be a graded A-module. Then, identifying A with B via ψ, M is A-isomorphic to $M' = M \otimes_A B \cong \{\sum_{i=-m}^{m} b_i \mathsf{T}^i \mid b_i \in M_i,\ m \geq 0\}$, which is a submodule of $M \otimes_A A[\mathsf{T}, \mathsf{T}^{-1}]$.

Chapter 4

Cohomology of Sheaves

§4.1 Injective Sheaves

a. Throughout this section, let $(X, \mathcal{O}_X)$ be a ringed space, and let $\mathcal{F}$, $\mathcal{G}$, $\mathcal{H}$ denote $\mathcal{O}_X$-modules.

Definition. An $\mathcal{O}_X$-module $\mathcal{S}$ is said to be *injective* if whenever any sequence of $\mathcal{O}_X$-modules of the form

$$0 \to \mathcal{F} \to \mathcal{G},$$

is exact, then

$$\mathrm{Hom}_{\mathcal{O}_X}(\mathcal{G}, \mathcal{S}) \to \mathrm{Hom}_{\mathcal{O}_X}(\mathcal{F}, \mathcal{S}) \to 0$$

is also exact.

Recalling Proposition 1.17.(iii), $\mathcal{S}$ is injective if and only if the functor $\mathrm{Hom}_{\mathcal{O}_X}(\,\cdot\,, \mathcal{S})$, defined by $\mathcal{F} \mapsto \mathrm{Hom}_{\mathcal{O}_X}(\mathcal{F}, \mathcal{S})$, is exact.

Remark. Occasionally, an injective $\mathcal{O}_X$-module is referred to as an *injective sheaf* on X.

We say that $\mathcal{F}$ is extended to $\mathcal{G}$, if $\mathcal{F}$ is an $\mathcal{O}_X$-submodule of $\mathcal{G}$.

Proposition 4.1. *Any $\mathcal{F}$ can be extended to some injective sheaf $\mathcal{S}$.*

PROOF. As is well known in homological algebra ([G, I.1.4.1, p. 6]), an $\mathcal{O}_{X,x}$-module $\mathcal{F}_x$ can be extended to an injective $\mathcal{O}_{X,x}$-module, which we denote I_x for any point $x \in X$. Define a presheaf $\mathcal{S}$ by $\mathcal{S}(U) = \prod_{x \in U} I_x$ for any open subset U of X. It is easy to see that $\mathscr{S}$ is an $\mathcal{O}_X$-module. We claim that for any $\mathscr{G}$, .

$$\mathrm{Hom}_{\mathcal{O}_X}(\mathscr{G}, \mathscr{S}) \cong \prod_{x \in X} \mathrm{Hom}_{\mathcal{O}_{X,x}}(\mathscr{G}_x, I_x).$$

To show this, let $p_{U,x}\colon \mathscr{S}(U) \to I_x$ be the projection, whenever $x \in U$. The direct limit of $p_{U,x}$ where $x \in U$ is denoted by $p_{\langle x \rangle}$, which is an $\mathcal{O}_{X,x}$-homomorphism from $\mathscr{S}_x$ to I_x. For any $\theta \in \mathrm{Hom}_{\mathcal{O}_X}(\mathscr{G}, \mathscr{S})$, one has $p_{\langle x \rangle} \circ \theta_x \in \mathrm{Hom}_{\mathcal{O}_{X,x}}(\mathscr{G}_x, I_x)$; hence we define $F(\theta)$ to be $\{p_{\langle x \rangle} \circ \theta_x \mid x \in X\}$. We leave it to the reader to verify that F is bijective.

We want to show that $\mathscr{S}$ is an injective sheaf. Given an injective $\mathcal{O}_X$-homomorphism $\varphi\colon \mathscr{F} \to \mathscr{G}$, and $\eta \in \mathrm{Hom}_{\mathcal{O}_X}(\mathscr{F}, \mathscr{S})$, one has $p_{\langle x \rangle} \circ \eta_x$: $\mathscr{F}_x \to I_x$ and $\theta_x\colon \mathscr{G}_x \to I_x$ such that $\theta_x \circ \varphi_x = p_{\langle x \rangle} \circ \eta_x$ for all $x \in X$, since I_x is injective. $\bar{\theta} = F^{-1}(\{\theta_x \mid x \in X\})$ satisfies $\bar{\theta} \circ \varphi = \eta$. □

The injective sheaf $\mathscr{S}$ in Proposition 4.1 is said to be an *injective extension* of $\mathscr{F}$.

b. Given $\mathscr{F}$, one has an injective extension $\mathscr{S}^0$ of $\mathscr{F}_0 = \mathscr{F}$. $\mathscr{F}_1 = \mathscr{S}^0/\mathscr{F}$ has an injective extension $\mathscr{S}^1$ and one has an injective extension $\mathscr{S}^2$ of $\mathscr{F}_2 = \mathscr{S}^1/\mathscr{F}_1$. Repeating this process, one has exact sequences

$$0 \longrightarrow \mathscr{F}_m \xrightarrow{i_m} \mathscr{S}^m \xrightarrow{p_m} \mathscr{F}_{m+1} \longrightarrow 0,$$

where $\mathscr{S}^m$ is an injective sheaf for all $m \geq 0$.

Composing $i_m\colon \mathscr{F}_m \to \mathscr{S}^m$ with $p_{m-1}\colon \mathscr{S}^{m-1} \to \mathscr{F}_m$, one has δ^{m-1}: $\mathscr{S}^{m-1} \to \mathscr{S}^m$. Since $\mathrm{Ker}\,\delta^{m-1} = \mathrm{Ker}\,p_{m-1}$ and $\mathrm{Im}\,\delta^{m-2} = \mathrm{Im}\,i_{m-1}$, the following sequence is exact

$$0 \longrightarrow \mathscr{F} \xrightarrow{i_0} \mathscr{S}^0 \xrightarrow{\delta^0} \mathscr{S}^1 \xrightarrow{\delta^1} \cdots.$$

Definition. The above sequence is said to be an *injective resolution* of $\mathscr{F}$.

Thus, one obtains the following result.

Lemma 4.1. *Any $\mathcal{O}_X$-module has an injective resolution.*

§4.2 Fundamental Theorems

a. As was shown in §4.1, every $\mathscr{F}$ has an injective resolution

$$0 \longrightarrow \mathscr{F} \xrightarrow{i} \mathscr{S}^0 \xrightarrow{\delta^0} \mathscr{S}^1 \xrightarrow{\delta^1} \mathscr{S}^2 \longrightarrow \cdots.$$

For any open subset U of X and any $q \geq 0$, define $H^q(U, \mathscr{F})$ to be $\operatorname{Ker} \delta_U^q / \operatorname{Im} \delta_U^{q-1}$, where $\delta^{-1} = 0$ and $\delta_U^q : \mathscr{S}^q(U) \to \mathscr{S}^{q+1}(U)$.

The next result is the fundamental theorem in the theory of sheaf cohomology.

Theorem 4.1

(*I*) *The groups $H^q(U, \mathscr{F})$ do not depend on the choice of the injective resolution of $\mathscr{F}$.*

(*II*) *To any $\theta : \mathscr{F} \to \mathscr{G}$, there are associated homomorphisms $H^q(U, \theta)$: $H^q(U, \mathscr{F}) \to H^q(U, \mathscr{G})$ for all q, which satisfy*

(*i*) $H^q(U, \mathrm{id}) = \mathrm{id}$.

(*ii*) *If $\eta : \mathscr{G} \to \mathscr{H}$, then*

$$H^q(U, \eta \circ \theta) = H^q(U, \eta) \circ H^q(U, \theta) \quad \textit{for} \quad q \geq 0.$$

(*III*) *With any exact sequence*

$$0 \longrightarrow \mathscr{F} \xrightarrow{\ \theta\ } \mathscr{G} \xrightarrow{\ \eta\ } \mathscr{H} \longrightarrow 0$$

there are associated homomorphisms $\mathfrak{d}^q : H^q(U, \mathscr{H}) \to H^{q+1}(U, \mathscr{F})$, $q \geq 0$, and these together with $H^q(U, \theta)$ and $H^q(U, \eta)$ form the long exact sequence

$$\begin{array}{l} 0 \longrightarrow H^0(U, \mathscr{F}) \longrightarrow H^0(U, \mathscr{G}) \longrightarrow H^0(U, \mathscr{H}) \\ \xrightarrow{\ \mathfrak{d}^0\ } H^1(U, \mathscr{F}) \longrightarrow \cdots \\ \xrightarrow{\ \mathfrak{d}^{q-1}\ } H^q(U, \mathscr{F}) \longrightarrow H^q(U, \mathscr{G}) \longrightarrow H^q(U, \mathscr{H}) \\ \longrightarrow \cdots . \end{array}$$

($\mathfrak{d}^q$ *is said to be the q-th connecting homomorphism.*)

(*IV*) *If*

$$\begin{array}{ccccccccc} 0 & \longrightarrow & \mathscr{F} & \longrightarrow & \mathscr{G} & \longrightarrow & \mathscr{H} & \longrightarrow & 0 \\ & & \downarrow & & \downarrow & & \downarrow & & \\ 0 & \longrightarrow & \mathscr{F}' & \longrightarrow & \mathscr{G}' & \longrightarrow & \mathscr{H}' & \longrightarrow & 0, \end{array}$$

is a commutative diagram with exact rows, then the square

$$\begin{array}{ccc} H^q(U, \mathscr{H}) & \xrightarrow{\ \mathfrak{d}^q\ } & H^{q+1}(U, \mathscr{F}) \\ \downarrow & & \downarrow \\ H^q(U, \mathscr{H}') & \xrightarrow{\ \mathfrak{d}'^q\ } & H^{q+1}(U, \mathscr{F}') \end{array}$$

commutes where $\mathfrak{d}'^q$ is the q-th connecting homomorphism for $0 \to \mathscr{F}' \to \mathscr{G}' \to \mathscr{H}' \to 0$.

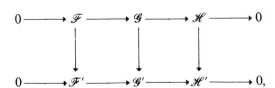

PROOF. We refer the reader to [H, pp. 204–205]. □

To explain the meaning of this theorem, we show some of its easy consequences.

Corollary

(i) $H^0(U, \mathscr{F}) \cong \Gamma(U, \mathscr{F})$.
(ii) *If* $\mathscr{S}$ *is injective, then* $H^q(U, \mathscr{S}) = 0$ *for all* $q > 0$.

PROOF. (i) By definition, $H^0(U, \mathscr{F}) = \Gamma(U, \mathscr{F})$, since $\delta^{-1} = 0$ and $\operatorname{Ker} \delta^0 = \mathscr{F}$.

(ii) One has the following injective resolution of $\mathscr{S}$:

$$0 \longrightarrow \mathscr{S} \xrightarrow{\text{id}} \mathscr{S} \longrightarrow 0 \longrightarrow \cdots.$$

Then $H^q(U, \mathscr{S}) = 0$, for all $q > 0$. □

§4.3 Flabby Sheaves

a. Definition. For an $\mathcal{O}_X$-module $\mathscr{F}$ and an open subset U of X, define a presheaf $\mathscr{G}$ as follows: $\mathscr{G}(W) = \mathscr{F}(W)$ if W is an open subset of X with $W \subseteq U$; and $\mathscr{G}(W) = 0$ if $W \nsubseteq U$. The sheafification ${}^a\mathscr{G}$ of $\mathscr{G}$ is denoted by $\mathscr{F}_U$.

One has a presheaf homomorphism $\{\theta_W\}: \mathscr{G} \to \mathscr{F}$ defined as follows: $\theta_W = \text{id}$ if $W \subseteq U$, and $\theta_W = 0$ if $W \nsubseteq U$. From this, an injective $\mathcal{O}_X$-homomorphism $i: {}^a\mathscr{G} = \mathscr{F}_U \to \mathscr{F}$ is induced by §1.10.**f**. For any $\mathcal{O}_X$-module $\mathscr{L}$, one has the isomorphism $\operatorname{Hom}_{\mathcal{O}_X}(\mathscr{G}, \mathscr{L}) \cong \operatorname{Hom}_{\mathcal{O}_X}(\mathscr{F}_U, \mathscr{L})$ by Lemma 1.14; hence $\operatorname{Hom}_{\mathcal{O}_X}(\mathcal{O}_{X,U}, \mathscr{L}) \cong \mathscr{L}(U)$.

If $\mathscr{L}$ is injective, then the homomorphism $\operatorname{Hom}_{\mathcal{O}_X}(\mathcal{O}_X, \mathscr{L}) \to \operatorname{Hom}_{\mathcal{O}_X}(\mathcal{O}_{X,U}, \mathscr{L})$ is surjective, i.e., the restriction map $r_{U,X}: \mathscr{L}(X) \to \mathscr{L}(U)$ is surjective.

Definition. An $\mathcal{O}_X$-module $\mathscr{L}$ is said to be *flabby* if the restriction map $\mathscr{L}(X) \to \mathscr{L}(U)$ is surjective for every open subset U of X.

From the above discussion, any injective $\mathcal{O}_X$-module is flabby.

EXAMPLE 4.1. If X is an irreducible space, then the constant sheaf G_X is flabby.

Lemma 4.2. *Let*

$$0 \longrightarrow \mathscr{L}' \xrightarrow{\alpha} \mathscr{L} \xrightarrow{\beta} \mathscr{L}'' \longrightarrow 0$$

be an exact sequence of $\mathcal{O}_X$-modules. If $\mathcal{L}'$ is flabby, then the sequence

$$0 \longrightarrow \mathcal{L}'(X) \xrightarrow{\alpha_X} \mathcal{L}(X) \xrightarrow{\beta_X} \mathcal{L}''(X) \longrightarrow 0$$

is exact.

PROOF. By Lemma 1.19, it suffices to show that β_X is surjective. Given any section $s'' \in \mathcal{L}''(X)$, define the set $\mathbb{M}$ to be $\{(t, W) \mid W \subseteq X,\ W \text{ is open},\ t \in \mathcal{L}(W), \text{ and } \beta_W(t) = s''|_W\}$. We introduce the order $\le$ on $\mathbb{M}$ by

$$(t, W) \le (t', W') \Leftrightarrow W \subseteq W' \quad \text{and} \quad t'|_W = t.$$

It is easy to see that $\mathbb{M}$ is inductively ordered. By Zorn's lemma, there exists a maximal element (s^*, W^*) of $\mathbb{M}$. If $W^* \neq X$, take $x \in X \backslash W^*$. Then since β is surjective, there exists $(s_x, U_x) \in \mathbb{M}$ such that $x \in U_x$. If one lets $B = W^* \cap U_x$, then $\beta_B(s^*|_B - s_x|_B) = 0$. Hence, one has $\tau \in \mathcal{L}'(B)$ such that $\alpha_B(\tau) = s^*|_B - s_x|_B$. Now, since $\mathcal{L}'$ is flabby, one has $\tau' \in \mathcal{L}'(X)$ with $\tau'|_B = \tau$. Thus, letting $\tau^* = \alpha_X(\tau')$, $(s_x + \tau^*|_{U_x})|_B = s^*|_B$. By condition (5) for a sheaf (§.1.10), letting W be $W^* \cup U_x$, one has $\sigma \in \mathcal{L}(W)$ with $\sigma|_{U_x} = s_x + \tau^*|_{U_x}$ and $\sigma|_{W^*} = s^*$. Then $\beta_W(\sigma) = s''|_W$, i.e., $(\sigma, W) \in \mathbb{M}$, which is greater than (s^*, W^*). This contradicts the choice of (s^*, W^*). Thus, $W^* = X$ and so β_X is surjective. □

Remark. If $\mathcal{L}'$ is flabby, then so is $\mathcal{L}'|_U$ for any open subset U.

Proposition 4.2. *With the same notation and assumptions as in Lemma 4.2, $\mathcal{L}$ is flabby if and only if $\mathcal{L}''$ is flabby.*

PROOF. This follows immediately from the following commutative diagram for any open subset U of X:

$$\begin{array}{ccccccccc}
0 & \longrightarrow & \mathcal{L}'(X) & \longrightarrow & \mathcal{L}(X) & \longrightarrow & \mathcal{L}''(X) & \longrightarrow & 0 \\
 & & \downarrow & & \downarrow & & \downarrow & & \\
0 & \longrightarrow & \mathcal{L}'(U) & \longrightarrow & \mathcal{L}(U) & \longrightarrow & \mathcal{L}''(U) & \longrightarrow & 0 \\
 & & \downarrow & & \downarrow & & \downarrow & & \\
 & & 0 & & 0 & & 0 & &
\end{array}$$

□

b. A flabby sheaf need not be injective. However, the following vanishing theorem holds.

Theorem 4.2. *If $\mathcal{L}$ is flabby, then $H^q(X, \mathcal{L}) = 0$ for any $q > 0$.*

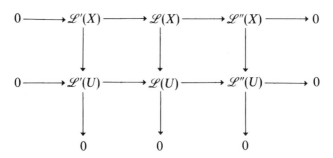

PROOF. By Proposition 4.1, $\mathscr{L}$ can be extended to an injective sheaf $\mathscr{S}^0$. Then by Proposition 4.2, $\mathscr{L}_1 = \mathscr{S}^0/\mathscr{L}$ is also flabby. Applying Theorem 4.1.(III) to the exact sequence

$$0 \to \mathscr{L} \to \mathscr{S}^0 \to \mathscr{L}_1 \to 0,$$

one has the exact sequence

$$0 \to \mathscr{L}(X) \to \mathscr{S}^0(X) \to \mathscr{L}_1(X) \to H^1(X, \mathscr{L}) \to H^1(X, \mathscr{S}^0).$$

By Corollary (ii) to Theorem 4.1, $H^1(X, \mathscr{S}^0) = 0$, and by Lemma 4.1 $\mathscr{S}^0(X) \to \mathscr{L}_1(X)$ is surjective. Hence, $H^1(X, \mathscr{L}) = 0$ for any flabby sheaf $\mathscr{L}$. Now, if $q \geq 2$, one has the exact sequence

$$\cdots\cdot \to H^{q-1}(X, \mathscr{S}^0) \to H^{q-1}(X, \mathscr{L}_1) \to H^q(X, \mathscr{L}) \to H^q(X, \mathscr{S}^0) \to \cdots.$$

Since $H^{q-1}(X, \mathscr{S}^0) = H^q(X, \mathscr{S}^0) = 0$ by Corollary (ii) to Theorem 4.1, it follows that $H^{q-1}(X, \mathscr{L}_1) \cong H^q(X, \mathscr{L})$. By induction on q, one obtains the result. □

By the same technique as in the proof of Theorem 4.2, one obtains the following result.

Proposition 4.3. *If a sequence of $\mathscr{O}_X$-modules*

$$0 \to \mathscr{F} \to \mathscr{G}^0 \to \mathscr{G}^1 \to \mathscr{G}^2 \to \cdots$$

is exact and $H^q(X, \mathscr{G}^i) = 0$ whenever $q > 0$ and $i \geq 0$, then each $H^q(X, \mathscr{F})$ is isomorphic to the q-th cohomology group of the complex $\{0 \to \mathscr{G}^0(X) \to \mathscr{G}^1(X) \to \mathscr{G}^2(X) \to \cdots\}$.

§4.4 Cohomology of Affine Schemes

a. The following result is the most basic result in the theory of cohomology of schemes.

Theorem 4.3 (Serre, Grothendieck). *If X is an affine scheme and $\mathscr{F}$ is a quasi-coherent $\mathscr{O}_X$-module, then*

$$H^q(X, \mathscr{F}) = 0 \qquad \textit{for any} \qquad q > 0.$$

PROOF. We refer the reader to [EGA, III.1.3.1]. □

To prove the converse of the above result, we need the next lemma.

Lemma 4.3. *Let $(X, \mathscr{O}_X)$ be a ringed space and let $\mathscr{F}$ be an $\mathscr{O}_X$-module of finite type. If* supp $\mathscr{F}$ *is a finite set $\{x_1, \ldots, x_r\}$, then $\Gamma(X, \mathscr{F}) \cong \bigoplus_{j=1}^r \mathscr{F}_{x_j}$.*

PROOF. Let $(\mathscr{F}^{\#}, \pi, X)$ denote the étale space associated with $\mathscr{F}$ (cf. §1.10.**d**). Then $\mathscr{F}^{\#} = \coprod_{x \in X} \mathscr{F}_x \cong \coprod_{j=1}^{r} \mathscr{F}_{x_j}$. Hence, $\Gamma(X, \mathscr{F}^{\#}) \cong \bigoplus_{j=1}^{r} \mathscr{F}_{x_j}$ by definition. □

b. Let X be a scheme.

Theorem 4.4. *If X is separated and quasi-compact, then the following are equivalent.*

(*a*) *X is affine.*
(*b*) *There exist $f_1, \ldots, f_r \in A = A(X)$ such that $A = \sum_{j=1}^{r} Af_j$ and the X_{f_j} are affine schemes.*
(*c*) *$H^q(X, \mathscr{F}) = 0$ for any quasi-coherent $\mathscr{O}_X$-module $\mathscr{F}$ and any positive integer q.*
(*d*) *$H^1(X, \mathscr{F}) = 0$ for any quasi-coherent $\mathscr{O}_X$-ideal $\mathscr{F}$.*

PROOF. (a) ⇒ (b) is obvious. (a) ⇒ (c) follows from Theorem 4.3.

(c) ⇒ (d) This is clear. Hence, it suffices to show (d) ⇒ (b) and (b) ⇒ (a).

(d) ⇒ (b) For any closed point x of X, take an affine open neighborhood U of x. Then $X \backslash U$ is a closed subset, which is defined by a quasi-coherent $\mathscr{O}_X$-ideal $\mathscr{S}$ by Proposition 1.27. If the reduced scheme $\{x\}$ is defined by a quasi-coherent $\mathscr{O}_X$-ideal $\mathscr{M}$, then $\{x\} \cup (X \backslash U)$ is defined by $\mathscr{J} = \mathscr{S} \cap \mathscr{M}$. But $\mathscr{S} + \mathscr{M} = \mathscr{O}_X$, since $\{x\} \cap (X \backslash U) = \varnothing$. Hence, $\mathscr{S}/\mathscr{J} = \mathscr{S}/\mathscr{S} \cap \mathscr{M} \cong (\mathscr{S} + \mathscr{M})/\mathscr{M} = \mathscr{O}_X/\mathscr{M}$, whose support is $\{x\}$. Thus by Lemma 4.2, $(\mathscr{S}/\mathscr{J})(X) = k(x)$. From the exact sequence

$$0 \to \mathscr{J} \to \mathscr{S} \to \mathscr{S}/\mathscr{J} \to 0,$$

one has by condition (d) the following exact sequence:

$$0 \to \mathscr{J}(X) \to \mathscr{S}(X) \to (\mathscr{S}/\mathscr{J})(X) \to H^1(X, \mathscr{J}) = 0.$$

Hence, there exists $f \in \mathscr{S}(X) \subseteq A = A(X)$ with $\bar{f}(x) = 1$ as an element of $k(x)$. However, $\bar{f}(y) = 0$ for all $y \in X \backslash U$; thus $X_f \subseteq U$ and $x \in X_f$. Since U is affine, $X_f = D(f|_U)$ is also an affine scheme. The f is denoted $f_{\{x\}}$. Clearly, $\{X_{f\{x\}} \mid x \text{ is a closed point of } X\}$ is an open cover of X. Since X is quasi-compact, there exist $x_1, \ldots, x_r \in X$ with $\bigcup_{j=1}^{r} X_{f\{x_j\}} = X$. Letting $f_j = f_{\{x_j\}} \in A(X)$, one defines an $\mathscr{O}_X$-homomorphism $\Phi\colon \mathscr{O}_X^r \to \mathscr{O}_X$ by $\Phi_U(\alpha_1, \ldots, \alpha_r) = \sum_{j=1}^{r} \alpha_j \cdot f_j|_U$, where U is any open subset of X.

We claim that Φ is surjective and, moreover, that $\Phi_X\colon \mathscr{O}_X^r(X) \to A(X) = \mathscr{O}_X(X)$ is also surjective. At any point $x \in X$, there exists j such that $\bar{f}_j(x) \neq 0$, i.e., $f_{j,x}$ is a unit. Hence, Φ is surjective. Let $\mathscr{F} = \operatorname{Ker} \Phi$, which is a sub-$\mathscr{O}_X$-module of $\mathscr{O}_X^r = \mathscr{O}_X \mathbf{e}_1 + \cdots + \mathscr{O}_X \mathbf{e}_r$ where the $\mathbf{e}_j$ form a base of the direct sum $\mathscr{O}_X^r$. $\mathscr{F}_j = \mathscr{F} \cap (\mathscr{O}_X \mathbf{e}_1 + \cdots + \mathscr{O}_X \mathbf{e}_j)$ for $1 \leq j \leq r$ is a quasi-coherent $\mathscr{O}_X$-module and each $\mathscr{F}_{j+1}/\mathscr{F}_j$ is also a quasi-coherent $\mathscr{O}_X$-module isomorphic to some $\mathscr{O}_X$-ideal. By (d), $H^1(X, \mathscr{F}_{j+1}/\mathscr{F}_j) = H^1(X, \mathscr{F}_1) = 0$ for all j. Hence from the exact sequences

$$H^1(X, \mathscr{F}_j) \to H^1(X, \mathscr{F}_{j+1}) \to H^1(X, \mathscr{F}_{j+1}/\mathscr{F}_j) = 0$$

for all j, we have $H^1(X, \mathscr{F}) = H^1(X, \mathscr{F}_r) = 0$. Thus the sequence

$$A(X)^r = \mathcal{O}_X^r(X) \to A(X) = \mathcal{O}_X(X) \to H^1(X, \mathscr{F}) = 0$$

is exact. Hence, Φ_X is surjective, i.e., $A(X) = \sum_{j=1}^r A(X)f_j$.

(b) $\Rightarrow$ (a) Let Ψ: $X \to \operatorname{Spec} A(X)$ be the canonical morphism for X defined in §1.15.**e**. For any $f \in A(X)$, $A(X_f) = A(X)_f$ by the Corollary to Theorem 1.15. If X is affine, then $\Psi|_{X_f}$: $X_f \to \operatorname{Spec} A(X_f) \cong D(f)$ is an isomorphism. Hence Ψ restricts to an isomorphism from each X_{f_j} onto $D(f_j)$, and so must be an isomorphism. Thus, X is an affine scheme. □

§4.5 Finiteness Theorem

a. The following fundamental result will be proved in §7.6.**b**. using the theory of cohomology groups of projective schemes.

Theorem 4.5. *Let X be a k-complete scheme, where k is a field, and let $\mathscr{F}$ be a coherent $\mathcal{O}_X$-module. Then for any $q \geq 0$, the k-vector space $H^q(X, \mathscr{F})$ is finite-dimensional.*

Let V be a normal variety and let D be a divisor on V. For any open subset U of V, define $\mathscr{F}(U)$ to be $\{\varphi \in \operatorname{Rat}(V) \mid \varphi = 0 \text{ or } \operatorname{div}(\varphi|_U) + D|_U \geq 0\}$. Then clearly $\{\mathscr{F}(U)\}$ with the natural restriction maps becomes an $\mathcal{O}_V$-module, denoted by $\mathcal{O}(D)$. If D is locally principal, this $\mathcal{O}(D)$ coincides with the $\mathcal{O}(D)$ as defined in §2.19.

Lemma 4.4. *$\mathcal{O}(D)$ is a coherent $\mathcal{O}_V$-module.*

PROOF. One may assume that V is an affine normal variety $\operatorname{Spec} A$. $\mathscr{F} = \mathcal{O}(D)$ satisfies condition (i) of Theorem 1.11.(c). To verify condition (ii), take $\varphi \in \mathscr{F}(V_f)$ for $f \in A$. Let $\operatorname{div}(f) = \sum_{i=1}^s m_i W_i$, and let $D = \sum_{i=1}^s n_i W_i + \sum_{j=1}^r \nu_j \Gamma_j$ be the irreducible decompositions where $m_i > 0$ and $n_i, \nu_j \in \mathbb{Z}$. Since $\operatorname{ord}_{\Gamma_j}(f^N\varphi) = \operatorname{ord}_{\Gamma_j}(\varphi)$ for all j, and since $\operatorname{ord}_{W_i}(f^N\varphi) = Nm_i + \operatorname{ord}_{W_i}(\varphi)$ for all i, it follows that $\operatorname{div}(f^N\varphi) + D \geq 0$ for $N \gg 0$; i.e., $f^N\varphi \in \mathscr{F}(V)$. Thus by Theorem 1.11, $\mathscr{F}$ is quasi-coherent. Clearly, we can choose ψ in A such that $D \leq \operatorname{div}(\psi)$. Then $\psi \cdot \mathscr{F} \subseteq \mathcal{O}_V$ and so $\mathscr{F} \subseteq \mathcal{O}_V \cdot \psi^{-1}$. Since V is Noetherian, $\mathscr{F}$ is of finite type, and is coherent (cf. Theorem 1.13). □

Since $\mathbf{L}_V(D) = \Gamma(V, \mathcal{O}(D))$, it is finite-dimensional as a k-vector space by Theorem 4.5. In the following chapters, we will use this result and obtain many important results.

b. The following result is a very basic fact, but will not be proved in this book.

Theorem 4.6 (Grothendieck). *Let X be a Noetherian space and let $\mathscr{F}$ be a sheaf. If $q > \dim X$, then $H^q(X, \mathscr{F}) = 0$.*

PROOF. We refer the reader to [H, III. Theorem 2.7] □

Thus, if V is a k-complete scheme and $\mathscr{F}$ is a coherent $\mathcal{O}_V$-module, then $\sum_{q=0}^{\infty}(-1)^q \dim_k H^q(V, \mathscr{F})$ is a finite sum of integers, called the (arithmetic) *Euler-Poincaré characteristic* of $\mathscr{F}$ on V, written as $\chi_V(\mathscr{F})$ or $\chi(\mathscr{F})$.

Proposition 4.4. *If a sequence of coherent* $\mathcal{O}_V$*-modules*

$$0 \to \mathscr{F}' \to \mathscr{F} \to \mathscr{F}'' \to 0$$

is exact, then $\chi_V(\mathscr{F}) = \chi_V(\mathscr{F}') + \chi_V(\mathscr{F}'')$.

PROOF. This follows from the long exact sequence obtained from the given exact sequence by Theorem 4.1.(III)

$$\cdots \to H^q(V, \mathscr{F}') \to H^q(V, \mathscr{F}) \to H^q(V, \mathscr{F}'')$$
$$\to H^{q+1}(V, \mathscr{F}') \to H^{q+1}(V, \mathscr{F}) \to H^{q+1}(V, \mathscr{F}'') \to \cdots.$$ □

Definition. If V is a complete normal variety, then $\chi_V = \chi_V(\mathcal{O}_V)$ is said to be the *arithmetic genus of V in the sense of Hirzebruch.* Moreover, when V is nonsingular with $\dim V = n$,

$$\dim_k H^n(V, \mathcal{O}_V) - \dim_k H^{n-1}(V, \mathcal{O}_V) + \cdots + (-1)^{n-1}\dim_k H^1(V, \mathcal{O}_V)$$

is said to be the *arithmetic genus* of V *in the sense of Severi*, denoted by $p_a(V)$.

Thus, one has $p_a(V) = (-1)^n(\chi_V - 1)$.

Let X be a scheme and let $\mathscr{F}$ be a quasi-coherent $\mathcal{O}_X$-module. Define $\eta: \mathcal{O} \to \mathrm{Hom}_{\mathcal{O}_X}(\mathscr{F}, \mathscr{F})$ by $\eta_U(\alpha)_W(\beta) = \alpha|_W \cdot \beta$ for any open subset W of an open set U, and for any $\alpha \in \mathcal{O}_X(U)$, $\beta \in \mathscr{F}(W)$. Then Ker η determines a closed subscheme Y of X. $\mathscr{F}$ has naturally a structure of $\mathcal{O}_Y$-module and then by definition $H^q(X, \mathscr{F}) = H^q(Y, \mathscr{F})$ for all q.

Furthermore, if $\mathscr{F}$ is coherent, then $\mathscr{H}om_{\mathcal{O}}(\mathscr{F}, \mathscr{F})_x \cong \mathrm{Hom}_{\mathcal{O}_{X,x}}(\mathscr{F}_x, \mathscr{F}_x)$ by Corollary 2 to Theorem 1.12, and so $\mathrm{supp}(\mathscr{H}om_{\mathcal{O}}(\mathscr{F}, \mathscr{F})) = \mathrm{supp}\, \mathscr{F}$. Hence, Theorem 4.6 can be restated in this case as follows:

Theorem 4.6*. *If* $\mathscr{F}$ *is coherent, then* $H^q(X, \mathscr{F}) = 0$ *for all* $q > \dim(\mathrm{supp}(\mathscr{F}))$.

§4.6 Leray's Spectral Sequence

a. Let $f: X \to Y$ be a morphism of ringed spaces and let $\mathscr{F}$ be an $\mathcal{O}_X$-module. Choosing an injective resolution of $\mathscr{F}$ by Lemma 4.1

$$0 \longrightarrow \mathscr{F} \longrightarrow \mathscr{I}^0 \xrightarrow{\delta^0} \mathscr{I}^1 \xrightarrow{\delta^1} \mathscr{I}^2 \longrightarrow \cdots$$

we have a complex of $\mathcal{O}_X$-modules $\{\cdots \to f_*(\mathcal{S}^{q-1}) \to f_*(\mathcal{S}^q) \to \cdots\}$. The q-th cohomology sheaf $\mathrm{Ker}(f_*(\delta^q))/\mathrm{Im}(f_*(\delta^{q-1}))$ is denoted by $R^q f_*(\mathcal{F})$, where $f_*(\delta^q)$ is defined by $f_*(\delta^q)_U = \delta^q_{f^{-1}(U)}$, for any q.

The next result is a direct consequence of the theory of derived functors.

Theorem 4.7

(*I*) *The $\mathcal{O}_Y$-modules $R^q f_*(\mathcal{F})$ do not depend on the choice of the injective resolution of $\mathcal{F}$. Furthermore, $R^0 f_*(\mathcal{F}) = f_* \mathcal{F}$.*

(*II*) *With any $\theta\colon \mathcal{F} \to \mathcal{G}$, $\mathcal{O}_Y$-homomorphisms $R^q f_*(\theta)\colon (R^q f_*)(\mathcal{F}) \to (R^q f_*)(\mathcal{G})$ are associated, satisfying:*

(*i*) $R^q f_*(\mathrm{id}) = \mathrm{id}$,
(*ii*) *If $\eta\colon \mathcal{G} \to \mathcal{H}$, then $(R^q f_*)(\eta \circ \theta) = R^q f_*(\eta) \circ R^q f_*(\theta)$.*

(*III*) *With any exact sequence*

$$0 \longrightarrow \mathcal{F} \xrightarrow{\theta} \mathcal{G} \xrightarrow{\eta} \mathcal{H} \longrightarrow 0,$$

there are associated $\mathcal{O}_Y$-homomorphisms $\mathfrak{d}^q\colon R^q f_*(\mathcal{H}) \to R^{q+1} f_*(\mathcal{F})$ and these, together with $R^q f_*(\theta)$ and $R^q f_*(\eta)$, form the long exact sequence

$$\begin{aligned} 0 &\longrightarrow f_*(\mathcal{F}) \longrightarrow f_*(\mathcal{G}) \longrightarrow f_*(\mathcal{H}) \\ &\xrightarrow{\mathfrak{d}^0} R^1 f_*(\mathcal{F}) \longrightarrow \cdots \\ &\xrightarrow{\mathfrak{d}^{q-1}} R^q f_*(\mathcal{F}) \longrightarrow R^q f_*(\mathcal{G}) \longrightarrow R^q f_*(\mathcal{H}) \\ &\longrightarrow \cdots. \end{aligned}$$

(*IV*) *Similar to* (*IV*) *of Theorem* 4.1.

Remark. If Y consists of a single point, then $f_*\mathcal{F} = \Gamma(X, \mathcal{F})$.

b. Now, let $\mathcal{L}$ be a flabby $\mathcal{O}_X$-module. Then $f_*\mathcal{L}$ is also flabby by definition. By Theorem 4.2, $H^q(Y, f_*\mathcal{L}) = 0$ for any $q > 0$. Choosing an injective resolution of a given $\mathcal{F}$

$$0 \to \mathcal{F} \to \mathcal{S}^0 \to \mathcal{S}^1 \to \cdots,$$

one has a complex $\{\cdots \to f_*\mathcal{S}^{q-1} \to f_*\mathcal{S}^q \to f_*\mathcal{S}^{q+1} \to \cdots\}$, which satisfies $H^p(Y, f_*\mathcal{S}^q) = 0$ for any $p > 0$ and $q \geq 0$. From this, applying the general theory of spectral sequences (cf. [G, II 4]), one obtains the next spectral sequence connecting $H^p(Y, R^q f_*(\mathcal{F}))$ with $H^n(X, \mathcal{F})$.

Theorem 4.8 (Leray). *There exists a spectral sequence*

$$\{E_r^{p,q}, d_r^{p,q}, E^n, \beta^{p,q}, F^p(E^n)\}_{p,q \geq 0, r \geq 2, n \geq 0},$$

such that $E_2^{p,q} = H^p(Y, R^q f_*(\mathcal{F}))$ and $E^n = H^n(X, \mathcal{F})$.

Proof. We refer the reader to [G, II.4.17.1]. □

The above spectral sequence is called the *Leray spectral sequence*. From the general theory of spectral sequences, one obtains the next corollary.

Corollary. *If* $R^q f_*(\mathscr{F}) = 0$ *for all* $q > 0$, *then* $H^p(X, \mathscr{F}) \cong H^p(Y, f_*\mathscr{F})$ *for all* $p \geq 0$.

§4.7 Cohomology of Affine Morphisms

a. Let $f\colon X \to Y$ be a morphism of schemes. If Y has an open cover $\{Y_\lambda \mid \lambda \in \Lambda\}$ such that the Y_λ and the $f^{-1}(Y_\lambda)$ are quasi-compact for all λ, then f is said to be a *quasi-compact morphism*. For example, any affine morphism or any morphism of finite type are quasi-compact.

Proposition 4.5. *If* f *is separated and quasi-compact, and if* $\mathscr{F}$ *is a quasi-coherent* $\mathscr{O}_X$*-module, then for any affine open subset* U *of* Y,

$$R^q f_*(\mathscr{F})|_U \cong H^q(f^{-1}(U), \mathscr{F})^\sim,$$

where $H^q(f^{-1}(U), \mathscr{F})$ *is regarded as an* $\mathscr{O}_Y(U)$*-module.*

PROOF. We refer the reader to [H, III Proposition 8.5]. □

By this and Theorem 4.3, we have the next result.

Theorem 4.9. *If* f *is an affine morphism and* $\mathscr{F}$ *is a quasi-coherent* $\mathscr{O}_X$*-module, then* $R^q f_*(\mathscr{F}) = 0$ *for any* $q > 0$.

Corollary. *With the same hypothesis as above,*

$$H^p(Y, f_*\mathscr{F}) \cong H^p(X, \mathscr{F}) \quad \textit{for any} \quad p \geq 0.$$

b. Let Z be a closed subscheme of X defined by a quasi-coherent $\mathscr{O}_X$-ideal $\mathscr{I}$. Denoting by $j\colon Z \to X$ the closed immersion, we have $j_*\mathscr{O}_Z = \mathscr{O}_X/\mathscr{I}$. Furthermore, if $\mathscr{L}$ is a locally free $\mathscr{O}_X$-module of finite type, then by the projection formula (Lemma 2.29),

$$j_*(j^*\mathscr{L}) \cong j_*(\mathscr{O}_Z) \otimes_{\mathscr{O}_X} \mathscr{L} \cong \mathscr{L}/\mathscr{I}\mathscr{L}.$$

Hence, we have the exact sequence

$$0 \to \mathscr{I} \otimes_{\mathscr{O}_X} \mathscr{L} \to \mathscr{L} \to j_*(j^*\mathscr{L}) \to 0.$$

Now, supposing that X is k-complete, one has $\chi_X(j_*j^*\mathscr{L}) = \chi_X(\mathscr{L}) - \chi_X(\mathscr{I} \otimes_{\mathscr{O}_X} \mathscr{L})$ by Proposition 4.4. On the other hand, $\chi_X(j_*j^*\mathscr{L}) = \chi_Z(j^*\mathscr{L})$ by the Corollary to Theorem 4.8. Thus, we obtain the next result.

Proposition 4.6. *For the above $\mathscr{S}$ and $\mathscr{L}$, it follows that*

$$\chi_Z(\mathscr{L}|_Z) = \chi_X(\mathscr{L}) - \chi_X(\mathscr{S} \otimes_{\mathscr{O}_X} \mathscr{L}),$$

where $\mathscr{L}|_Z$ denotes $j^*\mathscr{L}$.

To simplify the notation, we write $\mathscr{F}$ instead of $j_*\mathscr{F}$, if $\mathscr{F}$ is a coherent $\mathscr{O}_Z$-module.

§4.8 Riemann–Roch Theorem (Weak Form) on a Curve

a. Let Γ be a k-complete curve such that $\mathrm{Rat}(\Gamma)/k$ is an algebraically closed extension of fields.

Definition. If $\mathscr{L}$ is an invertible sheaf on Γ, define $\deg \mathscr{L}$ to be $\chi_\Gamma - \chi_\Gamma(\mathscr{L}^{-1})$.

Now, let (C, μ) be the normalization of Γ. Then $\mu_*(\mathscr{O}_C) = (\mathscr{O}_\Gamma)'$ which was defined in §2.14.**a** and $(\mathscr{O}_\Gamma)'$ is coherent by Theorem 2.1. $\mathscr{Z} = (\mathscr{O}_\Gamma)'/\mathscr{O}_\Gamma$ is also coherent. Define $\eta: \mathscr{O}_\Gamma \to \mathscr{H}om_{\mathscr{O}_\Gamma}(\mathscr{Z}, \mathscr{Z})$ as in the proof of Proposition 2.19 or as in §4.5.**b**. Then Ker η determines a closed subscheme Z of Γ. Clearly, $p \in Z \Leftrightarrow \mathscr{Z}_p \neq 0 \Leftrightarrow \mathscr{O}'_{\Gamma,p} \neq \mathscr{O}_{\Gamma,p} \Leftrightarrow p \in \Gamma \backslash \mathrm{Nor}(\Gamma)$. Hence, $\dim Z = 0$ and so $\mathscr{O}_{Z,p}$ is an Artinian ring by Theorem 1.10. Therefore, $\mathscr{Z}_p = (\mathscr{O}_{\Gamma,p})'/\mathscr{O}_{\Gamma,p}$ is a finite-dimensional vector space over k, since $\mathscr{Z}_p$ is finitely generated over $\mathscr{O}_{Z,p}$.

Definition. Define $\delta(\Gamma)_p$ to be $\dim_k(\mathscr{O}_{\Gamma,p})'/\mathscr{O}_{\Gamma,p}$.

Now, let $\mathscr{L}$ be an invertible sheaf on Γ, and consider the exact sequence

$$0 \to \mathscr{L} \to (\mathscr{O}_\Gamma)' \otimes_{\mathscr{O}_\Gamma} \mathscr{L} \to \mathscr{Z} \otimes_{\mathscr{O}_\Gamma} \mathscr{L} \to 0$$

Applying Proposition 4.4, we obtain

$$\chi_\Gamma((\mathscr{O}_\Gamma)' \otimes_{\mathscr{O}_\Gamma} \mathscr{L}) = \chi_\Gamma(\mathscr{L}) + \chi_\Gamma(\mathscr{Z} \otimes_{\mathscr{O}_\Gamma} \mathscr{L}).$$

Since $(\mathscr{O}_\Gamma)' \otimes_{\mathscr{O}_\Gamma} \mathscr{L} = \mu_*(\mu^*\mathscr{L})$ by Lemma 2.29 and μ is affine, it follows that $\chi_\Gamma((\mathscr{O}_\Gamma)' \otimes_{\mathscr{O}_\Gamma} \mathscr{L}) = \chi_C(\mu^*\mathscr{L})$.

For any p, $(\mathscr{Z} \otimes_{\mathscr{O}_\Gamma} \mathscr{L})_p \cong \mathscr{Z}_p$ and so $\mathrm{supp}(\mathscr{Z} \otimes_{\mathscr{O}_\Gamma} \mathscr{L}) = Z$. By Theorem 4.6*, we have

$$\chi_\Gamma(\mathscr{Z} \otimes_{\mathscr{O}_\Gamma} \mathscr{L}) = \dim_k \Gamma(Z, \mathscr{Z} \otimes_{\mathscr{O}_\Gamma} \mathscr{L}) = \sum_{p \in Z} \delta(\Gamma)_p,$$

which we denote by $\delta(\Gamma)$. Thus we obtain the formula

$$\chi_C(\mu^*\mathscr{L}) - \chi_\Gamma(\mathscr{L}) = \delta(\Gamma).$$

Applying this to the case of $\mathscr{L} = \mathcal{O}_\Gamma$, we have $\chi_C - \chi_\Gamma = \delta(\Gamma)$. Hence, we obtain the following result.

Theorem 4.10

(*i*) $\dim_k H^1(\Gamma, \mathcal{O}_\Gamma) - \dim_k H^1(C, \mathcal{O}_C) = p_a(\Gamma) - p_a(C) = \chi_C - \chi_\Gamma = \delta(\Gamma)$.
(*ii*) $\deg(\mu^*(\mathscr{L})) = \deg(\mathscr{L})$ *for any invertible sheaf* $\mathscr{L}$.

PROOF. (i) follows from $A(\Gamma) = A(C) = k$ by Lemma 1.53.
(ii) Since $\mathscr{L}^{-1}$ is also invertible by Proposition 2.22, we have $\chi_C(\mu^*\mathscr{L}^{-1}) - \chi_\Gamma(\mathscr{L}^{-1}) = \delta(\Gamma) = \chi_C - \chi_\Gamma$ and so

$$\deg \mu^*(\mathscr{L}) = \chi_C - \chi_C(\mu^*(\mathscr{L}^{-1})) = \chi_\Gamma - \chi_\Gamma(\mathscr{L}^{-1})$$

which is $\deg \mathscr{L}$. □

b. C is a complete normal curve, which is therefore locally factorial by Theorem 2.13.

Lemma 4.5. *Let* $\mathscr{L}$ *be an invertible sheaf on* C *and let* p *be a closed point on* C. *Then,*

(*i*) $\chi_C(\mathscr{L}) = \chi_C(\mathscr{L} \otimes \mathcal{O}(-p)) + \deg p$, *where* $\deg p$ *is* $\dim_k k(p)$.
(*ii*) $\chi_C(\mathscr{L}) = \chi_C(\mathscr{L} \otimes \mathcal{O}(-mp)) + m \deg p$ *for any* $m \in \mathbb{Z}$.

PROOF. (i) This follows from Proposition 4.6 and $\chi_p(\mathscr{L}|_p) = \dim_k k(p) = \deg p$.

(ii) If $m > 0$, then since $\mathscr{L} \otimes \mathcal{O}(-(m-1)p)$ is also invertible, (ii) follows from (i) by induction on m. If $m < 0$, apply the previous case to $\mathscr{L} \otimes \mathcal{O}(-mp)$. Then

$$\chi_C(\mathscr{L} \otimes \mathcal{O}(-mp)) = \chi_C(\mathscr{L} \otimes \mathcal{O}(-mp) \otimes \mathcal{O}(-(-m)p)) - m \deg p$$

and so

$$\chi_C(\mathscr{L}) = \chi_C(\mathscr{L} \otimes \mathcal{O}(-mp)) + m \deg p.$$

□

For a divisor $D = \sum_{i=1}^r m_i p_i$ on C, define $r(D)$ to be r, and $\deg D$ to be $\sum_{i=1}^r m_i \deg p_i$.

Theorem 4.11. *Let* D *be a divisor on* C. *Then*

$$\chi_C(\mathcal{O}(D)) = \deg D + \chi_C.$$

PROOF. We prove this by induction on $r(D)$. If $r(D) = 0$, then $D = 0$ and the equality is obvious. D is written as $mp + D_1$ for some $m \in \mathbb{Z}$, $p \in C$, and D_1 with $r(D_1) = r(D) - 1$. By Lemma 4.5.(ii), $\chi_C(\mathcal{O}(D)) = \chi_C(\mathcal{O}(D_1)) + m \deg p$. Hence by induction, $\chi(\mathcal{O}(D_1)) = \deg D_1 + \chi_C$ and so $\chi(\mathcal{O}(D)) = \deg D + \chi_C$. □

Corollary 1

(*i*) *If* $D \sim \Delta$, *then* $\deg D = \deg \Delta$.
(*ii*) $\deg D = \deg \mathcal{O}(D)$ *for any divisor* D *on* C.

PROOF. (i) This follows from $\mathcal{O}(D) \cong \mathcal{O}(\Delta)$.
(ii) Since $\deg \mathcal{O}(D) = \chi_C - \chi_C(\mathcal{O}(D)^{-1})$, the assertion follows from the last theorem. □

Corollary 2. *Let* $f\colon \Gamma \to Y$ *be a surjective morphism between complete curves, and let* $\mathcal{L}$ *and* $\mathcal{M}$ *be invertible sheaves on* Y. *Then* $\deg f^*(\mathcal{L}) = \deg f \cdot \deg \mathcal{L}$. *Furthermore,* $\deg(\mathcal{L} \otimes \mathcal{M}) = \deg \mathcal{L} + \deg \mathcal{M}$.

PROOF. By Theorem 4.10.(ii), one can assume that both curves are normal (cf. Theorem 2.25). Since any $\mathcal{L}$ on a normal curve can be written as $\mathcal{O}(D)$ for some divisor D, the assertion of the addivity is clear. By Lemma 2.26, $\deg f^*D = \deg f \cdot \deg D$. Thus, by Corollary 1(ii), one has the assertion. □

c. Note that $\dim_k H^1(C, \mathcal{O}(D))$ is denoted also by $i(D)$. Then Theorem 4.11 can be written as follows: recalling $p_a(C) = \dim_k H^1(C, \mathcal{O}_C)$, one has

$$l(D) - i(D) = \deg D + 1 - p_a(C).$$

This is called the *weak form of the Riemann–Roch Theorem* on a nonsingular complete curve C.

Furthermore, if $\mathcal{L}$ is an invertible sheaf on a complete curve Γ, one has by definition $\chi_\Gamma(\mathcal{L}) = -\deg(\mathcal{L}^{-1}) + \chi_\Gamma$ and so

$$\dim_k H^0(\Gamma, \mathcal{L}) - \dim_k H^1(\Gamma, \mathcal{L}) = -\deg \mathcal{L}^{-1} + 1 - p_a(\Gamma).$$

But, by (iii) one has $\deg \mu^*(\mathcal{L}^{-1}) = -\deg(\mu^*\mathcal{L})$. Then by Theorem 4.10.(ii), $\deg(\mu^*\mathcal{L}) = \deg \mathcal{L}$ and $\deg \mu^*(\mathcal{L}^{-1}) = \deg(\mathcal{L}^{-1})$. Hence, $-\deg \mathcal{L}^{-1} = \deg \mathcal{L}$. Thus,

$$\dim_k H^0(\Gamma, \mathcal{L}) - \dim_k H^1(\Gamma, \mathcal{L}) = \deg \mathcal{L} + 1 - p_a(\Gamma)$$

has been established.

This is said to be the *Riemann–Roch Theorem on a singular curve* Γ.

EXERCISE

4.1. Let X be a quasi-compact separated scheme. Show that the following conditions are equivalent.
(a) X is affine.
(b) For any open set U and any finite set $\{x_1, \ldots, x_r\} \subseteq U$, there exists $f \in A(X)$ such that $\{x_1, \ldots, x_r\} \subseteq X_f \subseteq U$.
(c) For any point x and any open neighborhood U of x, there exists $f \in A(X)$ with $x \in X_f \subseteq U$.

Chapter 5

Regular Forms and Rational Forms on a Variety

§5.1 Modules of Regular Forms and Canonical Derivations

a. Let A be a ring and let B be an A-algebra. Define $\mu_B: B \otimes_A B \to B$ by $\mu_B(\sum_{i=1}^n b_i \otimes c_i) = \sum_{i=1}^n b_i c_i$ and set $J_B = \operatorname{Ker} \mu_B$. Regarding $B \otimes_A B$ as a B-module by $b_1 \cdot b \otimes c = b_1 b \otimes c$ for $b_1, b, c \in B$, J_B also becomes a B-module. Define $\tilde{d}: B \to J_B$ by $\tilde{d}(b) = 1 \otimes b - b \otimes 1$ for $b \in B$, which is an A-homomorphism.

Lemma 5.1

(i) $\tilde{d}(A \cdot 1_B) = 0$.
(ii) $J_B = \sum_{b \in B} B\tilde{d}(b)$.
(iii) $\tilde{d}(bc) = \tilde{d}(b)c + b\tilde{d}(c) + \tilde{d}(b)\tilde{d}(c)$ *for any* $b, c \in B$.

PROOF. (i) is obvious.

(ii) If $\beta = \sum_{i=1}^n b_i \otimes c_i \in J_B$, then $\sum_{i=1}^n b_i c_i = 0$; hence

$$\beta = \sum_{i=1}^n \{b_i \cdot (1 \otimes c_i - c_i \otimes 1) + b_i c_i \otimes 1\}$$

$$= \sum_{i=1}^n b_i \tilde{d}(c_i) + \left(\sum_{i=1}^n b_i c_i\right) \otimes 1$$

$$= \sum_{i=1}^n b_i \tilde{d}(c_i) \in \sum_{b \in B} B\tilde{d}(b).$$

$$\begin{aligned}\text{(iii) } \tilde{d}(b)\tilde{d}(c) &= (1 \otimes b - b \otimes 1) \cdot (1 \otimes c - c \otimes 1)\\ &= bc \otimes 1 - b \otimes c - c \otimes b + 1 \otimes bc\\ &= \tilde{d}(bc) + 2bc \otimes 1 - b(\tilde{d}(c) + c \otimes 1) - c(\tilde{d}(b) + b \otimes 1)\\ &= \tilde{d}(bc) - b\tilde{d}(c) - c\tilde{d}(b). \qquad \square\end{aligned}$$

b. J_B/J_B^2 is a $B \otimes_A B/J_B$-module. But since $B \otimes_A B/J_B \cong B$, J_B/J_B^2 can be regarded as a B-module.

Definition. The B-module J_B/J_B^2 is said to be the *module of regular* (*differential*) *forms* of B over A, denoted by $\Omega_{B/A}$. $d: B \to \Omega_{B/A}$ is defined to be $\pi \circ \tilde{d}$, where π is the canonical homomorphism: $J_B \to J_B/J_B^2$. d is said to be the *canonical derivation* of B over A, and $d(b)$ is called the *differential* of b.

Then d is an A-homomorphism and satisfies the *derivation rule*: $d(bc) = bd(c) + cd(b)$ for any $b, c \in B$, since $\tilde{d}(b)\tilde{d}(c) \in J_B^2$.

Definition. Let M be a B-module. An A-homomorphism $\mathfrak{d}: B \to M$ is said to be an *A-derivation* of B into M, if

$$\mathfrak{d}(bc) = b\mathfrak{d}(c) + c\mathfrak{d}(b) \quad \text{for any} \quad b, c \in B.$$

Then the set $\mathrm{Der}_A(B, M)$ of all A-derivations $\mathfrak{d}: B \to M$ becomes a B-module if we define $(b_1\mathfrak{d}_1 + b_2\mathfrak{d}_2)(b) = b_1 \cdot \mathfrak{d}_1(b) + b_2 \cdot \mathfrak{d}_2(b)$ for $b_1, b_2, b \in B$; $\mathfrak{d}_1, \mathfrak{d}_2 \in \mathrm{Der}_A(B, M)$.

EXAMPLE 5.1. Let $B = k[x_1, \ldots, x_n]$ and let $\mathfrak{m} = (x_1 - a_1, \ldots, x_n - a_n)$, where $a_i \in k$. Then $B/\mathfrak{m} \cong k$. Since $B/\mathfrak{m}$ is a B-module, k becomes a B-module via the above isomorphism. Namely, if $\varphi(x_1, \ldots, x_n) \in B$, then $\varphi(x_1, \ldots, x_n) \cdot 1_k = \varphi(a_1, \ldots, a_n) \in k$. Let $\mathfrak{d} \in \mathrm{Der}_k(B, k)$. Then $\mathfrak{d}(x_j) = \lambda_j$ and $\mathfrak{d}(\varphi(x_1, \ldots, x_n)) = \sum_{j=1}^n (\partial_j \varphi)(a_1, \ldots, a_n)\lambda_j$. Furthermore, if $f(x_1, \ldots, x_n) = 0$ for some polynomial $f \in k[\mathrm{X}_1, \ldots, \mathrm{X}_n]$, then $\sum_{j=1}^n \lambda_j \cdot \partial_j f(a_1, \ldots, a_n) = 0$.

Now, let $f_1, \ldots, f_m$ be a basis of the ideal which is the kernel of ψ: $k[\mathrm{X}_1, \ldots, \mathrm{X}_n] \to B$ where $\psi(\mathrm{X}_i) = x_i$ for all i. Define $\mathbf{T}$ to be $\{(\lambda_1, \ldots, \lambda_n) \in k^n \mid \sum_{j=1}^n \lambda_j(\partial_j f_i)(a_1, \ldots, a_n) = 0$ for all $1 \le i \le m\}$. If $\lambda = (\lambda_1, \ldots, \lambda_n) \in \mathbf{T}$, one can define $\mathfrak{d}_\lambda$ by $\mathfrak{d}_\lambda(\varphi) = \sum_{j=1}^n \lambda_j(\partial_j \varphi)(a_1, \ldots, a_n)$, which is a k-derivation of B. Thus one has the k-linear isomorphism:

$$\begin{array}{ccc}\mathbf{T} & \cong & \mathrm{Der}_k(B, k)\\ \cup\!\shortmid & & \cup\!\shortmid\\ \lambda & \leftrightarrow & \mathfrak{d}_\lambda\end{array}$$

EXAMPLE 5.2. Let B be a polynomial ring $k[\mathrm{X}_1, \ldots, \mathrm{X}_n]$. Then $\partial_j = \partial/\partial \mathrm{X}_j$: $B \to B$ is a k-derivation of B into B. Furthermore, ∂_j can be regarded as a k-derivation of $Q(B)$ into $Q(B)$ for all j.

c. If $\eta: \Omega_{B/A} \to M$ is a B-homomorphism, then $\eta \circ d$ is an A-derivation. Letting $H(\eta) = \eta \circ d$, we have a B-homomorphism $H: \mathrm{Hom}_B(\Omega_{B/A}, M) \to \mathrm{Der}_A(B, M)$.

Proposition 5.1. *H is an isomorphism.*

PROOF. (1) H is injective. Actually, supposing $H(\eta) = 0$, we have $\eta(d(b)) = 0$ for all $b \in B$. But since $\Omega_{B/A} = \sum_{b \in B} Bd(b)$ by Lemma 5.1.(ii), this implies $\eta = 0$.

(2) H is surjective. Given $\mathfrak{d} \in \mathrm{Der}_A(B, M)$, we set $\psi(b, c) = b\mathfrak{d}(c)$ for any $b, c \in B$. Then ψ is A-bilinear and it defines an A-homomorphism $B \otimes_A B \to M$, which is denoted by the same letter ψ, i.e., $\psi(b \otimes c) = b\mathfrak{d}(c)$. Thus, ψ is a B-homomorphism and $\psi(\tilde{d}(b)) = \mathfrak{d}(b)$. Furthermore, $\psi(\tilde{d}(b)\tilde{d}(c)) = \psi(\tilde{d}(bc)) - \psi(b\tilde{d}(c)) - \psi(c\tilde{d}(b)) = \mathfrak{d}(bc) - b\mathfrak{d}(c) - c\mathfrak{d}(b) = 0$, and hence $\psi|_{J_B^2} = 0$. This implies that ψ is derived from a B-homomorphism $\eta: J_B/J_B^2 \to M$ such that $\psi = \eta \circ \pi$, i.e., $\psi(\tilde{d}b) = \eta(db) = \mathfrak{d}(b)$; thus $H(\eta) = \mathfrak{d}$. □

Remark. $d: B \to \Omega_{B/A}$ is often denoted by d_B or $d_{B/A}$, to avoid confusion.

EXAMPLE 5.3. Let B be a polynomial ring $k[\mathsf{X}_1, \ldots, \mathsf{X}_n]$. Then $\Omega_{B/k} = \sum_{j=1}^n Bd\mathsf{X}_j$, since $dF = \sum_{j=1}^n \partial_j F d\mathsf{X}_j$. Thus the map $\gamma: B^n \to \Omega_{B/k}$ defined by $\gamma(\psi_1, \ldots, \psi_n) = \sum_{j=1}^n \psi_j d\mathsf{X}_j$ is surjective. If $\gamma(\psi_1, \ldots, \psi_n) = 0$, then by using $\eta_i \in \mathrm{Hom}_B(\Omega_{B/k}, B)$ defined by $H(\eta_i) = \partial_i$, one has $0 = \eta_i(\sum_{j=1}^n \psi_j\, dX_j) = \psi_i$; hence $\psi_1 = \cdots = \psi_n = 0$. Therefore, $\gamma: B^n \cong \Omega_{B/k}$.

d. Let C be an A-algebra and let $\varphi: B \to C$ be an A-algebra homomorphism. $d_C \circ \varphi: B \to \Omega_{C/A}$ is an A-derivation and so, regarding $\Omega_{C/A}$ as a B-module via φ by Proposition 5.1, there exists a B-homomorphism $\bar{\varphi}: \Omega_{B/A} \to \Omega_{C/A}$ such that $\bar{\varphi} \circ d_B = d_C \circ \varphi$. $\bar{\varphi}$ induces a C-homomorphism $\delta: \Omega_{B/A} \otimes_B C \to \Omega_{C/A}$ such that $\delta(\omega \otimes \gamma) = \bar{\varphi}(\omega)\gamma$.

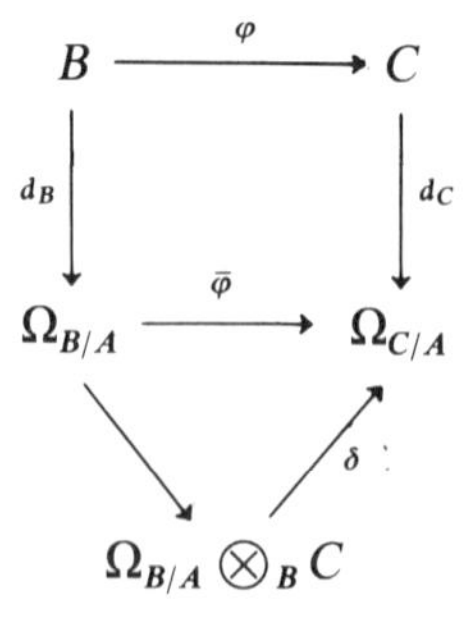

EXAMPLE 5.4. Let $\varphi: k[\mathsf{X}_1, \ldots, \mathsf{X}_n] \to k[\mathsf{Y}_1, \ldots, \mathsf{Y}_m]$ be a k-algebra homomorphism between polynomial rings. Since $\varphi(\mathsf{X}_i)$ is a polynomial $F_i(\mathsf{Y}_1, \ldots,$

Y_m) for $1 \le i \le n$, one has

$$\bar{\varphi}(d\mathsf{X}_i) = \sum_{j=1}^{m} \frac{\partial F_i(\mathsf{Y}_1, \ldots, \mathsf{Y}_m)}{\partial \mathsf{Y}_j} d\mathsf{Y}_j.$$

e. Proposition 5.2. *Let S be a multiplicative subset of an A-algebra B. Then* $S^{-1}\Omega_{B/A} \cong \Omega_{S^{-1}B/A}$.

PROOF. Letting $d\colon B \to \Omega_{B/A}$ be the canonical derivation, define ${}_S d\colon S^{-1}B \to S^{-1}\Omega_{B/A}$ by ${}_S d(b/s) = d(b)/s - bd(s)/s^2$ for any $b/s \in S^{-1}B$. Clearly, ${}_S d$ is well-defined and is an A-derivation of $S^{-1}B$. We shall prove that $(S^{-1}\Omega_{B/A}, {}_S d)$ is the universal element among pairs $(M, \mathfrak{d})$ where M is an $S^{-1}B$-module and $\mathfrak{d}$ is an A-derivation of $S^{-1}B$ into M.

For any such $\mathfrak{d}$, define $\Delta\colon B \to M$ by $\Delta(b) = \mathfrak{d}(b/1)$, which becomes an A-derivation of B. By Proposition 5.1, one has a unique $\eta\colon \Omega_{B/A} \to M$ with $\eta \circ d = \Delta$. Then η induces an $S^{-1}B$-homomorphism $\eta^{\#}\colon S^{-1}\Omega_{B/A} \to M$ such that $\eta^{\#}(\omega/1) = \bar{\eta}(\omega)$ for any ω. For $b/s \in S^{-1}B$, $\mathfrak{d}(b/s) = \eta^{\#} \circ {}_S d(b/s)$; hence $\mathfrak{d} = \eta^{\#} \circ {}_S d$. Thus, $(S^{-1}\Omega_{B/A}, {}_S d)$ is universal and so by Proposition 5.1, $S^{-1}\Omega_{B/A} \cong \Omega_{S^{-1}B/A}$. □

EXAMPLE 5.5. Let B be a polynomial ring $k[\mathsf{X}_1, \ldots, \mathsf{X}_n]$ and let S be $B\setminus\{0\}$. Then $S^{-1}B$ is the field $\mathfrak{K} = k(\mathsf{X}_1, \ldots, \mathsf{X}_n)$. Hence by Example 5.3,

$$\Omega_{\mathfrak{K}/k} = S^{-1}(\Omega_{B/k}) = S^{-1}\left(\sum_{i=1}^{n} Bd\mathsf{X}_i\right) = \sum_{i=1}^{n} \mathfrak{K}d\mathsf{X}_i \cong \mathfrak{K}^n.$$

§5.2 Lemmas

a. Lemma 5.2. *Let $\mathfrak{K}/L$ be a finite separable extension of fields with $L \supseteq k$, and let M be a $\mathfrak{K}$-module. Then any $\mathfrak{d} \in \mathrm{Der}_k(L, M)$ can be extended uniquely to a $\mathfrak{D} \in \mathrm{Der}_k(\mathfrak{K}, M)$.*

PROOF. For any $\alpha \in \mathfrak{K}$, there exists an irreducible separable polynomial $\Phi(\mathsf{X}) \in L[\mathsf{X}]$ such that $\Phi(\alpha) = 0$. Then letting $\Phi(\mathsf{X}) = a_0\mathsf{X}^n + \cdots + a_n$, $(\mathfrak{d}\Phi)(\mathsf{X}) = \mathfrak{d}(a_0)\mathsf{X}^n + \cdots + \mathfrak{d}(a_n)$, $\in M \otimes_k k[\mathsf{X}]$, and $\Phi'(\mathsf{X}) = na_0\mathsf{X}^{n-1} + \cdots + a_{n-1}$, one has

$$(\mathfrak{d}\Phi)(\alpha) + \Phi'(\alpha)\mathfrak{D}(\alpha) = \mathfrak{D}(\Phi(\alpha)) = 0,$$

if $\mathfrak{D}$ is an extension of $\mathfrak{d}$. Since $\Phi'(\alpha) \ne 0$, the uniqueness of $\mathfrak{D}$ is clear. To show the existence of $\mathfrak{D}$, for $F(\alpha) \in L(\alpha)$, define $\mathfrak{D}(F(\alpha))$ to be $(\mathfrak{d}F)(\alpha) - F'(\alpha) \cdot (\mathfrak{d}\Phi)(\alpha)/\Phi'(\alpha)$. Then $\mathfrak{D}$ is well-defined and is a k-derivation of $L(\alpha)$. Continuing this process, one can extend $\mathfrak{d}$ to a k-derivation of $\mathfrak{K}$. □

Let $\mathfrak{K}/k$ be a finitely generated and separable extension of fields, i.e., there exist $x_1, \ldots, x_n \in \mathfrak{K}$ such that (i) $L = k(x_1, \ldots, x_n)$ is a purely transcendental extension, i.e., $x_1, \ldots, x_n$ are algebraically independent, and (ii) $\mathfrak{K}/L$ is a finite separable extension. There exists $\partial_i \in \mathrm{Der}_k(L, L)$ such that $\partial_i(x_j) = \delta_{ij}$ for all i, j. For each i, the composition of ∂_i with the inclusion $L \subseteq \mathfrak{K}$ can be extended to $\mathfrak{D}_i \in \mathrm{Der}_k(\mathfrak{K}, \mathfrak{K})$ by the last lemma, which determines $\eta_i \in \mathrm{Hom}_{\mathfrak{K}}(\Omega_{\mathfrak{K}/K}, \mathfrak{K})$ by Proposition 5.1. Then $\eta_i(dx_j) = \delta_{ij}$ for all i, j.

Proposition 5.3. *With the notation as above,* $\Omega_{\mathfrak{K}/k} = \sum_{i=1}^n \mathfrak{K}dx_i \cong \mathfrak{K}^n$.

PROOF. For any $\alpha \in \mathfrak{K}$, one has an irreducible separable polynomial $\Phi(\mathsf{X}) \in L[\mathsf{X}]$ with $\Phi(\alpha) = 0$. Then $(d\Phi)(\alpha) + \Phi'(\alpha)d\alpha = 0$; hence $d\alpha \in \mathfrak{K}\, dL = \sum_{i=1}^n \mathfrak{K}dx_i$.

Suppose that $\sum_{i=1}^n \varphi_i\, dx_i = 0$. Then $0 = \eta_j(\sum_{i=1}^n \varphi_i\, dx_i) = \varphi_j$; thus $\varphi_1 = \cdots = \varphi_n = 0$. □

Remark. $(x_1, \ldots, x_n)$ is said to be a *separable transcendence basis* of $\mathfrak{K}/k$.

b. Now, let L/k and $\mathfrak{K}/L$ be finitely generated and separable extensions of fields. There exist $x_1, \ldots, x_n \in L$ and $y_1, \ldots, y_m \in \mathfrak{K}$ such that $(x_1, \ldots, x_n)$ and $(x_1, \ldots, x_n, y_1, \ldots, y_m)$ are separable transcendence bases of L and $\mathfrak{K}$ over k, respectively. By Proposition 5.3, $\Omega_{L/k} = \sum_{i=1}^n L d_L x_i$, and letting $d = d_{\mathfrak{K}}$, one has $\Omega_{\mathfrak{K}/k} = \sum_{i=1}^n \mathfrak{K}dx_i + \sum_{j=1}^m \mathfrak{K}dy_j$. The inclusion map $\imath$: $L \to \mathfrak{K}$ induces $\bar{\imath}$: $\Omega_{L/k} \to \Omega_{\mathfrak{K}/k}$ and $\bar{\imath}(d_L x_i) = dx_i$. Since $\Omega_{\mathfrak{K}/k} \cong \mathfrak{K}^{n+m}$, it follows that $\bar{\imath}$ is one-to-one. Thus via $\bar{\imath}$ one can identify $\Omega_{L/k}$ with a subspace of $\Omega_{\mathfrak{K}/k}$.

Definition. Let $\mathfrak{K}/k$ be a finitely generated and separable extension of fields with $(x_1, \ldots, x_n)$ a separable transcendence basis. For $F \in \mathfrak{K}$, dF can be written uniquely as $\sum_{i=1}^n A_i dx_i$. A_i is denoted by $\partial F/\partial x_i$. Furthermore, for $F_1, \ldots, F_n \in \mathfrak{K}$, the matrix $[\partial F_i/\partial x_j]$ is said to be the *Jacobian matrix* and its determinant is said to be the *Jacobian* of $F_1, \ldots, F_n$, denoted by $\partial(F_1, \ldots, F_n)/\partial(x_1, \ldots, x_n)$.

Remark. If $\mathfrak{K}$ is generated by $x_1, \ldots, x_n, x_{n+1}, \ldots, x_l$ such that $(x_1, \ldots, x_n)$ is a separable transcendence basis of $\mathfrak{K}/k$, any element F of $\mathfrak{K}$ is a rational function $Q(x_1, \ldots, x_l)$. Then one has the derivation of Q with respect to X_j obtained as $(\partial Q/\partial \mathsf{X}_j)(x_1, \ldots, x_l)$, which does not always agree with $\partial Q(x_1, \ldots, x_l)/\partial x_j$ defined above. For example, let $l = 2$, $n = 1$, and let x_1, x_2 satisfy $x_1^2 = x_2$. If $Q(\mathsf{X}_1, \mathsf{X}_2)$ is $\mathsf{X}_1^2 - \mathsf{X}_2$, then $Q(x_1, x_2) = 0$; hence $\partial Q(x_1, x_2)/\partial x_1 = 0$. But $(\partial Q/\partial \mathsf{X}_1)(x_1, x_2) = 2x_1$.

c. Let A be a k-algebra and let $\mathfrak{m}$ be a maximal ideal of A such that $A/\mathfrak{m} \cong k$. Hence, for any $x \in A$, one has $x \cdot 1 \in k$, which is denoted $x(0)$; i.e., $x - x(0) \in \mathfrak{m}$.

Lemma 5.3. *With the same hypothesis as above, one has*

$$\operatorname{Der}_k(A, k) \cong \operatorname{Hom}_k(\mathfrak{m}/\mathfrak{m}^2, k).$$

PROOF. If $\mathfrak{d} \in \operatorname{Der}_k(A, k)$, then $\mathfrak{d}(\mathfrak{m}^2) = 0$, since $\mathfrak{d}(xy) = x \cdot \mathfrak{d}(y) + y \cdot \mathfrak{d}(x) = 0$ for any $x, y \in \mathfrak{m}$. Hence $\mathfrak{d}$ induces a k-linear map $\mu: \mathfrak{m}/\mathfrak{m}^2 \to k$, denoted $J(\mathfrak{d})$. If $J(\mathfrak{d}) = 0$, then $\mathfrak{d}(\mathfrak{m}) = 0$, and for any $x \in A$, one has $\mathfrak{d}(x) = \mathfrak{d}(x - x(0)) = 0$; thus $\mathfrak{d} = 0$.

Given a k-linear map $\mu: \mathfrak{m}/\mathfrak{m}^2 \to k$, define $\mathfrak{d}^*: \mathfrak{m} \to k$ to be the composition of μ with the natural projection $\pi: \mathfrak{m} \to \mathfrak{m}/\mathfrak{m}^2$. Then define $\mathfrak{d}: A \to k$ by $\mathfrak{d}(x) = \mathfrak{d}^*(x - x(0))$. It is easy to see that $\mathfrak{d}$ is a k-derivation and $J(\mathfrak{d}) = \mu$.

□

By Proposition 5.1, $\operatorname{Hom}_A(\Omega_{A/k}, k) \cong \operatorname{Der}_k(A, k)$. Since $\operatorname{Hom}_A(\Omega_{A/k}, k) \cong \operatorname{Hom}_k(\Omega_{A/k} \otimes_A k, k)$, it follows that $\operatorname{Hom}_k(\Omega_{A/k} \otimes_A k, k) \cong \operatorname{Hom}_k(\mathfrak{m}/\mathfrak{m}^2, k)$. If A is Noetherian, then $\mathfrak{m}/\mathfrak{m}^2$ is finite-dimensional, and so $\Omega_{A/k} \otimes_A k \cong \mathfrak{m}/\mathfrak{m}^2$. Take a basis $\xi_1, \ldots, \xi_m$ of $\mathfrak{m}/\mathfrak{m}^2$ and let x_i be a representative of ξ_i for each i. Then for any element $\omega = \sum_{j=1}^r \alpha_j da_j$ of $\Omega_{A/k}$ with $\alpha_j, a_j \in A$, one has $\omega \otimes_A 1 = \sum_{j=1}^r \alpha_j(0) d(a_j - a_j(0))$, which is expressible as $\sum_{i=1}^m c_i dx_i$, $c_i \in k$. Thus, one has the isomorphism $u_A: \Omega_{A/k} \otimes_A k \to \mathfrak{m}/\mathfrak{m}^2$ defined by $u_A(\sum_{j=1}^m c_j dx_j) = \sum_{i=1}^m c_i \xi_i$. Hence, one obtains the next result.

Lemma 5.4. *Let A be a Noetherian k-algebra and let $\mathfrak{m}$ be a maximal ideal of A with $A/\mathfrak{m} \cong k$. Then $u_A: \Omega_{A/k} \otimes_A k \to \mathfrak{m}/\mathfrak{m}^2$ defined by $u_A(\alpha\, da) = \alpha(0)\pi(a - a(0))$ is an isomorphism, where $\pi: \mathfrak{m} \to \mathfrak{m}/\mathfrak{m}^2$ is the natural projection.*

Furthermore, let B be a Noetherian k-algebra and let $\mathfrak{n}$ be a maximal ideal of B with $B/\mathfrak{n} \cong k$. Suppose that a k-algebra homomorphism $\varphi: A \to B$ satisfies $\varphi^{-1}(\mathfrak{n}) = \mathfrak{m}$. Then φ induces a k-linear map $\varphi_*: \mathfrak{m}/\mathfrak{m}^2 \to \mathfrak{n}/\mathfrak{n}^2$. Furthermore, $\bar{\varphi}: \Omega_{A/k} \to \Omega_{B/k}$ induces $\delta: \Omega_{A/k} \otimes_A B \to \Omega_{B/k}$ and $\bar{\varphi}_* = \delta \otimes_B 1_k: \Omega_{A/k} \otimes_A k \to \Omega_{B/k} \otimes_B k$.

For any $x \in \mathfrak{m}$, $\varphi_*(x \bmod \mathfrak{m}^2) = \varphi(x) \bmod \mathfrak{n}^2$ and $\bar{\varphi}_*(dx \otimes 1) = d(\varphi(x)) \otimes 1$. Hence, $u_B \bar{\varphi}_*(dx \otimes 1) = \varphi(x) \bmod \mathfrak{n}^2 = \varphi_*(x \bmod \mathfrak{m}^2) = \varphi_*(u_A(dx \otimes 1))$, i.e., the next square is commutative:

$$\begin{array}{ccc}
\Omega_{A/k} \otimes_A k & \xrightarrow{\ \bar{\varphi}_*\ } & \Omega_{B/k} \otimes_B k \\
{\scriptstyle u_A}\, \wr\| & & \wr\| \,{\scriptstyle u_B} \\
\mathfrak{m}/\mathfrak{m}^2 & \xrightarrow[\ \varphi_*\ ]{} & \mathfrak{n}/\mathfrak{n}^2
\end{array}$$

Letting $T({}^a\varphi)_\mathfrak{n}$ be the dual map of φ_*, $T({}^a\varphi)_\mathfrak{n}$ is the linear map from $T_\mathfrak{n}(B)$ into $T_\mathfrak{m}(A)$.

§5.3 Sheaves of Regular Forms

a. In §5.1, we discussed the A-module $\Omega_{A/R}$ associated with any R-algebra A. We define the sheaf (of germs) of regular forms on Spec A over R to be $(\Omega_{A/R})^\sim$. For any $a \in A$, one has by Proposition 5.2

$$(\Omega_{A/R})^\sim|_{D(a)} \cong (\Omega_{A/R} \otimes_A A_a)^\sim \cong (\Omega_{A_a/R})^\sim.$$

Definition. Let X be a separated scheme over R and let $\mathscr{S}$ be the $\mathcal{O}_{X \times_R X}$-ideal defining the diagonal $\Delta_{X/R}(X)$, i.e., the closed subscheme of $X \times_R X$. $\mathscr{S}/\mathscr{S}^2$, regarded as an $\mathcal{O}_X$-module via the natural isomorphism $\mathcal{O}_{X \times_R X}/\mathscr{S} \cong \mathcal{O}_X$, is said to be the *$\mathcal{O}_X$-module of (germs of) regular forms* of X over R, denoted by $\Omega_{X/R}$.

It is easy to see that if U is an affine open subset $\cong$ Spec A, then $\Omega_{X/R}|_U \cong \Omega_{U/R} \cong (\Omega_{A/R})^\sim$.

Definition. For any $q \geq 0$, define the $\mathcal{O}_X$-module $\Omega^q_{X/R}$ to be $\wedge^q \Omega_{X/R}$, the q-th exterior power as an $\mathcal{O}_X$-module. Furthermore, if A is an R-algebra, define $\Omega^q_{A/R}$ to be $\wedge^q \Omega_{A/R}$. Then if $X = \operatorname{Spec} A$, one has $\Omega^q_{X/R} = (\Omega^q_{A/R})^\sim$.

When V is a variety over a field k, Ω^q_V stands for $\Omega^q_{V/k}$.

b. The next result is one of the upper semicontinuity theorems and is a consequence of Nakayama's Lemma (Lemma 2.8).

Lemma 5.5. *Let X be a scheme and $\mathscr{F}$ be an $\mathcal{O}_X$-module of finite type. Then the function $\delta_{\mathscr{F}}$ defined by $\delta_{\mathscr{F}}(x) = \dim_{k(x)} \mathscr{F}_x \otimes_{\mathcal{O}_{X,x}} k(x)$ for any $x \in X$ is upper semicontinuous.*

Proof. Take an arbitrary point $x \in X$ and put $\delta = \delta_{\mathscr{F}}(x)$. By Lemma 2.8, $\mathscr{F}_x$ has generators $\alpha_1, \ldots, \alpha_\delta$ as an $\mathcal{O}_{X,x}$-module. There exist an open neighborhood U of x and $a_1, \ldots, a_\delta \in \mathscr{F}(U)$ such that $a_{i,x} = \alpha_i$ for all i. Define $\mathscr{G}$ to be $(\mathscr{F}|_U)/\sum_{i=1}^{\delta} \mathcal{O}_U a_i$, which is of finite type. Hence, supp($\mathscr{G}$) is a closed subset and $x \notin$ supp($\mathscr{G}$); thus $U_0 = U \setminus$supp($\mathscr{G}$) is an open neighborhood of x and for any $y \in U_0$, $\mathscr{F}_y$is generated by $a_{1,y}, \ldots, a_{\delta,y}$ Hence, $\delta_{\mathscr{F}}(y) \leq \delta$. □

Let V be a variety over an algebraically closed field k. Then Rat(V)/k is finitely generated and separable. $\Omega_{V/k}$ is an $\mathcal{O}_V$-module of finite type and $(\Omega_{V/k})_* \cong \Omega_{\mathrm{Rat}(V)/k}$ with $\dim_{\mathrm{Rat}(V)} \Omega_{\mathrm{Rat}(V)/k} = \dim V = n$, where $*$ is generic point of V, by Propositions 5.2 and 5.3.

δ_Ω is upper semicontinuous on V, where Ω is Ω_V, $U = \{x \in V | \delta_\Omega(x) = n\}$ is an open dense subset by Lemma 5.5, and $\delta_\Omega(*) = n$ by Proposition 5.3. Furthermore, $U \cap \operatorname{spm}(V) = \operatorname{Reg}(U) \cap \operatorname{spm}(V)$ by Lemma 5.4. For any $p \in U$, let V_1 be the closure of p in U and take q in $V_1 \cap \operatorname{spm}(V)$. Then q belongs to Reg(V) and p is a generalization of q; hence, $p \in$ Reg(V) by

Theorem 2.12. Conversely, for a point $p \in \mathrm{Reg}(V)$, one has $x_1, \ldots, x_r$ such that $(x_1, \ldots, x_r)\mathcal{O}_{V,p} = \mathfrak{m}_p$, where $r = \dim \mathcal{O}_{V,p}$. One can choose an affine open subset W such that $p \in W$ and $f_1, \ldots, f_r \in A(W)$ such that $x_1 = f_1/1, \ldots, x_r = f_r/1$ in $\mathcal{O}_{V,p} = A(W)_{\mathfrak{p}}$, where $\mathfrak{p}$ corresponds to p. Moreover, we can assume that $f_1, \ldots, f_r$ generate $\mathfrak{p}$. $\mathrm{Spec}(A(W)/\mathfrak{p})$ is considered as a closed subvariety of W and so it has a closed regular point q. There exist $x_{r+1}, \ldots, x_n \in \mathcal{O}_{V,q}$ whose images in $\mathcal{O}_{W,q}$ form a regular system of parameters of $\mathcal{O}_{W,q}$. Then $f_1, \ldots, f_r, x_{r+1}, \ldots, x_n$ form a regular system of parameters of $\mathcal{O}_{V,q}$, i.e., $q \in \mathrm{Reg}(V) \cap \mathrm{spm}(V) = U \cap \mathrm{spm}(V)$. Thus $p \in U$, since U is open. Hence, we establish the following result.

Theorem 5.1. *If V is a variety, then* $\mathrm{Reg}(V)$ *is a dense open subset.*

c. With the same notation as in Theorem 5.1, suppose that V is an affine variety Spec A, i.e., there exist $x_1, \ldots, x_N \in A$ such that $A = k[x_1, \ldots, x_N]$. Take a nonsingular point p of V. Without loss of generality, one may assume p is the origin, i.e., $\bar{x}_i(p) = 0$, for all i. Now, define $\psi\colon k[\mathsf{X}_1, \ldots, \mathsf{X}_N] \to A$ by $\psi(\mathsf{X}_i) = x_i$ for all i and $\mathfrak{p}$ to be $\mathrm{Ker}\,\psi$; hence, $k[\mathsf{X}_1, \ldots, \mathsf{X}_N]/\mathfrak{p} \cong A$. Since $T_p(V)$ is a vector space of dimension $n = \dim V$, by definition, there exist $f_1, \ldots, f_{N-n} \in \mathfrak{p}$ and $\mathsf{X}_1, \ldots, \mathsf{X}_{N-n}$, after a permutation of the indices of the X_i, such that

$$\frac{\partial(f_1, \ldots, f_{N-n})}{\partial(\mathsf{X}_1, \ldots, \mathsf{X}_{N-n})}(0) \neq 0.$$

Setting $J = \partial(f_1, \ldots, f_{N-n})/\partial(\mathsf{X}_1, \ldots, \mathsf{X}_{N-n}) \bmod \mathfrak{p} \in A$, we have $p \in D(J)$. Defining $\partial_j f_i$ to be $\partial f_i/\partial \mathsf{X}_j \bmod \mathfrak{p}$, we have $\sum_{j=1}^{N} (\partial_j f_i) dx_j = 0$ for all $1 \leq i \leq N - n$; hence $J = \det[\partial_j f_i]_{1 \leq i, j \leq N-n}$. Then there exist $\beta_{ij} \in A_J$ with

$$dx_i = \sum_{j=N-n+1}^{N} \beta_{ij} dx_j \in \sum_{j=N-n+1}^{N} A_J dx_j \quad \text{if} \quad 1 \leq i \leq N - n.$$

Hence, $\Omega_{A_J/k} = \sum_{j=N-n+1}^{N} A_J dx_j \cong A_J^n$ and so $D(J) \subseteq \mathrm{Reg}\, V$.

Letting $z^1 = x_{N-n+1}, \ldots, z^n = x_N$, one says that $D(J)$ is a coordinate neighborhood of p with local coordinate system $(z^1, \ldots, z^n)$.

Definition. An affine open neighborhood U_λ of p in V with $\dim V = n$ is said to be a *coordinate neighborhood* of p, if there exist $z_\lambda^1, \ldots, z_\lambda^n \in \mathcal{O}_V(U_\lambda)$ with the property that $\Omega_V|_{U_\lambda} = \sum_{i=1}^{n} \mathcal{O}_{U_\lambda} dz_\lambda^i$. $(z_\lambda^1, \ldots, z_\lambda^n)$ is said to be a *local coordinate system* at p if $z_\lambda^i(p) = 0$ for all i.

Clearly, $\Omega_V|_{U_\lambda} \cong \mathcal{O}_{U_\lambda}^n$, and $(z_\lambda^1, \ldots, z_\lambda^n)$ is a separable transcendence basis of $\mathrm{Rat}(V)/k$.

Proposition 5.4. *If p is a nonsingular point of V, then there exists a coordinate neighborhood U_λ of p. Thus if V is nonsingular, then there exists an affine open cover $\{U_\lambda | \lambda \in \Lambda\}$ such that all U_λ are coordinate neighborhoods.*

Such an open cover $\{U_\lambda | \lambda \in \Lambda\}$ is said to be a *coordinate cover* of V. In this case, $\Omega_V|_{U_\lambda} \cong \mathcal{O}_{U_\lambda}^n$ for all $\lambda \in \Lambda$; thus Ω_V is a locally free $\mathcal{O}_V$-module of rank n.

Definition. If V is a nonsingular variety, any element of $\Gamma(V, \Omega_V^q)$ is said to be a *regular (differential) q-form* on V. Furthermore, any element of $\Gamma_{\text{rat}}(V, \Omega_V^q)$ is said to be a *rational q-form* on V.

Letting $\mathfrak{K}$ be Rat(V), we have the $\mathfrak{K}$-modules $\Omega_{\mathfrak{K}/k}^q$. If $U = \operatorname{Spec} A$ is an open dense subset of V, then $\Omega_{A/k}^q \otimes_A \mathfrak{K} \cong \Omega_{\mathfrak{K}/k}^q$ by Proposition 5.2, and so $\Omega_V^q \otimes_{\mathcal{O}_V} \mathcal{R}at_V$ is a constant sheaf on V; i.e., $\Gamma_{\text{rat}}(U, \Omega_V^q) \cong \Omega_{\mathfrak{K}/k}^q$ for any nonempty open subset U of V. In particular, any rational q-form ω on U can be regarded as a rational q-form on V.

Further, any rational q-form ω on V can be written as $\omega|_U = \omega_0/J$ for some $J \in A(U)\backslash\{0\}$, $\omega_0 \in \Omega_{A/k}^q = \Gamma(U, \Omega_V^q)$, where U is an affine open subset of V. Hence ω is regular on some dense open subset $D(J)$ of V.

d. Let C be an affine curve defined by an irreducible polynomial $f \in k[\mathsf{X}, \mathsf{Y}]$. Then $x = \mathsf{X} \bmod(f)$ and $y - \mathsf{Y} \bmod(f)$ satisfy $f(x, y) = 0$ and $A - A(C) = k[x, y] = k[\mathsf{X}, \mathsf{Y}]/(f)$. Suppose that C is nonsingular. If $f_x = \partial f/\partial \mathsf{X} \bmod(f)$, $f_y = \partial f/\partial \mathsf{Y} \bmod(f)$, then $C = D(f_x) \cup D(f_y)$. Since $f(x, y) = 0$, it follows that $f_x\,dx + f_y\,dy = 0$; hence,

$$dx/f_y = -dy/f_x.$$

Any $\omega \in \Gamma_{\text{rat}}(C, \Omega_C)$ can be written as $h\,dx/f_y$ for some $h \in Q(A) = \operatorname{Rat}(C)$. Since $h\ dx/f_y = -h\ dy/f_x$, ω is regular whenever $h \in \Gamma(D(f_x) \cup D(f_y), \mathcal{O}_C) = A$. Thus,

$$\Gamma(C, \Omega_C) = A\ dx/f_y.$$

§5.4 Birational Invariance of Genera

a. Let R be a ring, and let A, B, C be R-algebras. If $\varphi\colon A \to B$ is an R-algebra homomorphism, then by §5.1.**d**, one has an A-homomorphism $\bar{\varphi}$: $\Omega_{A/R} \to \Omega_{B/R}$ such that $\bar{\varphi} \circ d_A = d_B \circ \varphi$, which induces A-homomorphisms $\bar{\varphi}_q = \wedge^q \bar{\varphi}\colon \Omega_{A/R}^q \to \Omega_{B/R}^q$ for any $q \geq 0$.

Furthermore, if $\psi\colon B \to C$ is another R-homomorphism, then $\overline{(\psi \circ \varphi)}_q = \bar{\psi}_q \circ \bar{\varphi}_q\colon \Omega_{A/R}^q \to \Omega_{B/R}^q \to \Omega_{C/R}^q$. In particular, one obtains the next result.

Lemma 5.6. *Let A and B be k-algebras without zero divisors. If $\varphi\colon A \to B$ is one-to-one, then $Q(A) \to Q(B)$ is derived and the following square is commutative:*

$$\begin{array}{ccc} \Omega^q_{A/k} & \xrightarrow{\bar{\varphi}_q} & \Omega^q_{B/k} \\ \downarrow & & \downarrow \\ \Omega^q_{Q(A)/k} & \longrightarrow & \Omega^q_{Q(B)/k} \end{array}$$

b. Let $f\colon X \to Y$ be an R-morphism of separated R-schemes. Then we can define an $\mathcal{O}_X$-homomorphism $\Psi_f\colon f^*\Omega_Y \to \Omega_X$ such that each $\Psi_{f,x}$: $(f^*\Omega_Y)_x = \Omega_{Y,y} \otimes_{\mathcal{O}_{Y,y}} \mathcal{O}_{X,x} \to \Omega_{X,x}$, where $y = f(x)$, coincides with δ which was defined for $\theta(f)^\#_x\colon \mathcal{O}_{Y,y} \to \mathcal{O}_{X,x}$ in §5.1.**d**. Note that if $f = {}^a\varphi$: Spec $B \to$ Spec A, then $\Psi_f\colon (\Omega_{A/R} \otimes_A B)^\sim \to (\Omega_{B/R})^\sim$ is derived from $\delta\colon \Omega_{A/R} \otimes_A B \to \Omega_{B/R}$, i.e., $\Psi_f = \tilde{\delta}$.

One has the canonical homomorphism $\sigma\colon \Omega^q_Y \to f_* f^*\Omega^q_Y$ of Ω^q_Y and $f^\# = \sigma_Y\colon \Gamma(Y, \Omega^q_Y) \to \Gamma(Y, f_*f^*\Omega^q_Y) = \Gamma(X, f^*\Omega^q_Y)$ (see §2.20.**d**). Composing this with $\Psi_{f,X}\colon \Gamma(X, f^*\Omega^q_Y) \to \Gamma(X, \Omega^q_X)$, one obtains a homomorphism: $\Gamma(Y, \Omega^q_Y) \to \Gamma(X, \Omega^q_X)$, also denoted by f^*.

By definition, if $f = {}^a\varphi$: Spec $B \to$ Spec A, then $f_q^*\colon \Gamma(Y, \Omega^q_Y) = \Omega^q_{A/R} \to \Omega^q_{B/R} = \Gamma(X, \Omega^q_X)$ agrees with $\bar{\varphi}_q$.

If $g\colon Y \to Z$ is another R-morphism of schemes, then $(g \circ f)^* = f^* \circ g^*$: $\Gamma(Z, \Omega^q_Z) \to \Gamma(Y, \Omega^q_Y) \to \Gamma(X, \Omega^q_X)$.

Furthermore, let $f\colon V \to W$ be a dominating morphism of varieties over a field k. Then by Lemma 5.6, the following square is commutative

$$\begin{array}{ccc} \Gamma(W, \Omega^q_W) & \xrightarrow{f^*} & \Gamma(V, \Omega^q_V) \\ \downarrow & & \downarrow \\ \Omega^q_{\mathrm{Rat}(W)/k} & \xrightarrow{\bar{f}_q^*} & \Omega^q_{\mathrm{Rat}(V)/k}. \end{array}$$

c. Let V be a nonsingular variety over an algebraically closed field k. Then for any $q > 0$, $\Omega^q_{V/k}$ is a locally free $\mathcal{O}_V$-module of rank $\binom{n}{q}$, where $n = \dim V$. Thus by Lemma 2.32, one obtains the next result.

Theorem 5.2. *Let F be a closed subset of a nonsingular variety V with $F \neq V$.*

(i) *If ω_1 and ω_2 are rational q-forms such that $\omega_1|_{V \setminus F} = \omega_2|_{V \setminus F}$, then $\omega_1 = \omega_2$.*

(ii) *If* $\operatorname{codim}(F) \geq 2$, *then*

$$\Gamma(V, \Omega^q_V) = \Gamma(V \setminus F, \Omega^q_V).$$

By this theorem, one can define the pullback of regular forms by a strictly rational map $f\colon V \to W$, where both V and W are nonsingular as follows.

Let $\varphi = f|_{\mathrm{dom}(f)}\colon \mathrm{dom}(f) \to W$. Then $\varphi^*\colon \Gamma(W, \Omega^q_W) \to \Gamma(\mathrm{dom}(f), \Omega^q_V)$ is obtained. By Theorem 2.19, $V \setminus \mathrm{dom}(f)$ is a closed subset of codimension

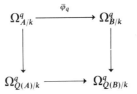

≥ 2. Hence, by Theorem 5.2.(ii), $\Gamma(V, \Omega_V^q) = \Gamma(\mathrm{dom}(f), \Omega_V^q)$. Composing this with φ^*, one obtains the linear map $\Gamma(W, \Omega_W^q) \to \Gamma(V, \Omega_V^q)$, denoted by f^*.

If a strictly rational map f is dominating, one has the injection f^*: $\mathrm{Rat}(W) \to \mathrm{Rat}(V)$. From this, one obtains the k-linear map $\bar{f}_q^*$: $\Omega^q_{\mathrm{Rat}(W)/k} \to \Omega^q_{\mathrm{Rat}(V)/k}$. Since $\Gamma_{\mathrm{rat}}(V, \Omega_V^q) = \Omega^q_{\mathrm{Rat}(V)/k}$ and $\Gamma_{\mathrm{rat}}(W, \Omega_W^q) = \Omega^q_{\mathrm{Rat}(W)/k}$, $\bar{f}_q^*$ gives rise to the pullback of rational q-forms.

Definition. A dominating rational map f: $V \to W$ is said to be *separable*, if f^*: $\mathrm{Rat}(W) \to \mathrm{Rat}(V)$ gives rise to a separable extension of fields.

Hence, if f is separable, $\bar{f}^*$: $\Omega_{\mathrm{Rat}(W)/k} \to \Omega_{\mathrm{Rat}(V)/k}$ is one-to-one by §5.2.**b**; thus $\bar{f}_q^*$: $\Omega^q_{\mathrm{Rat}(W)/k} \to \Omega^q_{\mathrm{Rat}(V)/k}$ is one-to-one for all q, and we obtain the following result.

Proposition 5.5. *If f is a dominating, separable, and strictly rational map, then f^*: $\Gamma(W, \Omega_W^q) \to \Gamma(V, \Omega_V^q)$ and $\bar{f}_q^*$: $\Gamma_{\mathrm{rat}}(W, \Omega_W^q) \to \Gamma_{\mathrm{rat}}(V, \Omega_V^q)$ are both one-to-one for any $q > 0$.*

Further, if g: $W \to U$ is a strictly rational map, where U is nonsingular, such that the composition $g \circ f$: $V \to U$ is defined, then $(g \circ f)^* = f^* \circ g^*$: $\Gamma(U, \Omega_U^q) \to \Gamma(W, \Omega_W^q) \to \Gamma(V, \Omega_V^q)$. In particular, if f is a proper birational map, in other words, if f and f^{-1} are both strictly rational by Proposition 2.17, then f^*: $\Gamma(W, \Omega_W^q) \to \Gamma(V, \Omega_V^q)$ is an isomorphism.

d. The above results hold for more general regular forms which will be defined below.

Definition. For any n-tuple $M = (m_1, \ldots, m_n)$ of nonnegative integers, define Ω_V^M to be $(\Omega_V^1)^{\otimes m_1} \otimes \cdots \otimes (\Omega_V^n)^{\otimes m_n}$. Furthermore, define $T_M(V)$ to be $\Gamma(V, \Omega_V^M)$.

Note that if $m_j > 0$ for some $j > \dim V$, then $\Omega_V^M = 0$ and so $T_M(V) = 0$. Since the Ω_V^M is locally free, one can prove the next result.

Theorem 5.3. *Let f: $V \to W$ be a strictly rational map between nonsingular varieties and let M be an n-tuple of nonnegative integers.*

(i) *The map f induces a k-linear map f^*: $T_M(W) \to T_M(V)$.*
(ii) *If f is dominating and separable, then f^* is one-to-one.*
(iii) *If $h : W \to U$ is a strictly rational map into a nonsingular variety U such that the composition $h \circ f$ is defined, then $(h \circ f)^* = f^* \circ h^*$.*
(iv) *If f is a proper birational map, then f^* is an isomorphism.*

If V is complete, then by Theorem 4.5, $T_M(V)$ is a finite-dimensional vector space.

Definition. In this case, $\dim_k T_M(V)$ is denoted by $P_M(V)$, and is called the *M-genus* of V.

e. If $M = (0, \ldots, 0, 1, 0, \ldots, 0)$ (the i-th component is 1), then $T_M(V)$ is denoted by $T_i(V)$ and $q_i(V)$ is used to denote $P_M(V) = \dim_k T_i(V)$. $q_i(V)$ is called the i-th *irregularity* of V. In particular, the first irregularity is said to be the *irregularity* of V, denoted by $q(V)$. If $n = \dim V$, then $q_n(V)$ is said to be the *geometric genus* of V, denoted by $p_g(V)$.

Furthermore, if $n = \dim V$ and $M = (0, \ldots, 0, m)$, then $T_M(V)$ is written as $T_{\{m\}}(V)$ and $P_m(V)$ is used to denote $P_M(V)$. $P_m(V)$ is said to be the *m-genus* of V. Hence $P_1(V) = p_g(V)$. $P_2(V)$ is also said to be the *bigenus* of V.

All the $P_M(V)$ are birational invariants by Theorem 5.3.

Given a variety V, when k is of characteristic zero, there exists a complete nonsingular variety V^* birationally equivalent to V by Hironaka's Theorem (§7.30). Since V^* is unique up to (proper) birational equivalence, $P_M(V^*)$ is independent of the choice of V^*.

Definition. The *M-genus* of V is defined to be $P_M(V^*)$, also denoted by $P_M(V)$.

Remark. In the case of $k = \mathbb{C}$ (more generally, if the characteristic of k is zero), one obtains the next interpretation of $q_i(V)$: if V is a complete nonsingular variety, then $\dim_k H^i(V, \mathcal{O}_V) = q_i(V)$. In particular, $q(V) = \dim_k H^1(V, \mathcal{O}_V)$ and $p_g(V) = \dim_k H^n(V, \mathcal{O}_V)$, where n is $\dim V$ (see [G–H, p. 116]).

f. Let V be a complete nonsingular variety of dimension n, which is covered by coordinate neighborhoods U_λ with local coordinate systems $(z_\lambda^1, \ldots, z_\lambda^n)$. In this case, Ω_V^n is an invertible sheaf; i.e., if dz_λ stands for $dz_\lambda^1 \wedge \cdots \wedge dz_\lambda^n$, then $\Omega_V^n|_{U_\lambda} = \mathcal{O}_{U_\lambda} dz_\lambda \cong \mathcal{O}_{U_\lambda}$.

Definition. Any divisor defined by a nonzero rational section of Ω_V^n is said to be a *canonical divisor* on V.

All canonical divisors on V are linearly equivalent. $K(V)$ *denotes one of the canonical divisors on* V.

By definition, $\mathcal{O}(K(V)) \cong \Omega_V^n$; hence one has

$$T_{\{m\}}(V) = \Gamma(V, (\Omega_V^n)^{\otimes m}) \cong \mathbf{L}_V(mK(V)),$$

$$P_m(V) = l(mK(V)) \quad \text{if } m \geq 1.$$

EXAMPLE 5.6. Let $\mathbf{P}_k^n = \operatorname{Proj} k[\mathsf{T}_0, \ldots, \mathsf{T}_n]$ be the projective space and let $z_0^1 = \mathsf{T}_1/\mathsf{T}_0, \ldots, z_0^n = \mathsf{T}_n/\mathsf{T}_0$; $z_1^1 = \mathsf{T}_0/\mathsf{T}_1, \ldots, z_1^n = \mathsf{T}_n/\mathsf{T}_1, \ldots$. Then $z_0^1 = 1/z_1^1$, $z_0^2 = z_1^2/z_1^1, \ldots, z_0^n = z_1^n/z_1^1, \ldots$. Hence $dz_0 = -dz_1/(z_1^1)^{n+1}, \ldots$. Recalling that $\mathsf{T}_0 = z_1^1 \mathsf{T}_1, \ldots$ and $H = \operatorname{div}(\mathsf{T}_0)$, one concludes that $K(\mathbf{P}_k^n) \sim -(n + 1)H$.

Whenever $P_m(V) > 0$, one has a rational map $\Phi\colon V \to \mathbf{P}((\Omega^n)^{\otimes m}) = \mathbf{P}(mK(V))$ associated with the invertible sheaf $(\Omega^n)^{\otimes m}$, which is denoted by $\Phi_{V,m}$.

Definition. $\Phi_{V,m}$ is said to be the m-th *canonical rational map* of V.

Proposition 5.6. *Let V and W be complete nonsingular varieties of dimension n and suppose $P_m(V) = P_m(W)$ for some $m > 0$. If $f\colon V \to W$ is a dominating separable rational map, then $\Phi_{W,m} \circ f = \Phi_{V,m}$.*

Proof. Since $f^*\colon T_{\{m\}}(W) \to T_{\{m\}}(V)$ is an isomorphism, one has $\Phi_{V,m} = (f^*\omega_0 : \cdots : f^*\omega_l) = (\omega_0 : \cdots : \omega_l) \circ f$, where $(\omega_0, \ldots, \omega_l)$ is a basis of $T_{\{m\}}(W)$, where $l + 1 = P_m(V) = P_m(W)$. □

When $V = W$, one has $f^*\omega_i \in T_{\{m\}}(V)$ and so there exist $c_{ij} \in k$ with $f^*\omega_i = \sum_{j=0}^{l} c_{ij}\omega_j$ for all i. Corresponding to the matrix $[c_{ij}]$, one obtains a ring homomorphism $\varphi\colon k[\mathsf{T}_0, \ldots, \mathsf{T}_l] \to k[\mathsf{T}_0, \ldots, \mathsf{T}_l]$ such that $\varphi(\mathsf{T}_i) = \sum_{j=0}^{l} c_{ji}\mathsf{T}_j$. Furthermore, Proj $\varphi\colon \mathbf{P}_k^l \to \mathbf{P}_k^l$ is an automorphism. Hence, setting $p_m(f) = \text{Proj}\ \varphi$, one obtains the next commutative square.

$$\begin{array}{ccc} V & \xrightarrow{\ f\ } & V \\ {\scriptstyle \Phi_{V,m}}\big\downarrow & & \big\downarrow{\scriptstyle \Phi_{V,m}} \\ \mathbf{P}_k^l & \xrightarrow[p_m(f)]{} & \mathbf{P}_k^l \end{array}$$

Corollary. *If $K(V)$ is ample, then any dominating separable rational map $f\colon V \to V$ is an isomorphism.*

Proof. One can choose $m > 0$ such that $\Phi_{V,m}\colon V \to \mathbf{P}_k^l$ is a closed immersion. Then, identifying V with $\Phi_{V,m}(V)$, one has $f = p_m(f)|_V$, which is an isomorphism. □

§5.5 Adjunction Formula

a. Let X be a nonsingular variety of dimension $n + 1$, which is covered by coordinate neighborhoods U_λ with local coordinate systems $(z_\lambda^1, \ldots, z_\lambda^{n+1})$ for all $\lambda \in \Lambda$. If V is a prime divisor on X, then one may assume that there exist $R_\lambda \in \Gamma(U_\lambda, \mathcal{O}_X)$ such that $V \cap U_\lambda = \text{div}(R_\lambda)$ for all λ, i.e., one has $s_{\lambda\mu} \in \Gamma(U_\lambda \cap U_\mu, \mathcal{O}_X^*)$ with $R_\lambda = s_{\lambda\mu} R_\mu$. Note that $\mathcal{O}(V)|_{U_\lambda} = \mathcal{O}_{U_\lambda} \cdot R_\lambda^{-1}$ (cf. §2.19.**c**).

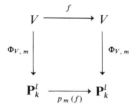

Suppose V is nonsingular. Then one may assume that $R_\lambda = z_\lambda^{n+1}$ for all $\lambda \in \Lambda$ with $V \cap U_\lambda \neq \varnothing$. Take a rational n-form ω on X such that the pullback $i^*\omega$ is defined and is not zero, where $i\colon V \to X$ is the inclusion map. (For example, a rational form ω defined by $dz_l^1 \wedge \cdots \wedge dz_l^n$ for some $l \in \Lambda$, has the required property.) The pullback $i^*\omega$ is often denoted by $\omega|_V$.

On each U_λ, $\omega|_{U_\lambda}$ is written as $\varphi_\lambda dz_\lambda^1 \wedge \cdots \wedge dz_\lambda^n + \psi_\lambda \wedge dz_\lambda^{n+1}$ for $\varphi_\lambda \in \mathrm{Rat}(X)$ and $\psi_\lambda \in \Gamma_{\mathrm{rat}}(U_\lambda, \Omega_X^{n-1})$. Then since $\omega \wedge dz_\lambda^{n+1} = \varphi_\lambda dz_\lambda$, $\omega|_{V \cap V_\lambda} = \varphi'_\lambda \cdot (dz_\lambda^1 \wedge \cdots \wedge dz_\lambda^n)|_{V \cap V_\lambda}$, in which φ'_λ denotes the restriction of φ_λ to $U_\lambda \cap V$. Hence, $K(V) \sim \mathrm{div}(i^*\omega) = \mathrm{div}\{\varphi'_\lambda\}$.

On the other hand,

$$\begin{aligned} \omega \wedge dz_\lambda^{n+1} &= (\varphi_\mu dz_\mu^1 \wedge \cdots \wedge dz_\mu^n + \psi_\mu \wedge dz_\mu^{n+1}) \\ &\quad \wedge (s_{\lambda\mu} dz_\mu^{n+1} + z_\mu^{n+1} ds_{\lambda\mu}) \\ &= s_{\lambda\mu}\varphi_\mu dz_\mu + z_\mu^{n+1}(\cdots). \end{aligned}$$

Hence,

$$\varphi'_\lambda(\partial z_\lambda/\partial z_\mu)' = s'_{\lambda\mu}\varphi'_\mu$$

for all λ and μ, where ε' denotes the restriction of ε to $V \cap U_\lambda \cap U_\mu$. Since $\{\varphi_\lambda'\}$ satisfies $\varphi'_\lambda = (s_{\lambda\mu} \cdot (dz_\lambda/\partial z_\mu)^{-1})'\varphi'_\mu$, $\{\varphi'_\lambda\}$ corresponds to a rational section of $\mathcal{O}(V) \otimes \Omega_X^{n+1}|_V \cong \mathcal{O}(K(X) + V)|_V$. Thus, the next formula has been established.

Theorem 5.4 (Adjunction formula). *Let V be a nonsingular prime divisor on a nonsingular variety X. Then,*

$$\mathcal{O}(K(V)) \cong \mathcal{O}(K(X) + V)|_V.$$

b. EXAMPLE 5.7. Let V be a nonsingular hypersurface of $\mathbf{P}_k^{n+1}$. Then $K(V) \sim \mathcal{O}_{\mathbf{P}_k^{n+1}}(r - n - 2)|_V$, where $r = \deg V$ is the degree of the homogeneous polynomial defining V.

For any $m \geq 1$, $T_{\{m\}}(V) \cong \Gamma(V, \mathcal{O}_V(m(r - n - 2)))$, where $\mathcal{O}_{V(1)} = \mathcal{O}_{\mathbf{P}_k^{n+1}}(1)|_V$. As was seen in Example 2.20, $\Gamma(V, \mathcal{O}_V(m(r - n - 2))) \cong$ the space of elements of $k[\mathrm{T}_0, \ldots, \mathrm{T}_{n+1}]/(F)$ represented by homogeneous polynomials of degree $m(r - n - 2)$.

In particular, if $r < n + 2$, then $P_m(V) = 0$ for all $m > 0$; if $r = n + 2$, then $P_m(V) = 1$ for all $m > 0$.

Let $x_1 = (\mathrm{T}_1/\mathrm{T}_0)|_V, \ldots, x_{n+1} = (\mathrm{T}_{n+1}/\mathrm{T}_0)|_V$. Then $f(x_1, \ldots, x_{n+1}) = F(1, x_1, \ldots, x_{n+1}) = 0$, and so $\sum_{i=1}^{n+1} \partial_i f dx_i = 0$. Hence,

$$\begin{aligned} \frac{dx_2 \wedge \cdots \wedge dx_{n+1}}{\partial_1 f} &= -\frac{dx_1 \wedge dx_3 \wedge \cdots \wedge dx_{n+1}}{\partial_2 f} = \cdots \\ &= (-1)^n \frac{dx_1 \wedge \cdots \wedge dx_n}{\partial_{n+1} f}, \end{aligned}$$

which we denote by ω_0. The set $\{\varphi(x)\omega_0 \mid \varphi(X)$ is a monic monomial, $\deg \varphi \leq r - n - 2\}$ is a basis of $T_n(V)$.

By the above observation, one has the formulas:

$$p_g(V) = P_m(V) = 0 \quad \text{if } r < n + 2;$$
$$p_g(V) = P_m(V) = 1 \quad \text{if } r = n + 2;$$
$$p_g(V) = \frac{(r-1)(r-2)\cdots(r-n-1)}{(n+1)!},$$

and

$$P_m(V) = \binom{m(r-n-2)+n+1}{n+1} - \binom{m(r-n-2)+n-r+1}{n+1}$$
$$= \frac{r}{n!}m^n + \cdots \qquad \text{if } r \geq n+2,\ m \geq 2.$$

§5.6 Ramification Formula

a. Let $f: V \to W$ be a morphism between nonsingular varieties. For closed points $x \in V$ and $y = f(x) \in W$, choose coordinate neighborhoods U_λ and W_α with local coordinate systems $(z_\lambda^1, \ldots, z_\lambda^n)$ and $(w_\alpha^1, \ldots, w_\alpha^m)$, respectively, such that $f(U_\lambda) \subseteq W_\alpha$. Then $f^*(dw_\alpha^i) = d(f^*w_\alpha^i) = \sum_{j=1}^n (\partial(f^*w_\alpha^i)/\partial z_\lambda^j) \cdot dz_\lambda^j$. Since $k(x) = k(y) = k$, $\theta(f)_x^\#: \mathcal{O}_{V,y} \to \mathcal{O}_{V,x}$ induces the linear maps $\mathfrak{m}_y/\mathfrak{m}_y^2 \to \mathfrak{m}_x/\mathfrak{m}_x^2$ and $T(f)_x: T_x(V) \to T_y(W)$. By the commutative square in §5.2.**c**, one obtains the next result.

Proposition 5.7. *The dual map of $T(f)_x$ is represented by the matrix $[(\partial f^*w_\alpha^i/\partial z_\lambda^j)(x)]_{1 \leq i \leq m,\, 1 \leq j \leq n}$. In particular, if $n = m$ and $(\partial(f^*w_\alpha^1, \ldots, f^*w_\alpha^n)/\partial(z_\lambda^1, \ldots, z_\lambda^n))(x) \neq 0$, then $T(f)_x: T_x(V) \cong T_y(W)$, i.e., $\mathfrak{m}_y/\mathfrak{m}_y^2 \cong \mathfrak{m}_x/\mathfrak{m}_x^2$.*

b. Now, suppose that $n = m$ and f is dominating. Then $\psi_{\alpha\lambda} = \partial(f^*w_\alpha^1, \ldots, f^*w_\alpha^n)/\partial(z_\lambda^1, \ldots, z_\lambda^n)$ is regular on U_λ and $f^*dw_\alpha = \psi_{\alpha\lambda}\, dz_\lambda$, where $dw_\alpha = dw_\alpha^1 \wedge \cdots \wedge dw_\alpha^n$ and $dz_\lambda = dz_\lambda^1 \wedge \cdots \wedge dz_\lambda^n$. Therefore, $\psi_{\alpha\lambda} \cdot \psi_{\beta\mu}^{-1} \in \Gamma(U_\lambda \cap U_\mu, \mathcal{O}_V^*)$ where $f(U_\mu) \subseteq W_\beta$. Hence, $\operatorname{div}\{\psi_{\alpha\lambda}\}$ is an effective divisor independent of the choice of coordinate covers of V and W.

Definition. The divisor $\operatorname{div}\{\psi_{\alpha\lambda}\}$ is said to be the *ramification divisor for f*, denoted by R_f.

Theorem 5.5 (Ramification formula). *If f is a dominating morphism, then*

$$K(V) \sim f^*(K(W)) + R_f.$$

Proof. Take a nonzero rational n-form ω on W. Then $\omega|_{W_\alpha} = \varphi_\alpha dw_\alpha$ and $f^*\omega|_{U_\lambda} = f^*(\varphi_\alpha)f^*(dw_\alpha) = f^*(\varphi_\alpha)\cdot\psi_{\alpha\lambda}dz_\lambda$. By definition, $K(W) \sim \operatorname{div}(\omega) = \operatorname{div}\{\varphi_\alpha\}$ and $K(V) \sim \operatorname{div}(f^*\omega) = \operatorname{div}\{f^*(\varphi_\alpha)\cdot\psi_{\alpha\lambda}\} = f^* \operatorname{div}\{\varphi_\alpha\} + R_f \sim f^*K(W) + R_f$. □

Let $x \in V$ be a closed point. If $x \notin R_f$, then by Proposition 5.7, $T(f)_x$: $T_x(V) \to T_{f(x)}(W)$ is an isomorphism. Hence $\mathfrak{m}_{f(x)}/\mathfrak{m}_{f(x)}^2 \cong \mathfrak{m}_x/\mathfrak{m}_x^2$, i.e., $\mathfrak{m}_x = \mathfrak{m}_{f(x)}\mathcal{O}_{V,x} + \mathfrak{m}_x^2$. By Lemma 2.8, this implies that $\mathfrak{m}_x = \mathfrak{m}_{f(x)}\mathcal{O}_{V,x}$.

Lemma 5.7. *Let $f: X \to Y$ be a morphism of schemes and let $x \in X$, $y = f(x)$. With the fiber $f^{-1}(y)$ being regarded as a scheme, one has*

$$\mathcal{O}_{f^{-1}(y),x} \cong \mathcal{O}_{X,x} \otimes_{\mathcal{O}_{Y,y}} k(y).$$

Proof. Left to the reader. □

Therefore, in our case, i.e., if $\mathfrak{m}_{f(x)}\mathcal{O}_{V,x} = \mathfrak{m}_x$, then $\mathcal{O}_{f^{-1}(y),x} \cong k$ and the fiber $f^{-1}(y)$ has an isolated component $\{x\}$.

Thus, if f is proper and y is a closed point of $W \backslash f(R_f)$, then $f^{-1}(y)$ is a finite set $\{x_1, \ldots, x_n\}$ such that $\mathcal{O}_{f^{-1}(y),x_i} \cong k$, i.e., $f^{-1}(y)$ is nonsingular and 0-dimensional. It will be said that $f|_{V_0}: V_0 \to W_0$, where $W_0 = W \backslash f(R_f)$ and $V_0 = f^{-1}(W_0)$, is an étale covering (see §7.8.**a**).

Proposition 5.8. *If $f: V \to W$ is proper birational, where V and W are nonsingular, then* $\operatorname{dom}(f^{-1}) = W \backslash f(R_f)$, *i.e.,* $f^{-1}|_{\operatorname{dom}(f^{-1})}: \operatorname{dom}(f^{-1}) \cong V \backslash R_f$.

Proof. If $y \in W \backslash f(R_f)$, then $f^{-1}(y)$ is a finite set; hence f^{-1} is defined at y by Theorem 2.22; thus $\operatorname{dom}(f^{-1}) \supseteq W \backslash f(R_f)$. Clearly, $\operatorname{dom}(f^{-1}) \subseteq W \backslash f(R_f)$; hence $\operatorname{dom}(f^{-1}) = W \backslash f(R_f)$. □

c. Let C and Y be complete nonsingular curves over an algebraically closed field of characteristic zero. Let $f: C \to Y$ be a surjective morphism and let p be a closed point of C. Taking a local coordinate t at $f(p)$, define $e(p)$ to be $\operatorname{ord}_p(f^*(t))$ where $f^* = \theta(f)_p^\#: \mathcal{O}_{Y,f(p)} \to \mathcal{O}_{C,p}$ and ord_p denotes the valuation at p. Then by Lemma 2.26, for a fixed point q of Y, $\sum_{f(p)=q} e(p)$ is $\deg f$. Moreover, by definition, R_f is written as $\sum_{p \in C}(e(p) - 1)p$.

Definition. $e(p)$ is said to be the *ramification index* of f (or of the covering f: $C \to Y$) at p.

For example, let $C = Y = \mathbf{P}_k^1$ and let $f: C \to Y$ be defined by $\varphi: k[\mathsf{T}_0, \mathsf{T}_1] \to k[\mathsf{T}_0, \mathsf{T}_1]$, where $\varphi(\mathsf{T}_i) = \mathsf{T}_i^n$, i.e., $f = \operatorname{Proj}\varphi$. Then $R_f = (n-1)((0) + (\infty))$ and so $\deg R_f = 2n - 2$.

§5.7 Generalized Adjunction Formula and Conductors

a. Any variety V of dimension n is birationally equivalent to a hypersurface of $\mathbf{P}_k^{n+1}$, if k has characteristic zero. Thus, it is desirable to establish an adjunction formula for a singular hypersurface. In general, let X be a nonsingular variety of dimension $n+1$, and let V be a complete subvariety of X with $\operatorname{codim}(V) = 1$. Given a proper separable surjective morphism f: $W \to V$, where W is a complete nonsingular variety with $\dim W = \dim V = n$, we shall obtain a relation between $\mathcal{O}(K(W))$ and $f^*(\mathcal{O}(K(X) + V)|_V)$, that will be called the *generalized adjunction formula.*

b. Since V is a locally principal divisor, $\mathcal{O}(V)|_{U_\lambda} = \mathcal{O}_{U_\lambda} \cdot R_\lambda^{-1}$ for some $R_\lambda \in \Gamma(U_\lambda, \mathcal{O}_X)$. Putting $\omega_\lambda^i = dz_\lambda^1 \wedge \cdots \wedge \widehat{dz_\lambda^i} \wedge \cdots \wedge dz_\lambda^{n+1}$ for all i, where $\frown$ indicates the deletion of dz_λ^i, we define Dz_λ^i by

$$Dz_\lambda^i = \left(\omega_\lambda^i\Big|_V\right) \Big/ \left(\frac{\partial R_\lambda}{\partial z_\lambda^i}\right)\Big|_V .$$

Since $dR_\lambda|_V = 0$, one has

$$\sum_{j=1}^{n+1} \left(\frac{\partial R_\lambda}{\partial z_\lambda^j} \cdot dz_\lambda^j\right)\Big|_{V \cap U_\lambda} = 0,$$

and hence

$$Dz_\lambda^1 = (-1)^j Dz_\lambda^{j+1} \quad \text{for all } j.$$

Take a rational n-form ω on X such that $\omega|_V$ is defined and $\omega|_V \neq 0$. (Such a form ω can be explicitly constructed as in §5.5.) Then ω can be written as $\sum_{i=1}^{n+1} \varphi_\lambda^i \omega_\lambda^i$ with $\varphi_\lambda^i \in \operatorname{Rat}(X)$, and so

$$\omega|_V = \sum_{j=1}^{n+1} \varphi_\lambda^i|_V \cdot \omega_\lambda^i|_V = \sum_{j=1}^{n+1} (-1)^{j-1}\left(\varphi_\lambda^j \frac{\partial R_\lambda}{\partial z_\lambda^j}\right)\Big|_V \cdot Dz_\lambda^1 .$$

Defining vectors $[\varphi_\lambda] = [\varphi_\lambda^1, -\varphi_\lambda^2, \ldots, (-1)^n \varphi_\lambda^{n+1}]$ and $[\partial R_\lambda] = [\partial R_\lambda/\partial z_\lambda^1, \ldots, \partial R_\lambda/\partial z_\lambda^{n+1}]$, we have $\omega|_V = \psi_\lambda|_V \cdot Dz_\lambda^1$, where ψ_λ is defined to be $[\varphi_\lambda] \cdot {}^t[\partial R_\lambda]$.

For simplicity, let ε' denote $\varepsilon|_V$ for a function or a matrix ε of X. Letting $e_{\lambda\mu} = R_\lambda/R_\mu$ and $j_{\lambda\mu} = \partial(z_\mu^1, \ldots, z_\mu^{n+1})/\partial(z_\lambda^1, \ldots, z_\lambda^{n+1})$, we shall prove that $\psi'_\lambda = e'_{\lambda\mu} \cdot j'_{\lambda\mu} \cdot \psi'_\mu$, for all λ, μ.

From $R_\lambda = e_{\lambda\mu} \cdot R_\mu$, it follows that

$$\left(\sum_{i=1}^{n+1} \frac{\partial R_\lambda}{\partial z_\lambda^i} \cdot \frac{\partial z_\lambda^i}{\partial z_\mu^j}\right)\Big|_V = \frac{\partial R_\lambda}{\partial z_\mu^j}\Big|_V = \left(e_{\lambda\mu} \cdot \frac{\partial R_\mu}{\partial z_\mu^j}\right)\Big|_V .$$

Letting $J_{\mu\lambda}$ be the matrix $[\partial z_\lambda^i/\partial z_\mu^j]$, one can rewrite the above equality as

$$[\partial R_\lambda]' \cdot J'_{\mu\lambda} = e'_{\lambda\mu}[\partial R_\mu]'.$$

On the other hand, defining $A^{i,j}_{\lambda\mu}$ by $\omega^i_\lambda = \sum_{j=1}^{n+1} A^{i,j}_{\lambda\mu}\omega^j_\mu$, one defines a matrix $A_{\lambda\mu} = [(-1)^{i+j}A^{i,j}_{\lambda\mu}]$. By definition, one has clearly ${}^tJ_{\mu\lambda}\cdot A_{\lambda\mu} = j_{\mu\lambda}\cdot 1_{n+1}$; hence ${}^tJ'_{\mu\lambda}\cdot A'_{\lambda\mu} = j'_{\mu\lambda}\cdot 1_{n+1}$. Since $\sum_{i=1}^{n+1}\varphi^i_\lambda\omega^i_\lambda = \sum_{j=1}^{n+1}\varphi^j_\mu\,\omega^j_\mu$, it follows that $[\varphi_\lambda]\cdot A_{\lambda\mu} = [\varphi_\mu]$. Thus $[\varphi_\lambda]' = [\varphi_\mu]' j'_{\lambda\mu}\cdot {}^tJ'_{\mu\lambda}$; hence

$$\psi'_\lambda = [\varphi_\lambda]'\cdot {}^t[\partial R_\lambda]' = [\varphi_\mu]' j'_{\lambda\mu}\cdot {}^tJ'_{\mu\lambda}\cdot e'_{\lambda\mu}\cdot {}^tJ'^{-1}_{\mu\lambda}[\partial R_\mu]' = j'_{\lambda\mu}\cdot e'_{\lambda\mu}\psi'_\mu.$$

These imply that $\{\psi'_\lambda\}$ determines a rational section of $\mathcal{O}(K(X) + V)|_V$.

Now, choosing a suitable coordinate cover $\{W_\alpha \mid \alpha \in \Phi\}$ of W with local coordinate system $(w^1_\alpha, \ldots, w^n_\alpha)$ on each W_α such that $f(W_\alpha) \subseteq U_\lambda$, define ξ_α to be a rational function on W_α by $f^*(Dz^1_\lambda) = \xi_\alpha\, dw_\alpha$. Clearly $\{\xi_\alpha\}$ satisfies $\xi_\alpha/\xi_\beta \in \Gamma(W_\alpha \cap W_\beta, \mathcal{O}^*_W)$ for all $\alpha, \beta \in \Phi$, and $f^*(\omega)|_{W_\alpha} = f^*(\psi'_\lambda)\cdot \xi_\alpha\, dw_\alpha$.

Definition. $\operatorname{div}\{\xi_\alpha^{-1}\}$ is said to be the *conductor* of f, denoted by cond_f.

By this, we obtain the next formula.

Theorem 5.6 (Generalized adjunction formula).

$$K(W) \sim f^*((K(X) + V)|_V) - \operatorname{cond}_f,$$

where $(K(X) + V)|_V$ *denotes the divisor* $\operatorname{div}(\zeta)$ *of a nonzero rational section* ζ *of* $\mathcal{O}(K(X) + V)|_V$.

Remark. If V is nonsingular, $-\operatorname{cond}_f = R_f$. Furthermore, if $W = V$ and is nonsingular, then $\operatorname{cond}_f = 0$ and so $\mathcal{O}(K(V)) \cong \mathcal{O}(K(X) + V)|_V$.

c. Now, we compute cond_f, where $\dim X = 2$ and f is a birational morphism. Then $Dz^1_\lambda = (dz^2_\lambda/\partial_1 R_\lambda)|_V$ and so $\xi_\alpha^{-1} = (\partial_1 R_\lambda/(z^2_\lambda)^\cdot)$, where $(z^2_\lambda)^\cdot$ denotes dz^2_λ/dw_α. Hence if $w_\alpha(p) = 0$, $z^1_\lambda(f(p)) = z^2_\lambda(f(p)) = 0$ and if $\operatorname{cond}_f = \sigma_p\cdot p + \cdots$, then

$$\sigma_p = \operatorname{ord}_w(\partial_1 R_\lambda/(z^2_\lambda)^\cdot) = \operatorname{ord}_w(\partial_1 R_\lambda) - \operatorname{ord}_w(z^2_\lambda) + 1,$$

where $w = w_\alpha$.

Example 5.8. If $V = V(\mathsf{X}^\nu - \mathsf{Y}^n) \subset \mathbf{A}^2_k$ for relatively prime numbers n, ν with $n > \nu \geq 2$, then $f^{-1}(0)$ is a set $\{p\}$ consisting of a single point. The computation of σ_p is easily performed as follows: $\partial_1 R_\lambda = \partial_{\mathsf{X}}(\mathsf{X}^\nu - \mathsf{Y}^n)|_V = \nu\mathsf{X}^{\nu-1}|_V = \nu x^{\nu-1}$, where x and y denote $\mathsf{X}|_V$ and $\mathsf{Y}|_V$, respectively. Hence,

$$\xi_\alpha^{-1} = \nu x^{\nu-1}/\nu w^{\nu-1},$$

where

$$w^n = x, \qquad w^\nu = y,$$

i.e.,

$$\xi_\alpha^{-1} = w^{(n-1)(\nu-1)}.$$

Therefore, $\sigma_p = (n-1)(\nu-1)$.

Definition. In the case when $\dim X = 2$ and $\dim V = 1$, for a closed point p of V, define σ_p to be $\sum_{f(p_j)=p} \sigma_{p_j}$.

Thus we obtain the formula

$$\deg(K(V)) = \deg(\mathcal{O}(K(X) + V)|_V) - \sum_{p \in V} \sigma_p,$$

since $\deg f^*(\mathcal{O}(K(X) + V)|_V) = \deg(\mathcal{O}(K(X) + V)|_V)$ by Corollary 2 to Theorem 4.11.

A genus formula for an embedded curve V will follow (cf. 9.2.**a**) from this.

§5.8 Serre Duality

a. The following very important duality theorem was proved first by Serre in a wider framework ("Un théorème de dualite", *Comment. Math. Helv.* 29, (1955)). However, an algebraic proof requires the theory of homological algebra which is omitted in this book. Hence, we refer the reader to [H, III 7] or [A-K] for an algebraic proof.

Theorem 5.7 (Serre duality). *Let V be a complete nonsingular variety of dimension n, and let D be a divisor on V. Then*

(*i*) $H^n(V, \Omega_V^n) \cong k$.
(*ii*) *The natural pairing*

$$H^p(V, \Omega_V^q \otimes \mathcal{O}(D)) \times H^{n-p}(V, \Omega_V^{n-q} \otimes \mathcal{O}(-D)) \to H^n(V, \Omega_V^n) \cong k$$

is nondegenerate for all $0 \le p, q \le n$.

Corollary.

(*i*) $\mathbf{L}_V(\Omega_V^n + D) \cong H^n(V, \mathcal{O}(-D))$.
(*ii*) $\dim_k H^p(V, \mathcal{O}(D)) = \dim_k H^{n-p}(V, \mathcal{O}(K(V) - D))$ *for all* p.
(*iii*) $q_s(V) = \dim_k \Gamma(V, \Omega_V^s) = \dim_k H^n(V, \Omega_V^{n-s})$ *for all* s.
(*iv*) *Let* $h^{p,q}(V) = \dim_k H^q(V, \Omega_V^p)$. *Then*

$$h^{p,q}(V) = h^{n-p,n-q}(V).$$

(*v*) *If V is a curve, i.e., $n = 1$, then*

$$i(D) = \dim H^1(V, \mathcal{O}(D)) = l(K(V) - D);$$

in particular, $p_a(V) = g(V)$.

EXERCISES

5.1. Let p, q be the projection morphisms of a k-product of k-schemes X and Y, where k is a field. If $\mathscr{F}$ and $\mathscr{G}$ are coherent sheaves on X and Y, respectively, let $\mathscr{F} \boxtimes \mathscr{G}$ denote $(p^*\mathscr{F}) \otimes (q^*\mathscr{G})$. Show that $\Gamma(X \times Y, \mathscr{F} \boxtimes \mathscr{G}) \cong \Gamma(X, \mathscr{F}) \otimes_k \Gamma(Y, \mathscr{G})$.

5.2. Let k be an algebraically closed field and let V, W be nonsingular varieties.

(i) Show that $V \times W$ is also nonsingular.

(ii) Letting p, q be the projection morphisms of $V \times W$, show that $\Omega_{V \times W} = p^*(\Omega_V) \oplus q^*(\Omega_W)$.

(iii) For any s, show that

$$\Omega^s_{V \times W} = \sum_{i=0}^{s} (\Omega^i_V \boxtimes \Omega^{s-1}_W).$$

(iv) Let D and Δ be divisors on V and W, respectively. Then $D \times W + V \times \Delta$ is a divisor on $V \times W$. Show that $\mathcal{O}(D \times W + V \times \Delta) \cong \mathcal{O}(D) \boxtimes \mathcal{O}(\Delta)$.

(v) Show that $K(V \times W)$ is linearly equivalent to $K(V) \times W + V \times K(W)$.

Chapter 6

Theory of Curves

§6.1 Riemann–Roch Theorem

a. Throughout this chapter, we study curves over an algebraically closed field k of characteristic zero. Points on curves mean closed points.

The following result is the most important theorem in the theory of curves.

Theorem 6.1 (Riemann–Roch Theorem). *Let C be a complete nonsingular curve and let D be a divisor on C. If K denotes the canonical divisor of C, then*

$$l(D) = l(K - D) + 1 - g + \deg D,$$

where g is the genus of C.

PROOF. This follows immediately from Theorem 4.11 and Corollary (v) to Theorem 5.7. □

Corollary. $l(K) = p_a(C) = g$ *and* $\deg K = 2g - 2$.

PROOF. The first assertion is obvious. Applying Theorem 6.1 to $D = K$, one has

$$g = l(K) = l(0) + 1 - g + \deg K.$$

Hence, $\deg K = 2g - 2$. □

b. Now, we prove some elementary lemmas concerning divisors D on a complete nonsingular curve C of genus g.

Lemma 6.1

(i) *If* $l(D) > 0$, *then* $\deg D \geq 0$.
(ii) *If* $l(D) > 0$ *and* $\deg D = 0$, *then* $D \sim 0$.

PROOF. Taking $D' \in |D|$, one has $\deg D = \deg D' \geq 0$ by Corollary 1 to Theorem 4.11. Furthermore, if $\deg D' = 0$, then $D' = 0$; thus $D \sim 0$. □

Lemma 6.2. *If* $\deg D \geq 2g - 1$, *then* $i(D) = 0$ *and so*

$$l(D) = 1 - g + \deg D \geq g.$$

PROOF. Since $i(D) = l(K - D)$ by Corollary (v) to Theorem 5.7, $\deg(K - D) = 2g - 2 - \deg D < 0$ implies $l(K - D) = 0$ by the last lemma. Furthermore, applying Theorem 6.1, one obtains the equality. □

Remark. Expressing D as $K + F$ for a divisor F, the last lemma can be restated as follows: if $\deg F > 0$, then $H^1(C, \mathcal{O}(K + F)) = 0$. This is a prototype of a famous vanishing theorem of cohomology groups by Kodaira (cf. [G–H p.148]).

c. Here we are concerned with the simplest possible curve, i.e., the projective line $\mathbf{P}_k^1$ with $(\mathsf{T}_0, \mathsf{T}_1)$ a homogeneous coordinate system. $z = \mathsf{T}_1/\mathsf{T}_0$ is a rational function, and for any $\lambda \in k$, $\operatorname{div}(z - \lambda) = (\lambda) - (\infty)$, where (λ) denotes the prime divisor determined by a point $\lambda \in k \subset \mathbf{P}_k^1$, and (∞) is the homogeneous prime ideal (T_0) (cf. Example 3.1). Hence any two points are linearly equivalent, i.e., $l((\lambda)) = l((\infty)) = 2$. On the other hand, $g(\mathbf{P}_k^1) = 0$ and $K(\mathbf{P}_k^1) \sim -2(\infty)$ by Example 5.6 in §5.4.f. Furthermore, $l(m(\infty)) = m + 1$ by Example 2.13 in §2.18; thus $\mathbf{L}(m(\infty))$ is spanned by $1, z, \ldots, z^m$, if $m \geq 0$. Furthermore, $\mathbf{L}(\Omega_{\mathbf{P}^1} + m(\infty)) = \{\varphi(z)dz | \varphi \in k[z]$ and $\deg \varphi \leq m - 2\}$, if $m \geq 2$ (cf. §2.20.**a**). Therefore, $l(K(\mathbf{P}_k^1) + m(\infty)) = m - 1$.

d. Let $\mathscr{L}$ be an invertible sheaf on a nonsingular complete curve C and let $\mathbf{L}$ be a nonzero subspace of $\Gamma(C, \mathscr{L})$. One has the linear system $\Lambda = \Lambda(\mathbf{L})$ defined by $\mathbf{L}$. A point $p \in C$ is a base point of Λ if and only if $\varphi(p) = 0$ for every $\varphi \in \mathbf{L}$.

Definition. For any $p \in C$, define $\Lambda(p)$ to be $\Lambda(\mathbf{L}(p))$, where $\mathbf{L}(p) = \{\varphi \in \mathbf{L} \mid \varphi(p) = 0\}$.

Hence, $p \in \operatorname{Bs} \Lambda$ if and only if $\Lambda = \Lambda(p)$.

Definition. If $\mathbf{L}(p) \neq \{0\}$, define $\Lambda - p$ to be $\{D - p \mid D \in \Lambda(p)\}$. This then is $\Lambda(\mathbf{L}(p)/\varphi)$, where $\varphi \in \Gamma(C, \mathcal{O}(p))$ with $\operatorname{div}(\varphi) = p$, and $\mathbf{L}(p)/\varphi \subseteq \Gamma(C, \mathscr{L} \otimes \mathcal{O}(-p))$.

Lemma 6.3

(i) *If* $p \notin \mathrm{Bs}\,\Lambda$, *then* $\dim \Lambda(p) = \dim \Lambda - 1$.
(ii) *If* $p \in \mathrm{Bs}\,\Lambda$, *then* $\Lambda = \Lambda(p)$ *and* $\Lambda = (\Lambda - p) + p$.

PROOF. (i) Take a basis $\varphi_1, \ldots, \varphi_l$ of $\mathbf{L}$. Then $\mathbf{L}(p) = \{\sum_{i=1}^{l} \lambda_i \varphi_i | \sum_{i=1}^{l} \lambda_i \varphi_i(p) = 0\}$, which is $(l-1)$-dimensional if and only if $\varphi_j(p) \neq 0$ for some j.

Theorem 6.2. *If there exists a divisor D on a complete nonsingular curve C with* $\deg D = 1$ *and* $l(D) \geq 2$, *then* $C \cong \mathbf{P}^1$ *and* $\mathrm{Rat}(C) = k(\varphi)$ *for any* $\varphi \in \mathbf{L}(D) \backslash k$.

PROOF. Since $\dim|D| = l(D) - 1 \geq 1 > 0$, we may assume D is effective. Then $D = p$ for some $p \in C$, since $\deg D = 1$. If $\mathrm{Bs}|D| \neq \varnothing$, then $\mathrm{Bs}|D| = p$ and so $l(D) = l(D - p) = l(0) = 1$. By hypothesis, however, $l(D) \geq 2$. Hence, $\mathrm{Bs}|D| = \varnothing$; thus $\Phi_D\colon C \to \mathbf{P}_k^1$ is a surjective morphism with $p \sim \Phi_D^*(a)$ for any $a \in \mathbf{P}_k^1$. By Theorem 4.10.(ii), $1 = \deg p = \deg \Phi_D^*(a) = \deg \Phi_D \cdot 1$, i.e., Φ_D is birational. In view of Theorem 2.22, Φ_D is an isomorphism. Since $\Phi_D = (1 : \varphi)$, one has $\mathrm{Rat}(C) = k(\varphi)$. □

Corollary.

(i) *If* $g(C) = 0$, *then* $C \cong \mathbf{P}_k^1$.
(ii) *If* $g(C) \geq 1$, *then* $l(p) = 1$ *for any point* $p \in C$.

PROOF. (i) follows from Theorems 6.1 and 6.2.
(ii) is easily proved by the last theorem. □

Note that if Λ is a linear system on a nonsingular complete curve C, then Λ_{fix} consists of points and so the support of Λ_{fix} coincides with $\mathrm{Bs}\,\Lambda$. In particular, if $\Lambda_{\mathrm{fix}} = 0$, then $\mathrm{Bs}\,\Lambda = \varnothing$.

Let D be a divisor on C. For a point $p \in C$, $|D - p| = |D|(p) - p$ if $|D|(p) \neq \varnothing$. Furthermore, $l(D - p) = l(D)$ if and only if $p \in \mathrm{Bs}|D|$.

Proposition 6.1. *Let C be a nonsingular complete curve with* $g = g(C) \geq 1$. *Then the canonical linear system* $|K(C)|$ *has no base points.*

PROOF. If $g = 1$, then $K(C) \sim 0$; hence the assertion is trivial. If $g \geq 2$, then take any point $p \in C$ and compute $l(K(C) - p)$ by Theorem 6.1. Thus $l(K(C) - p) = 1 - g + 2g - 3 + l(p) = g - 1$, since $l(p) = 1$ follows from Corollary (ii) to Theorem 6.2. Therefore, $l(K(C) - p) = l(K(C)) - 1$ for any $p \in C$; thus $|K(C)|$ has no base points. □

§6.2 Fujita's Invariant $\Delta(C, D)$

a. Definition. For a divisor D with $\deg D > 0$ on a complete nonsingular curve C, define $\Delta(C, D)$ to be $1 + \deg D - l(D)$.

The invariant $\Delta(C, D)$ is a special case of Fujita's invariant $\Delta(V, \mathscr{F})$ for a complete variety V and an invertible sheaf $\mathscr{F}$, which is called the *total deficiency* by T. Fujita.

$\Delta(C, D)$ behaves like a genus as if every divisor D' had $i(D') = 0$. Indeed, one can rewrite the definition as $l(D) = 1 - \Delta(C, D) + \deg D$.

Theorem 6.3. *If* $\deg D > 0$, *then* $\Delta(C, D) \geq 0$. *Furthermore,* $\Delta(C, D) = 0$ *if and only if* $C \cong \mathbf{P}^1$.

Proof. If $\deg D = 1$, the result follows immediately from Theorem 6.2. We now use induction on $\deg D$. If $\deg D \geq 2$ and $l(D) \geq 1$, then taking a point $p \in C \backslash \mathrm{Bs}|D|$, one has $l(D - p) = l(D) - 1$ and $\deg(D - p) = \deg D - 1 > 0$; thus $\Delta(C, D - p) = \Delta(C, D)$. Hence by the induction hypothesis, $\Delta(C, D) = \Delta(C, D - p) \geq 0$. Furthermore, if $\Delta(C, D) = 0$, then $\Delta(C, D - p) = 0$, which implies $C \cong \mathbf{P}^1$ by the induction hypothesis. On the other hand, if $\deg D \geq 2$ and $l(D) = 0$, then $\Delta(C, D) = 1 + \deg D - l(D) = 1 + \deg D \geq 3$. □

Corollary. *If* $\deg D > 0$, *then* $\Delta(C, D) \leq g$ *and* $l(K - D) \leq l(K) = g(C)$. *Furthermore, in the second inequality, equality holds if and only if* $g(C) = 0$.

Proof. Since $\Delta(C, D) = g - l(K - D)$ by Theorem 6.1, the assertion follows from the last theorem, where g is $g(C)$. □

The corollary can be restated as follows:

If $\deg D \leq 2g - 3$, *then* $l(D) \leq g$.
If $\deg D \leq 2g - 3$ *and* $g > 0$, *then* $l(D) \leq g - 1$.

Theorem 6.4

(i) *If* $\deg D = 2g - 2$ *and* $g > 0$, *then* $l(D) \leq g$.
(ii) *If* $\deg D = 2g - 2$ *and* $l(D) = g$, *then* $D \sim K$.

Proof. (i) For any point $p \in C$, one has $\deg(D - p) = 2g - 3$; hence $l(D - p) \leq g - 1$. By Lemma 6.3, $l(D) \leq l(D - p) + 1 \leq g$.

(ii) If $\deg D = 2g - 2$ and $l(D) = g$, then by Theorem 6.1 $g = l(D) = 1 - g + l(K - D) + \deg D = l(K - D) + g - 1$; hence $l(K - D) = 1$ and $\deg(K - D) = 0$. Thus $K - D \sim 0$ by Lemma 6.1. □

Whenever $l(K-D)=0$, $l(D)=1-g+\deg D$. It is rather difficult to know $l(D)$ when both $l(D)$ and $i(D)=l(K-D)$ are positive. Such an effective divisor D is called a *special divisor*, because $i(D)$ is said to be the *index of speciality* of D.

If a special divisor D is neither 0 nor linearly equivalent to K, then $\deg D \le 2g-3$ and $l(D) \le g-1$ by Theorem 6.4. We shall obtain an inequality between $l(D)$ and $\deg(D)$, when D is a special divisor.

b. In general, let V be a complete normal variety and let $D_1 = \sum_{i=1}^s n_i\Gamma_i$, $D_2 = \sum_{i=1}^s m_i\Gamma_i$ be effective divisors on V with Γ_i prime divisors such that $\Gamma_i \neq \Gamma_j$, if $i \neq j$.

Definition. Define the two divisors:

$$D_1 \wedge D_2 = \sum_{i=1}^{s} \min\{n_i, m_i\}\Gamma_i \quad \text{and}$$

$$D_1 \vee D_2 = \sum_{i=1}^{s} \max\{n_i, m_i\}\Gamma_i.$$

Then one has the inclusion relations as subspaces of $\mathrm{Rat}(V)$:

$$\mathbf{L}(D_1) \cap \mathbf{L}(D_2) = \mathbf{L}(D_1 \wedge D_2) \quad \text{and}$$

$$\mathbf{L}(D_1) \cup \mathbf{L}(D_2) \subseteq \mathbf{L}(D_1 \vee D_2).$$

By a theorem on vector spaces, one has

$$\begin{aligned} l(D_1) + l(D_2) &= \dim_k(\mathbf{L}(D_1) \cap \mathbf{L}(D_2)) + \dim_k(\mathbf{L}(D_1) + \mathbf{L}(D_2)) \\ &\le l(D_1 \wedge D_2) + l(D_1 \vee D_2); \end{aligned}$$

thus the next lemma has been obtained.

Lemma 6.4. *If D_1 and D_2 are effective divisors, then*

$$l(D_1) + l(D_2) \le l(D_1 \wedge D_2) + l(D_1 \vee D_2).$$

Now, let C be a complete nonsingular curve and let $\mathbf{L}$ be a subspace of $\Gamma(C, \mathscr{L})$, where $\mathscr{L}$ is an invertible sheaf. Take a basis $\varphi_1, \ldots, \varphi_l$ of $\mathbf{L}$ and define φ_λ to be $\sum_{i=1}^l \lambda_i\varphi_i$ for $\lambda = (\lambda_1, \ldots, \lambda_l) \in k^l$.

Lemma 6.5. *If E is an effective divisor with $E \cap \mathrm{Bs}\,\Lambda = \varnothing$, then there is a closed proper subset F of $\mathbf{A}_k^l$ such that if $\lambda \in \mathrm{spm}(\mathbf{A}_k^l \backslash F)$ then $E \wedge \mathrm{div}(\varphi_\lambda) = 0$.*

Proof. Suppose $\mathrm{supp}\, E = \{p_1, \ldots, p_r\}$. Define F_j to be $V(\sum_{i=1}^l \bar{\varphi}_i(p_j)\mathrm{X}_i)$ and F to be $F_1 \cup \cdots \cup F_r$ as subsets of $\mathbf{A}_k^l = \mathrm{Spec}\, k[\mathrm{X}_1, \ldots, \mathrm{X}_l]$. Then by hypothesis, the F_j are proper subsets, and identifying $\mathrm{spm}(\mathbf{A}_k^l)$ with k^l, one obtains the assertion. □

Occasionally, if $\lambda \in \operatorname{spm}(\mathbf{A}_k^l \backslash F)$ for some closed proper subset F, then $\operatorname{div}(\varphi_\lambda)$ is said to be a *general member* of Λ (see §7.9.**b**, for a precise discussion of general members). Hence, Lemma 6.5 can be restated as follows: if $E \wedge \operatorname{Bs} \Lambda = \varnothing$, and D is a general member of Λ, then $E \wedge D = 0$.

Lemma 6.6. *Let D_1 and D_2 be effective divisors on C. Then $l(D_1) + l(D_2) \leq 1 + l(D_1 + D_2)$. Furthermore, if $l(D_1) + l(D_2) = 1 + l(D_1 + D_2)$, then $D_1 = 0$ or $|D_1|_{\text{fix}} \leq |D_1 + D_2|_{\text{fix}}$.*

Proof. Let $B = |D_1|_{\text{fix}}$ and $D = D_1 - B$. Then $l(D) = l(D_1)$, and replacing D by a general member of $|D|$, one may assume that $D \wedge D_2 = 0$. By Lemma 6.4, one has

$$l(D) + l(D_2) \leq l(0) + l(D + D_2) \leq 1 + l(D_1 + D_2 - B) \leq 1 + l(D_1 + D_2).$$

Hence, if $l(D_1) + l(D_2) = 1 + l(D_1 + D_2)$, then $l(D_1 + D_2 - B) = l(D_1 + D_2)$, which implies that $B \leq |D_1 + D_2|_{\text{fix}}$. □

c. Theorem 6.5 (Clifford). *If $l(D) > 0$ and $i(D) > 0$, then $l(D) \leq \deg D/2 + 1$. If equality holds, $D \sim 0$ or $\operatorname{Bs}|D| = \varnothing$.*

Proof. Applying the last lemma to D and $K - D$, one has

$$l(D) + l(K - D) \leq 1 + l(K) = 1 + g.$$

If equality holds, then $|D|_{\text{fix}} \leq |K|_{\text{fix}} = 0$, i.e., $\operatorname{Bs}|D| = \varnothing$. Now, rewriting the above inequality by Theorem 6.1, one obtains the required inequality. □

§6.3 Degree of a Curve

a. When V is a closed subvariety of a projective space $\mathbf{P}_k^n$ one has the fundamental sheaf $\mathcal{O}_V(1)$ defined as $\mathcal{O}_{\mathbf{P}_k^n}(1)|_V$. The degree of V in $\mathbf{P}_k^n$ (or, with respect to the immersion $V \to \mathbf{P}_k^n$) will be defined in §8.5.**a**. We consider here a 1-dimensional V.

Definition. Let Γ be a curve in $\mathbf{P}_k^n$. $\deg(\mathcal{O}_\Gamma(1))$ is said to be the *degree of the curve* Γ in $\mathbf{P}_k^n$.

Proposition 6.2. *Let D be a divisor on a nonsingular complete curve C with $l(D) \geq 2$. $\Gamma = \Phi_D(C)$ is a complete curve in $\mathbf{P}|D|$ and $\deg \Gamma$ satisfies the formula $\deg(D - |D|_{\text{fix}}) = \deg \Phi_D \cdot \deg \Gamma$, with respect to the immersion $\Gamma \to \mathbf{P}|D|$.*

Proof. Since $l(D) = l(D - |D|_{\text{fix}})$ and $\Phi_D = \Phi_{D'}$, where D' is $D - |D|_{\text{fix}}$, one can assume $|D|_{\text{fix}} = 0$, i.e., $\operatorname{Bs}|D| = \varnothing$. Letting f be Φ_D and $\mathbf{P}$ be $\mathbf{P}|D|$, one has $f^*\mathcal{O}_{\mathbf{P}}(1) \cong \mathcal{O}(D)$ and $f^*(\mathcal{O}_{\mathbf{P}}(1)|_\Gamma) \cong \mathcal{O}(D)$. Hence, $\deg D = \deg \mathcal{O}(D) = \deg f \cdot \deg \mathcal{O}_{\mathbf{P}}(1) = \deg f \cdot \deg \Gamma$, by Corollary 2 to Theorem 4.11. □

b. Definition. Let $\mathscr{L}$ be an invertible sheaf on a complete curve Γ with $\deg \mathscr{L} > 0$. Define $\Delta(\Gamma, \mathscr{L})$ to be $1 + \deg \mathscr{L} - \dim_k H^0(\Gamma, \mathscr{L})$.

If D is a divisor on a nonsingular complete curve C, then $\Delta(C, \mathscr{O}(D))$ agrees with $\Delta(C, D)$ defined in §6.2.**a**.

Proposition 6.3. $\Delta(\Gamma, \mathscr{L}) \geq 0$. *If* $\Delta(\Gamma, \mathscr{L}) = 0$, *then* $\Gamma \cong \mathbf{P}_k^1$.

PROOF. Let (C, μ) be the normalization of Γ. Then by Lemma 2.35, $\dim_k H^0(C, \mu^*\mathscr{L}) \geq \dim_k H^0(\Gamma, \mathscr{L})$; and by Theorem 4.10.(ii), $\deg \mu^*\mathscr{L} = \deg \mathscr{L} > 0$. Hence, $\Delta(\Gamma, \mathscr{L}) \geq \Delta(C, \mu^*\mathscr{L})$. By Theorem 6.3, $\Delta(\Gamma, \mathscr{L}) \geq \Delta(C, \mu^*\mathscr{L}) \geq 0$. If $\Delta(\Gamma, \mathscr{L}) = 0$, then $\Delta(C, \mu^*\mathscr{L}) = 0$ and $H^0(C, \mu^*\mathscr{L}) \cong H^0(\Gamma, \mathscr{L})$. By Theorem 6.3, $C \cong \mathbf{P}_k^1$; hence $\Phi_{\mu^*\mathscr{L}}$ is a closed immersion into $\mathbf{P}|\mu^*\mathscr{L}| = \mathbf{P}|\mathscr{L}|$. Since $H^0(C, \mu^*\mathscr{L}) \cong H^0(\Gamma, \mathscr{L})$, one has $\Phi_{\mu^*\mathscr{L}} = \Phi_{\mathscr{L}} \circ \mu$. Since $\Phi_{\mu^*\mathscr{L}}$ is a closed immersion, it follows that μ is a closed immersion by Theorem 1.23.(iv); hence μ is an isomorphism. □

c. Now, let D be a divisor on C with $\mathrm{Bs}|D| = \varnothing$. Then $f = \Phi_D$ is a morphism onto a complete curve $\Gamma = \Phi_D(C)$ in $\mathbf{P}|D|$. Letting $\mathscr{O}_\Gamma(1) = (\mathscr{O}_{\mathbf{P}|D|}(1))|_\Gamma$, one has $\Delta(\Gamma, \mathscr{O}_\Gamma(1)) = 1 + \deg(\mathscr{O}_\Gamma(1)) - l(D)$.

By Proposition 6.3, one has $\Delta(\Gamma, \mathscr{O}_\Gamma(1)) \geq 0$ and so $\deg \Gamma = \deg \mathscr{O}_\Gamma(1) \geq l(D) - 1$. Since $\deg D = \deg f \cdot \deg \Gamma$, it follows that $\deg D \geq \deg f \cdot (l(D) - 1)$, which is rewritten as $l(D) \leq \deg D/\deg f + 1$. Whenever $l(D) = \deg D/\deg f + 1$, one has $\Delta(\Gamma, \mathscr{O}_\Gamma(1)) = 0$ and then $\Gamma \cong \mathbf{P}^1$ by Proposition 6.3.

§6.4 Hyperplane Section Theorem

a. Throughout this section, let D denote a divisor with $\deg D > 0$ on a nonsingular complete curve C and let $g = g(C)$.

Proposition 6.4.

(i) *If* $\deg D \geq 2\Delta(C, D) - 1$, *then* $l(D) \geq 1$.
(ii) *If* $\deg D \geq 2\Delta(C, D)$, *then* $\mathrm{Bs}|D| = \varnothing$.
(iii) *If* $\deg D = 2\Delta(C, D) - 1$ *and* $\mathrm{Bs}|D| \neq \varnothing$, *then* $\deg(|D|_{\mathrm{fix}}) = 1$.
(iv) *If* $\deg D \geq 2\Delta(C, D) + 1$, *then* $\Delta(C, D) = g$.

PROOF. (i) This is clear by definition and hypothesis.

(iv) From the hypothesis, it follows that $l(D) \geq \frac{1}{2}\deg D + \frac{3}{2}$. By Theorem 6.5, this implies that $l(K - D) = 0$; hence $\Delta(C, D) = 1 + \deg D - (l(K - D) + 1 - g + \deg D) = g$.

(ii) Assume $\mathrm{Bs}|D| \neq \varnothing$ and take $p \in \mathrm{Bs}|D|$. Then, letting $D_1 = D - p$, one has $\deg D_1 = \deg D - 1 > 0$ and $l(D_1) = l(D)$; thus $\Delta(C, D_1) = \Delta(C, D) - 1$ and so $\deg D_1 \geq 2\Delta(C, D_1) + 1$. By the assertion (iv), one has $\Delta(C, D_1) = g$ and $\Delta(C, D) = 1 + g$. But $\Delta(C, D) \leq g$ by the Corollary to Theorem 6.3, and hence a contradiction.

(iii) Let p be a base point of $|D|$. $D_1 = D - p$ satisfies $\deg D_1 = 2\Delta(C, D_1)$; hence $|D_1|_{\mathrm{fix}} = 0$ by (ii). Thus, $|D|_{\mathrm{fix}} = p$. □

Lemma 6.7. *If $l(D - p) = l(D) - 1$ and $l(D - p - q) = l(D) - 2$ for any p, $q \in C$, then D is very ample, i.e., $\mathrm{Bs}|D| = \varnothing$ and $\Phi_D\colon C \to \mathbf{P}|D|$ is a closed immersion.*

Proof. It is clear that $\mathrm{Bs}|D| = \varnothing$. Take a basis $\varphi_1, \ldots, \varphi_l$ of $\Gamma(C, \mathcal{O}(D))$. First, we shall prove $f = \Phi_D = (\varphi_1 : \cdots : \varphi_l)\colon C \to \mathbf{P}|D| = \mathbf{P}_k^{l-1}$ is one-to-one. Suppose that $f(p) = f(q)$ for some points p and q. Then $\varphi_i(p) = \lambda\varphi_i(q)$ for $i = 1, \ldots, l$, where $\lambda \in k^*$; thus any $\varphi = \sum_{i=1}^{l} c_i\varphi_i$ satisfying $\varphi(p) = 0$ vanishes at q, i.e., $\mathrm{Bs}|D - p| \ni q$. This contradicts the hypothesis. For a fixed $p \in C$, one has $l(D - p) = l - 1$ and $l(D - 2p) = l - 2$; i.e., $\Gamma(C, \mathcal{O}(D)) = \mathbf{L}(\mathcal{O}(D) + 0) \supset \mathbf{L}(\mathcal{O}(D) - p) \supset \mathbf{L}(\mathcal{O}(D) - 2p)$ (cf. § 2.20.**a**). Hence, choosing $\psi_1 \in \mathbf{L}(\mathcal{O}(D) + 0)\backslash\mathbf{L}(\mathcal{O}(D) - p)$, $\psi_2 \in \mathbf{L}(\mathcal{O}(D) - p)\backslash\mathbf{L}(\mathcal{O}(D) - 2p)$, and $\psi_3, \ldots, \psi_l \in \mathbf{L}(\mathcal{O}(D) - 2p)$, one has a basis $\psi_1, \ldots, \psi_l$ of $\Gamma(C, \mathcal{O}(D))$. $f = \Phi_D = (\psi_1 : \cdots\cdots : \psi_l)\colon C \to \mathbf{P}|D|$ is represented by a morphism $F = (\psi_2/\psi_1, \ldots, \psi_l/\psi_1)_k$ from $U = \{x \in C \mid \psi_1(x) \neq 0\}$ to $\mathbf{A}_k^{l-1}$, since the ψ_j/ψ_1 are regular on U. Since U is $f^{-1}(D_+(\mathsf{T}_0))$, F is obtained from f by base extension. On the other hand, f is a finite morphism, since C is a curve. Thus F is a finite morphism; thus the homomorphism $\rho = \Gamma(F)$ from $A = A(\mathbf{A}_k^{l-1}) = k[\mathsf{X}_1, \ldots, \mathsf{X}_{l-1}]$ into $B = A(U)$ is a finite homomorphism, where $\rho(\mathsf{X}_i) = \psi_{i+1}/\psi_1$ for all $i \geq 1$. Letting $\mathfrak{m} = (\mathsf{X}_1, \ldots, \mathsf{X}_{l-1})$ and letting $\mathfrak{p}$ be the prime ideal of B corresponding to p, one sees that $A_\mathfrak{m} = \mathcal{O}_{\mathbf{A}^{l-1}, x}$, where $x = \Phi_D(p)$, $B_\mathfrak{p} = \mathcal{O}_{C,p}$, and $\mathfrak{m}B_\mathfrak{p} = \mathfrak{m}_p = \psi_2/\psi_1\, B_\mathfrak{p}$. From the exact sequence of A-modules

$$A \xrightarrow{\ \rho\ } B \longrightarrow \operatorname{Coker}\rho \longrightarrow 0,$$

one has $A \otimes_A (A/\mathfrak{m}) \to B \otimes_A (A/\mathfrak{m}) \to (\operatorname{Coker}\rho) \otimes_A (A/\mathfrak{m}) \to 0$, which is exact. Since $A/\mathfrak{m} \cong k$ and $B \otimes_A (A/\mathfrak{m}) \cong B/\mathfrak{m}B = B_\mathfrak{p}/\mathfrak{m}B_\mathfrak{p} \cong k$, it follows that $(\operatorname{Coker}\rho) \otimes_A (A/\mathfrak{m}) = 0$. By Lemma 2.8, $(\operatorname{Coker}\rho)_\mathfrak{m} = 0$; thus there exists $a \in A\backslash\mathfrak{m}$ with $(\operatorname{Coker}\rho)_a = 0$. Hence, $A_a \to B_{\rho(a)}$ is surjective, i.e., $\Phi_D\colon C \to \mathbf{P}_k^{l-1}$ is a closed immersion on $D(a)$. Therefore, Φ_D is a closed immersion, since it is a closed immersion on some neighborhood of every point of C. □

b. Theorem 6.6 (Hyperplane Section Theorem). *If $\deg D \geq 2g + 1$, then D is very ample, i.e., D is a hyperplane section divisor.*

PROOF. By Lemma 6.3.(i), $l(D - p) = l(D) - 1$ and $l(D - p - q) = l(D) - 2$ for any $p, q \in C$. Hence, by Lemma 6.7, one obtains the assertion. □

Corollary 1. *Any nonsingular complete curve is projective.*

Corollary 2

(i) *If $g \geq 3$, then $mK(C)$ is very ample for any $m \geq 2$.*
(ii) *If $g = 2$, then $mK(C)$ is very ample for any $m \geq 3$, and $\Phi_{2K(C)}: C \to \mathbf{P}_k^1$ is a double covering; i.e., $\deg \Phi_{2K(C)} = 2$.*

PROOF. Since $\deg(mK(C)) = 2m(g - 1)$, every assertion is clear except for the last part of (ii). If $g = 2$, then $l(2K(C)) = 1 - 2 + 4 = 3$ and $l(K(C)) = 2$. Letting φ_0, φ_1 be a basis of $\Gamma(C, \Omega_C)$, one has a basis $\varphi_0^2, \varphi_0\varphi_1, \varphi_1^2$ of $\Gamma(C, \Omega_C^{\otimes 2})$. Hence, $\Phi_{2K(C)}(C)$ is a conic, which is isomorphic to $\mathbf{P}_k^1$. □

c. Suppose that $g(C) = 1$. Then for any point p on C, $D = 3p$ becomes very ample by Theorem 6.6. Hence, $\Gamma = \Phi_D(C)$ is a nonsingular curve on $\mathbf{P}_k^2$ of degree 3, which is said to be a *nonsingular cubic*. Since $\Phi_D: C \to \Gamma$ is an isomorphism, one can identify C with Γ. Then $3p$ is obtained from a line on $\mathbf{P}_k^2$, say $\{p\} = V_+(\mathrm{T}_0) \cap \Gamma$. One can choose a homogeneous coordinate system $(\mathrm{T}_0, \mathrm{T}_1, \mathrm{T}_2)$ such that $\Gamma = V_+(\mathrm{T}_0\mathrm{T}_2^2 - \mathrm{T}_1(\mathrm{T}_1 - \mathrm{T}_0)(\mathrm{T}_1 - \lambda\mathrm{T}_0))$, where $\lambda \neq 0, 1$; in other words, $\mathrm{Rat}(C) = k(z, w)$ with $w^2 = z(z - 1)(z - \lambda)$.

Now, suppose that $g(C) = 2$. Since $l(K(C)) = 2$, it follows that $\mathbf{P}|K(C)| = \mathbf{P}_k^1$ and $f = \Phi_{K(C)}: C \to \mathbf{P}_k^1$ satisfies $\deg f \geq 2$. Hence, by §6.3.**c**, $2 = l(K(C)) \leq (\deg K(C)/\deg f) + 1 = (2/\deg f) + 1 \leq 2$. From this, $\deg f = 2$ follows. Therefore, $\mathrm{Rat}(C)$ is a quadratic extension of $\mathrm{Rat}(\mathbf{P}_k^1) = k(z)$. Since k has characteristic zero, there exists w such that $\mathrm{Rat}(C) = k(z, w)$ and $w^2 = z(z - 1)(z - \lambda_1)(z - \lambda_2)(z - \lambda_3)$ for some $\lambda_1, \lambda_2, \lambda_3 \in k$ with $\lambda_i \neq \lambda_j, 0, 1$ for all $i \neq j$ (see §6.5.**b**).

Definition. If a surjective morphism $f: C \to Y$ between complete curves has $\deg f = r$, f is said to be an *r-sheeted covering*. A 2-sheeted covering is often also called a *double covering*.

§6.5 Hyperelliptic Curves

a. Applying the ramification formula (Theorem 5.5) to a morphism between nonsingular complete curves, one obtains the next formula (see §5.6.**c**).

Theorem 6.7 (Riemann–Hurwitz Formula). *If there exists an r-sheeted covering $f: C \to Y$, then*

$$2g(C) - 2 = r \cdot (2g(Y) - 2) + \sum_{p \in C} (e(p) - 1).$$

Corollary. *Let $\varphi \in \operatorname{Rat}(C)\backslash\{0\}$. Then, letting $n = \deg \varphi$, one has*

$$g(C) = \frac{1}{2} \sum_{p \in C} (e(p) - 1) - n + 1.$$

This is called *the genus formula of Riemann.*

b. Definition. If there exists a double covering $f\colon C \to \mathbf{P}_k^1$ with $g(C) \geq 2$, then C is said to be a *hyperelliptic curve.*

In this case, by Theorem 6.7, $\sum_{p \in C} (e(p) - 1) = 2g + 2$, where $g = g(C)$. Note that $e(p) \geq 2$ if and only if $f^*f(p) = 2p$; in which case $e(p) = 2$. Such a point p is called a *ramification point* on C with respect to f. Thus there exist $2g + 2$ ramification points.

Proposition 6.5. *If there exists a double covering $f\colon C \to \mathbf{P}_k^1$ with $g = g(C) \geq 2$, then for any divisor $\mathfrak{d}$ with $0 < \deg \mathfrak{d} < g$, $l(f^*(\mathfrak{d})) = l(\mathfrak{d})$ and $\deg(f^*(\mathfrak{d})) = 2 \deg \mathfrak{d}$. In particular, if $\deg \mathfrak{d} = g - 1$, then $K(C) \sim f^*\mathfrak{d} \sim (g-1)f^*(p)$ for any $p \in \mathbf{P}_k^1$.*

Proof. If $\deg \mathfrak{d}_0 = g - 1$, then $l(f^*\mathfrak{d}_0) \geq l(\mathfrak{d}_0) = g$ and $\deg(f^*\mathfrak{d}_0) = 2g - 2$. By Theorem 6.4, $f^*\mathfrak{d}_0 \sim K(C)$. For any effective divisor $\mathfrak{d}$ on $\mathbf{P}_k^1$ with $\deg \mathfrak{d} \leq g - 1$, there is an effective divisor $\mathfrak{d}'$ with $\deg(\mathfrak{d} + \mathfrak{d}') = g - 1$. Hence, $l(f^*(\mathfrak{d} + \mathfrak{d}')) = l(\mathfrak{d} + \mathfrak{d}')$ and $l(f^*(\mathfrak{d})) \geq l(\mathfrak{d})$. From these, it follows that $l(f^*(\mathfrak{d})) = l(\mathfrak{d}) = 1 + \deg \mathfrak{d}$. □

Corollary. $\Phi_{K(C)} = \Phi_{(g-1)p} \circ f$ *for any point $p \in \mathbf{P}^1$.*

Thus, any double covering $f\colon C \to \mathbf{P}_k^1$ is obtained as the covering associated with the canonical morphism $\Phi_{K(C)}\colon C \to \Phi_{K(C)}(C) \cong \mathbf{P}_k^1$. The ramification points of f are said to be the *canonical ramification points* of C.

Theorem 6.8. *If C is not a hyperelliptic curve and $g = g(C) \geq 2$, then $K = K(C)$ is very ample.*

Proof. Since $|K|$ has no base points by Proposition 6.1, if K is not very ample, then there exist $p, q \in C$ such that $l(K - p - q) = l(K - p) = g - 1$ by Lemma 6.7. By Theorem 6.1,

$$l(p + q) = 1 - g + 2 + l(K - p - q) = 2,$$

which implies that Φ_{p+q} is the double covering; hence, C is hyperelliptic. □

c. Suppose that there is an effective special divisor D on C such that $D \neq 0$, $D \not\sim K$, and $2(l(D) - 1) = \deg D$. Then we shall prove that C is hyperelliptic, i.e., there exists a double covering $f\colon C \to \mathbf{P}_k^1$, and that $D \sim f^*(\mathfrak{d})$ for some effective divisor $\mathfrak{d}$. First, we show the next lemma.

Lemma 6.8. *Let D and E be effective divisors on C with $g(C) \geq 1$. If $\mathrm{Bs}\,|D - p| = \varnothing$ for some $p \leq E$, then*

$$l(D) + l(E) \leq 1 + l(D + E - p).$$

PROOF. Take a general member D' of $|D - p|$. Then defining E' by $E = p + E'$, one has $E \vee (D' + p) = D' + E' + p \sim D + E - p$ and $E \wedge (D' + p) = p$. By Lemma 6.4 and $l(p) = 1$ (cf. Corollary (ii) to Theorem 6.2), one obtains the assertion. □

Now, taking a suitable point p of $\Gamma = \Phi_D(C)$, one has $\Phi_D^{-1}(p) = \{p_1, \ldots, p_r\}$, where $r = \deg \Phi_D$. Take $E' \in |K - D|(p_1)$ and apply the above lemma. Since $l(D) + l(K - D) = g + 1$, one concludes that $\mathrm{Bs}\,|D - p_1| \neq \varnothing$. Taking $p_2 \in \mathrm{Bs}\,|D - p_1|$, one defines D_1 to be a member of $|D - p_1 - p_2|$. Then $l(D_1) = l(D) - 1$ and so $2(l(D_1) - 1) = \deg D_1$. Hence, by Lemma 6.4, $|D_1|$ has no base points. This implies $\Phi_D^{-1}(p) = \{p_1, p_2\}$; i.e., $r = 2$. Since $l(D) = \frac{1}{2} \deg D + 1 = (\deg D/\deg \Phi_D) + 1$, one sees that $\Delta(\Gamma, \mathcal{O}_\Gamma(1)) = 0$ and $\Gamma \cong \mathbf{P}^1$ by §6.3.**c**. Therefore, identifying Γ with $\mathbf{P}^1$ and letting f be Φ_D we obtain the following assertion.

Theorem 6.9. *If D is an effective special divisor on C with $g(C) \geq 2$ such that $2(l(D) - 1) = \deg D$, then $D \sim 0$, $D \sim K(C)$, or there exists a double covering $f: C \to \mathbf{P}_k^1$ with $D = f^*(\mathfrak{d})$ for some $\mathfrak{d}$ on $\mathbf{P}_k^1$.*

d. We now strengthen the hyperplane section theorem (Theorem 6.6).

Theorem 6.10. *If $\deg D \geq 2g$, then D is very ample unless $D \sim K + p + q$ for some $p, q \in C$. Furthermore, $\Phi_{K+p+q}: C \to \Gamma = \Phi_{K+p+q}(C)$ is not birational if and only if C is hyperelliptic and $\Phi_K(p) = \Phi_K(q)$.*

PROOF. By Theorem 6.6, we may assume $\deg D = 2g$. Since $|D|$ has no base points by Proposition 6.4.(ii), and D is not very ample, then there exist p, $q \in C$ with $l(D - p - q) = l(D - p) = l(D) - 1 = g$ by Lemma 6.2. On the other hand, $l(D - p - q) = 1 - g + \deg D - 2 + l(K + p + q - D) = l(K + p + q - D) + g - 1$. Hence, $l(K + p + q - D) = 1$ and $\deg(K + p + q - D) = 0$; thus $K + p + q \sim D$ by Lemma 6.1.(ii).

Suppose that $f = \Phi_{K+p+q}: C \to \Gamma$ is not birational. Then from the formula $g + 1 = l(K + p + q) \leq \deg(K + p + q)/2 + 1 = g + 1$ in §6.3.**c**, one has $\deg f = 2$ and $\Delta(\Gamma, \mathcal{O}_\Gamma(1)) = 0$. Thus $\Gamma \cong \mathbf{P}_k^1$ and $K + p + q = f^*(\mathfrak{d})$ for some $\mathfrak{d}$ on $\mathbf{P}_k^1$; hence C is hyperelliptic, f is the double covering of C, and $f(p) = f(q)$. □

e. Let C be a hyperelliptic curve of genus g and let $f: C \to \mathbf{P}_k^1$ be the double covering. For $1 \leq i \leq g$, $|f^*(ip)|$ has no base points and $l(f^*(ip)) = i + 1$.

Hence, for a canonical ramification point p_1 of C, one has $f^*(f(p_1)) = 2p_1$ and $l(2ip_1) = l((2i + 1)p_1) = i + 1$, if $1 \le i \le g$. Furthermore by Lemma 6.2, $l(jp_1) = j + 1 - g$, if $j \ge 2g$. We shall show that if p' is not a canonical ramification point, then $l(ip') = 1$ for $1 \le i \le g$; thus $l(gp) = 1$ if and only if p is not a canonical ramification point (see §6.7.**e**).

§6.6 Λ-Gap Sequences and Weierstrass Points

a. If C is a nonhyperelliptic curve with $g = g(C) \ge 3$, C has no canonical ramification points. But there exist a finite number of specified points similar to canonical ramification points on a hyperelliptic curve, as will be seen in subsection **c**.

Definition. For $p \in C$, define $SG_p(C)$ to be $\{-\mathrm{ord}_p(\varphi) \mid \varphi \in \mathrm{Rat}(C)^*$ and $\varphi \in A(C \setminus \{p\})\}$. $SG_p(C)$ is a semigroup, called the *Weierstrass semigroup* of C at p.

EXAMPLE 6.1. If p is a canonical ramification point of a hyperelliptic curve C, then

$$SG_p(C) = \{0, 2, 4, \ldots, 2g - 2, 2g, 2g + 1, 2g + 2, \ldots\}.$$

Note that $m \in SG_p(C)$ if and only if $\mathbf{L}(mp) \supset \mathbf{L}((m - 1)p)$. Hence, $m \in SG_p(C)$ if and only if $l(mp) = l((m - 1)p) + 1$. By Theorem 6.1,

$$m \in SG_p(C) \Leftrightarrow l(K(C) - mp) = l(K(C) - (m - 1)p)$$
$$\Leftrightarrow p \in \mathrm{Bs}\,|K(C) - (m - 1)p|.$$

b. Let Λ be a linear system on a nonsingular complete curve C with $\dim \Lambda = l - 1$.

Definition. For any point p, define the linear system $\Lambda(ip)$ to be $\{D \in \Lambda \mid D \ge ip\}$ (see §6.1.**d**).

Definition. If a is the coefficient at p of the fixed part of Λ, define $a_p(\Lambda)$ to be a.

Then $a_p(\Lambda) = \max\{i \mid \Lambda(ip) = \Lambda\}$; hence $a = a_p(\Lambda)$ satisfies $\Lambda((a + 1)p) \subset \Lambda(ap) = \Lambda$ and so $\dim \Lambda((a + 1)p) = \dim \Lambda(ap) - 1 = l - 2$.

First, we put $a_p(\Lambda) = a_1$ and further define a_2 by $a_p(\Lambda((a_1 + 1)p)) = a_2$. Inductively, we define a_j for $j = 3, 4, \ldots$, by

$$a_p(\Lambda((a_j + 1)p)) = a_{j+1} \quad \text{if } \Lambda((a_j + 1)p) \ne \varnothing.$$

Since $\dim \Lambda((a_j + 1)p) = \dim \Lambda(a_j p) - 1 = \cdots = l - j - 1$, there exist exactly l numbers $a_1, \ldots, a_l$. The sequence $\{a_1, \ldots, a_l\}$ is said to be the *Λ-gap sequence* at p. Since Λ has a member $a_l p + \Delta$ for some effective divisor Δ, it follows that $a_l \leq \deg D$ for any $D \in \Lambda$.

Definition. The Λ-*index* $\rho_p(C, \Lambda)$ at p is defined to be $a_1 + a_2 - 1 + a_3 - 2 + \cdots + a_l - (l - 1)$, where $\{a_1, \ldots, a_l\}$ is the Λ-gap sequence at p.

Clearly, $\rho_p(C, \Lambda) \geq 0$; thus $\rho_p(C, \Lambda) = 0$ if and only if $a_1 = 0$, $a_2 = 1, \ldots, a_l = l - 1$.

c. If Λ is a complete linear system, i.e., $\Lambda = |D|$ for some D, then the $|D|$-gap sequence, $|D|$-index, and $\rho_p(C, |D|)$ are referred to as the *D-gap sequence*, *D-index*, and $\rho_p(C, D)$, respectively.

Letting $\{a_1, \ldots, a_l\}$ be the D-gap sequence at p, one has $l(D - a_j p) = l - (j - 1)$ and $l(D - (a_j + 1)p) = l - j$.

Proposition 6.6. *Let* $\{a_1, \ldots, a_g\}$ *be the* $K(C)$*-gap sequence at* p. *Then* $\mathbb{N} \backslash SG_p(C) = \{a_1 + 1, a_2 + 1, \ldots, a_g + 1\}$.

PROOF. Since $m \notin SG_p(C)$ if and only if $l(K - mp) = l(K - (m - 1)p) - 1$, the assertion follows. □

The sequence $\{a_1 + 1, \ldots, a_g + 1\}$ is called the *gap sequence* at p and $\rho_p(C, K(C)) = \sum_{j=1}^{g} (a_j - (j - 1))$ is denoted by $w_p(C)$ or w_p.

$w_p(C) > 0$, if and only if $l(gp) \geq 2$, which is equivalent to $l(K - gp) \geq 1$.

Definition. If $w_p(C) > 0$, then p is said to be a *Weierstrass point* on C. $W(C)$ denotes the set of all Weierstrass points of C.

We shall prove that $W(C)$ is a finite set (Corollary to Theorem 6.11).

d. EXAMPLE 6.2. Let C denote a Fermat curve of degree n, i.e., $C = V_+(\mathrm{T}_2^n + \mathrm{T}_1^n - \mathrm{T}_0^n)$, for $n \geq 4$, and let $x = (\mathrm{T}_1/\mathrm{T}_0)|_C$, $y = (\mathrm{T}_2/\mathrm{T}_0)|_C$. Then $x^n + y^n = 1$. Recalling §5.3.**d** and Example 5.7 in §5.5.**b**, one has

$$\Gamma(C, \Omega_C) = \left\{ Q(x, y) \frac{dx}{y^{n-1}} \,\middle|\, Q \text{ is a polynomial and } \deg Q \leq n - 3 \right\}.$$

If $p = (1 : 0 : 1) \in C$, then $\mathrm{ord}_p(x) = 1$, $\mathrm{ord}_p(y - 1) = n$. For a pair of non-negative integers (i, j) with $i + j \leq n - 3$, define

$$\omega_{i,j} = x^i (y - 1)^j \cdot \frac{dx}{y^{n-1}}.$$

Then $\mathrm{div}(\omega_{i,j}) \in |K(C)|$ and $\mathrm{ord}_p(\omega_{i,j}) = i + nj$. It is easy to see that these $\omega_{i,j}$ form a basis of $\Gamma(C, \Omega_C)$. Hence, the gap sequence at p is $\{i + nj + 1 \mid 0 \leq i, j \text{ and } i + j \leq n - 3\}$. Thus $n - 1, n \in SG_p(C)$ and hence,

$SG_p(C) \supseteq \{\lambda n + \mu(n-1) \mid \lambda, \mu \geq 0\}$. From this, we infer readily that $SG_p(C)$ is the semigroup generated by $n-1$ and n.

If $n = 4, 5, 6$, we can compute the gap sequences and w_p:

n	g	gap sequence	w_p
4	3	$\{1, 2, 5\}$	2
5	6	$\{1, 2, 3, 6, 7, 11\}$	9
6	10	$\left\{\begin{matrix}1, 2, 3, 4, 7, 8, \\ 9, 13, 21, 19\end{matrix}\right\}$	25

§6.7 Wronski Forms

a. Let U be a coordinate neighborhood of p on a nonsingular curve C with z the local coordinate. If $\varphi \in R = A(U)$, then $d\varphi = \alpha\, dz$ for some $\alpha \in R$, since $\Omega_{R/k} = R\, dz$. α is denoted $d\varphi/dz$ or φ', and is called the *derived function from* φ *with respect* to z. Further, we put $\varphi^{\langle s \rangle} = (\varphi^{\langle s-1 \rangle})'$ with $\varphi^{\langle 0 \rangle} = \varphi$ for $s \geq 1$.

Definition. Given $\varphi_1, \ldots, \varphi_l \in R$, we define $W_z(\varphi_1, \ldots, \varphi_l)$ by

$$W_z(\varphi_1, \ldots, \varphi_l) = \det \begin{bmatrix} \varphi_1, & \varphi_2, & \ldots, & \varphi_l \\ \varphi_1', & \varphi_2', & \ldots, & \varphi_l' \\ \vdots & & & \\ \varphi_1^{\langle l-1 \rangle}, & \varphi_2^{\langle l-1 \rangle}, & \ldots, & \varphi_l^{\langle l-1 \rangle} \end{bmatrix}.$$

Further, we define $W(\varphi_1, \ldots, \varphi_l)$ to be $W_z(\varphi_1, \ldots, \varphi_l)(dz)^{l(l-1)/2}$, where $(dz)^m$ stands for $(dz)^{\otimes m} \in \Gamma(U, (\Omega_U^1)^{\otimes m})$.

Remark. If ζ is another local coordinate of U, then $W_\zeta(\varphi_1, \ldots, \varphi_l) = W_z(\varphi_1, \ldots, \varphi_l)(dz/d\zeta)^{l(l-1)/2}$. Thus, $W(\varphi_1, \ldots, \varphi_l)$ does not depend on the choice of the local coordinate.

$W(\varphi_1, \ldots, \varphi_l)$ is said to be the *Wronski form* of $\varphi_1, \ldots, \varphi_l$.

Proposition 6.7

(i) *If* $\psi_i = \sum_{j=1}^{l} c_{ij}\varphi_j$ *for* $1 \leq i \leq l$, $c_{i,j} \in k$, *then* $W(\psi_1, \ldots, \psi_l) = c \cdot W(\varphi_1, \ldots, \varphi_l)$, *where* $c = \det[c_{i,j}]$.
(ii) *If* $\varphi_1, \ldots, \varphi_l$ *are k-linearly dependent, then* $W(\varphi_1, \ldots, \varphi_l) = 0$.
(iii) *If* $a_1 = \operatorname{ord}_p \varphi_1 < a_2 = \operatorname{ord}_p \varphi_2 < \cdots < a_l = \operatorname{ord}_p \varphi_l$ *then* $\operatorname{ord}_p W(\varphi_1, \ldots, \varphi_l) = \sum_{j=1}^{l} (a_j - (j-1))$.

(iv) *If $\varphi_1, \ldots, \varphi_l$ are k-linearly independent, then $W(\varphi_1, \ldots, \varphi_l) \neq 0$.*
(v) *For any $\psi \in R$,*

$$W(\psi\varphi_1, \ldots, \psi\varphi_l) = \psi^l W(\varphi_1, \ldots, \varphi_l).$$

PROOF. (i) This follows immediately from the definition.
(ii) By (i), one can assume $\varphi_1 = 0$.
(iii) This is left to the reader as an exercise.
(iv) This follows from (iii) after a linear transformation.
(v) This can be easily proved by the theory of determinants. □

b. Let $\mathscr{L}$ be an invertible sheaf on C which is covered by coordinate neighborhoods $\{U_\lambda\}$ with $\mathscr{L}|_{U_\lambda} = \mathscr{O}_{U_\lambda} \cdot \beta_\lambda$. If $\mathbf{L}$ is an l-dimensional subspace of $\Gamma(C, \mathscr{L})$ with basis $(\varphi_1, \ldots, \varphi_l)$, such that $\varphi_i|_{U_\lambda} = \varphi_\lambda^{(i)}\beta_\lambda$, then define ω_λ by

$$\omega_\lambda = W(\varphi_\lambda^{(1)}, \ldots, \varphi_\lambda^{(l)})\beta_\lambda^l.$$

Then by Proposition 6.7.(v), one can check that $\omega_\lambda|_{U_\lambda \cap U_\mu} = \omega_\mu|_{U_\lambda \cap U_\mu}$. Thus $\{\omega_\lambda\}$ determines a section of $\Gamma(C, (\Omega)^{\otimes l(l-1)/2} \otimes \mathscr{L}^{\otimes l})$, which is denoted by $\omega(\mathbf{L})$ or $\omega(C, \Lambda)$, where Λ denotes $\Lambda(\mathbf{L})$.

Definition. $\omega(C, \Lambda)$ is said to be the *Weierstrass form associated with* Λ. $W(C, \Lambda)$ denotes $\operatorname{div}(\omega(C, \Lambda)) \in |\frac{1}{2}l(l-1) \cdot K(C) + l \cdot \operatorname{div}(\varphi_1)|$, called the *Weierstrass divisor associated with* Λ.

Note that $W(C, \Lambda)$ does not depend on the choice of the basis $(\varphi_1, \ldots, \varphi_l)$ of $\mathbf{L}$, by Proposition 6.7.(i).

Lemma 6.9. $\rho_p(C, \Lambda) = \operatorname{ord}_p \omega(C, \Lambda)$.

PROOF. Letting $\{a_1, \ldots, a_l\}$ be the Λ-gap sequence at p, one has $\Lambda(a_1 p) \supset \Lambda(a_2 p) \supset \cdots \supset \Lambda(a_l p)$. Corresponding to this, one can choose a basis $\varphi_1, \ldots, \varphi_l$ of $\mathbf{L}$ such that $\operatorname{ord}_p \varphi_j = a_j$. Hence by Proposition 6.7.(ii), $\operatorname{ord}_p \omega(C, \Lambda) = \sum_{j=1}^{l} (a_j - (j-1))$, i.e., $\rho_p(C, \Lambda)$. □

Theorem 6.11. *If Λ is a linear system and $D \in \Lambda$, then $\sum_{p \in C} \rho_p(C, \Lambda) = l \deg D + l(l-1)(g-1)$, where $l = \dim \Lambda + 1$ and $g = g(C)$.*

PROOF. By Lemma 6.9 and the Corollary to Theorem 6.1, $\sum_{p \in C} \rho_p(C, \Lambda) = \sum_{p \in C} \operatorname{ord}_p \omega(C, \Lambda) = \deg W(C, \Lambda) = \deg(\frac{1}{2}l(l-1)K(C) + lD) = l(l-1)(g-1) + l \deg D$. □

Corollary.

(i) *If $g = g(C) \geq 2$, then $\sum_{p \in C} w_p(C) = (g^2 - 1)g$. Hence, $W(C)$ is a finite set.*
(ii) *If C is a hyperelliptic curve, then $W(C)$ consists of the canonical ramification points.*

PROOF. (i) This is clear, since $l(K(C)) = g$.

(ii) If p is a canonical ramification point, then the gap sequence at p is $\{1, 3, \ldots, 2g - 1\}$; hence $w_p = g(g - 1)/2$. Since there exist $2g + 2$ canonical ramification points, one obtains $\sum_{i=1}^{2g+2} w_{p_i} = g(g^2 - 1)$. Hence by (i), there do not exist any other Weierstrass points. □

EXAMPLE 6.3. If C is a Fermat curve of degree 4, then $g = 3$ and $\sum_{p \in C} w_p = (9 - 1) \cdot 3 = 24$. If $p = (1 : 0 : 1)$, then $w_p = 2$. One has 12 such points on C. Thus these points exhaust the Weierstrass points on C.

c. Let C be a nonsingular complete curve of genus $g \geq 3$, and let $K = K(C)$ as usual. For the next result, see [G-H, p. 275].

Theorem 6.12. *Suppose that C is not hyperelliptic and $p \in C$.*

(i) *If $g \geq 5$, then $w_p(C) \leq (g - 1)(g - 2)/2$.*
(ii) *If $g = 4$, then $w_p(C) \leq 4$. If $g = 3$, then $w_p(C) \leq 2$.*
(iii) *There exist at least $2g + 6$ Weierstrass points on C.*

PROOF. Letting $\{a_1, \ldots, a_g\}$ be the K-gap sequence at p, one has $l(K - a_j p) = g - j + 1$ for all j by §6.6.c. Since $a_1 = 0$, we let $D = K - a_j p$ for $j \geq 2$. Then D is a special divisor and so, if $D \not\sim 0$, one can apply Theorems 6.5 and 6.9. Thus $l(D) < (\deg D)/2 + 1$; hence, $2(g - j) < 2g - 2 - a_j$, i.e., $a_j < 2j - 2$.

Since $a_{g-1} < a_g \leq \deg K = 2g - 2$, one can conclude that

(α) $a_1 = 0$, $a_j \leq 2j - 3$ for $j = 2, \ldots, g$; or
(β) $a_1 = 0$, $a_j \leq 2j - 3$ for $j = 2, \ldots, g - 1$ and $a_g = 2g - 2$.

The latter case occurs only when $K \sim (2g - 2)p$. Hence, if $\sum_{j=1}^{g} (a_j - (j - 1)) > (g - 1)(g - 2)/2$, then $a_1 = 0$, $a_2 = 1$, $a_3 = 3$, $\ldots$, $a_{g-1} = 2g - 5$, and $a_g = 2g - 2$. Therefore, $\{1, 2, 4, \ldots, 2g - 4, 2g - 1\}$ is the gap sequence at p and

$$SG_p(C) = \{0, 3, 5, \ldots, 2g - 3, 2g - 2, 2g, 2g + 1, \ldots\}$$

which includes $\{3n + 5m \mid n \geq 0 \text{ and } m \geq 0\} = \{0, 3, 5, 6, 8, 9, 10, 11, \ldots\}$. Since $2g - 1 \notin SG_p(C)$, it follows that $2g - 1 \leq 7$; hence $g \leq 4$. If $g = 4$, then the gap sequence is $\{1, 2, 4, 7\}$; hence $w_p = 4$. If $g = 3$, then the gap sequence is $\{1, 2, 5\}$ hence $w_p = 2$.

(iii) Let w be $\max\{w_p(C) \mid p \in W(C)\}$. Then letting $\#(S)$ denote the cardinal number of a set S, by Corollary (i) to Theorem 6.11, one has

$$\#(W(C)) \geq \frac{g(g^2 - 1)}{w} \geq 2g + 6 + \frac{12}{g - 2}, \quad \text{if } g \geq 5.$$

If $g = 3$, then $\#(W(C)) \geq 12 = 2 \cdot 3 + 6$. If $g = 4$, then $\#(W(C)) \geq 15 > 2 \cdot 4 + 6 = 14$. □

EXAMPLE 6.4. As in Example 6.2, let C be a Fermat curve of degree n and let $p = (1 : 0 : 1)$. If $\omega = (y-1)^{n-3}\,dx/y^{n-1}$, then $\mathrm{ord}_p(\omega) = n(n-3) = 2g-2$. Hence, $\mathrm{div}(\omega) = (2g-2)p \in |K(C)|$.

§6.8 Theorems of Hurwitz and Automorphism Groups of Curves

a. Throughout this section, let C denote a nonsingular complete curve of genus g.

Theorem 6.13 (H. Weber). *If $g \geq 2$ and f: $C \to C$ is a surjective morphism, then f is an isomorphism.*

PROOF. By Theorem 6.7, one has

$$2g-2 = \deg f \cdot (2g-2) + \deg R_f.$$

Since $2g-2 > 0$, it follows that $R_f = 0$ and $\deg f = 1$. Hence, f is birational; and hence by Theorem 2.19, $f \in \mathrm{Aut}(C)$. □

Definition. If $f \in \mathrm{Aut}(C)$ with $f \neq \mathrm{id}$, then $\mathbb{F}(f) = \mathrm{Ker}(f, \mathrm{id}) \subset C$ is a closed set. Define $v^*(f)$ to be $\#(\mathbb{F}(f))$.

EXAMPLE 6.5. Let $\varphi: C \to \mathbf{P}_k^1$ be the canonical double covering for a hyperelliptic curve C. Then $\mathrm{Rat}(C) = k(w, z)$ where $z = \varphi^*(z_1)$ for an inhomogeneous coordinate $z_1 = \mathsf{T}_1/\mathsf{T}_0$ and $w^2 = R(z)$. The involution i: $\mathrm{Rat}(C) \to \mathrm{Rat}(C)$ defined by $i(w) = -w$ and $i(z) = z$ determines a morphism f: $C \to C$ such that $\varphi \circ f = \varphi$ and $f^2 = \mathrm{id}$. f is said to be the *canonical involution* of C. Then $\mathbb{F}(f) = \{p_1, \ldots, p_{2g+2}\}$ consists of the canonical ramification points; hence $v^*(f) = 2g+2$.

Theorem 6.14 (A. Hurwitz). *If $f \in \mathrm{Aut}(C)$ is not the identity and $g \geq 2$, then*

(i) $v^*(f) \leq 2g+2$.
(ii) *If $v^*(f) = 2g+2$, then C is hyperelliptic and f is the canonical involution.*
(iii) *If C is nonhyperelliptic, then $v^*(f) \leq 2g$.*
(iv) *If C is hyperelliptic, and f is not the canonical involution, then $v^*(f) \leq 4$.*

PROOF. (iv) Supposing that $\mathbb{F}(f) \supseteq \{p_1, \ldots, p_5\}$, we shall prove that f fixes all the canonical ramification points. Let p_c be a canonical ramification point such that $f(p_c) \neq p_c$. Take $\varphi \in \mathbf{L}(2p_c)\backslash k$ and define ψ to be $\varphi - f^*(\varphi)$, which is not zero, since $f(p_c) \neq p_c$. Then $(\psi)_0 \geq p_1 + \cdots + p_5$ and $(\psi)_\infty = 2p_c + 2f^{-1}(p_c)$; hence $0 = \deg(\mathrm{div}\,\psi) \geq 5 - 4 > 0$, a contradiction. Then $\Phi_K \circ f = \Phi_K$, which implies that f is the canonical involution.

(iii) Supposing that $\mathbb{F}(f) \supseteq \{p_1, \ldots, p_{2g}, p_{2g+1}\}$, we shall show that f fixes all Weierstrass points p_w of C. Let $p_w \in W(C)$ be such that $f(p_w) \neq p_w$. Then from $\varphi \in \mathbf{L}(gp_w)\backslash k$, one forms $\psi = \varphi - f^*(\varphi)$. Then $(\psi)_0 \geq p_1 + \cdots + p_{2g+1}$ and $(\psi)_\infty = gp_w + gf^{-1}(p_w)$; $0 = \deg(\operatorname{div}(\psi)) \geq 2g + 1 - 2g > 0$.

Thus f fixes every point of $W(C)$. Letting $W(C)$ be $\{p_1, \ldots, p_N\}$ and choosing φ from $\mathbf{L}((g+1)p)\backslash k$, one defines ψ to be $\varphi - f^*(\varphi)$. Then if $f(p) \neq p$, one has $\psi \neq 0$; thus $(\psi)_0 \geq p_1 + \cdots + p_N$ and $(\psi)_\infty = (g+1)p + (g+1)f^{-1}(p)$. By Theorem 6.12.(iii), $N \geq 2g + 6$. Hence, $0 = \deg(\psi)_0 - \deg(\psi)_\infty \geq 2g + 6 - 2g - 2 > 0$. Therefore, f fixes all points p of C.

(ii) By (iii), C is hyperelliptic. Since $2g + 2 \geq 6 > 4$, f is the canonical involution by (iv).

(i) This follows from (ii) and (iii). □

b. Theorem 6.15 (Hurwitz). *If $g \geq 2$, then* Aut(C) *is a finite group.*

Proof. Since $l(gp) = l(gf(p))$ for all $f \in \operatorname{Aut}(C)$, Aut$(C)$ acts on the finite set $W(C)$ (see §6.6.**b**). Thus it suffices to prove that a subgroup G of Aut(C) acting trivially on $W(C)$ is a finite group. If C is a hyperelliptic curve, then $W(C)$ consists of the $2g + 2$ canonical ramification points of C; hence, any nontrivial $f \in G$ must be the canonical involution and so $\#(G) = 2$. If C is not hyperelliptic, then $\#(W(C)) \geq 2g + 6$ and hence $v^*(f) \geq 2g + 6$ for any nontrivial $f \in G$. This contradicts the assertion (iii) of Theorem 6.14; hence $G = \{\mathrm{id}\}$. □

Exercises

6.1. (Hurwitz.) Let C be a nonsingular complete curve of genus g. Assume that $g \geq 2$. If $f \in \operatorname{Aut}(C)$ satisfies that $f^*\omega = \omega$ for all regular 1-forms ω, then show that $f = \mathrm{id}$.

6.2. Let Γ be a nonsingular complete curve and let $f, g\colon \Gamma \to C$ be surjective morphisms. If $f^*\omega = g^*\omega$ for all regular 1-forms ω, then show that $f = g$.

6.3. Let C be a hyperelliptic curve, and let a, b be positive integers. Multiplication induces a k-linear map $\mu\colon T_{\{a\}}(C) \otimes T_{\{b\}}(C) \to T_{\{a+b\}}\{C\}$. Compute $\dim_k(\operatorname{Im} \mu)$.

6.4. Let R be a DVR with maximal ideal $\mathfrak{m}$. Suppose that $k = R/\mathfrak{m}$ is contained in R and that k is of characteristic zero. If G is a finite subgroup of $\operatorname{Aut}_k(R)$, then show that G is a finite cyclic group.

Chapter 7

Cohomology of Projective Schemes

§7.1 The Homomorphism $\boldsymbol{\alpha}_M$

a. Let R be a ring and let A be a finitely generated graded R-algebra. Let $X = \operatorname{Proj} A$, which is a quasi-compact scheme. If M is a graded A-module, then the $\mathcal{O}_X$-module $\tilde{M}$ has been defined in §3.4. If $\alpha \in A_d$, $d > 0$, then $\tilde{M}|_{D+(\alpha)} \cong \widetilde{M_{[\alpha]}}$ (see §3.4.**c**).

Now, let N be a graded A-module and let $\varphi\colon M \to N$ be an A-homomorphism preserving degree. Then for any $\alpha \in A_d$, an $A_{[\alpha]}$-homomorphism $\varphi_{[\alpha]}\colon M_{[\alpha]} \to N_{[\alpha]}$ is induced by $\varphi_{[\alpha]}(x/\alpha^m) = \varphi(x)/\varphi(\alpha)^m$ for $x \in M_{md}$. Then $\{\widetilde{\varphi_{[\alpha]}}\colon \widetilde{M_{[\alpha]}} \to \widetilde{N_{[\alpha]}}\}_\alpha$ determines an $\mathcal{O}_X$-homomorphism $\tilde{\varphi}$: $\tilde{M} \to \tilde{N}$.

Definition. A graded A-module N is said to be *of TN*, if there exists m_0 such that $N_m = 0$ for all $m \geq m_0$.

Obviously, if N is a graded A-module of TN, then $\tilde{N} = 0$.

Proposition 7.1

(*i*) *If a sequence of graded A-modules*

$$0 \longrightarrow M' \xrightarrow{\varphi} M \xrightarrow{\psi} M'' \longrightarrow 0$$

is exact and φ, ψ *preserve degree, then the sequence*

$$0 \longrightarrow \tilde{M}' \xrightarrow{\tilde{\varphi}} \tilde{M} \xrightarrow{\tilde{\psi}} \tilde{M}'' \longrightarrow 0$$

is also exact.

(ii) *If N is a graded A-module finitely generated over A such that $\tilde{N} = 0$, then N is of TN.*

PROOF. Left to the reader as an exercise. □

b. If a graded A-module M is finitely generated, there exist a finite number of homogeneous generators $x_1 \in M_{l_1}, \ldots, x_r \in M_{l_r}$. Hence, if $L = A(-l_1) \oplus \cdots \oplus A(-l_r)$ and $\varphi: L \to M$ is defined by $\varphi(a_1, \ldots, a_r) = a_1 x_1 + \cdots + a_r x_r$, then φ preserves degree and is surjective.

Definition. A direct sum of the form $A(l_1) \oplus \cdots \oplus A(l_r)$ for some $l_1, \ldots, l_r \in \mathbb{Z}$ is said to be a *graded free A-module*.

If R is a Noetherian ring, then so is A by the Hilbert basis theorem. Thus, if M is a finitely generated graded A-module, then there exist graded free A-modules L', L, and A-homomorphisms φ, ψ preserving degree such that the following sequence is exact:

$$L' \xrightarrow{\varphi} L \xrightarrow{\psi} M \longrightarrow 0.$$

In this case, we say that M has a *finite presentation*.

c. Let A be a finitely generated graded R-algebra, and let $X = \operatorname{Proj} A$. If $A_1 = R\alpha_0 + \cdots + R\alpha_r$, then $\tilde{A}(m)|_{D_+(\alpha_i)} = \widetilde{A_{[\alpha_i]}} \cdot \alpha_i^m$ for all i; hence $\tilde{A}(m) \cong \tilde{A}(1)^{\otimes m}$. $\tilde{A}(m)$ is denoted by $\mathcal{O}_X(m)$.

Definition. If $\mathscr{F}$ is an $\mathcal{O}_X$-module, define $\mathscr{F}(m)$ to be $\mathscr{F} \otimes_{\mathcal{O}_X} \mathcal{O}_X(m)$.

It is easy to see that if M is a graded A-module, then $(M(m))^\sim \cong \tilde{M}(m)$.

d. Let M be a graded A-module and let $\alpha \in A_d$, $\beta \in A_e$. Then one has the canonical homomorphisms $\psi_\alpha: M_0 \to M_{[\alpha]}$ and $\psi_\beta: M_0 \to M_{[\beta]}$ defined by $\psi_\alpha(x) = x/1$ and $\psi_\beta(x) = x/1$. Then the following diagram is clearly commutative:

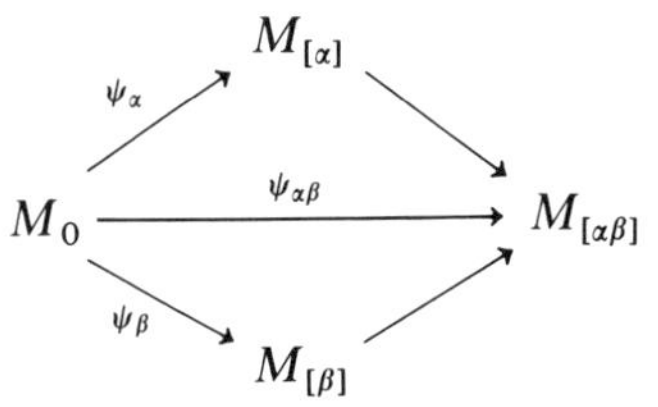

Since $M_{[\alpha]} \cong \Gamma(D_+(\alpha), \tilde{M})$, $\{\psi_\alpha\}$ determines an R-homomorphism $M_0 \to \Gamma(X, \tilde{M})$, denoted by $\boldsymbol{\alpha}_0$, i.e., $\boldsymbol{\alpha}_0(x)|_{D_+(\alpha)} = \psi_\alpha(x) = x/1$. Replacing M by $M(n)$, one obtains $\boldsymbol{\alpha}_n: M_n = M(n)_0 \to \Gamma(X, \tilde{M}(n))$, and defines $\boldsymbol{\alpha}_M$ to be $\bigoplus_{n \in \mathbb{Z}} \boldsymbol{\alpha}_n: M \to \bigoplus_{n \in \mathbb{Z}} \Gamma(X, \tilde{M}(n))$.

§7.2 The Homomorphism $\beta_{\mathscr{F}}$

a. Here, in general, let X denote a quasi-compact scheme over R and let $\mathscr{L}$ be an invertible sheaf on X.

Given two $\mathscr{O}_X$-modules $\mathscr{F}$ and $\mathscr{G}$, one has the natural homomorphism $\Gamma(X, \mathscr{F}) \otimes_R \Gamma(X, \mathscr{G}) \to \Gamma(X, \mathscr{F} \otimes_{\mathscr{O}_X} \mathscr{G})$. The image of $\varphi \otimes \psi$, under this homomorphism, is denoted by $\varphi \cdot \psi$. With this multiplication, $\Gamma_*(X, \mathscr{O}_X)$, which is defined to be $\bigoplus_{m=-\infty}^{\infty} \Gamma(X, \mathscr{L}^{\otimes m})$, becomes a graded ring with positive and negative gradation. Letting $\mathscr{F}(m)$ be $\mathscr{F} \otimes \mathscr{L}^{\otimes m}$, $\Gamma_*(X, \mathscr{F})$, which is defined to be $\bigoplus_{m=-\infty}^{\infty} \Gamma(X, \mathscr{F}(m))$, has a structure of a graded $\Gamma_*(X, \mathscr{O}_X)$-module.

Recall that in §2.18.**c** for any $\alpha \in \Gamma(X, \mathscr{L})$, an open subset X_α is defined which consists of the points x such that $\bar{\alpha}(x) \neq 0$.

b. Lemma 7.1. *The homomorphism* $r_m\colon \Gamma(X_\alpha, \mathscr{F}) \to \Gamma(X_\alpha, \mathscr{F}(m))$ *defined by* $r_m(\varphi) = (\alpha|_{X_\alpha})^m \varphi$ *is an isomorphism for any* $m \geq 0$.

Proof. We choose an affine open cover $\{U_\lambda | \lambda \in \Lambda\}$ such that $\mathscr{L}|_{U_\lambda} = \mathscr{O}_{U_\lambda} \cdot \beta_\lambda$ and $\alpha|_{U_\lambda} = a_\lambda \cdot \beta_\lambda$ for all $\lambda \in \Lambda$. Let $U_{\lambda\alpha} = U_\lambda \cap X_\alpha$.

(1) r_m is one-to-one. If $r_m(\varphi) = 0$, then $r_m(\varphi)|_{U_{\lambda\alpha}} = 0$ and so $(a_\lambda/1)^m \cdot \varphi = 0$ in $\Gamma(D(a_\lambda), \mathscr{F})$, i.e., $\varphi|_{U_{\lambda\alpha}} = 0$ for all λ. Hence $\varphi = 0$.

(2) r_m is surjective. Given $\psi \in \Gamma(X_\alpha, \mathscr{F}(m))$, one has $\psi|_{U_{\lambda\alpha}} = \psi_\lambda \cdot \beta_{\lambda\alpha}^m$ where $\psi_\lambda \in \Gamma(U_{\lambda\alpha}, \mathscr{F})$ and $\beta_{\lambda\alpha} = \beta_\lambda|_{U_{\lambda\alpha}}$. Letting $\sigma_\lambda = \psi_\lambda/a_\lambda^m$, one has $\psi_\lambda \cdot \beta_{\lambda\alpha}^m = \sigma_\lambda \cdot (a_\lambda \beta_{\lambda\alpha})^m = \sigma_\lambda \cdot \alpha^m|_{U_{\lambda\alpha}}$. Since $(\psi|_{U_{\lambda\alpha}})|_{U_{\lambda\mu\alpha}} = (\psi|_{U_{\mu\alpha}})|_{U_{\lambda\mu\alpha}}$, where $U_{\lambda\mu\alpha}$ denotes $U_\lambda \cap U_\mu \cap X_\alpha$, it follows that $\sigma_\lambda|_{U_{\lambda\mu\alpha}} = \sigma_\mu|_{U_{\lambda\mu\alpha}}$ for all $\lambda, \mu \in \Lambda$; hence there exists $\sigma \in \Gamma(X_\alpha, \mathscr{F})$ with $\sigma|_{U_{\lambda\alpha}} = \sigma_\lambda$ for all λ. Then $r_m(\sigma) = (\alpha|_{X_\alpha})^m \sigma = \psi$. □

Now, noting $r_{m+l}(\sigma) = (\alpha|_{X_\alpha})^l \cdot r_m(\sigma)$ for any l, we define $\tau_\alpha\colon \Gamma_*(X, \mathscr{F})_{[\alpha]} \to \Gamma(X_\alpha, \mathscr{F})$ by $\tau_\alpha(\varphi/\alpha^m) = r_m^{-1}(\varphi|_{X_\alpha})$, where $\varphi \in \Gamma(X, \mathscr{F}(m))$.

Lemma 7.2. *Suppose that* $\mathscr{F}$ *is quasi-coherent.*

(i) *If* $\varphi \in \Gamma(X, \mathscr{F})$ *satisfies* $\varphi|_{X_\alpha} = 0$, *then there exists* $m \geq 0$ *with* $\alpha^m \varphi = 0$.
(ii) *If* X *is separated, then for any* $\psi \in \Gamma(X_\alpha, \mathscr{F})$, *there exist* m *and* $\varphi \in \Gamma(X, \mathscr{F}(m))$ *such that* $\psi \cdot (\alpha|_{X_\alpha})^m = \varphi|_{X_\alpha}$.

Proof. Since X is quasi-compact, one can choose a finite affine open cover $\{U_\lambda | \lambda \in \Lambda\}$ which has the property used in the proof of Lemma 7.1.

(i) Since $\varphi|_{U_{\lambda\alpha}} = (\varphi|_{U_\lambda})|_{D(a_\lambda)} = 0$, there exists m_λ such that $\alpha^{m_\lambda}\varphi|_{U_\lambda} = 0$. Since Λ is a finite set, take $m = \max\{m_\lambda | \lambda \in \Lambda\}$. Then $\alpha^m \varphi = 0$.

(ii) This is left to the reader. □

Corollary. *If X is separated, then $\tau_\alpha\colon \Gamma_*(X, \mathscr{F})_{[\alpha]} \to \Gamma(X_\alpha, \mathscr{F})$ is an isomorphism.*

For $\mathscr{L} = \mathscr{O}_X$, the corollary asserts that $\Gamma(X, \mathscr{F})_\alpha \cong \Gamma(X_\alpha, \mathscr{F})$, if $\mathscr{F}$ is quasi-coherent. In particular $A(X)_\alpha \cong A(X_\alpha)$ (cf. Corollary to Theorem 1.15).

c. We shall apply the above result to the case where $X = \operatorname{Proj} A$ and $\mathscr{L} = \widetilde{A(1)}$.

Lemma 7.3. *For any $\gamma \in A_1$, $X_{\boldsymbol{\alpha}(\gamma)} = D_+(\gamma)$.*

Proof. On each $D_+(\alpha_i)$, one has

$$X_{\boldsymbol{\alpha}(\gamma)} \cap D_+(\alpha_i) = D_+(\alpha_i)_{\gamma/1} = D(\gamma/\alpha_i) \subseteq \operatorname{Spec} A_{[\alpha_i]}$$

and

$$D_+(\gamma) \cap D_+(\alpha_i) = D(\gamma/\alpha_i) \subseteq \operatorname{Spec} A_{[\alpha_i]}.$$

Hence, $X_{\boldsymbol{\alpha}(\gamma)} = D_+(\gamma)$. □

For any quasi-coherent $\mathscr{O}_X$-module $\mathscr{F}$, the $\tau_{\boldsymbol{\alpha}(\alpha_i)}\colon \Gamma_*(X, \mathscr{F})_{[\alpha_i]} \xrightarrow{\sim} \Gamma(X_{\alpha(\alpha_i)}, \mathscr{F})$ satisfy $\tilde{\tau}_{\boldsymbol{\alpha}(\alpha_i)}|_{D_+(\alpha_i\alpha_j)} = \tilde{\tau}_{\boldsymbol{\alpha}(\alpha_j)}|_{D_+(\alpha_i\alpha_j)}$ for all i, j. Hence, these $\{\tilde{\tau}_{\boldsymbol{\alpha}(\alpha_i)}\}$ define an $\mathscr{O}_X$-isomorphism: $\Gamma_*(X, \mathscr{F})^\sim \xrightarrow{\sim} \mathscr{F}$, denoted by $\boldsymbol{\beta}_{\mathscr{F}}$. Thus, one obtains the next result.

Proposition 7.2. *For any quasi-coherent $\mathscr{O}_X$-module $\mathscr{F}$, $\boldsymbol{\beta}_{\mathscr{F}}\colon \Gamma_*(X, \mathscr{F})^\sim \to \mathscr{F}$ is an isomorphism. Hence, $\mathscr{F} \cong \tilde{M}$ for some graded A-module M.*

Now, suppose that R is Noetherian. If $\mathscr{F}$ is a quasi-coherent $\mathscr{O}_X$-module of finite type, then $M = \Gamma_*(X, \mathscr{F})$ satisfies $\tilde{M} \cong \mathscr{F}$, each $\tilde{M}|_{D_+(\alpha_i)}$ is of finite type, and hence $\Gamma(D_+(\alpha_i), \tilde{M}) = M_{[\alpha_i]}$ is finitely generated as an $A_{[\alpha_i]}$ module by Theorem 1.13. Thus, there exist m, and $x_{ij} \in M_m$ such that $M_{[\alpha_i]}$ is generated by $x_{i,j}/\alpha_i^m$ for $j = 1, \ldots, t$, where t and m are chosen independently of i. Define M' to be $\sum_{i,j} Ax_{i,j}$, which is a graded A-submodule. Then $\tilde{M}'|_{D_+(\alpha_i)} = \tilde{M}|_{D_+(\alpha_i)}$ for all i. Hence $\tilde{M}' \cong \tilde{M} \cong \mathscr{F}$. Thus, we obtain the next result.

Proposition 7.3. *If R is Noetherian and $\mathscr{F}$ is a quasi-coherent $\mathscr{O}_X$-module of finite type, then there exists a finitely generated graded A-module M' with $\tilde{M}' \cong \mathscr{F}$. Hence, there exist free graded A-modules L' and L, and A-homomorphisms φ, ψ preserving degree such that the sequence*

$$\tilde{L}' \xrightarrow{\tilde{\varphi}} \tilde{L} \xrightarrow{\tilde{\psi}} \tilde{M}'(\cong \mathscr{F}) \longrightarrow 0$$

is exact.

§7.3 Cohomology Groups of Coherent Sheaves on $\mathbf{P}_R^n$

a. In what follows in this chapter, we let R denote a Noetherian ring. Let A be a finitely generated graded R-algebra with $A_1 = \sum_{i=0}^{r} R\alpha_i$. Then $X = \operatorname{Proj} A$ is a closed subscheme of $\mathbf{P}_R^r$ (cf. §3.3.**a**).

Lemma 7.4. *Let $j\colon X \to Y$ be a finite morphism between Noetherian schemes. If $\mathscr{F}$ is a coherent $\mathscr{O}_X$-module, then $j_*\mathscr{F}$ is also coherent.*

Proof. One can assume X and Y are affine schemes. Then the assertion is obvious, by Theorem 1.13. □

Now, let $j\colon X \to \mathbf{P} = \mathbf{P}_R^r$ be a closed immersion and let $\mathscr{F}$ be a coherent $\mathscr{O}_X$-module. Then $\mathscr{O}_X(m) = j^*(\mathscr{O}_{\mathbf{P}}(m))$ by definition, and letting $\mathscr{F}(m)$ be $\mathscr{F} \otimes_{\mathscr{O}_X} \mathscr{O}_X(m)$ and $(j_*\mathscr{F})(m)$ be $(j_*\mathscr{F}) \otimes_{\mathscr{O}_{\mathbf{P}}} \mathscr{O}_{\mathbf{P}}(m)$, one has by Lemma 2.29

$$j_*(\mathscr{F}(m)) \cong (j_*\mathscr{F})(m).$$

Further, since j is affine, one obtains the next result by the Corollaries to Theorems 4.8 and 4.9.

Lemma 7.5. *With the notation above,*

$$H^q(X, \mathscr{F}(m)) \cong H^q(\mathbf{P}_R^r, (j_*\mathscr{F})(m)) \quad \textit{for any} \quad q \geq 0.$$

Thus, the computation of cohomology groups of coherent sheaves on X is reduced to that on $\mathbf{P}_R^r$. However, our definition of $H^q(\mathbf{P}_R^r\, \mathscr{O}_{\mathbf{P}}(m))$ is too abstract to allow us to compute them quickly. In order to simplify the discussion, we assume the next basic fact.

Theorem 7.1. *If $q > 0$, then $H^q(\mathbf{P}_R^r, \mathscr{O}_{\mathbf{P}}(m)) = 0$ for $m \gg 0$.*

Proof. We refer the reader to [H, pp. 225–228]. □

b. Proposition 7.4. *If S is the polynomial ring $R[\mathrm{T}_0, \ldots, \mathrm{T}_r]$ with $r > 0$, then $\boldsymbol{\alpha}_S\colon S \to \Gamma_*(\mathbf{P}_R^r, \mathscr{O}_{\mathbf{P}})$ is an isomorphism.*

Proof. Since $\mathscr{O}_{\mathbf{P}}(m)|_{D_+(\mathrm{T}_i)} = \mathscr{O}_{\mathbf{P}}|_{D_+(\mathrm{T}_i)} \cdot \mathrm{T}_i^m$, $\varphi \in \Gamma(\mathbf{P}_R^r, \mathscr{O}_{\mathbf{P}}(m))$ corresponds to $\{\varphi_i \in R[\mathrm{T}_0/\mathrm{T}_i, \ldots, \mathrm{T}_r/\mathrm{T}_i] \mid \varphi_i \mathrm{T}_i^m = \varphi_0 \mathrm{T}_0^m \text{ for all } i\}$. Thus $\varphi_0 \mathrm{T}_0^m$ is a homogeneous polynomial F of degree m and φ_i is obtained as F/T_i^m. By definition, $\boldsymbol{\alpha}_m(F) = \{\varphi_i \cdot \mathrm{T}_i^m\}$, i.e., $\boldsymbol{\alpha}_m(F) = \varphi$. Hence, $\boldsymbol{\alpha}_m\colon S_m \to \Gamma(\mathbf{P}_R^r, \mathscr{O}_{\mathbf{P}}(m))$ is an isomorphism for every m. □

Theorem 7.2

(*i*) $H^q(\mathbf{P}_R^r, \mathscr{O}_{\mathbf{P}}(m)) = 0$, *if* ($\alpha$) $0 < q < r$, or (β) $r > 0$, $q = 0$ *and* $m < 0$, *or* (γ) $q = r > 0$ *and* $m > -(r+1)$, *or* (δ) $q > r$.

(ii) $H^q(\mathbf{P}_R^r\ \mathcal{O}_{\mathbf{P}}(m))$ *is finitely generated over* R *for all* q *and* m.
(iii) *If* R *is a field* k, *then for any* m

$$\dim_k H^r(\mathbf{P}_k^r, \mathcal{O}_{\mathbf{P}}(-m)) = \dim_k \Gamma(\mathbf{P}_k^r, \mathcal{O}_{\mathbf{P}}(m-r-1)).$$

PROOF. We prove this by induction on r. Since the case of $r = 0$ is trivial, we assume $r > 0$. Let $H = V_+(\mathsf{T}_0) \cong \mathbf{P}_R^{r-1}$. Then $\mathcal{O}_{\mathbf{P}}(-H) \cong \mathcal{O}_{\mathbf{P}}(-1)$ and we obtain the following exact sequence (cf. §4.7.**b**):

$$0 \longrightarrow \mathcal{O}_{\mathbf{P}}(m-1) \longrightarrow \mathcal{O}_{\mathbf{P}}(m) \longrightarrow \mathcal{O}_H(m) \longrightarrow 0$$

for all m. By applying Theorem 4.1.(III) to the above sequence, we obtain the following long exact sequence:

$$\longrightarrow H^{q-1}(H, \mathcal{O}_H(m)) \xrightarrow{\mathfrak{d}^{q-1}} H^q(\mathbf{P}_R^r, \mathcal{O}_{\mathbf{P}}(m-1)) \longrightarrow H^q(\mathbf{P}_R^r, \mathcal{O}_{\mathbf{P}}(m))$$
$$\longrightarrow H^q(H, \mathcal{O}_H(m)) \longrightarrow \cdots.$$

Since $H \cong \mathbf{P}_R^{r-1}$, we assume the assertion holds for H. Thus, if $r \geq 4$ and $2 \leq q \leq r-2$, then $H^{q-1}(H, \mathcal{O}_H(m)) = H^q(H, \mathcal{O}_H(m)) = 0$; hence $H^q(\mathbf{P}_R^r, \mathcal{O}_{\mathbf{P}}(m-1)) \cong H^q(\mathbf{P}_R^r, \mathcal{O}_{\mathbf{P}}(m))$ for any m. Thus choosing $l \gg 0$, one has $H^q(\mathbf{P}_R^r, \mathcal{O}_{\mathbf{P}}(m+l)) = 0$ by Theorem 7.1; hence $H^q(\mathbf{P}_R^r, \mathcal{O}_{\mathbf{P}}(m)) = 0$ for any m. If $q = 1 < r$, then from the long exact sequence

$$0 \longrightarrow \Gamma(\mathbf{P}_R^r, \mathcal{O}_{\mathbf{P}}(m-1)) \longrightarrow \Gamma(\mathbf{P}_R^r, \mathcal{O}_{\mathbf{P}}(m)) \longrightarrow \Gamma(H, \mathcal{O}_H(m))$$
$$\xrightarrow{\mathfrak{d}^0} H^1(\mathbf{P}_R^r, \mathcal{O}_{\mathbf{P}}(m-1)) \longrightarrow H^1(\mathbf{P}_R^r, \mathcal{O}_{\mathbf{P}}(m)) \longrightarrow H^1(H, \mathcal{O}_H(m)),$$

one has $H^1(H, \mathcal{O}_H(m)) \cong H^1(\mathbf{P}_R^{r-1}, \mathcal{O}_{\mathbf{P}}(m-1)) = 0$ by the induction hypothesis, and $S_m \cong \Gamma(\mathbf{P}_R^r, \mathcal{O}_{\mathbf{P}}(m))$, one has $(S/\mathsf{T}_0 S)_m \cong \Gamma(\mathbf{P}_R^{r-1}, \mathcal{O}_{\mathbf{P}}(m)) \cong \Gamma(H, \mathcal{O}_H(m))$ by Proposition 7.4. Hence, $\mathfrak{d}^0$ is the zero map and we obtain an isomorphism

$$H^1(\mathbf{P}_R^r, \mathcal{O}_{\mathbf{P}}(m-1)) \cong H^1(\mathbf{P}_R^r, \mathcal{O}_{\mathbf{P}}(m)) \quad \text{for any } m.$$

Thus, again from Theorem 7.1, the vanishing of $H^1(\mathbf{P}_R^r, \mathcal{O}_{\mathbf{P}}(m))$ for any m follows. If $q = r-1 \geq 2$, we observe the long exact sequence:

$$\longrightarrow H^{r-2}(H, \mathcal{O}_H(m)) \xrightarrow{\mathfrak{d}^{r-2}} H^{r-1}(\mathbf{P}_R^r, \mathcal{O}_{\mathbf{P}}(m-1)) \longrightarrow H^{r-1}(\mathbf{P}_R^r, \mathcal{O}_{\mathbf{P}}(m))$$
$$\longrightarrow H^{r-1}(H, \mathcal{O}_H(m)) \longrightarrow \cdots$$

Since $r-2 \geq 1$, it follows that $H^{r-2}(H, \mathcal{O}_H(m)) = 0$ for any m by the induction hypothesis. Thus each homomorphism $H^{r-1}(\mathbf{P}_R^r, \mathcal{O}_{\mathbf{P}}(m-1)) \to H^{r-1}(\mathbf{P}_R^r, \mathcal{O}_{\mathbf{P}}(m))$ is an isomorphism. By Theorem 7.1, $H^{r-1}(\mathbf{P}_R^r, \mathcal{O}_{\mathbf{P}}(m+l)) = 0$ for $l \gg 0$; hence $H^{r-1}(\mathbf{P}_R^r, \mathcal{O}_{\mathbf{P}}(m)) = 0$ for any m. Thus, the vanishing of the cohomology groups for the case (α) has been established. For the case (β), the assertion follows from Proposition 7.4.

Let $q = r \geq 2$ and assume $m \geq -(r+1)$. We have again the long exact sequence

$$\cdots \longrightarrow H^{r-1}(H, \mathcal{O}_H(m)) \longrightarrow H^r(\mathbf{P}^r_R, \mathcal{O}_{\mathbf{P}}(m-1)) \longrightarrow H^r(\mathbf{P}^r_R, \mathcal{O}_{\mathbf{P}}(m))$$
$$\longrightarrow H^r(H, \mathcal{O}_H(m)).$$

By the induction hypothesis, $H^{r-1}(H, \mathcal{O}_H(m)) = H^r(H, \mathcal{O}_H(m)) = 0$. Hence, $H^r(\mathbf{P}^r_R, \mathcal{O}_{\mathbf{P}}(m-1)) \cong H^r(\mathbf{P}^r_R, \mathcal{O}_{\mathbf{P}}(m))$ for any $m \geq -(r+1)$. Thus, using the same techniques as before, we have

$$H^r(\mathbf{P}^r_R, \mathcal{O}_{\mathbf{P}}(m-1)) = 0 \quad \text{if} \quad m - 1 \geq -(r+2),$$

and this completes the proof for the case (γ). The proof for the case (δ) is easy and is left to the reader.

(ii) This follows from the next lemma applied to the long exact sequence.

Lemma 7.6. *Let R be a Noetherian ring and let*

$$M' \xrightarrow{\ \varphi\ } M \xrightarrow{\ \psi\ } M''$$

be an exact sequence of R-modules. If both M' and M'' are finitely generated, then so is M.

PROOF. Since $\operatorname{Im}\psi$ is a submodule of the finitely generated module M'', it is also finitely generated. Furthermore, $\operatorname{Im}\varphi$ is finitely generated, since it is a quotient module of M'. From the isomorphism $\operatorname{Im}\psi \cong M/\operatorname{Im}\varphi$, one sees that M is finitely generated. □

We proceed with the proof of (iii). Let $\psi_r(m) = \dim_k H^r(\mathbf{P}^r_k, \mathcal{O}_{\mathbf{P}}(-m))$. If $r > 0$ and $m \leq r$, then $\psi_r(m) = 0$ by case (γ). By the long exact sequence, one has $\psi_r(m+1) - \psi_r(m) = \psi_{r-1}(m)$, which is $\dim_k \Gamma(\mathbf{P}^{r-1}_k, \mathcal{O}_{\mathbf{P}}(m-r)) = \binom{m-1}{m-r}$ if $m \geq r$. Hence, $\psi_r(m) = \binom{m-1}{m-r-1} + \text{const}$. But since $\psi_r(r+1) - \psi_r(r) = \psi_{r-1}(r) = 1$, it follows that $\psi_r(r+1) = 1$, i.e., $\psi_r(m) = \binom{m-1}{m-r-1}$. On the other hand, $\dim_k \Gamma(\mathbf{P}^r_k, \mathcal{O}_{\mathbf{P}}(m-r-1)) = \binom{m-1}{m-r-1}$; hence one obtains the assertion. □

c. Theorem 7.3. *Let F be an irreducible homogeneous polynomial in $k[\mathsf{T}_0, \ldots, \mathsf{T}_{n+1}]$ with $\delta = \deg F \geq 1$. $V = V_+(F)$ has the following cohomology groups:*

(i) *$H^q(V, \mathcal{O}_V(m)) = 0$ for any $0 < q < n$ and any m.*
(ii) *$\dim_k H^n(V, \mathcal{O}_V(m)) = \dim_k \Gamma(V, \mathcal{O}_V(-n-2+\delta-m))$.*
(iii) *$\boldsymbol{\alpha}_V : A = S/(F) \xrightarrow{\sim} \Gamma_*(V, \mathcal{O}_V)$ is an isomorphism.*

PROOF. Since one has the exact sequence (cf. §4.7.**b**)

$$0 \to \mathcal{O}_{\mathbf{P}}((m-\delta)) \to \mathcal{O}_{\mathbf{P}}(m) \to \mathcal{O}_V(m) \to 0, \text{ where } \mathbf{P} = \mathbf{P}^{n+1}_k,$$

one obtains the long exact sequence involving $H^q(\mathbf{P}_k^{n+1}, \mathcal{O}_{\mathbf{P}}(m))$ and $H^q(V, \mathcal{O}_V(m))$. By Theorem 7.2.(i) and (iii), one can obtain the assertion. □

Corollary

$$\dim_k H^n(V, \mathcal{O}_V) = \binom{\delta-1}{\delta-1-n-1} = \frac{(\delta-1)(\delta-2)\cdots(\delta-n-1)}{(n+1)!}.$$

d. Proposition 7.5. *Let $X = \mathbf{P}_R^r$ and let $\mathcal{F}$ be a coherent $\mathcal{O}_X$-module. Then*

(i) *$H^q(X, \mathcal{F}) = 0$ for $q > r$.*
(ii) *$H^q(X, \mathcal{F})$ is finitely generated over R for any q.*

PROOF. First assume R is a local ring. Then $\dim X = r + \dim R$, since $\dim R[\mathsf{X}_1, \ldots, \mathsf{X}_r] = r + \dim R$ by [N2, Theorem (9.10)]. By Proposition 7.3, there exist a free graded A-module $L = A(l_1) \oplus \cdots \oplus A(l_s)$ and an A-homomorphism $\psi\colon L \to \Gamma_*(X, \mathcal{F})$ preserving degree such that $\tilde{\psi}$ is surjective. Hence letting $\mathcal{R} = \operatorname{Ker} \tilde{\psi}$, one has the exact sequence

$$0 \to \mathcal{R} \to \tilde{L} \to \mathcal{F} \to 0.$$

If $q > r$, then $0 = H^q(X, \tilde{L}) = \oplus_{i=1}^s H^q(X, \mathcal{O}_X(l_i))$ by Theorem 7.2.(i), and so $H^q(X, \mathcal{F}) \cong H^{q+1}(X, \mathcal{R})$. Thus, we shall show (i) by induction on q. If $q \gg 0$, $H^q(X, \mathcal{F}) = 0$ follows from Theorem 4.6. Hence assuming the case $q + 1$ applied to $\mathcal{R}$, one has $H^q(X, \mathcal{F}) \cong H^{q+1}(X, \mathcal{R}) = 0$. For general R, we use the next simple result.

Lemma 7.7. *Let S be a multiplicative subset of R and let X be a separated scheme algebraic over R. Then for any quasi-coherent $\mathcal{O}_X$-module $\mathcal{F}$ and any $q \geq 0$, $H^q(X \otimes_S S^{-1}R, \mathcal{F} \otimes_R S^{-1}R) \cong H^q(X, \mathcal{F}) \otimes_R S^{-1}R$, where $\mathcal{F} \otimes_R S^{-1}R$ is $p^*\mathcal{F}$ and p is the projection: $X \otimes_R S^{-1}R \to X$.*

PROOF. We refer the reader to [H, III Proposition 9.3.] Note that $S^{-1}R$ is R-flat, by Lemma 1.2. □

Returning to the proof of (i), we let $\mathfrak{m}$ be a maximal ideal of R. Then $\dim R_{\mathfrak{m}} < \infty$ and $H^q(\mathbf{P}_R^r, \mathcal{F}) \otimes_R R_{\mathfrak{m}} \cong H^q(\mathbf{P}_{R_{\mathfrak{m}}}^r, \mathcal{F} \otimes_R R_{\mathfrak{m}}) = 0$, if $q > r$. Thus, by Example 1.7 one has $H^q(\mathbf{P}_R^r, \mathcal{F}) = 0$ if $q > r$.

PROOF OF (ii) We induce on $r - q + 1$. If $q = r + 1$, the assertion is clear by (i). From the long exact sequence

$$\cdots \longrightarrow H^q(X, \tilde{L}) \longrightarrow H^q(X, \mathcal{F}) \longrightarrow H^{q+1}(X, \mathcal{R}),$$

one concludes that $H^q(X, \mathcal{F})$ is finitely generated by Lemma 7.6, since $H^q(X, \tilde{L}) = \oplus_{i=1}^s H^q(X, \mathcal{O}_X(l_i))$ is finitely generated by Theorem 7.2.(ii), and since $H^{q+1}(X, \mathcal{R})$ is finitely generated by the induction hypothesis. □

Theorem 7.4. *Let* $X = \operatorname{Proj} A \subseteq \mathbf{P}_R^r$ *with the notation in* §7.3.**a** *and let* $\mathscr{F}$ *be a coherent* $\mathscr{O}_X$*-module. Then* $H^q(X, \mathscr{F}) = 0$ *for* $q > r$ *and* $H^q(X, \mathscr{F})$ *is finitely generated for any* q.

PROOF. This follows from Lemma 7.4 and Proposition 7.5 applied to $j_*\mathscr{F}$. □

e. Further, by the same technique, one can prove the so-called Theorem B for $X = \operatorname{Proj} A$.

Theorem 7.5. *Let* $X = \operatorname{Proj} A$ *and let* $\mathscr{F}$ *be a coherent* $\mathscr{O}_X$*-module as before. Then* $H^q(X, \mathscr{F}(m)) = 0$ *for any* $q > 0$ *and* $m \gg 0$.

PROOF. By Lemma 7.5, one can assume $X = \mathbf{P}_R^r$. If $\mathscr{F} = \mathscr{O}_X(l_1) \oplus \cdots \oplus \mathscr{O}_X(l_s)$, then $H^q(\mathbf{P}_R^r, \mathscr{F}(m)) = \bigoplus_{i=1}^s H^q(\mathbf{P}_R^r, \mathscr{O}_X(l_i + m)) = 0$ for $q > 0$ and m with $l_i + m \geq -r - 1$, i.e., $m \geq \max\{-l_i - r - 1\}$. Now, we use induction on $r - q + 1$, since $H^{r+1}(\mathbf{P}_R^r, \mathscr{F}(m)) = 0$. By making use of the long exact sequence in the proof of Proposition 7.5, one has the exact sequence

$$H^q(\mathbf{P}_R^r, \tilde{L}(m)) \longrightarrow H^q(\mathbf{P}_R^r, \mathscr{F}(m)) \longrightarrow H^{q+1}(\mathbf{P}_R^r, \mathscr{R}(m)).$$

From this, the result follows immediately. □

Corollary. *In addition, if the sequence of coherent* $\mathscr{O}_X$*-modules*

$$\mathscr{F}' \xrightarrow{\ \theta\ } \mathscr{F} \xrightarrow{\ \eta\ } \mathscr{F}'' \longrightarrow 0$$

is exact, then

$$\Gamma(X, \mathscr{F}'(m)) \to \Gamma(X, \mathscr{F}(m)) \to \Gamma(X, \mathscr{F}''(m)) \to 0$$

is exact for $m \gg 0$.

PROOF. It suffices to note that $H^1(X, (\operatorname{Ker} \eta)(m)) = 0$ for $m \gg 0$. □

f. Let A be a finitely generated graded R-algebra and let M be a finitely generated graded A-module.

Proposition 7.6. $\alpha_{M,m}\colon M_m \xrightarrow{\sim} \Gamma(X, \tilde{M}(m))$ *for* $m \gg 0$.

PROOF. By Proposition 7.3, one has a finite presentation of M by graded free A-modules L' and L:

$$L' \to L \to M \to 0.$$

Then, by the last corollary, the following sequence is exact for $m \gg 0$:

$$\Gamma(X, \tilde{L}'(m)) \to \Gamma(X, \tilde{L}(m)) \to \Gamma(X, \tilde{M}(m)) \to 0,$$

and one has the commutative diagram:

$$\begin{array}{ccccccc} L'_m & \longrightarrow & L_m & \longrightarrow & M_m & \longrightarrow & 0 \\ \downarrow{\scriptstyle \alpha_{L',m}} & & \downarrow{\scriptstyle \alpha_{L,m}} & & \downarrow{\scriptstyle \alpha_{M,m}} & & \\ \Gamma(X, \tilde{L}'(m)) & \longrightarrow & \Gamma(X, \tilde{L}(m)) & \longrightarrow & \Gamma(X, \tilde{M}(m)) & \longrightarrow & 0 \end{array}$$

Thus, it suffices to show that $\boldsymbol{\alpha}_{L,m}$ and $\boldsymbol{\alpha}_{L',m}$ are isomorphisms for $m \gg 0$. In other words, one may assume M is free. Thus, it suffices to prove this result for $M = A(l)$. This implies that one can assume M to be A.

Now, let S be $R[\mathsf{T}_0, \ldots, \mathsf{T}_r]$ and let $\psi\colon S \to A$ be defined by $\psi(\mathsf{T}_i) = \alpha_i$. Then A is a graded S-module and $j = \operatorname{Proj} \psi\colon X \to \mathbf{P}_R^r$ is a closed immersion. By Lemma 7.5, $\Gamma(X, \mathcal{O}_X(m)) = \Gamma(\mathbf{P}_R^r, (j_*\mathcal{O}_X)(m))$. Thus applying the first part of the proof to A, one can reduce the proof for A to that of S because $j_*(\mathcal{O}_X) = \tilde{A}$ as a graded S-module. Then, $\boldsymbol{\alpha}_S$ is an isomorphism by Proposition 7.4. □

Corollary. *If R is a field k, then*

$$\chi_X(\tilde{M}(m)) = \dim_k \Gamma(X, \tilde{M}(m)) = \dim_k M_m \quad \textit{for} \quad m \gg 0.$$

PROOF. This follows from Theorem 7.5 and the last proposition. □

Remark. Let B be a Noetherian local ring with maximal ideal $\mathfrak{n}$. Then letting $k = B/\mathfrak{n}$, the graded ring $\operatorname{gr}_{\mathfrak{n}}(B) = \bigoplus_{j=0}^{\infty} \mathfrak{n}^j/\mathfrak{n}^{j+1}$ is finitely generated as a k-algebra. Hence, $X = \operatorname{Proj}(\operatorname{gr}_{\mathfrak{n}}(B))$ is a projective scheme and

$$\chi_X(\mathcal{O}_X(m)) = \dim_k(\mathfrak{n}^m/\mathfrak{n}^{m+1}) \quad \text{for } m \gg 0.$$

§7.4 Ample Sheaves

a. Let X be a Noetherian R-scheme and let $\mathscr{L}$ be an invertible sheaf. Given a finite subset $\Sigma = \{f_1, \ldots, f_l\}$ of $\Gamma(X, \mathscr{L})$, we define X_Σ to be $\bigcup_{j=1}^{l} X_{f_j}$, and $\Phi_j\colon X_{f_j} \to D_+(\mathsf{T}_j) \subset \mathbf{P}_R^{l-1} = \operatorname{Proj} R[\mathsf{T}_1, \ldots, \mathsf{T}_l]$ by $\Phi_j = (f_1/f_j, \ldots, f_l/f_j)_R$ (cf. §2.18.**c**, Remark). As morphisms into $\mathbf{P}_R^{l-1}$, $\Phi_j|_{U_{i,j}} = \Phi_i|_{U_{j,i}}$, where $U_{i,j} = X_{f_i} \cap X_{f_j}$ for all i, j. Hence, $\{\Phi_i\}$ determines an R-morphism: $X_\Sigma \to \mathbf{P}_R^{l-1}$, which is denoted by Φ_Σ or $(f_1 : \cdots : f_l)$.

Definition. An invertible sheaf $\mathscr{L}$ on X is said to be *R-very ample*, if there exists a finitely generated graded R-algebra A such that $X = \operatorname{Proj} A$ and $\mathscr{L} \cong A(1)^\sim$.

In this case, one has $f_i = \boldsymbol{\alpha}_1(\alpha_i) \in \Gamma(X, \mathscr{L})$ with the identification $\mathscr{L} =$

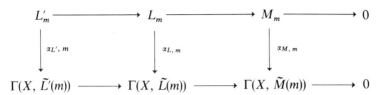

$A(1)^{\sim}$, for $i = 0, \ldots, r$. If Σ denotes $\{f_0, \ldots, f_r\}$, then $X_\Sigma = X$ and Φ_Σ is a closed immersion $X \to \mathbf{P}_R^r$.

Proposition 7.7. *If a finite subset $\Xi = \{\varphi_1, \ldots, \varphi_l\}$ of $\Gamma(X, \mathscr{L})$ generates $\mathscr{L}$ as an $\mathscr{O}_X$-module and Φ_Ξ is a closed immersion into $\mathbf{P} = \mathbf{P}_R^{l-1}$, then $\mathscr{L}$ is R-very ample. Furthermore, any finite subset Σ between Ξ and $\Gamma(X, \mathscr{L})$ satisfies $X_\Sigma = X$ and Φ_Σ is a closed immersion.*

PROOF. Define A_Ξ to be the graded R-subalgebra of $\Gamma_*(X, \mathscr{O}_X) = \sum_{m \in \mathbb{Z}} \Gamma(X, \mathscr{L}^{\otimes m})$ generated by R and Ξ. Then $\Phi_\Xi^*(\mathscr{O}_\mathbf{P}(1)) \cong \mathscr{L}$, i.e., $\mathscr{L} \cong (A_\Xi(1))^{\sim}$. The proof of the rest is left to the reader. □

EXAMPLE 7.1. Let $X = \operatorname{Spec}(R/I)$, which is a closed subscheme of Spec R. Then $\mathscr{O}_X$ is R-very ample.

Remark. Let V be a complete normal variety over a field k and D be a locally principal divisor on V. If a vector subspace $\mathbf{L}$ of $\Gamma(V, \mathscr{O}(D)) \cong \mathbf{L}_V(D)$ has a finite basis $\Sigma = \{\varphi_1, \ldots, \varphi_l\}$, then $V \backslash V_\Sigma = \operatorname{Bs}|D|$ and Φ_Σ agrees with $\Phi_\mathbf{L}$. Thus, D is very ample if and only if $\mathscr{O}(D)$ is k-very ample.

b. Definition. An invertible sheaf $\mathscr{L}$ on X is said to be *R-ample*, if there exists $m \geq 0$ such that $\mathscr{L}^{\otimes m}$ is R-very ample.

By Theorem 3.1, the existence of an R-ample sheaf on X implies that X is proper over R.

Theorem 7.6. *Let $\mathscr{L}$ be an R-ample sheaf on X. For any coherent $\mathscr{O}_X$-module $\mathscr{F}$, define $\mathscr{F}(m)$ to be $\mathscr{F} \otimes \mathscr{L}^{\otimes m}$. Then*

$$H^q(X, \mathscr{F}(m)) = 0 \quad \text{for any } q > 0 \text{ and } m \gg 0.$$

PROOF. Choose e such that $\mathscr{O}_X(e)$ $(= \mathscr{L}^{\otimes e})$ is R-very ample. For each $0 \leq i < e$, by Theorem 7.5, one has μ_i such that $H^q(X, \mathscr{F}(i)(me)) = 0$ for any $q > 0$ and $m \geq \mu_i$. Letting $m_0 = \max\{\mu_0, \ldots, \mu_{e-1}\}$ and $m_1 = (m_0 + 1)e$, we claim that $H^q(X, \mathscr{F}(m)) = 0$ for any $q > 0$ and $m \geq m_1$. In fact, dividing m by e, i.e., $m = ve + i$ with $0 \leq i < e$, one has $ve + i \geq (m_0 + 1)e$, i.e., $v \geq m_0$. Thus

$$H^q(X, \mathscr{F}(m)) = H^q(X, \mathscr{F}(i)(ve)) = 0.$$

□

A vanishing theorem of the above type is called *Theorem* B. The second part of the next assertion may be called *Theorem* A.

c. Let X be an R-scheme which is proper over Spec R, and let $\mathscr{L}$ be an invertible sheaf on X. As in §7.2.**a**, we put $\mathscr{F}(m) = \mathscr{F} \otimes_{\mathscr{O}_X} \mathscr{L}^{\otimes m}$ for any $\mathscr{O}_X$-module $\mathscr{F}$.

Theorem 7.7. *Suppose that, given any coherent $\mathcal{O}_X$-module $\mathcal{F}$, there exists m_0 such that $H^q(X, \mathcal{F}(m)) = 0$ for all $q > 0$ and $m \geq m_0$. Then, for a given coherent $\mathcal{O}_X$-module $\mathcal{G}$, one has m_1 such that $\mathcal{G}(m)$ is generated as an $\mathcal{O}_X$-module by $\Gamma(X, \mathcal{G}(m))$ where $m \geq m_1$.*

PROOF. Take an arbitrary closed point x of X and let $\mathcal{M}$ be the $\mathcal{O}_X$-ideal which defines the reduced scheme $\{x\}$. From the exact sequence

$$0 \to \mathcal{M}\mathcal{G} \to \mathcal{G} \to \mathcal{G}/\mathcal{M}\mathcal{G} \to 0$$

one obtains the following exact sequence for any m:

$$0 \to (\mathcal{M}\mathcal{G})(m) \to \mathcal{G}(m) \to (\mathcal{G}/\mathcal{M}\mathcal{G})(m) \to 0.$$

Hence, $(\mathcal{G}/\mathcal{M}\mathcal{G})(m) \cong \mathcal{G}(m)/(\mathcal{M}\mathcal{G})(m)$ and $\operatorname{supp}(\mathcal{G}/\mathcal{M}\mathcal{G})(m) \subseteq \{x\}$. First, assume $\operatorname{supp}(\mathcal{G}/\mathcal{M}\mathcal{G})(m) \neq \varnothing$, and so one has by Lemma 4.3,

$$\begin{aligned}\Gamma(X, (\mathcal{G}/\mathcal{M}\mathcal{G})(m)) &\cong \Gamma(X, \mathcal{G}(m)/\mathcal{M}\mathcal{G}(m)) \\ &\cong \mathcal{G}(m)_x/\mathcal{M}\mathcal{G}(m)_x \cong \mathcal{G}(m)_x \otimes_{\mathcal{O}_{X,x}} k(x).\end{aligned}$$

Applying the hypothesis to $\mathcal{M}\mathcal{G}$, one has $m > 0$ such that $H^1(X, (\mathcal{M}\mathcal{G})(m)) = 0$. Thus one has the natural surjection $p_x \colon \Gamma(X, \mathcal{G}(m)) \to \Gamma(X, (\mathcal{G}/\mathcal{M}\mathcal{G})(m))$. Since there is $\gamma_x \colon \Gamma(X, \mathcal{G}(m)) \to \mathcal{G}(m)_x$ defined by $\gamma_x(\varphi) = \varphi_x$, one has $p_x(\varphi) = \gamma_x(\varphi) \bmod (\mathcal{M}\mathcal{G})(m)_x$. Since p_x is surjective, so is γ_x by Lemma 2.8, i.e., $\mathcal{G}(m)_x$ is generated as an $\mathcal{O}_{X,x}$-module by $\Gamma(X, \mathcal{G}(m))$. We use the next result.

Lemma 7.8. *Let X be a Noetherian scheme, let x be a closed point, and let $\mathcal{F}$ be a coherent $\mathcal{O}_X$-module such that $\mathcal{F}_x$ is generated as an $\mathcal{O}_{X,x}$-module by $\Gamma(X, \mathcal{F})$. Then there exists an affine open neighborhood U of x such that $\mathcal{F}|_U$ is generated as an $\mathcal{O}_U$-module by $\Gamma(X, \mathcal{F})$.*

PROOF. Take an affine open neighborhood U of x. Then letting $B = A(U)$, $M = \Gamma(U, \mathcal{F})$, M is finitely generated as a B-module, i.e., there exist $\psi_1, \ldots, \psi_r \in M$ with $M = \sum_{i=1}^r B\psi_i$. By assumption, one has $s_{ij} \in \Gamma(X, \mathcal{F})$ and $a_{ij} \in \Gamma(U_x, \mathcal{O}_X)$, where U_x is an open neighborhood of x, such that

$$\psi_{ix} = \sum_{j=1}^{u} a_{ijx} s_{ijx} = \left(\sum_{j=1}^{u} a_{ij} s_{ij}\right)_x.$$

Hence, there exists an affine open neighborhood W of x with $W \subseteq U_x$ such that

$$\psi_i\Big|_W = \left(\sum_{j=1}^{u} a_{ij} s_{ij}\right)\Big|_W \quad \text{for all } i. \qquad \square$$

Now, we continue the proof of the theorem. Thus by the last lemma, there exists an affine open neighborhood U such that $\mathscr{G}(m)|_U$ is generated by $\Gamma(X, \mathscr{G}(m))$. However, U may depend on m.

Note that in the case of $\operatorname{supp}(\mathscr{G}/\mathscr{M}\mathscr{G}) = \varnothing$, one has $\mathscr{G}_x = \mathfrak{m}_x \mathscr{G}_x$; hence $\mathscr{G}_x = 0$ by Lemma 2.8. Thus there exists an open neighborhood U with $\mathscr{G}(m)|_U \cong 0$. Hence in this case also, it is trivially generated by $\Gamma(X, \mathscr{G}(m))$.

Applying these arguments to $\mathscr{G} = \mathscr{O}_X$, we have m and an open neighborhood U' of x such that $\mathscr{O}_X(m)|_{U'}$ is generated by $\Gamma(X, \mathscr{O}_X(m))$. Furthermore, we have r such that $H^1(X, \mathscr{M}\mathscr{G}(i)(r)) = 0$ for $i = 0, \ldots, m-1$. Then by the preceding arguments, there exists an open neighborhood U'' of x such that the $\mathscr{G}(i)(r)|_{U''}$ are generated by $\Gamma(X, \mathscr{G}(i)(r)) = \Gamma(X, \mathscr{G}(i+r))$.

Now, let $U\langle x\rangle$ be an open neighborhood of x contained in $U' \cap U''$ such that $\mathscr{L}|_{U\langle x\rangle} \cong \mathscr{O}_{U\langle x\rangle}$, i.e., $\mathscr{L}|_{U\langle x\rangle} = \mathscr{O}_{U\langle x\rangle} \cdot \beta$.

We claim that if $n > 0$, then $\mathscr{G}(i+r+mn)|_{U\langle x\rangle}$ is generated by $\Gamma(X, \mathscr{G}(i+r+mn))$. Since $\mathscr{G}(i+r+mn)|_{U\langle x\rangle} = \mathscr{G}(i+r)|_{U\langle x\rangle} \cdot \beta^{mn}$, for each point y of $U\langle x\rangle$ and for every $\psi = \varphi \cdot \beta_y^{mn} \in \mathscr{G}(i+r)_y \cdot \beta_y^{mn} = \mathscr{G}(i+r+mn)_y$, one has $\sigma \in \Gamma(X, \mathscr{O}_X(m))$ with $\sigma_y = \beta_y^m$, and $\tau \in \Gamma(X, \mathscr{G}(i+r))$ with $\tau_y = \varphi$; hence $(\tau\sigma^n)_y = \psi$.

If $N > r + m$, then N can be written as $i + r + mn$ with $n > 0$, $0 \le i < m$. Therefore $\mathscr{G}(N)|_{U\langle x\rangle}$ is generated by $\Gamma(X, \mathscr{G}(N))$. Since $\{U\langle x\rangle \mid x$ is a closed point of $X\}$ is an open cover, one has $x_1, \ldots, x_v \in X$ such that $\bigcup_{i=1}^{v} U\langle x_i\rangle = X$. Letting $N_0 = \max\{r + m \mid r + m$ is the number corresponding to x_j, $i \le j \le v\}$, we conclude that $\mathscr{G}(N)$ is generated by $\Gamma(X, \mathscr{G}(N))$, if $N \ge N_0$. □

Remark. Let $\mathscr{L}$ be an invertible sheaf on X and $\Sigma = \{f_1, \ldots, f_l\}$. Then $x \in X_\Sigma$ if and only if $\mathscr{L}_x$ is generated by $\Gamma(X, \mathscr{L})$ as an $\mathscr{O}_{X,x}$-module.

d. Proposition 7.8. *Let $\mathscr{L}$ be an invertible sheaf on X which is proper over R. Suppose that there exists a finite subset $\Sigma = \{f_1, \ldots, f_l\}$ of $\Gamma(X, \mathscr{L}^{\otimes v})$ for some $v > 0$ such that $X = X_\Sigma$ and all the X_{f_i} are affine. Then $\mathscr{L}$ is R-ample.*

Proof. Clearly, one can assume $v = 1$. $\Phi_\Sigma = (f_1 : \cdots : f_l)$: $X \to \mathbf{P}(\mathscr{L}) = \operatorname{Proj} R[\mathrm{T}_1, \ldots, \mathrm{T}_l]$ is a morphism and $\Phi_\Sigma^{-1}(D_+(\mathrm{T}_i)) = X_{f_i}$, which is of finite type over R by Lemma 1.46. Hence, there exist $\varphi_{i1}, \ldots, \varphi_{is} \in \Gamma(X_{f_i}, \mathscr{O}_X)$ such that $\Gamma(X_{f_i}, \mathscr{O}_X) = R[\varphi_{i1}, \ldots, \varphi_{is}]$. Then by Lemma 7.2.(ii), one has $m > 0$ and $\psi_{ij} \in \Gamma(X, \mathscr{L}^{\otimes m})$ such that $\psi_{ij}|_{X_{f_i}} = \varphi_{ij} \cdot (f_i|_{X_{f_i}})^m$ for all i, j. Thus, define Ξ to be $\{f_1^m, f_2^m, \ldots, f_l^m, \psi_{11}, \ldots, \psi_{ij}, \ldots, \psi_{ls}\}$. Then $X_\Xi = X$ and Φ_Ξ is a closed immersion; hence $\mathscr{L}^{\otimes m}$ is R-very ample. □

Theorem 7.8. *If the conclusion of Theorem B holds for $\mathscr{L}$, then $\mathscr{L}$ is R-ample.*

Proof. Take a closed point x of X and let $\mathscr{M}$ be the $\mathscr{O}_X$-ideal defining the reduced subscheme $\{x\}$. Take an affine open neighborhood U of x and let the closed subscheme $X \backslash U$ be defined by an $\mathscr{O}_X$-ideal $\mathscr{S}$. $\mathscr{J} = \mathscr{S}\mathscr{M}$ has the properties $\operatorname{supp}(\mathscr{O}_X/\mathscr{J}) = (X \backslash U) \coprod \{x\}$ and $\mathscr{S}/\mathscr{J} \cong (\mathscr{S} + \mathscr{M})/\mathscr{M} = \mathscr{O}_X/\mathscr{M}$.

Thus one has the exact sequences for all $m > 0$:

$$0 \longrightarrow \mathscr{J}(m) \longrightarrow \mathscr{S}(m) \xrightarrow{p} (\mathscr{S}/\mathscr{J})(m) \longrightarrow 0.$$

There exists m_1 such that $H^1(X, \mathscr{J}(m)) = 0$ for all $m \geq m_1$. Hence p^*: $\Gamma(X, \mathscr{S}(m_1)) \to \Gamma(X, (\mathscr{S}/\mathscr{J})(m_1)) = k(x)$ is surjective. By the same argument as in the proof of Theorem 4.4, one has $h \in \Gamma(X, \mathscr{S}(m_1))$ $(\subseteq \Gamma(X, \mathscr{O}(m_1)))$ with $\bar{h}(x) = 1$. Thus $X_h \subseteq U$ and $x \in X_h$. Since U is affine, $X_h = D(h|_U)$ is also affine. Applying Proposition 7.8, one can conclude that $\mathscr{O}(m_1)$ is R-ample; hence $\mathscr{O}_X(1) = \mathscr{L}$ is R-ample. □

Corollary. *Let h: $X \to Y$ be a finite R-morphism and let $\mathscr{L}$ be an invertible sheaf on Y. If $\mathscr{L}$ is R-ample, then so is $h^*\mathscr{L}$.*

Proof. Given any coherent $\mathscr{O}_X$-module $\mathscr{F}$, $h_*(\mathscr{F})$ is coherent by Lemma 7.4. Then by Theorem 7.6, there exists m_1 such that $H^q(Y, h_*(\mathscr{F}) \otimes \mathscr{L}^{\otimes m}) = 0$ for all $q > 0$ and $m \geq m_1$. But by the Corollary to Theorem 4.8, $H^q(X, \mathscr{F} \otimes (h^*\mathscr{L})^{\otimes m}) \cong H^q(Y, h_*(\mathscr{F} \otimes (h^*\mathscr{L})^{\otimes m})) \cong H^q(Y, h_*(\mathscr{F}) \otimes \mathscr{L}^{\otimes m}) = 0$. Thus $h^*\mathscr{L}$ is R-ample by the last theorem. □

§7.5 Projective Morphisms

a. Let Y be a Noetherian scheme and let f: $X \to Y$ be a morphism of schemes.

Definition. An invertible sheaf $\mathscr{L}$ on X is said to be *f-ample* if there exists an affine open cover $\{U_\lambda \mid \lambda \in \Lambda\}$ of Y such that the $\mathscr{L}|_{f^{-1}(U_\lambda)}$ are $A(U_\lambda)$-ample for all $\lambda \in \Lambda$.

Theorems 7.4, 7.6, and 7.7 are restated as follows: let f: $X \to Y$ be a morphism between Noetherian schemes and let $\mathscr{L}$ be an f-ample invertible sheaf on X. Given a coherent $\mathscr{O}_X$-module $\mathscr{F}$, we put $\mathscr{F}(m) = \mathscr{F} \otimes_{\mathscr{O}_X} \mathscr{L}^{\otimes m}$.

Theorem 7.9 (Finiteness Theorem). *Given a coherent $\mathscr{O}_X$-module $\mathscr{F}$, the $R^qf_*(\mathscr{F})$ are coherent $\mathscr{O}_Y$-modules for all $q \geq 0$.*

Theorem 7.10

I (*Theorem A*). *Given a coherent $\mathscr{O}_X$-module $\mathscr{F}$, there exists m_1 such that $\rho_{\mathscr{F}}: f^*(f_*\mathscr{F}(m)) \to \mathscr{F}(m)$ is surjective for all $m \geq m_1$.*

II (*Theorem B*). *Given a coherent $\mathscr{O}_X$-module $\mathscr{F}$, there exists m_0 such that $R^qf_*(\mathscr{F}(m)) = 0$ for $q > 0$ and any $m \geq m_0$.*

For proofs, we note Proposition 4.5 which says, if U is an open subscheme Spec R of Y, then $R^q f_*(\mathscr{F})|_U = H^q(f^{-1}(U), \mathscr{F})^\sim$ for any q. The result follows immediately from the corresponding theorems in §7.4. □

EXAMPLE 7.2. If $f: X \to Y$ is a finite morphism, then $\mathcal{O}_X$ is f-ample.

Definition. $f: X \to Y$ is said to be a *projective morphism*, if there exists an f-ample invertible sheaf on X.

A projective morphism is always locally projective.

b. The next lemma is easy to prove; hence the proof is left to the reader.

Lemma 7.9. *Let $f: X \to$ Spec R be a morphism and let $\mathscr{L}$ be an invertible sheaf on X. If there exist $a_1, \ldots, a_r \in R$ such that $\bigcup_{j=1}^r D(a_j) =$ Spec R and the $\mathscr{L}_j$ are R_{a_j}-ample, where $\mathscr{L}_j$ denotes $\mathscr{L}|_{X_j}$ and $X_j = f^{-1}(D(a_j))$ for all j, then $\mathscr{L}$ is also R-ample.*

By this one obtains the following result.

Proposition 7.9. *Let $f: X \to Y$ be a morphism and let $\mathscr{L}$ be an invertible sheaf which is f-ample. Then for any affine open subset U of Y, $\mathscr{L}|_{f^{-1}(U)}$ is $A(U)$-ample.*

Theorem 7.11. *Let $f: X \to Y$ and $g: Y \to$ Spec R be morphisms, and let $\mathscr{L}$, $\mathscr{M}$ be invertible sheaves on X and Y, respectively. If $\mathscr{M}$ is R-ample and $\mathscr{L}$ is f-ample, then there exists $v > 0$ such that $\mathscr{L} \otimes f^*(\mathscr{M}^{\otimes v})$ is R-ample.*

PROOF. One can assume $\mathscr{M}$ is R-very ample, i.e., there exist a finitely generated graded R-algebra A with $A_1 = \sum_{i=0}^r A\alpha_i$, $Y =$ Proj A, and an isomorphism $\mathscr{M} \cong A(1)^\sim$. $\mathscr{M}$ is denoted by $\mathcal{O}_Y(1)$. Corresponding to α_i, one has $\varphi_i \in \Gamma(Y, \mathcal{O}_Y(1))$. Then letting ψ_i be $f^*(\varphi_i) \in \Gamma(X, f^*(\mathcal{O}_Y(1)))$ and Y_i be Y_{φ_i}, one has $X_{\psi_i} = f^{-1}(Y_i)$, denoted by X_i. Since $\mathscr{L}|_{X_i}$ is $A(Y_i)$-ample, there exists a finite set $\Sigma_i = \{\sigma_{i1}, \ldots, \sigma_{is}\} \subset \Gamma(X_i, \mathscr{L}^{\otimes m})$ for some m such that $X_i = X_{i\Sigma_i}$ and Φ_{Σ_i} is a closed immersion, where s and m can be chosen independently of i. Note that each $X_{\sigma_{ij}}$ is affine.

By Lemma 7.2.(ii), one can choose $v > 0$ such that there exists $\tau_{ij} \in \Gamma(X, (\mathscr{L} \otimes f^*\mathcal{O}_Y(v))^{\otimes m})$ with $\tau_{ij}|_{X_i} = \psi_i^{mv}|_{X_i} \cdot \sigma_{ij}$ for all i, j. Let $\rho_{ij} = \tau_{ij} \cdot \psi_i$ for all i, j and define Σ to be $\{\rho_{01}, \ldots, \rho_{rs}\}$. Then $X = X_\Sigma$ and the $X_{\rho_{ij}} = X_{\tau_{ij}} \cap X_i = X_{\sigma_{ij}}$ are affine. Hence by Proposition 7.8, $\mathscr{L}^{\otimes m} \otimes f^*(\mathcal{O}_Y(vm))$ is R-ample; thus $\mathscr{L} \otimes f^*\mathcal{O}_Y(v)$ is R-ample. □

Corollary. *Let $f: X \to Y$ and $g: Y \to$ Spec R be as above. Assume g is a separated morphism. If an invertible sheaf $\mathscr{L}$ on X is R-ample, then $\mathscr{L}$ is also f-ample.*

PROOF. We use the notation of Lemma 1.40 (§1.23.**b**). Then $f = q \circ \Gamma_f$ and Γ_f is the closed immersion. It is easy to see that $p^*\mathscr{L}$ is q-ample. Since $\mathcal{O}_X$ is Γ_f-ample, one has $m > 0$ such that $\mathcal{O}_X \otimes \Gamma_f^*(p^*\mathscr{L}^{\otimes m})$ is $q \circ \Gamma_f$-ample by the last theorem, i.e., $\mathscr{L}^{\otimes m}$ is f-ample.

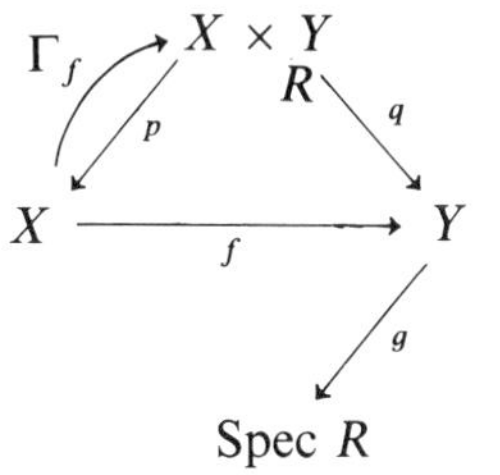

□

c. Theorem 7.12

(i) *Any closed immersion (into a Noetherian scheme) is projective.*
(ii) *The composition of any two projective morphisms is also projective.*
(iii) *The property of being a projective morphism is stable under base extension by a Noetherian scheme.*
(iv) *Let f: $X \to Y$ and g: $Y \to Z$ be morphisms where X, Y, Z are Noetherian schemes. If g is separated and $g \circ f$ is projective, then f is also projective.*

PROOF. (i) is obvious. (ii) follows from Theorem 7.11. (iii) is also obvious. (iv) follows from the Corollary to Theorem 7.11. □

Remark. The property of f being projective is not local with respect to Y, i.e., there exists a locally projective morphism which is not projective.

§7.6 Unscrewing Lemma and Its Applications

a. In order to prove the finiteness theorem for proper morphisms, we have to show a very tricky lemma, called the unscrewing lemma. We start by proving the following simple facts.

Lemma 7.10 (Noetherian Induction). *Let X be a Noetherian T_0-space, and let $\mathbb{P}$ be some property concerning closed subspaces of X. Suppose that the following hold for $\mathbb{P}$:*

(α) *Any closed point satisfies $\mathbb{P}$.*
(β) *Given any closed subspace Z, Z satisfies $\mathbb{P}$, if all closed proper subspaces of Z satisfy $\mathbb{P}$.*

Then X itself satisfies $\mathbb{P}$.

PROOF. Put $\mathscr{Z} = \{Z \mid Z$ is a closed subspace of X that does not satisfy $\mathbb{P}\}$. If $\mathscr{Z} \neq \varnothing$, then it has a minimal element Z^*, since X is Noetherian. By (α), Z^* cannot be a point. Since X is a T_0-space, so is Z^*. Hence, Z^* has closed proper subsets by Lemma 1.5. Then by (β), $Z^* \notin \mathscr{Z}$, a contradiction. □

Lemma 7.11. *Let X be a Noetherian scheme, and let $\mathscr{F}$, $\mathscr{G}$ be coherent $\mathscr{O}_X$-modules. Suppose that $\mathscr{F}_x \cong \mathscr{G}_x$ for some point $x \in X$. Then there exist a coherent $\mathscr{O}_X$-module $\mathscr{H}$ and $\mathscr{O}_X$-homomorphisms $\theta\colon \mathscr{H} \to \mathscr{F}$ and $\eta\colon \mathscr{H} \to \mathscr{G}$ such that θ_x and η_x are isomorphisms and $\eta_x \circ \theta_x^{-1}$ is the given isomorphism.*

PROOF. Since $\operatorname{Hom}(\mathscr{F}_x, \mathscr{G}_x) = (\mathscr{H}om_{\mathscr{O}_X}(\mathscr{F}, \mathscr{G}))_x$ by Corollary 2 to Theorem 1.12, one has $\varphi \in \operatorname{Hom}(\mathscr{F}|_U, \mathscr{G}|_U)$ for some affine open neighborhood U of x such that φ_x is the given isomorphism. The graph $\Gamma_\varphi\colon \mathscr{F}|_U \to (\mathscr{F} \oplus \mathscr{G})|_U$ is one-to-one, and letting $j\colon U \to X$ be the open immersion, define $\mathscr{H}$ to be $\sigma^{-1}(j_* \Gamma_\varphi(\mathscr{F}|_U))$ where $\sigma\colon \mathscr{F} \oplus \mathscr{G} \to j_*((\mathscr{F} \oplus \mathscr{G})|_U)$ was defined in §2.16.**a**. Then since $\mathscr{H}$ is an $\mathscr{O}_X$-submodule of the coherent sheaf $\mathscr{F} \oplus \mathscr{G}$, it is also coherent. Clearly, $\mathscr{H}$ and the homomorphisms: $\mathscr{H} \subset \mathscr{F} \oplus \mathscr{G} \to \mathscr{F}$ and $\mathscr{H} \subset \mathscr{F} \oplus \mathscr{G} \to \mathscr{G}$ have the required property. □

Lemma 7.12. *Let X be a reduced but reducible Noetherian scheme, i.e., there exist closed subschemes X' and X'' of X with $X' \cup X'' = X$. Given any coherent $\mathscr{O}_X$-module $\mathscr{F}$, there exist coherent $\mathscr{O}_X$-modules $\mathscr{F}'$, and $\mathscr{F}''$ with* $\operatorname{supp}(\mathscr{F}') \subseteq X'$, $\operatorname{supp}(\mathscr{F}'') \subseteq X''$ *such that there exists an $\mathscr{O}_X$-homomorphism $\xi\colon \mathscr{F} \to \mathscr{F}' \oplus \mathscr{F}''$, with* $\operatorname{supp}(\operatorname{Ker} \xi) \subseteq X' \cap X''$ *and* $\operatorname{supp}(\operatorname{Coker} \xi) \subseteq X' \cap X''$.

PROOF. Let $j'\colon X' \to X$ and $j''\colon X'' \to X$ be the closed immersions. Defining $\mathscr{F}' = j'_*(\mathscr{F}|_{X'})$ and $\mathscr{F}'' = j''_*(\mathscr{F}|_{X''})$, one has $\sigma'\colon \mathscr{F} \to \mathscr{F}'$, $\sigma''\colon \mathscr{F} \to \mathscr{F}''$ by §2.16.**a**. Then define $\xi = \sigma' \oplus \sigma''\colon \mathscr{F} \to \mathscr{F}' \oplus \mathscr{F}''$. ξ is an isomorphism outside $X' \cap X''$. Hence, ξ has the required property. □

Lemma 7.13. *Let X be a scheme, let $\mathscr{N}$ be an $\mathscr{O}_X$-ideal with $\mathscr{N}^m = 0$ for some $m > 0$, and let $\mathscr{F}$ be an $\mathscr{O}_X$-module. Then $\mathscr{N}^{m-1}\mathscr{F}$ and $\mathscr{F}_j = \mathscr{N}^j\mathscr{F}/\mathscr{N}^{j+1}\mathscr{F}$ are $\mathscr{O}_X/\mathscr{N}$-modules and the following sequences are exact:*

$$0 \to \mathscr{N}^{j+1}\mathscr{F} \to \mathscr{N}^j\mathscr{F} \to \mathscr{F}_j \to 0 \qquad (N_j)$$

for $j = 0, 1, \ldots, m-1$.

PROOF. Obvious. □

b. Let X be a Noetherian scheme and let $\mathbb{K}$ be a full subcategory of the category of coherent $\mathscr{O}_X$-module, in which $\operatorname{Hom}(\mathscr{F}, \mathscr{G})$ is $\operatorname{Hom}_{\mathscr{O}_X}(\mathscr{F}, \mathscr{G})$ for any coherent $\mathscr{O}_X$-modules $\mathscr{F}$ and $\mathscr{G}$.

Theorem 7.13 (Unscrewing Lemma of Grothendieck). *Suppose that* $\mathbb{K}$ *satisfies the following properties*:

(1) $\mathbb{K}$ *is an exact subcategory, i.e., for any exact sequence of coherent* $\mathcal{O}_X$*-modules*

$$0 \to \mathscr{F}' \to \mathscr{F} \to \mathscr{F}'' \to 0,$$

if any two of $\mathscr{F}'$, $\mathscr{F}$, $\mathscr{F}''$ *belong to* $\mathbb{K}$, *then the third belongs to* $\mathbb{K}$, *too.*

(2) *Given a coherent* $\mathcal{O}_X$*-module* $\mathscr{F}$,

$$\mathscr{F} \in \mathbb{K} \quad \textit{whenever} \quad \mathscr{F}^{\oplus m} \in \mathbb{K} \quad \textit{for some} \quad m > 0.$$

(3) *Given a closed integral subscheme* Z *of* X, *there exists* $\mathscr{G} \in \mathbb{K}$ *with* $\operatorname{supp}(\mathscr{G}) = Z$.

(4) *If* $\operatorname{supp}(\mathscr{F}) = \{x\}$ *for some coherent* $\mathscr{F}$, *then* $\mathscr{F} \in \mathbb{K}$.

Then $\mathbb{K}$ *coincides with the category of coherent* $\mathcal{O}_X$*-modules.*

Proof. We consider a property of subsets Z of X to the effect that if $\operatorname{supp}(\mathscr{F}) \subseteq Z$, then $\mathscr{F} \in \mathbb{K}$, where $\mathscr{F}$ is any coherent $\mathcal{O}_X$-module. This property satisfies (α) of Lemma 7.10 by condition (4). Thus, for the proof, it suffices to verify (β); in other words, we can also suppose that if $\operatorname{supp}(\mathscr{F}) \neq X$, then $\mathscr{F} \in \mathbb{K}$. Hence, consider $\mathscr{F}$ with $\operatorname{supp}(\mathscr{F}) = X$.

Now, first suppose that X is a reduced scheme. If X is irreducible, then X is an integral scheme. By (3), one has $\mathscr{G} \in \mathbb{K}$ with $\operatorname{supp}(\mathscr{G}) = X$. If $* \in X$ is the generic point of X, then $\mathscr{G}_* \neq 0$ and $\mathscr{G}_*$ is a free $k(*)\,(=\operatorname{Rat}(X))$-module, i.e., $\mathscr{G}_* \cong \mathcal{O}_{X,*}^{\oplus m}$ for some $m > 0$. On the other hand, $\operatorname{supp}(\mathscr{F}) = X$ implies that $\mathscr{F}_* \cong \mathcal{O}_{X,*}^{\oplus s}$ for some $s > 0$. Thus $\mathscr{G}_*^{\oplus s} \cong \mathscr{F}_*^{\oplus m}$. By Lemma 7.11, there exist $\mathscr{H}$ and $\theta\colon \mathscr{H} \to \mathscr{G}^{\oplus s}$, $\eta\colon \mathscr{H} \to \mathscr{F}^{\oplus m}$ such that θ_* and η_* are isomorphisms, i.e., $\operatorname{supp}(\operatorname{Ker}\theta)$, $\operatorname{supp}(\operatorname{Ker}\eta)$, $\operatorname{supp}(\operatorname{Coker}\theta)$, and $\operatorname{supp}(\operatorname{Coker}\eta)$ $\neq X$. By hypothesis, $\operatorname{Ker}\theta$, $\operatorname{Ker}\eta$, $\operatorname{Coker}\theta$, and $\operatorname{Coker}\eta \in \mathbb{K}$. Recalling that $\mathscr{G}^{\oplus s} \in \mathbb{K}$ by (1), one concludes $\mathscr{H} \in \mathbb{K}$ by (1) applied to the exact sequences derived from θ, and $\mathscr{F}^{\oplus m} \in \mathbb{K}$ by (1), again. Hence by (2), $\mathscr{F} \in \mathbb{K}$.

Second, suppose that X is reduced but reducible, i.e., $X = X' \cup X''$ with $X' \neq X$ and $X'' \neq X$. Then as in Lemma 7.12 one has $\mathscr{F}'$ and $\mathscr{F}''$ with $\operatorname{supp}(\mathscr{F}') \neq X$ and $\operatorname{supp}(\mathscr{F}'') \neq X$. Hence $\mathscr{F}', \mathscr{F}'' \in \mathbb{K}$, and so $\mathscr{F}' \oplus \mathscr{F}'' \in \mathbb{K}$ by (1). Applying (1) to the exact sequences derived from ξ in Lemma 7.12, one concludes that $\mathscr{F} \in \mathbb{K}$.

Finally, suppose that X is a scheme such that if a coherent $\mathcal{O}_X$-module $\mathscr{F}'$ can be regarded as an $\mathcal{O}_X/\mathcal{N}$-module, where $\mathcal{N}$ is the nilradical of X, then $\mathscr{F}' \in \mathbb{K}$. Now, given $\mathscr{F}$, by making use of (1) applied to $(N_{m-1}), \ldots,$ (N_0) in Lemma 7.13, one concludes $\mathscr{F} \in \mathbb{K}$. □

c. Now, we shall prove the finiteness theorem.

Theorem 7.14 (Finiteness Theorem). *Let* $f\colon X \to Y$ *be a proper morphism,* *where* Y *is a Noetherian scheme. If* $\mathscr{F}$ *is a coherent* $\mathcal{O}_X$*-module, then all the* $(R^qf_*)(\mathscr{F})$ *are coherent* $\mathcal{O}_Y$*-modules.*

PROOF. Define $\mathbb{K} = \{\mathscr{F} \mid \text{all } (R^qf_*)(\mathscr{F}) \text{ are coherent}\}$. We check conditions (1), (2), (3), (4) of the last theorem.

(1) This is easy by Lemma 7.6 applied to the long exact sequence derived from the short exact sequence in (1).

(2) This is clear.

(4) This is obvious.

(3) One may assume Y is an affine scheme Spec R. By Theorem 3.4, there exist a scheme X' and a proper morphism $\mu: X' \to X$ such that (i) $f \circ \mu$ is projective, i.e., there exists an R-ample sheaf $\mathscr{L}$ on X', and (ii) there exists a dense open subset U of X with the property that $\mu^{-1}(U) \cong U$. By the Corollary to Theorem 7.11, $\mathscr{L}$ is μ-ample. By Theorem 7.10.II there exists m with $(R^q\mu_*)(\mathscr{L}^{\otimes m}) = 0$ for all $q > 0$. Hence, by the Corollary to Theorem 4.8,

$$H^q(X', \mathscr{L}^{\otimes m}) \cong H^q(X, \mu_*(\mathscr{L}^{\otimes m})).$$

By Theorem 7.4, all the $H^q(X', \mathscr{L}^{\otimes m})$ are finitely generated. Hence, letting $\mathscr{G}$ be $\mu_*(\mathscr{L}^{\otimes m})$, $\mathscr{G}$ belongs to $\mathbb{K}$. On the other hand, $\mathscr{G}|_U$ is isomorphic to an invertible sheaf; hence $\mathrm{supp}(\mathscr{G}) \supseteq U$, i.e., $\mathrm{supp}(\mathscr{G}) = X$. □

d. Any divisor D is a difference of two effective divisors, i.e., $D = D_+ - D_-$. The following result is a similar fact to this for invertible sheaves.

Lemma 7.14. *Let* $\mathscr{L}$ *be an invertible sheaf on an integral scheme* X*. Then there exist two quasi-coherent* $\mathcal{O}_X$*-ideals* $\mathscr{S}$ *and* $\mathscr{J}$ *such that* $\mathscr{J} = \mathscr{S} \otimes \mathscr{L}^{-1}$*. If* Y *and* Z *are closed subschemes of* X *defined by* $\mathscr{S}$ *and* $\mathscr{J}$*, respectively, then* $Y \neq X$, $Z \neq Y$*, and the following sequences are exact*

$$\begin{array}{c} 0 \to \mathscr{S}(m) \to \mathcal{O}_X(m) \to \mathcal{O}_Y(m) \to 0 \\ \wr\| \\ 0 \to \mathscr{J}(m+1) \to \mathcal{O}_X(m+1) \to \mathcal{O}_Z(m+1) \to 0, \end{array}$$

where $\mathscr{F}(m) = \mathscr{F} \otimes \mathscr{L}^{\otimes m}$ *for any* $\mathcal{O}_X$*-module* $\mathscr{F}$*.*

PROOF. By the remark in §2.19.**c**, $\mathscr{L}$ can be identified with a Cartier sheaf, i.e., $\mathscr{L} \subset \mathscr{R}at_X$. $\mathscr{S} = \mathscr{L} \cap \mathcal{O}_X$ is a quasi-coherent $\mathcal{O}_X$-ideal and $\mathscr{J} = \mathscr{S} \otimes_{\mathcal{O}_X} \mathscr{L}^{-1} \subset \mathscr{L} \otimes_{\mathcal{O}_X} \mathscr{L}^{-1} = \mathcal{O}_X$, i.e., $\mathscr{J}$ is also a quasi-coherent $\mathcal{O}_X$-ideal. Clearly, $\mathscr{S} \neq 0$ and $\mathscr{J} \neq 0$, i.e., $Y \neq X$, and $Z \neq X$. □

Remark. If $X = \mathrm{Spec}\, A$ and $\mathscr{L}|_X = \mathcal{O}_X \cdot \alpha$, where $\alpha \in Q(A)$, then $\mathscr{S}|_X = \tilde{I}$, where I is $A \cap A\alpha$. Thus if $\alpha = a/b$ for $a, b \in A$, then $a \in I$. The ideal I need not be principal.

Theorem 7.15. *Let X be a complete scheme over a field k, let $\mathscr{L}$ be an invertible sheaf, and let $\mathscr{F}$ be a coherent $\mathscr{O}_X$-module. Then $\chi_X(\mathscr{F}(m))$ is a polynomial in m with* $\deg \chi_X(\mathscr{F}(m)) \leq \dim(\operatorname{supp} \mathscr{F})$, *where* $\mathscr{F}(m) = \mathscr{F} \otimes \mathscr{L}^{\otimes m}$.

Proof. Define a subcategory $\mathbb{K}$ consisting of coherent $\mathscr{O}_X$-modules having the required property. To apply the unscrewing lemma, we shall verify conditions (1), (2), (3), (4).

(1) This is easily checked. Given a short exact sequence as in (1), one has $\chi_X(\mathscr{F}(m)) = \chi_X(\mathscr{F}'(m)) + \chi_X(\mathscr{F}''(m))$ for any m by Proposition 4.4 and furthermore, $\operatorname{supp}(\mathscr{F}') \subseteq \operatorname{supp}(\mathscr{F})$, $\operatorname{supp}(\mathscr{F}'') \subseteq \operatorname{supp}(\mathscr{F}) \subseteq \operatorname{supp}(\mathscr{F}') \cup \operatorname{supp}(\mathscr{F}'')$. Hence, (1) is clearly satisfied.

(2) and (4) These are obvious.

(3) It suffices to show $\mathscr{O}_X \in \mathbb{K}$, if X is an integral scheme, and if any $\mathscr{F}$ satisfies the claim $\operatorname{supp}(\mathscr{F}) \neq X \Rightarrow \mathscr{F} \in \mathbb{K}$. By Lemma 7.14,

$$\begin{aligned}\chi_X(\mathscr{O}_X(m)) - \chi_Y(\mathscr{O}_Y(m)) &= \chi_X(\mathscr{S}(m)) = \chi_X(\mathscr{J}(m+1)) \\ &= \chi_X(\mathscr{O}_X(m+1)) - \chi_Z(\mathscr{O}_Z(m+1));\end{aligned}$$

hence,

$$\chi_X(\mathscr{O}_X(m+1)) - \chi_X(\mathscr{O}_X(m)) = \chi_Z(\mathscr{O}_Z(m+1)) - \chi_Y(\mathscr{O}_Y(m)).$$

Since $\operatorname{supp}(\mathscr{O}_Y) = Y \neq X$ and $\operatorname{supp}(\mathscr{O}_Z) = Z \neq X$, the right-hand side is a polynomial in m; hence so is $\chi_X(\mathscr{O}_X(m))$. Furthermore, $\deg(\chi_X(\mathscr{O}_X(m))) \leq \max\{\deg(\chi_Y(\mathscr{O}_Y(m))), \deg(\chi_Z(\mathscr{O}_Z(m)))\} + 1 \leq \dim X$. □

Similarly, one can prove a more general result.

Theorem 7.15*. *Let $\mathscr{L}_1, \ldots, \mathscr{L}_s$ be invertible sheaves on X and let $\mathscr{F}$ be a coherent $\mathscr{O}_X$-module. Then $\chi_X(\mathscr{F} \otimes \mathscr{L}_1^{\otimes m_1} \otimes \cdots \otimes \mathscr{L}_s^{\otimes m_s})$ is a polynomial (called the Snapper polynomial) in $m_1, \ldots, m_s$ with total degree* $\leq \dim(\operatorname{supp} \mathscr{F})$.

Proof. Left to the reader as an exercise. □

Finally, we prove the next result due to Hilbert.

Proposition 7.10. *Let A be a graded ring finitely generated over a field k and let M be a finitely generated graded A-module. Then $\varphi_M(m) = \dim_k M_m$ is a polynomial in m for $m \gg 0$ and* $\deg \varphi_M(m) = \dim(\operatorname{supp} \tilde{M})$, *where $\tilde{M}$ is the $\mathscr{O}_{\operatorname{Proj} A}$-module associated with M.*

Proof. Recalling Proposition 7.6, one has $\varphi_M(m) = \chi_X(\tilde{M}(m))$ for $m \gg 0$, where X is Proj A, that is a polynomial by Theorem 7.15. It suffices to show that deg $\chi_X(\mathscr{F}(m)) = \dim(\operatorname{supp} \mathscr{F})$ for any coherent sheaf $\mathscr{F}$, where $\mathscr{O}_X(1) = A(1)^\sim$. Leaving the general case to the reader, we shall prove here the result for $\mathscr{F} = \mathscr{O}_X$ by induction on $r = \dim X$. If $r = 0$, then $\mathscr{O}_X(m) = \mathscr{O}_X$ and so $\chi_X(\mathscr{O}_X(m))$ is a constant function, i.e., deg $\chi_X(\mathscr{O}_X(m)) = 0$. Assuming that $r > 0$, we choose $\alpha \in A_1$ such that $Y = V_+(\alpha)$ does not contain any generic points of the irreducible components of X. Then $Y \cong \operatorname{Proj}(A/(\alpha))$, $\dim Y = r - 1$ by Proposition 1.61, and $\mathscr{O}(-Y) \cong \mathscr{O}_X(-1)$ (cf. §2.19). Hence, one has an exact sequence

$$0 \to \mathscr{O}_X(m-1) \to \mathscr{O}_X(m) \to \mathscr{O}_Y(m) \to 0$$

for any m, where $\mathscr{O}_Y(1) = (A/(\alpha))(1)^\sim$. By Proposition 4.4, one obtains $\chi(\mathscr{O}_X(m)) - \chi(\mathscr{O}_X(m-1)) = \chi(\mathscr{O}_Y(m))$, which is of degree $r - 1$, by the induction hypothesis. Hence, deg $\chi(\mathscr{O}_X(m)) = r$. □

Corollary. *Let B be a local Noetherian ring with maximal ideal $\mathfrak{n}$. Then $\sigma_B(m) = \sum_{j=0}^{m} \dim_k(\mathfrak{n}^j/\mathfrak{n}^{j+1})$ is a polynomial of degree $1 + \dim X$ for $m \gg 0$, where $X = \operatorname{Proj}(\operatorname{gr}_{\mathfrak{n}}^{\cdot}(B))$ and $k = B/\mathfrak{n}$.*

Remark. From Exercise 7.3, it follows that $\dim X = \dim B - 1$. Thus deg $\sigma_B(m) = \dim B$. This establishes Theorem 2.5.

§7.7 Projective Normality

a. Let A be a finitely generated graded k-algebra, where k is a field. Suppose that A is an integral domain. Taking a nonzero element α of A_1, one has an isomorphism

$$\xi_A : A \xrightarrow{\sim} \sum_{m=0}^{\infty} A_m \cdot \frac{1}{\alpha^m} \cdot \mathsf{T}^m \subseteq \operatorname{Rat}(\operatorname{Proj} A)[\mathsf{T}].$$

defined by $\xi_A(x) = (x/\alpha^m) \cdot \mathsf{T}^m$, for $x \in A_m$, where T is transcendental over the field $\operatorname{Rat}(\operatorname{Proj} A)$.

Proposition 7.11. *Let D be a divisor on a complete normal variety V. Then the ring $R(V, D) = \sum_{m=0}^{\infty} \mathrm{L}(mD)\mathsf{T}^m \subset \operatorname{Rat}(V)[\mathsf{T}]$ is normal.*

Proof. Let $D = \sum_{j=1}^{s} e_j \Gamma_j$, where $\Gamma_i \neq \Gamma_j$ for $i \neq j$. The valuation associated with Γ_j is denoted by ord_j. For any $\varphi(\mathsf{T}) = \sum_{m=0}^{r} a_m T^m \in \operatorname{Rat}(V)[\mathsf{T}]$, one has $\varphi(\mathsf{T}) \in R(V, D)$, if and only if $\operatorname{ord}_j(a_m) + e_j m \geq 0$ for all j, and m with $a_m \neq 0$. Furthermore, if $\mathfrak{O}_j$ denotes the valuation ring of Γ_j and π_j is the regular parameter of $\mathfrak{O}_j$, then

$$\operatorname{ord}_j(a_m) + e_j m \geq 0 \Leftrightarrow a_m \in \mathfrak{O}_j \pi_j^{-e_j m}.$$

Since $\mathrm{Rat}(V)[\mathsf{T}]$ is normal, for the proof it suffices to show that if $\varphi(\mathsf{T})$ is integral over $R(V, D)$, then $\varphi(\mathsf{T}) \in R(V, D)$. Thus, suppose that there exist $\alpha_1(\mathsf{T}), \ldots, \alpha_r(\mathsf{T}) \in R(V, D)$ such that

$$\varphi(\mathsf{T})^r + \alpha_1(\mathsf{T})\varphi(\mathsf{T})^{r-1} + \cdots + \alpha_r(\mathsf{T}) = 0.$$

If $\alpha_i(\mathsf{T}) = \sum_{m=0}^{p} b_{im}\mathsf{T}^m$, then $b_{im} \in \mathfrak{O}_j \pi_j^{-e_j m}$, i.e., $b_{im}\pi_j^{e_j m} \in \mathfrak{O}_j$ for all j. Thus, $\alpha_i(\pi_j^{e_j}\mathsf{T}) \in \mathfrak{O}_j[\mathsf{T}]$. One has $\varphi(\pi_j^{e_j}\mathsf{T})^r + \alpha_1(\pi_j^{e_j}\mathsf{T})\varphi(\pi_j^{e_j}\mathsf{T})^{r-1} + \cdots + \alpha_r(\pi_j^{e_j}\mathsf{T}) = 0$. This implies $\varphi(\pi_j^{e_j}\mathsf{T})$ is integral over $\mathfrak{O}_j[\mathsf{T}]$. But $\mathfrak{O}_j[\mathsf{T}]$ is a UFD, since $\mathfrak{O}_j$ is a UFD (cf. Proposition 2.7.(ii)). Hence, $\varphi(\pi_j^{e_j}\mathsf{T}) \in \mathfrak{O}_j[\mathsf{T}]$ for all j, i.e., $\varphi(\mathsf{T}) \in R(V, D)$. □

b. Now, suppose that Proj A is a normal variety V, in which A is assumed to be an integral domain and $\dim V > 0$. Then for a nonzero α of A_1, one has $D = \mathrm{div}(\boldsymbol{\alpha}_1(\alpha))$ and so $\mathcal{O}(mD) \cong \mathcal{O}_V(m)$. Thus $\eta_m : \Gamma(V, \mathcal{O}_V(m)) \cong \mathbf{L}_V(mD)$ is defined by $\eta_m(\sigma) = \sigma/\alpha^m$. Hence $\eta_D : \Gamma_*(V, \mathcal{O}_V) \cong R(V, D)$ is defined by $\eta_D = \sum_{m=0}^{\infty} \eta_m \mathsf{T}^m$, i.e., $\eta_D(\sigma) = \eta_m(\sigma)\mathsf{T}^m$ if $\sigma \in \Gamma(V, \mathcal{O}_V(m))$. One thus obtains the next commutative square, and $\Gamma(V, \mathcal{O}_V(-m)) = 0$ for $m > 0$.

$$\begin{array}{ccc}
A & \xrightarrow[\sim]{\xi_A} & \displaystyle\sum_{m=0}^{\infty} A_m \cdot \frac{1}{\alpha^m} \cdot \mathsf{T}^m \\
{\scriptstyle \alpha_A}\Big\downarrow & & \cap\Big| \\
\Gamma_*(V, \mathcal{O}_V) & \xrightarrow[\eta_D]{\sim} & R(V, D)
\end{array}$$

Therefore, by Proposition 7.6, $A_m \cdot (1/\alpha^m) = \mathbf{L}_V(mD)$, if $m \gg 0$. Hence, $Q(\sum_{m=0}^{\infty} A_m \cdot (1/\alpha^m) \cdot \mathsf{T}^m) \cong Q(R(V, D))$.

Theorem 7.16 (Zariski). *If $V = \mathrm{Proj}\, A$ is a normal variety and α is an element of A_1, then*

(*i*) *$\Gamma_*(V, \mathcal{O}_V)$ is a normal ring.*
(*ii*) *$\Gamma_*(V, \mathcal{O}_V)$ is the integral closure of A in $Q(A)$.*
(*iii*) *Let Λ_m denote the linear system defined by $A_m \cdot (1/\alpha^m)$. Then Λ_m is complete for $m \gg 0$.*
(*iv*) *A is normal if and only if all Λ_m are complete.*

Definition. Proj A is said to be *projectively normal* if A is a normal ring.

By Theorem 2.1, $\Gamma_*(V, \mathcal{O}_V)$ is a finitely generated A-module, i.e., one has $\sigma_1 \in \Gamma(V, \mathcal{O}_V(l_1)), \ldots, \sigma_t \in \Gamma(V, \mathcal{O}_V(l_t))$ such that $\Gamma_*(V, \mathcal{O}_V) = \sum_{i=1}^{t} A(-l_i)\sigma_i$.

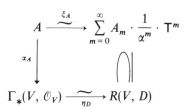

§7.8 Etale Morphisms

a. In this section, we assume k is an algebraically closed field.

Proposition 7.12. *Let $f\colon V \to W$ be a morphism between varieties and let p be a closed point on V. If $T(f)_p\colon T_p(V) \to T_{f(p)}(W)$ is one-to-one, then*

(i) $\mathfrak{m}_{f(p)}\mathcal{O}_{V,p} = \mathfrak{m}_p$.
(ii) $\hat{f}^*\colon \hat{\mathcal{O}}_{W,f(p)} \to \hat{\mathcal{O}}_{V,p}$ *is surjective, where* $f^* = \theta(f)_p^{\#}\colon \mathcal{O}_{W,f(p)} \to \mathcal{O}_{V,p}$.
(iii) *Let Z be the closed image of V under f. Then* $\hat{\mathcal{O}}_{Z,f(p)} \cong \hat{\mathcal{O}}_{V,p}$.

PROOF. By hypothesis and Lemma 2.8, (i) can be proved immediately. Since $\mathcal{O}_{Z,f(p)} \cong \mathcal{O}_{W,f(p)}/\operatorname{Ker} f^*$ and $\mathfrak{m}_{f(p)}\mathcal{O}_{V,p} = \mathfrak{m}_p$, one has $\hat{\mathcal{O}}_{Z,f(p)} \subseteq \hat{\mathcal{O}}_{V,p}$ with $\hat{\mathcal{O}}_{V,p} \otimes k(f(p)) \cong k(p) = k$, and hence by Theorem 2.6, $\hat{\mathcal{O}}_{Z,f(p)} = \hat{\mathcal{O}}_{V,p}$. This implies (ii) and (iii). □

Definition. f is said to be *étale* at p, if $\hat{f}^*\colon \hat{\mathcal{O}}_{W,f(p)} \to \hat{\mathcal{O}}_{V,p}$ is an isomorphism.

Then by (iii), $\operatorname{Ker} f^* = 0$ and so $\mathcal{O}_{Z,f(p)} \cong \mathcal{O}_{W,p}$, in other words, $Z = W$, i.e., $f\colon V \to W$ is dominating, and $\mathfrak{m}_{f(p)}\mathcal{O}_{V,p} = \mathfrak{m}_p$.

Definition. f is said to be an *étale covering* (*map*), if f is étale at every closed point and f is proper.

Proposition 7.13. *Any étale covering is a finite morphism.*

PROOF. If p is a closed point and q is $f(p)$, then the fiber $f^{-1}(q)$ satisfies $\mathcal{O}_{f^{-1}(q),p} = k$; hence p is an isolated component of $f^{-1}(q)$ (see §5.6.**b**). If f is not finite, there exists a closed irreducible subset F with $\dim F \geq 1$ such that $f(F) = \{q\}$ (see §2.14.**d**). Take a closed point p of F. Then $F \subseteq f^{-1}(q)$, which contradicts the fact that p is an isolated component of $f^{-1}(q)$. □

b. Now, suppose further that V and W are nonsingular varieties over an algebraically closed field k of characteristic zero. Then any morphism $f\colon V \to W$ is separable.

Proposition 7.14. *If $T(f)_p\colon T_p(V) \to T_{f(p)}(W)$ is surjective at a closed point p of V, then $f^{-1}(f(p))$ is nonsingular at p and f is dominating.*

PROOF. We use the notation used in §5.5.**a**. Then by hypothesis, after a permutation of indices and changing the local coordinate systems, one can assume that $z_\lambda^1 = f^*w_\alpha^1, \ldots, z_\lambda^m = f^*w_\alpha^m$, where $\dim V = n$ and $\dim W = m$. Since $p \in U_\lambda \cap f^{-1}(f(p)) = V(z_\lambda^1, \ldots, z_\lambda^m)$ in U_λ, p is a nonsingular point of $f^{-1}(f(p))$ and $\dim_p f^{-1}(f(p)) = n - m$. Furthermore, $f\colon U_\lambda \to W_\alpha$ is clearly dominating, i.e., f is dominating. □

Lemma 7.15. *If f: $V \to W$ is dominating, then there exists a closed point p such that $T(f)_p$ is surjective.*

PROOF. Since $\mathrm{Rat}(V)/\mathrm{Rat}(W)$ is separable, there exist separable transcendence bases $(x_1, \ldots, x_m)$ of $\mathrm{Rat}(W)$ and $(x_1, \ldots, x_m, y_1, \ldots, y_{n-m})$ of $\mathrm{Rat}(V)$. These $x_1, \ldots, y_{n-m}$ are regular functions on some affine open subset V_0 of V. One has an n-form $\omega = dx_1 \wedge \cdots \wedge dy_{n-m}$ and an effective divisor $D_0 = \mathrm{div}(\omega)$ on V_0. At any closed point p of $V_0 \backslash D_0$, $T(f)_p$: $T_p(V) \to T_{f(p)}(W)$ is surjective. □

Definition. Define a closed set Σ of V by

$$\Sigma \cap U_\lambda = \left\{ p \in U_\lambda \,\middle|\, \mathrm{rank}_p \left[\frac{\partial f^* w_\alpha^i}{\partial z_\lambda^j} \right] < m \right\},$$

on each U_λ where $\mathrm{rank}_p\, A$ denotes the rank of $\bar{A}(p)$, where $\bar{A}(p)$ denotes the matrix $[\bar{a}_{ij}(p)]$, and $A = [a_{ij}]$.

Σ is also denoted by R_f.

By the last lemma, if f is dominating, then R_f is not the whole space V.

If $\dim V = \dim W$ and f is dominating, then R_f coincides with the support of the ramification divisor of f (see §5.6.**a**).

Definition. If $R_f = \varnothing$, then f is said to be a *smooth* morphism.

Theorem 7.17. *Let f: $V \to W$ be a dominating morphism between nonsingular varieties. Then $f(R_f)$ is not dense in W.*

PROOF. Suppose that $f(R_f)$ is dense, i.e., there exists a closed subvariety Y of R_f such that $f|_Y$: $Y \to W$ is dominating. Then letting $\varphi = f|_{\mathrm{Reg}\, Y}$, R_φ is a closed proper subset of Reg Y by Lemma 7.15. Hence, take a closed point p of Reg $Y \backslash R_\varphi$. Then $T(\varphi)_p$ is surjective. But $T(f)_p$ is not surjective, since $p \in R_f$. Letting j denote the immersion of Reg Y into V, one has $T(\varphi)_p = T(f)_p \circ T(j)_p$, which is not surjective. This is a contradiction. □

If p is a closed point of W_0, where W_0 is the complement of the closure of $f(R_f)$ in W, then $f^{-1}(p)$ is a finite union of nonsingular varieties by Proposition 7.14. Let $V_0 = f^{-1}(W_0)$ and $f_0 = f|_{V_0}$. Then f_0: $V_0 \to W_0$ is a smooth morphism.

EXAMPLE 7.3. Let $F \in k[\mathsf{X}_1, \ldots, \mathsf{X}_n]$ be a polynomial and φ: $k[\mathsf{T}] \to k[\mathsf{X}_1, \ldots, \mathsf{X}_n]$ be a k-algebra homomorphism defined by $\varphi(\mathsf{T}) = F$. If f: $\mathbf{A}_k^n \to \mathbf{A}_k^1$ is the morphism associated with φ, then $R_f = V(\partial F/\partial \mathsf{X}_1, \ldots, \partial F/\partial \mathsf{X}_n)$. Hence $f(R_f) = \{p \in k \mid F(x_1, \ldots, x_n) = p,\ (\partial F/\partial \mathsf{X}_1)(q) = \cdots = (\partial F/\partial \mathsf{X}_n)(q) = 0$ for some $q = (x_1, \ldots, x_n) \in k^n\}$.

§7.9 Theorems of Bertini

a. We say that a *general point of a variety V has some property* $\mathbb{P}$, if there exists a closed proper subset F of V such that every (closed) point p of $V\backslash F$ has the property $\mathbb{P}$. Note that a finite union of closed proper subsets is also closed and proper. Thus, when a general point has a property $\mathbb{P}_1$, and at the same time, a general point has a property $\mathbb{P}_2$, a general point has the properties $\mathbb{P}_1$ and $\mathbb{P}_2$.

For example, a general point of a variety over an algebraically closed field k is a nonsingular point. In the situation of Theorem 7.17, a general fiber $f^{-1}(x)$ is understood as the fiber at a general point x of W. Thus, the theorem can be restated as follows: a *general fiber of f is a finite union of nonsingular varieties.*

b. Let Λ be a linear system on a normal variety V, i.e., there exist an invertible sheaf $\mathscr{L}$ on V and a finite-dimensional vector subspace $\mathbf{L}$ of $\Gamma(V, \mathscr{L})$ such that $\Lambda = \{\operatorname{div}(\varphi) \mid \varphi \in \mathbf{L}\backslash\{0\}\}$. Fixing a basis of $\mathbf{L}$, one has the bijection $\Lambda \cong \operatorname{spm}(\mathbf{P}^*(\Lambda))$. Any member of Λ corresponding to a general closed point of $\mathbf{P}^*(\Lambda)$ is said to be a *general member* of Λ.

Remark. It is meaningless to say that a point p of V or a member of Λ is general. But one can say that a general member of Λ has such and such property.

For simplicity, we assume V to be nonsingular and choose a finite coordinate cover $\{U_j \mid j \in J\}$ of V such that $\mathscr{L}|_{U_j} = \mathcal{O}_{U_j} \cdot \beta_j$. Let $\alpha_1, \ldots, \alpha_l$ be a basis of $\mathbf{L}$ such that $\alpha_i|_{U_j} = \varphi_i^{(j)}\beta_j$ for all i, j. If R_j is $A(U_j)$, then $U_j \times \mathbf{P}(\Lambda) \cong \mathbf{P}_{R_j}^{l-1} = \operatorname{Proj}(R_j[\mathrm{T}_1, \ldots, \mathrm{T}_l])$, where $\mathrm{T}_1, \ldots, \mathrm{T}_l$ are algebraically independent over $\operatorname{Rat}(V)$, and $\{U_j \times \mathbf{P}(\Lambda) \mid j \in \Lambda\}$ is an open cover of $V \times \mathbf{P}(\Lambda)$. The closed subschemes $Z_j = V_+(\sum_{i=1}^{l} \varphi_i^{(j)} \mathrm{T}_i)$ of $\mathbf{P}_{R_j}^{l-1}$ define a closed subscheme Z_Λ of $V \times \mathbf{P}(\Lambda)$. Composing $Z_\Lambda \to V \times \mathbf{P}(\Lambda)$ with the projection $V \times \mathbf{P}(\Lambda) \to \mathbf{P}(\Lambda)$, one obtains a morphism $f\colon Z_\Lambda \to \mathbf{P}(\Lambda)$. Then for a closed point $a = (a_1 : \cdots : a_l)$ of $\mathbf{P}(\Lambda)$, $f^{-1}(a) = \operatorname{div}(\sum_{i=1}^{l} a_i \alpha_i) \times \{a\}$.

Lemma 7.16. *If Λ is a reduced linear system, then Z_Λ is a variety and* $\operatorname{Sing} Z_\Lambda \subseteq \operatorname{Bs} \Lambda \times \mathbf{P}(\Lambda)$.

PROOF. Let $(z_j^1, \ldots, z_j^n)$ be a coordinate system on U_j and $\zeta_2 = \mathrm{T}_2/\mathrm{T}_1, \ldots, \zeta_l = \mathrm{T}_l/\mathrm{T}_1$. Then $W_j = U_j \times D_+(\mathrm{T}_1)$ is a coordinate neighborhood of $V \times \mathbf{P}(\Lambda)$ with $(z_j^1, \ldots, z_j^n, \zeta_2, \ldots, \zeta_l)$ a local coordinate system. $Z_\Lambda \cap W_j$ is defined by $\psi_j = \varphi_1^{(j)} + \varphi_2^{(j)}\zeta_2 + \cdots + \varphi_l^{(j)}\zeta_l$ in W_j; i.e., $Z_\Lambda \cap W_j = V(\psi_j)$. Hence, any irreducible component of Z_Λ has codimension 1. Since

$$d\psi_j = d\varphi_1^{(j)} + d\varphi_2^{(j)} \cdot \zeta_2 + \cdots + \varphi_l^{(j)} \cdot d\zeta_l,$$

it follows that $\operatorname{Sing}(V(\psi_j)) \subseteq V(\varphi_1^{(j)}, \varphi_2^{(j)}, \ldots, \varphi_l^{(j)}) \times D_+(\mathrm{T}_1)$. This implies that $\operatorname{Sing}(Z_\Lambda) \subseteq \operatorname{Bs}(\Lambda) \times \mathbf{P}(\Lambda)$. Hence, $Z_\Lambda\backslash\operatorname{Bs}(\Lambda) \times \mathbf{P}(\Lambda)$ is nonsingular. If $x \in$

$U_j \backslash \mathrm{Bs}(\Lambda)$, then ψ_j is irreducible and so (ψ_j) is a prime ideal of $\mathcal{O}_{V,x}[\zeta_2, \ldots, \zeta_l]$ by Proposition 2.7 and Theorem 2.11. Hence $Z_\Lambda \backslash \mathrm{Bs}(\Lambda) \times \mathbf{P}(\Lambda)$ is irreducible. By hypothesis, $\Lambda_{\mathrm{fix}} = 0$; hence $\mathrm{codim}(\mathrm{Bs}(\Lambda)) \geq 2$. Therefore, Z_Λ coincides with the closure of $Z_\Lambda \backslash (\mathrm{Bs}(\Lambda) \times \mathbf{P}(\Lambda))$, which is a variety. □

If Λ has a fixed part F, then one has the reduced linear system Λ_1 obtained from Λ and $Z_\Lambda = (F \times \mathbf{P}(\Lambda)) \cup Z_{\Lambda_1}$.

Theorem 7.18. *If Λ is a linear system on a nonsingular variety V, then any general member of Λ is nonsingular at any point $p \in V \backslash \mathrm{Bs}\,\Lambda$. In particular, if $\mathrm{Bs}(\Lambda) = \varnothing$, then any general member is nonsingular at any point, i.e., it is a disjoint union of nonsingular varieties.*

PROOF. Applying Theorem 7.17 to $f\colon Z_{\Lambda_1} \to \mathbf{P}(\Lambda)$, one obtains the result. □

c. Now, we shall prove that $f\colon Z_{\Lambda_1} \to \mathbf{P}(\Lambda)$ has an irreducible general fiber, if $\dim(\Phi_\Lambda(V)) \geq 2$. This results from the next lemma.

Lemma 7.17

(i) *Let K/k be an algebraically closed extension of fields and let t be a transcendental element over K. Then $K(t)/k(t)$ is also an algebraically closed extension.*

(ii) *Let $x_1, \ldots, x_n$ be elements of K such that* $\mathrm{tr.\,deg}_k\, k(x_1, \ldots, x_n) \geq 2$, *and $u_2, \ldots, u_n$ be elements algebraically independent over K. Suppose that the extension K/k is finitely generated and k is of characteristic zero. Then, if $z = x_1 + u_2 x_2 + \cdots + u_n x_n$, the field $k(u_2, \ldots, u_n, z)$ is algebraically closed in $\mathfrak{K} = K(u_2, \ldots, u_n)$.*

PROOF. (i) Left to the reader as an exercise.

(ii) By hypothesis, z is transcendental over $k(u_2, \ldots, u_n, x_i)$ for some $i \geq 2$. For $c \in L = k(u_2, \ldots, u_n)$, we define $t_c = z + c x_i$, which is transcendental over $L(x_i)$. Denote by H_c the algebraic closure of $L(t_c)$ in $\mathfrak{K}$. $H_c(x_i)/L(t_c, x_i)$ is algebraic and $L(t_c, x_i) = L(z, x_i)$. Hence, $H_c(x_i)$ is contained in the algebraic closure Σ of $L(z, x_i)$ in $\mathfrak{K}$.

$$\begin{array}{ccccccc} L(z, x_i) = L(t_c, x_i) & \subseteq & H_c(x_i) & \subseteq & \Sigma & \subseteq & \mathfrak{K} \\ \cup\!| & & \cup\!| & & & & \\ L(t_c) & \subseteq & H_c & & & & \end{array}$$

Since $\Sigma/L(z, x_i)$ is finite separable, the set of intermediate fields between these is finite by Galois theory. Hence one has γ and $\delta \in L$ with $\gamma \neq \delta$ such that $H_\gamma(x_i) = H_\delta(x_i)$. Since $t_\gamma - t_\delta = (\gamma - \delta)x_i$, it follows that $H_\gamma(t_\delta) = H_\gamma(x_i) = H_\delta(x_i) = H_\delta(t_\gamma)$. Clearly, t_δ is transcendental over H_γ.

On the other hand, L is algebraically closed in H_γ, since so is L in $\mathfrak{K}$.

Thus by assertion (i), $L(t_\delta)$ is algebraically closed in $H_\gamma(t_\delta)$. Hence, $L(t_\delta)$ is also algebraically closed in $H_\delta(t_\gamma) = H_\gamma(t_\delta)$. But $L(t_\delta)$ is algebraic in H_δ by definition. Thus $L(t_\delta) = H_\delta$, i.e., $L(t_\delta)$ is algebraically closed in $\mathfrak{K}$. Now, $t_\delta = x_1 + u_2 x_2 + \cdots + (u_i + \delta)x_i + \cdots + u_n x_n$ and $u_2, \ldots, u_i + \delta, \ldots, u_n$ are algebraically independent over K. Therefore, one knows that $L(x_1 + u_2 x_2 + \cdots + u_n x_n)$ is algebraically closed in $\mathfrak{K}$. □

In order to prove that $f^*(\text{Rat } \mathbf{P}(\Lambda))$ is algebraically closed in $\text{Rat}(Z_\Lambda)$, when $\dim(\Phi_\Lambda(V)) \geq 2$, we use the notation in §7.9.**b**. Then $\text{Rat}(Z_\Lambda) = \text{Rat }(V)(\zeta_2, \ldots, \zeta_{l-1})$ and $\text{Rat}(\mathbf{P}(\Lambda)) = k(\zeta_2, \ldots, \zeta_l)$, where $f^*(\zeta_l) = x_1 + x_2\zeta_2 + \cdots + x_{l-1}\zeta_{l-1}$, and x_i is $-\varphi_i^{(j)}/\varphi_l^{(j)}$ for $1 \leq i \leq l-1$. By hypothesis, $\text{tr. deg}_k\, k(x_1, \ldots, x_{l-1}) = \dim\, \Phi_\Lambda(V) \geq 2$ and $\zeta_2, \ldots, \zeta_{l-1}$ are algebraically independent over $\text{Rat}(V)$. Thus applying Lemma 7.17.(ii), one obtains the next result.

Theorem 7.19. *If Λ is a reduced linear system on a nonsingular complete variety V such that* $\dim\, \Phi_\Lambda(V) \geq 2$, *then any general member of Λ is irreducible and any member of Λ is connected.*

PROOF. Since the field extension $\text{Rat}(Z_\Lambda)/\text{Rat}(\mathbf{P}(\Lambda))$ is algebraically closed, any general member of Λ, i.e., any general fiber $f^{-1}(a)$, is irreducible and an arbitrary member of Λ is connected by Theorem 2.20. □

d. Lemma 7.18. *Let D and D' be divisors on a normal complete variety V. If* $\text{Bs}\,|D'| = \varnothing$ *and D is very ample, then $D + D'$ is also very ample.*

PROOF. One has morphisms $\Phi_D\colon V \to \mathbf{P}|D|$ and $\Phi_{D'}\colon V \to \mathbf{P}|D'|$. Then $h = (\Phi_D, \Phi_{D'})_k\colon V \to \mathbf{P}|D| \times \mathbf{P}|D'|$ is a closed immersion, since so is $\Phi_D = p \circ h$. One has the Segre morphism $\zeta\colon \mathbf{P}|D| \times \mathbf{P}|D'| \to \mathbf{P}^c$, where c is $l(D) \cdot l(D') - 1$. Then $\zeta \circ h$ becomes the closed immersion defined by $(\varphi_0\psi_0 : \cdots : \varphi_a\psi_0 : \cdots : \varphi_a\psi_b)$, where $\Phi_D = (\varphi_0 : \cdots : \varphi_a)$ and $\Phi_{D'} = (\psi_0 : \cdots : \psi_b)$. Since $\{\varphi_0\psi_0, \ldots, \varphi_a\psi_b\} \subset \mathbf{L}_V(D + D')$ and since clearly $\text{Bs}\,|D + D'| \subseteq \text{Bs}\,|D| \cap \text{Bs}\,|D'| = \varnothing$, $D + D'$ is very ample. □

Proposition 7.15. *Let D be an ample divisor and let Δ be an arbitrary divisor on a normal complete variety. Then $\Delta + \nu D$ is a very ample divisor for some* $\nu > 0$.

PROOF. One can assume D is very ample. Then $\mathcal{O}(D)$ is invertible. Applying Theorem 7.7 to $\mathcal{O}(\Delta)$ and $\mathcal{O}(D)$, one has $m > 0$ such that $\mathcal{O}(\Delta) \otimes \mathcal{O}(mD)$ is generated by global sections. Since $\mathcal{O}(\Delta) \otimes \mathcal{O}(mD) \cong \mathcal{O}(\Delta + mD)$, it follows that $\text{Bs}\,|\Delta + mD| = \varnothing$. Hence by the last lemma, $\Delta + (m+1)D$ is very ample. □

Corollary. *If D is a divisor on a projective nonsingular variety V with* $\dim V \geq 2$, *then D is linearly equivalent to $\Gamma_1 - \Gamma_2$, where the Γ_i are nonsingular varieties.*

PROOF. Take a very ample divisor H and choose m such that $D + mH$ is also very ample. Then general members Γ_1 of $|mH|$ and Γ_2 of $|D + mH|$ are nonsingular varieties by Theorems 7.18 and 7.19. □

§7.10 Monoidal Transformations

a. Let I be an ideal of the Noetherian ring R. Define a graded subalgebra $S(R, I)$ of a graded polynomial ring $R[\mathsf{T}]$ by $S(R, I) = \sum_{d=0}^{\infty} I^d \mathsf{T}^d$, which is isomorphic to $\oplus_{d=0}^{\infty} I^d$ (see Exercise 3.3).

Proposition 7.16

(i) *If I has a basis $(x_1, \ldots, x_n)$, then $S(R, I) = R[x_1\mathsf{T}, \ldots, x_n\mathsf{T}]$, i.e., it is finitely generated.*
(ii) *If $I \cong R$, i.e., $I = xR$ for some x which is not a zero divisor, then $R[\mathsf{T}] \cong S(R, I) = R[x\mathsf{T}]$.*
(iii) *If R is an integral domain, then so is $S(R, I)$.*
(iv) *$S(R, I)_{[x_i\mathsf{T}]} \cong R[x_1/x_i, \ldots, x_n/x_i]$ for any $1 \le i \le n$.*

PROOF. Left to the reader as an exercise. □

Associated with the graded R-algebra $S(R, I)$, one has the R-scheme Proj $S(R, I)$, whose structure morphism is denoted by μ: Proj $S(R, I) \to$ Spec R.

Proposition 7.17. *Let $X = \operatorname{Spec} R$, let $Z = \operatorname{Spec}(R/I)$ be identified with $V(I)$, and let $Y = \operatorname{Proj} S(R, I)$. Then*

(i) *$\mu^{-1}(\tilde{I})$ is an invertible sheaf; $\mu^{-1}(\tilde{I}) \cong \mathcal{O}_Y(1)$ (defined to be $S(R, I)(1)^{\sim}$).*
(ii) *μ induces an isomorphism $Y \setminus \mu^{-1}(Z) \xrightarrow{\sim} X \setminus Z$.*

PROOF. (i) Since $\mu^{-1}(\tilde{I})|_{D_+(x_i\mathsf{T})} = (IR[x_1/x_i, \ldots, x_n/x_i])^{\sim}$ for all i, with the identification $D_+(x_i\mathsf{T}) = \operatorname{Spec} R[x_1/x_i, \ldots, x_n/x_i]$, $\mu^{-1}(\tilde{I})$ is the sheaf associated with the homogeneous ideal $\sum_{d=0}^{\infty} I^{d+1}\mathsf{T}^d = IS(R, I)$. But $S(R, I)(1)$ is expressed as $\sum_{d=-1}^{\infty} I^{d+1}\mathsf{T}^d$ ($\subseteq R[\mathsf{T}, \mathsf{T}^{-1}]$), and $S(R, I)(1)/IS(R, I)$ is of TN; hence by Proposition 7.1.(ii), $\mu^{-1}(\tilde{I}) = S(R, I)(1)^{\sim}$, that is $\mathcal{O}_Y(1)$. Since $S(R, I)$ is finitely generated, it follows that $\mathcal{O}_Y(1)$ is invertible.

(ii) Take any $\mathfrak{p} \in X \setminus Z$. Then $IR_\mathfrak{p} = R_\mathfrak{p}$ and $S(R, I) \otimes_R R_\mathfrak{p} \cong S(R_\mathfrak{p}, IR_\mathfrak{p}) \cong R_\mathfrak{p}[\mathsf{T}]$ by Propositions 3.4.(iv) and 7.16.(ii). Hence, $\mu^{-1}(\operatorname{Spec} R_\mathfrak{p}) \cong \operatorname{Spec} R_\mathfrak{p}$, which implies that $\mu^{-1}(\mathfrak{p}) = \{\mathfrak{q}\}$ and μ is an isomorphism around $\mathfrak{q}$ by Proposition 1.50. □

Intuitively speaking, Y is obtained from X by replacing Z by $\mu^{-1}(Z)$ such that $\mu^{-1}(\tilde{I})$ becomes invertible. We say that (Y, μ) is the *monoidal transformation* of X with center $V(I)$. Y is denoted by $Q_{V(I)}(X)$.

b. Furthermore, let $\varphi\colon R \to R_1$ be a homomorphism of Noetherian rings and let I_1 be IR_1. Then one has the canonical homomorphism

$$\psi\colon S(R, I) \underset{R}{\otimes} R_1 = \left(\sum_{d=0}^{\infty} I^d \mathrm{T}^d\right) \underset{R}{\otimes} R_1 \to \sum_{d=0}^{\infty} I^d R_1 \mathrm{T}^d = S(R_1, I_1).$$

Thus $G(\psi) = Q_{V(I_1)}(\mathrm{Spec}\ R_1)$ and $h = \mathrm{Proj}\ \psi$ is a morphism into $(Q_{V(I)}(\mathrm{Spec}\ R)) \otimes_R R_1$. In particular, if $I_1 \cong R_1$, then $S(R, I) \otimes_R R_1 \cong S(R_1, I_1) \cong R_1[\mathrm{T}]$ and so $(Q_{V(I)}(\mathrm{Spec}\ R)) \otimes_R R_1 \cong Q_{V(I_1)}(\mathrm{Spec}\ R_1) \cong \mathrm{Spec}\ R_1$.

Now, let X be a Noetherian scheme and let Z be a closed subscheme defined by an $\mathcal{O}_X$-ideal $\mathcal{S}$. Let $\{U_\lambda \mid \lambda \in \Lambda\}$ be an affine open cover of X such that $\mathcal{S}|_{U_\lambda} = \tilde{I_\lambda}$ for some ideal I_λ of $R_\lambda = A(U_\lambda)$. Then, it is easy to see that $\{\mu_\lambda\colon Q_{V(I_\lambda)}(U_\lambda) \to U_\lambda \subset X\}$ can be glued, obtaining a morphism $\mu\colon Y \to X$ with $\mu^{-1}(U_\lambda) \cong Q_{V(I_\lambda)}(U_\lambda)$, and $\mu|_{\mu^{-1}(U_\lambda)} = \mu_\lambda$ for each λ.

Definition. Y is said to be the *monoidal transform* of X with center Z, and is denoted by $Q_Z(X)$. Furthermore, $\mu\colon Q_Z(X) \to X$ is said to be the monoidal transformation.

From Proposition 7.17, the next result follows.

Proposition 7.18

(i) *$\mu^{-1}\mathcal{S}$ is an invertible sheaf which is μ-ample.*
(ii) *μ induces an isomorphism $Y \backslash \mu^{-1}(Z) \cong X \backslash Z$.*

If $g\colon X_1 \to X$ is a morphism between Noetherian schemes and Z_1 is the closed subscheme defined by $\mu^{-1}(\mathcal{S})$, then one has a morphism $\tau\colon Q_{Z_1}(X_1) \to Q_Z(X) \times_X X_1$. Hence composing τ with the projection, one obtains $h\colon Q_{Z_1}(X_1) \to Q_Z(X)$ and the commutative diagram:

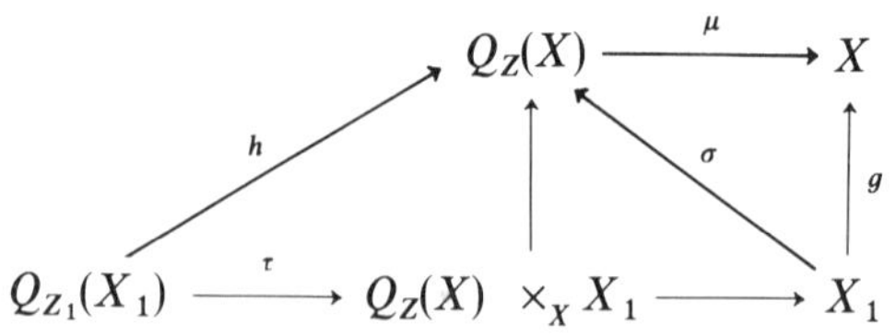

Note that if $g^{-1}\mathcal{S}$ is invertible, then $Q_{Z_1}(X_1) \cong X_1$ and so one obtains $\sigma\colon X_1 \to Q_Z(X)$ with $g = \mu \circ \sigma$. Therefore, $(Q_Z(X), \mu)$ is universal among pairs (X_1, g) such that the $g^{-1}\mathcal{S}$ are invertible.

EXAMPLE 7.4. If $\mathcal{S}$ is invertible, then $Q_Z(X) = X$. If $\mathcal{S}^m = 0$ for some $m > 0$, then $Q_Z(X) = \varnothing$.

c. If I is an ideal of R, then $S(R, I) \otimes_R R/I \cong \sum_{d=0}^{\infty} (I^d/I^{d+1})\mathsf{T}^d$, i.e., $\mathrm{gr}_I(R)$. Hence, letting $\mu: Q_{V(I)}(\mathrm{Spec}\ R) \to \mathrm{Spec}\ R$ be the monoidal transformation, one sees that $\mu^{-1}(V(I)) \cong \mathrm{Proj}(\mathrm{gr}_I(R))$. $E = \mu^{-1}(V(I))$ is defined by $(S(R, I)(1))^{\sim}$; hence $\mathcal{O}(-E) = (S(R, I)(1))^{\sim}$, and so $\mathcal{O}(-E)|_E \cong \mathcal{O}_E(1)$ which is the fundamental sheaf associated with the projective scheme $\mathrm{Proj}(\mathrm{gr}_I(R))$.

Suppose that V is a variety over a field k and p is a closed point of V with $k(p) = k$. If U is an affine open neighborhood of p, then $Q_p(V)|_{\mu^{-1}(U)} = Q_p(U)$ and so $\mu^{-1}(p) \cong \mathrm{Proj}(\mathrm{gr}_I(R))$, where I is a maximal ideal with $V(I) = \{p\}$. But since $I^d/I^{d+1} \otimes_R R_I \cong (IR_I)^d/(IR_I)^{d+1}$ and IR_I is the maximal ideal of $R_I = \mathcal{O}_{V,p}$, $\mu^{-1}(p) \cong \mathrm{Proj}(\mathrm{gr}_{\mathfrak{m}}(\mathcal{O}_{V,p}))$, where $\mathfrak{m}$ denotes $\mathfrak{m}_p$. Therefore, if p is a regular point of V and $n = \dim V$, then $\mathrm{gr}_I(R) \cong k[\mathsf{T}_1, \ldots, \mathsf{T}_n]$ as graded rings and so $\mu^{-1}(p) \cong \mathbf{P}_k^{n-1}$. Furthermore, $\mathcal{O}(-E)|_E$ is the invertible sheaf defined by a hyperplane. Hence, the next result is obtained.

Proposition 7.19. *If p is a nonsingular point of a variety V with $n = \dim V$, then $E = \mu^{-1}(p) \cong \mathbf{P}_k^{n-1} = \mathbf{P}$ and $\mathcal{O}(-E)|_E \cong \mathcal{O}_{\mathbf{P}}(1)$, where $\mu: Q_p(V) \to V$ is the monoidal transformation.*

d. The next fundamental result was proved by Hironaka.

Theorem 7.20 (Hironaka). *Given any variety V over a field k of characteristic zero, there exists a sequence of monoidal transformations $\{\mu_j: V_j = Q_{Z_{j-1}}(V_{j-1}) \to V_{j-1} | 1 \le j \le l\}$ such that*

(1) $V_0 = V$,
(2) *Z_j is a nonsingular closed subvariety contained in* $\mathrm{Sing}(V_j)$,
(3) *V_l is a nonsingular variety.*

Proof. We refer the reader to "Resolution of singularities of an algebraic variety over a field of characteristic zero" by H. Hironaka, in *Ann. of Math.* 79(1964), 109–326. □

Definition. Given a variety V, a pair of a nonsingular variety $V^{\#}$ and a proper birational morphism $\mu: V^{\#} \to V$ is said to be a *nonsingular model* of V.

Thus, by the last theorem, for any variety one has a nonsingular model $(V_l, \mu_l \circ \cdots \circ \mu_1)$. Note that each μ_j is a projective morphism by Proposition 7.18. Hence, $\mu = \mu_l \circ \cdots \circ \mu_1$ is also a projective morphism. In particular, if V is a projective variety, then V^* is a projective nonsingular variety by Theorem 7.11.

Corollary to Theorem 7.20. *Given a variety W, there exists a nonsingular projective variety V_l which is birationally equivalent to W.*

Proof. Let W_0 be a dense affine open subset of W, which is a closed subvariety of some affine space $\mathbf{A}_k^N$. Regarding $\mathbf{A}_k^N$ as an open subset of the projective space $\mathbf{P}_k^N$, take the closure scheme of W_0 in $\mathbf{P}_k^N$, which is denoted by V. Applying the last theorem to V, one obtains a nonsingular model (V_l, μ). Then V_l has the required property. □

The next simple example shows how singularities are resolved by monoidal transformations. Let C be an affine curve $V(\mathsf{Y}^2 - \mathsf{X}^{2r+1})$ on $\mathbf{A}_k^2$ for some $r > 0$. Let $\mu_1\colon C_1 \to C$ be the monoidal transformation with center $p = (0, 0)$. Then $R = A(C) = k[x, y]$, where $y = \mathsf{Y}|_C$ and $x = \mathsf{X}|_C$; hence C_1 is covered by $D_+(x\mathsf{T}) = \operatorname{Spec} R[y/x]$ and $D_+(y\mathsf{T}) = \operatorname{Spec} R[x/y]$ by Proposition 7.14.(iv). If $u = y/x$, $R[u] = k[x, u]$ and $u^2 = x^{2r-1}$. Furthermore, if $v = x/y$, $R[v] = k[y, v] = k[y, y^{-1}] = k[xu, (xu)^{-1}]$. Hence $D_+(y\mathsf{T}) \subset D_+(x\mathsf{T})$ and so $C_1 = D_+(x\mathsf{T}) = V(\mathsf{Y}^2 - \mathsf{X}^{2r-1})$. Thus after a succession of r monoidal transformations, the singularity of C at p is completely resolved.

e. Let V be a closed subvariety of a variety X and let Z be a closed subscheme of V with $Z \neq V$ defined by an $\mathcal{O}_X$-ideal $\mathcal{J}$. Then if $\mathcal{S} = \mathcal{J}|_V$, Z is defined by the $\mathcal{O}_V$-ideal $\mathcal{S}$. Let $\mu\colon Q_Z(X) \to X$ and $\mu'\colon Q_Z(V) \to V$ denote monoidal transformations.

Define Y to be the closure of $\mu^{-1}(V \backslash Z)$ which is isomorphic to $V \backslash Z$. The τ induced from the inclusion map $j\colon V \to X$ is a morphism τ_1 onto Y by Proposition 1.30, since τ gives rise to the isomorphism: $Q_Z(V) \backslash \mu'^{-1}(Z) \cong Y \backslash \mu^{-1}(Z)$.

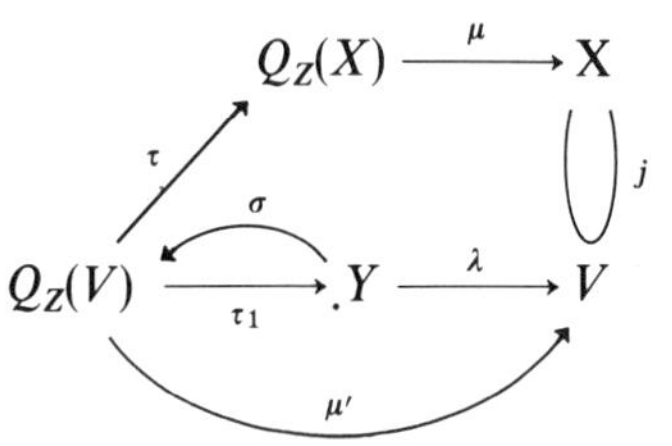

Since $\mu^{-1}\mathcal{J}$ is invertible, $\mu^{-1}(\mathcal{J})|_Y$ is invertible. But, letting $\lambda\colon Y \to V$ be the restriction of μ to Y, $\mu^{-1}(\mathcal{J})|_Y = \lambda^{-1}(\mathcal{S})$. Hence $\lambda^{-1}(\mathcal{S})$ is also an invertible sheaf. Therefore, from the universality of monoidal transformation (subsection **b**) one has $\sigma\colon Y \to Q_Z(V)$ such that $\lambda = \mu' \circ \sigma$. Therefore, $\lambda \circ \tau_1 = \mu'$ implies that $\lambda \circ \tau_1 \circ \sigma = \mu' \circ \sigma = \lambda$; hence $\tau_1 \circ \sigma = \mathrm{id}$, because λ is a birational morphism. Thus, τ_1 and σ are isomorphisms, i.e., $Q_Z(V) \cong Y$.

Definition. If V is a subvariety of X, then the closure of $\mu^{-1}(V \backslash Z)$ is said to be the *strict* (or, *proper*) *transform* of V by the monoidal transformation $\mu\colon Q_Z(X) \to X$.

Hence, by the above observation, the monoidal transform of V is obtained as the strict transform of V by $\mu: Q_Z(X) \to X$ if $Z \subset V \subset X$.

Suppose that X is a nonsingular variety and Z is a nonsingular closed subvariety of a closed subvariety V of X with $\operatorname{codim}(V) = 1$. Take a general closed point p of Z and a coordinate neighborhood U of p in X. Since Z is nonsingular, one has a local coordinate system $(z_1, \ldots, z_r, x_{r+1}, \ldots, x_n)$, where $r = \dim Z$ and $n = \dim X$, such that $Z \cap U = V(x_{r+1}, \ldots, x_n)$. Replacing U by X, and letting $R = A(U)$ and $I = (x_{r+1}, \ldots, x_n)$, $Q_Z(X)$ is covered by $\{D_+(x_i \mathrm{T}) \mid r+1 \le i \le n\}$. $Y_i = D_+(x_i \mathrm{T})$ ($\cong \operatorname{Spec} R_i$, $R_i = R[x_{r+1}/x_i, \ldots, x_n/x_i]$,) is a coordinate neighborhood with $(z_1, \ldots, z_r, u_{r+1}, \ldots, u_{i-1}, x_i, u_{i+1}, \ldots, u_n)$ the local coordinate system, where the $u_j = x_j/x_i$. Hence, $Q_Z(X)$ is also nonsingular.

If V is defined by $f \in R$, then $\mu^{-1}(V) \cap Y_i$ is defined by fR_i. Let $\nu = \nu_p(f)$, which is the order of f with respect to $\mathfrak{m}_p$. Note that $\nu = e(p, V)$ by Proposition 2.11. Since p is a general point of Z, it follows that $Z = V(I)$ and $f \in I^\nu \setminus I^{\nu+1}$. Then $IR_i = x_i R_i$ and so $f = x_i g_i$ in R_i, where g_i is not a multiple of x_i. Therefore, $\mu^{-1}(V) \cap Y_i = (\mu^{-1}(Z) \cap Y_i) \cup V(g_i)$ and $V(g_i) \not\supseteq \mu^{-1}(Z)$. Hence, $(\mu^{-1}(V) \setminus \mu^{-1}(Z)) \cap Y_i = V(g_i) \setminus \mu^{-1}(Z) \cap Y_i$; therefore $V(g_i)$ is the strict transform of V on Y_i. Thus, one obtains the next result.

Proposition 7.20. *Let V be a prime divisor on a nonsingular variety X and let Z be a closed nonsingular subvariety of X with $Z \neq X$, and $\mu: Q_Z(X) \to X$ the monoidal transformation with center Z.*

(i) *If $Z \subset V$ and $\nu = e(p, V)$, where p is a general point of Z, then $\mu^*(V) = \nu \cdot \mu^{-1}(Z) + V'$, where V' is the strict transform of V by μ.*
(ii) *If $Z \not\subseteq V$, then $\mu^*(V)$ is the strict transform of V by μ.*

If $\operatorname{codim}(Z) \ge 2$, then $\mu^{-1}(Z)$ is called the *exceptional divisor* of μ, or with respect to μ.

f. Let D be an effective divisor $\sum_{i=1}^r m_i \Gamma_i$, with $m_i > 0$, and $\Gamma_i \neq \Gamma_j$ for $i \neq j$, on a normal variety V.

Definition. $\sum_{i=1}^r \Gamma_i$ is said to be the *reduced divisor obtained from D*, denoted D_{red}. If $D = D_{\text{red}}$, it is said to be a *reduced divisor*.

Definition. If V is nonsingular, a reduced divisor $D = \sum_{i=1}^r \Gamma_i$ is said to have only *simple normal crossings* at a closed point p, if there exists a coordinate neighborhood U of p with $(z_1, \ldots, z_n)$ a local coordinate system such that $D \cap U$ is defined by $z_{j_1} \cdots z_{j_s}$ for $1 \le j_1 < \cdots < j_s \le n$.

If D has only simple normal crossings everywhere, D is said to be a *divisor with simple normal crossings*. Furthermore, D is said to be a *smooth boundary* of $V \setminus D$.

The next result is a consequence of Hironaka's Main Theorem II.

Theorem 7.21. *Let D be a reduced divisor on a nonsingular variety V over a field k with characteristic zero. There exists a sequence of monoidal transformations μ_j: $V_j = Q_{Z_{j-1}}(V_{j-1}) \to V_{j-1}$ and reduced divisors $D_{[j]}$ on V_j for $1 \le j \le l$ such that*

(1) $V_0 = V$, $D_{[0]} = D$,
(2) Z_j *is nonsingular and* $Z_j \subset D_j$,
(3) $D_{[j]} = \mu_j^{-1}(D_{[j-1]})$ *(defined to be* $\mu^*(D_{[j-1]})_{\text{red}}$*)*,
(4) $D_{[l]}$ *is a divisor with simple normal crossings.*

PROOF. We refer the reader to the paper cited in the proof of Theorem 7.20. □

The following is a typical example of Theorem 7.21. Let $V = \mathbf{A}_k^2$ and $D = V(\mathsf{Y}^2 - \mathsf{X}^3)$. Furthermore, D' denotes the strict transform of D and E_i denotes $\mu_i^{-1}(p_{i-1})$, where $p_{i-1} = Z_{i-1}$ for all i.

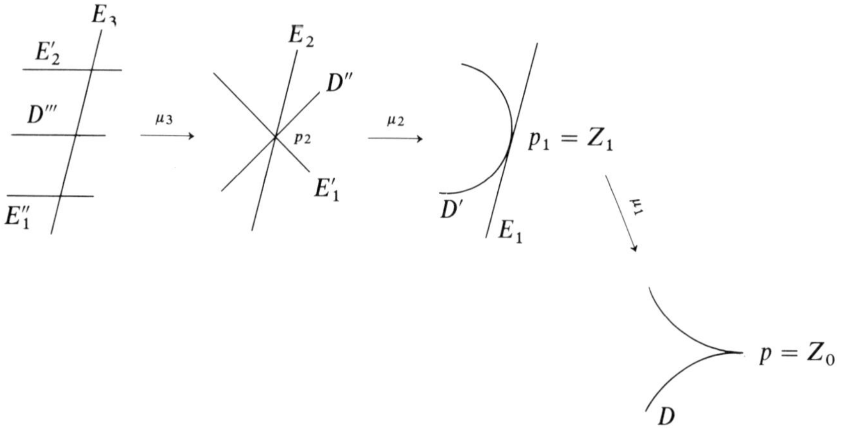

Now, if D is a closed proper subset of a nonsingular variety V, let μ: $Q_D(V) \to V$ be the monoidal transformation and let (W, λ) be a nonsingular model of $Q_D(V)$. Then if $\varphi = \mu \circ \lambda$, $\varphi^{-1}(D)$ is a support of some reduced divisor. But since $\operatorname{Sing}(Q_D(V)) \subseteq \mu^{-1}(D)$, one can assume that λ is an isomorphism outside $\mu^{-1}(D)$. Hence, $W \backslash \varphi^{-1}(D) \cong Q_D(V) \backslash \mu^{-1}(D) \cong V \backslash D$. Then applying Theorem 7.21, one obtains a nonsingular variety W_l and a proper birational morphism $\tilde{\mu}$: $W_l \to W$ such that $\tilde{\mu}^{-1}\varphi^{-1}(D)$ has only simple normal crossings at every point and $W_l \backslash \psi^{-1}(D) \cong V \backslash D$, where ψ is $\varphi \circ \tilde{\mu}$. The pair $(W_l, \psi^{-1}(D))$ is said to be a *smooth completion* of $V \backslash D$ *with smooth boundary* $\psi^{-1}(D)$ (see §11.1).

Let V be a nonsingular variety and X be a variety. Let f: $V \to X$ be a strictly rational map. Then, there exists a graph Γ_f and the proper birational morphism π: $\Gamma_f \to V$. By Theorem 7.20, one obtains a nonsingular model

(W, λ) of Γ_f. $(W, \pi \circ \lambda)$ is a reduction model of f. In this case, since W is nonsingular, one says that $(W, \pi \circ \lambda)$ is a *nonsingular reduction model of f*. Note that $\pi \circ \lambda$ is a proper birational morphism.

Exercises

7.1. Using Theorem 7.8, prove the following assertions. Let X be a scheme proper over Spec R, and let $\mathscr{L}$ be an invertible sheaf on X.
(i) If $\mathscr{L}/\mathscr{N}\mathscr{L}$ is R-ample, then $\mathscr{L}$ is also R-ample, where $\mathscr{N}$ is the nilradical of $\mathscr{O}_X$.
(ii) Let $X_1, \ldots, X_n$ be the irreducible components of X. If all the $\mathscr{L}|_{X_j}$ are R-ample, then $\mathscr{L}$ is also R-ample.

7.2. Let $h: X \to Y$ be a finite surjective R-morphism, and let $\mathscr{L}$ be an invertible sheaf on Y. If $h^*\mathscr{L}$ is R-ample, then show that $\mathscr{L}$ is also R-ample.

7.3. Let B be a Noetherian local ring with maximal ideal $\mathfrak{n}$. Let $X = \operatorname{Spec} B$, $p = V(\mathfrak{n})$, $Y = Q_p(X)$, and $E = \mu^{-1}(p)$, where $\mu: Y \to X$ is the monoidal transformation with center p. Prove that $\dim Y = \dim X$, $\dim E = \dim X - 1$, and finally that $\dim(\operatorname{Proj}(\operatorname{gr}_{\mathfrak{n}}(B))) = \dim B - 1$.

7.4. Let X be a projective scheme and let $\mathscr{L}$ be an ample sheaf on X. Given an open subset U and $x_1, \ldots, x_r \in U$, show that there exists $f \in \Gamma(X, \mathscr{L}^{\otimes m})$ for some $m > 0$ such that $\{x_1, \ldots, x_r\} \subseteq X_f \subseteq U$.

7.5. Let V and W be nonsingular varieties over $\mathbb{C}$, and let $f: V \to W$ be a strictly rational map. If $f(p)$ is not a point for a closed point p of V, then show that $f(p)$ contains a rational curve (cf. Theorem 7.21).

Chapter 8

Intersection Theory of Divisors

§8.1 Intersection Number of Curves on a Surface

a. Throughout this chapter, let S denote a projective nonsingular surface over an algebraically closed field k with characteristic zero. For any divisors D_1 and D_2, we wish to define an intersection number $(D_1, D_2)_S \in \mathbb{Z}$ which is required to have the following properties (see §2.8.**b**):

(i) If $D_1 \sim 0$, then $(D_1, D_2)_S = 0$, for any $D_2 \in \mathbb{G}(S)$.
(ii) $(D_1, D_2)_S = (D_2, D_1)_S$, for any $D_1, D_2 \in \mathbb{G}(S)$.
(iii) For any $a_1, a_2 \in \mathbb{Z}$ and $D_1, D_2, D_3 \in \mathbb{G}(S)$, it follows that $(a_1 D_1 + a_2 D_2, D_3)_S = a_1(D_1, D_3)_S + a_2(D_2, D_3)_S$.
(iv) For any nonsingular curve C on S and $D \in \mathbb{G}(S)$, $(D, C)_S = \deg(\mathcal{O}(D)|_C)$.

Since any divisor D_1 is linearly equivalent to a difference of nonsingular curves by the Corollary to Proposition 7.15, these properties determine $(\ ,\ )_S$ uniquely. But how can we define $(\ ,\ )_S$? To find the definition of $(\ ,\ )_S$, we consider condition (iv). From the exact sequence

$$0 \to \mathcal{O}(-D-C) \to \mathcal{O}(-D) \to \mathcal{O}(-D)|_C \to 0,$$

letting $\chi(D) = \chi_S(\mathcal{O}(D))$ and $\chi_S = \chi(0)$, one obtains

$$\chi_C(\mathcal{O}(-D)|_C) = \chi(-D) - \chi(-D-C).$$

Putting $D = 0$, one has

$$\chi_C = \chi_S - \chi(-C).$$

By (iv), one obtains $(-D, C)_S = \deg(\mathcal{O}(-D)|_C)$, i.e., $-\chi_C + \chi_C(\mathcal{O}(-D)|_C)$ by Theorem 4.11. Combining these, one has

$$(-D, C)_S = -\chi_C + \chi_C(\mathcal{O}(-D)|_C) = -\chi_S + \chi(-D) + \chi(-C) - \chi(-D-C).$$

From (iii), it follows that $(-D, C)_S = -(D, C)_S$; hence,

$$(D, C) = \chi_S - \chi(-D) - \chi(-C) + \chi(-D-C).$$

With the above observation in mind, one defines $(D_1, D_2)_S$ as follows.

Definition. For any D_1 and $D_2 \in \mathbb{G}(S)$, define $(D_1, D_2)_S$ to be

$$\chi_S - \chi(-D_1) - \chi(-D_2) + \chi(-D_1 - D_2).$$

By definition, $(D_1, D_2)_S$ has properties (i), (ii), and (iv). We shall prove that (iii) is also satisfied. First, we show this when D_3 is a nonsingular curve C. Then by (iv),

$$\begin{aligned}(a_1D_1 + a_2D_2, D_3)_S &= \deg(\mathcal{O}(a_1D_1 + a_2D_2)|_C)\\ &= \deg(\mathcal{O}(D_1)|_C^{\otimes a_1} \otimes_{\mathcal{O}_C} \mathcal{O}(D_2)|_C^{\otimes a_2}) = a_1 \deg(\mathcal{O}(D_1)|_C) + a_2 \deg(\mathcal{O}(D_2)|_C)\end{aligned}$$

by Corollary 2 to Theorem 4.11. Hence, (iii) is established in this case. For a general D_3, take a very ample divisor H. Then there exists m_0 such that $D_3 + mH$ is very ample for all $m \geq m_0$, by Proposition 7.15. Choosing a general member $C_{(m)}$ of $|D_3 + mH|$, $C_{(m)}$ is a nonsingular curve by Theorems 7.18 and 7.19, and so one has by the above discussion,

$$(a_1D_1 + a_2D_2, C_{(m)})_S = a_1(D_1, C_{(m)})_S + a_2(D_2, C_{(m)})_S.$$

By (i), $(a_1D_1 + a_2D_2, C_{(m)})_S = (a_1D_1 + a_2D_2, D_3 + mH)_S$ and $(D_i, C_{(m)})_S = (D_i, D_3 + mH)_S$. By Theorem 7.15, the function $\psi(m) = (a_1D_1 + a_2D_2, D_3 + mH)_S - a_1(D_1, D_3 + mH)_S - a_2(D_2, D_3 + mH)_S$ is a polynomial for all m, and $\psi(m) = 0$ for $m \geq m_0$. Hence $\psi(m) = 0$ for all m. Then $0 = \psi(0) = (a_1D_1 + a_2D_2, D_3)_S - a_1(D_1, D_3)_S - a_2(D_2, D_3)_S$.

b. For a closed point p of S, if φ_1 and φ_2 are elements of $\mathfrak{m}_p \subset \mathcal{O}_{S,p}$ without common prime factors, then $\dim(\mathcal{O}_{S,p}/(\varphi_1, \varphi_2)) = 0$; hence $\mathcal{O}_{S,p}/(\varphi_1, \varphi_2)$ is an Artinian ring by Theorem 1.10. Thus $\mathcal{O}_{S,p}/(\varphi_1, \varphi_2)$ is a finite-dimensional vector space over k. Letting (x, y) be a local coordinate system around p, one has $I_0(\varphi, x) = \operatorname{ord}_y(\varphi \bmod (x)) \geq \nu(\varphi)$ which is equal to $e(p, V(\varphi))$.

Definition. The *local intersection number* $I_0(\varphi_1, \varphi_2)$ is defined to be the dimension of the k-vector space $\mathcal{O}_{S,p}/(\varphi_1, \varphi_2)$. If D_1 and D_2 are effective divisors on S, and $p \in D_1 \cap D_2$ such that D_1 and D_2 have no common prime divisors containing p, then the *local intersection number* $I_p(D_1, D_2)$ is defined to be $I_0(\varphi_1, \varphi_2)$ where $\mathcal{O}(-D_i)_p = \varphi_i \mathcal{O}_{S,p}$.

Proposition 8.1. *Let D_1 and D_2 be effective divisors with $D_1 \wedge D_2 = 0$. Then $(D_1, D_2)_S = \sum_{p \in D_1 \cap D_2} I_p(D_1, D_2)$.*

PROOF. From the exact sequence

$$0 \to \mathcal{O}(-D_1) \to \mathcal{O}_S \to \mathcal{O}_{D_1} \to 0,$$

one obtains an exact sequence

$$0 \to \mathcal{O}(-D_1)|_{D_2} \to \mathcal{O}_{D_2} \to \mathcal{O}_{D_1} \otimes_{\mathcal{O}_S} \mathcal{O}_{D_2} \to 0 \tag{$*$}$$

by the next lemma.

Lemma 8.1. *Let R be a UFD and let φ, ψ be nonzero elements of R. If these are not units and have no common prime factors, then the natural homomorphism*

$$\varphi R \underset{R}{\otimes} R/\psi R \to R/\psi R$$

is injective.

PROOF. Any element of $\varphi R \otimes_R R/\psi R$ can be written as $\varphi \otimes (\omega \bmod \psi R)$ for some ω. If $\varphi\omega \in \psi R$, then $\omega \in \psi R$, by the Remark to Proposition 2.4, since R is a UFD. This implies $\varphi \otimes (\omega \bmod \psi R) = 0$. □

Now we return to the proof of Proposition 8.1. By §1.21.**d**, the scheme-theoretic intersection $D_1 \cap D_2$ is defined by an $\mathcal{O}_S$-ideal $\mathcal{O}(-D_1) + \mathcal{O}(-D_2)$; hence $\mathcal{O}_{D_1} \otimes_{\mathcal{O}_S} \mathcal{O}_{D_2} \cong \mathcal{O}_{D_1 \cap D_2}$. Thus by Lemma 4.3

$$\dim_k \Gamma(D_1 \cap D_2, \mathcal{O}_{D_1 \cap D_2}) = \sum_{p \in D_1 \cap D_2} I_p(D_1, D_2),$$

since $D_1 \cap D_2$ consists of finitely many (algebraic) points.

On the other hand, by the exact sequence $(*)$, one has

$$\begin{aligned}\dim_k \Gamma(D_1 \cap D_2, \mathcal{O}_{D_1 \cap D_2}) &= \chi(\mathcal{O}_{D_1} \otimes_{\mathcal{O}_S} \mathcal{O}_{D_2}) = \chi_{D_2} - \chi_{D_2}(\mathcal{O}(-D_1)|_{D_2}) \\ &= \chi_S - \chi(-D_2) - \chi(-D_1) + \chi(-D_1 - D_2).\end{aligned}$$

The last term is $(D_1, D_2)_S$, by definition. □

Corollary

(i) *Let C_1 and C_2 be distinct curves on S. Then $(C_1, C_2)_S \geq 0$. Furthermore, in this case $(C_1, C_2)_S = 0$ if and only if $C_1 \cap C_2 = \varnothing$.*
(ii) *If D is an effective divisor and C is a curve on S with $C \not\subseteq D$, then $(C, D)_S \geq 0$.*
(iii) *Let H be an effective divisor on S such that $S\backslash H$ is an affine surface* Spec *A. If D_1 and D_2 are effective divisors without common prime divisors, then*

$$\sum_{p \notin H} I_p(D_1, D_2) = \dim_k(A/(I_1, I_2)),$$

where $I_i = \Gamma(S \backslash H, \mathcal{O}(-D_i))$, *for* $i = 1, 2$. *In particular, if* $(D_1 \cap D_2) \cap H = \emptyset$ *and* $I_i = Af_i$, *then*

$$(D_1, D_2)_S = \dim_k(A/(f_1, f_2)).$$

PROOF. (i) This is clear from the last proposition. (ii) follows from (i). (iii) is left to the reader. □

c. EXAMPLE 8.1. Let $S = \mathbf{P}_k^2$ and let C, D be distinct curves on S with $n = \deg C$ and $m = \deg D$. Then letting H denote a line $V_+(\mathrm{T}_0)$ on $\mathbf{P}_k^2$, one has $C \sim nH$ and $D \sim mH$ by §2.8.**d**. By definition, $(C, D)_S = (nH, mH)_S = nm(H, H)_S$. Since $\mathcal{O}(H)|_H$ has degree 1, $(H, H)_S = 1$ by (iv). Thus $(C, D)_S = nm$. On the other hand, $(C, D)_S = \sum_{p \in C \cap D} I_p(C, D)$ by Proposition 8.1. Hence, $\sum_{p \in C \cap D} I_p(C, D) = nm$. This is the so called *Bézout's Theorem*.

EXAMPLE 8.2. Let $S = \mathbf{P}_k^1 \times \mathbf{P}_k^1$ and let $H_1 = \mathbf{P}_k^1 \times p$, $H_2 = p \times \mathbf{P}_k^1$ for a closed point p. Then $(H_i, H_i)_S = 0$ and $(H_1, H_2)_S = 1$ by applying Corollary (i) to the last proposition. If C and D are curves on S, then $C \sim a_1 H_1 + a_2 H_2$ and $D \sim b_1 H_1 + b_2 H_2$ for some $a_i, b_i \in \mathbb{Z}$. Thus $(C, D)_S = a_1 b_2 + a_2 b_1$.

d. If C is a curve on S and D is a divisor on S, then by rewriting the definition of $(D, C)_S$ one has

$$(D, C)_S = \chi_C - \chi_C(\mathcal{O}(-D)|_C).$$

The right-hand side is $-\deg(\mathcal{O}(-D)|_C)$ by the definition of deg (§4.8.**a**). Furthermore, by Corollary 2 to Theorem 4.11, one has $\deg(\mathcal{O}(-D)|_C) = -\deg(\mathcal{O}(D)|_C)$; hence,

$$(D, C)_S = \deg(\mathcal{O}(D)|_C).$$

This is a generalization of property (iv) into the case of singular curves. By this, one can prove the next proposition.

Proposition 8.2. *Let* W *be a nonsingular projective surface and let* $f\colon W \to S$ *be a surjective morphism. If* D_1 *and* D_2 *are divisors on* S, *then* $(f^*D_1, f^*D_2)_W = \deg f \cdot (D_1, D_2)_S$.

PROOF. One can assume that D_2 is a curve C. If $f^*C = \sum_{i=1}^s n_i \Gamma_i$ is the irreducible decomposition of the pullback, one has $\deg f = \sum_{i=1}^s n_i \deg f_i$, where f_i is $f|_{\Gamma_i}\colon \Gamma_i \to C$, by Lemma 2.26. Moreover, $(f^*D_1, f^*C)_W = \sum_{i=1}^s n_i (f^*D_1, \Gamma_i)_W$ by property (iii), and $(f^*D_1, \Gamma_i)_W = \deg(\mathcal{O}(f^*D_1)|_{\Gamma_i}) = \deg(f_i^*(\mathcal{O}(D_1)|_C)$. But by Corollary 2 to Theorem 4.11, one has $\deg f_i^*(\mathcal{O}(D_1)|_C) = \deg f_i \cdot \deg(\mathcal{O}(D_1)|_C) = \deg f_i \cdot (D_1, C)_S$. Thus,

$$\sum_{i=1}^s n_i (f^*D_1, \Gamma_i)_W = \sum_{i=1}^s n_i \deg f_i \cdot (D_1, C)_S = \deg f \cdot (D_1, C)_S. \qquad \square$$

Proposition 8.3. *With the same assumptions as above, let E be a curve on W such that $f(E)$ is a point p and let D be a divisor on S. Then $(f^*D, E)_W = 0$.*

PROOF. $(f^*D, E)_W = \deg(\mathcal{O}(f^*D)|_E)$ and $\mathcal{O}(f^*D)|_E \cong \mathcal{O}_E$, since $\mathcal{O}(f^*D)|_E = f^*(\mathcal{O}(D))|_E = f_1^*(\mathcal{O}(D)|_p) \cong \mathcal{O}_E$, where $f_1 = f|_E \colon E \to p$. Hence, $(f^*D, E)_W = 0$. □

Lemma 8.2. *Let p be a closed point of S and let $\mu \colon W = Q_p(S) \to S$ be the corresponding monoidal transformation. Then $E = \mu^{-1}(p)$ is a projective line with $(E, E)_W = -1$. Furthermore, $K(W) \sim \mu^*K(S) + E$.*

PROOF. The first part is a special case of Proposition 7.19. To prove the second part, it suffices to establish the formula $R_\mu = E$ by Theorem 5.5. But this is clear, since $dx_1 \wedge dx_2 = x_1 dx_1 \wedge du_2$, where $u_2 = x_2/x_1$, and (x_1, x_2) is a local coordinate system at p (cf. §7.10.**e**). □

e. If C is a curve on S passing through p, the multiplicity $e(p, C)$ of C at p has been defined in §2.5.**d**. Let $n = e(p, C)$ and let C^* be the strict (proper) transform of C by μ. Then $\mu^*C = C^* + nE$ by Proposition 7.20.(i). For simplicity, D^2 denotes $(D, D)_S$, if $D \in \mathbb{G}(S)$.

Theorem 8.1 (Max Noether)

(*i*) *If C and Γ are curves on S with $n = e(p, C)$ and $m = e(p, \Gamma)$, then $(C^*, \Gamma^*)_W = (C, \Gamma)_S - nm$. In particular, $(C^*)^2 = C^2 - n^2$, where C^* and Γ^* denote the strict transforms of C and Γ by μ, respectively.*

(*ii*) *$(K(W), C^*)_W = (K(S), C)_S + n$.*

PROOF. These follow immediately from the last proposition and Lemma 8.2. □

EXAMPLE 8.3. If p is a nonsingular point of C and Γ, and C, Γ have no common tangent at p, i.e., $T_p(C) \neq T_p(\Gamma)$ as subspaces of $T_p(S)$; then $(C, \Gamma)_S = (C^*, \Gamma^*)_W + 1$ and $I_p(C, \Gamma) = 1$. This is clear, since C^* and Γ^* have no common points on $\mu^{-1}(p)$.

§8.2 Riemann–Roch Theorem on an Algebraic Surface

a. If C is a nonsingular curve on a nonsingular projective surface S, then by Theorem 5.4,

$$\mathcal{O}(K(C)) \cong \mathcal{O}(K(S) + C)|_C.$$

Hence, deg $K(C) = (K(S) + C, C)_S$. On the other hand, by the Corollary to Theorem 6.1, $2g(C) - 2 = \deg K(C)$; hence, $g(C) = ((K(S) + C, C)_S + 2)/2$.

Definition. For any divisor D on S, the *virtual genus* $\pi(D)$ of D is defined to be $((K(S) + D, D)_S + 2)/2$.

Then, if C is a nonsingular curve, one has $g(C) = \pi(C)$. Since $g(C) = p_a(C) = 1 - \chi_C = 1 - \chi_S + \chi(-C)$, one obtains the formula:

$$\chi(-C) = \chi_S + \pi(C) - 1 = \frac{(K(S) + C, C)_S}{2} + \chi_S.$$

We prove that this holds for any $D \in \mathbb{G}(S)$.

Theorem 8.2 (Riemann–Roch Theorem). *For any* $D \in \mathbb{G}(S)$,

$$\chi(D) = \pi(-D) - 1 + \chi_S = \frac{(D - K(S), D)_S}{2} + \chi_S.$$

PROOF. Take a very ample divisor H of S. By Proposition 7.15, there exists m_0 such that $-D + mH$ is very ample for every $m \geq m_0$. A general member $C_{(m)}$ of $|-D + mH|$ is a nonsingular curve by Theorems 7.18 and 7.19. Hence,

$$\chi(-C_{(m)}) = \pi(C_{(m)}) - 1 + \chi_S$$

for $m \geq m_0$. But $\chi(-C_{(m)}) = \chi(D - mH)$ and $\pi(C_{(m)}) = \pi(-D + mH)$ are polynomials in m by Theorem 7.14. Hence, one has the equality for any m; thus $\chi(D - mH) = \pi(-D + mH) - 1 + \chi_S$. By putting $m = 0$, one obtains the required equality. □

Corollary 1. *If C is a curve on S, then* $\pi(C) = \dim_k H^1(C, \mathcal{O}_C)$.

PROOF. From the exact sequence

$$0 \to \mathcal{O}(-C) \to \mathcal{O}_S \to \mathcal{O}_C \to 0,$$

one obtains $\chi_C = \chi_S - \chi(-C)$, which is $1 - \pi(C)$ by the last theorem. But since $\dim_k \Gamma(C, \mathcal{O}_C) = 1$, one has the equality. □

Corollary 2 (Riemann–Roch Inequality due to Castelnuovo). *For any* $D \in \mathbb{G}(S)$, $l(D) + l(K(S) - D) \geq \pi(-D) - 1 + \chi_S$.

PROOF. By Theorem 5.7, $\dim_k H^2(S, \mathcal{O}(D)) = l(K(S) - D)$. Then since $\dim_k H^1(S, \mathcal{O}(D)) \geq 0$, one obtains the inequality. □

EXAMPLE 8.4. Let S be a nonsingular surface $V_+(F)$ in $\mathbf{P}_k^3$ with degree δ and let L be a line (i.e., an intersection of two distinct planes) which lies on S. Then $L^2 = 2 - \delta$.

To show this, let H be a divisor $V_+(T_0)|_S$. Then $K(S) \sim (\delta - 4)H$ and $1 = \deg L = \deg(\mathcal{O}(H)|_L) = (H, L)_S$. Since $L \cong \mathbf{P}^1_k$, it follows that $\pi(L) = g(L) = 0$. Thus $-2 = (K(S), L)_S + L^2$. Furthermore, $(K(S), L)_S = (\delta - 4)(H, L)_S = \delta - 4$. Hence $L^2 = 2 - \delta$. In particular, if $\delta = 3$ then $L^2 = -1$. Any curve L on S with $L^2 = (K(S), L)_S = -1$ will be called an *exceptional curve of the first kind* (see §9.5).

b. Let S be a nonsingular projective surface. Define a subgroup $\mathbb{G}(S)^\perp$ of the divisor group $\mathbb{G}(S)$ by

$$\mathbb{G}(S)^\perp = \{D \in \mathbb{G}(S) \mid (D, \Delta)_S = 0 \quad \text{for any} \quad \Delta \in \mathbb{G}(S)\}.$$

Definition. The quotient group $\mathbb{G}(S)/\mathbb{G}(S)^\perp$ is said to be the *group of numerical equivalence classes of divisors*, denoted $\mathrm{Num}(S)$.

The following result can be proved easily, if one applies the analytic theory of divisors.

Proposition 8.4. $\mathrm{Num}(S)$ *is a finitely generated Abelian group.*

Proof. We refer the reader to [G–H, p. 461]. □

Definition. The rank of the Abelian group $\mathrm{Num}(S)$ is said to be the *Picard number* of S (or the *base number* of S), denoted by $\rho(S)$.

Clearly, $\mathrm{Num}(S)$ is a free Abelian group and the pairing $I(\ ,\ )$ is induced from $(\ ,\ )_S$. I is symmetric and nondegenerate.

Furthermore, define a $\mathbb{Q}$-vector space $\mathbf{N}_0(S)$ to be $\mathrm{Num}(S) \otimes_{\mathbb{Z}} \mathbb{Q}$ and introduce the pairing $I(\xi, \eta)$ by $I(m\xi, n\eta)/mn$ where $m\xi$ and $n\eta \in \mathrm{Num}(S)$ for some m and $n \neq 0$. ξ^2 denotes $I(\xi, \xi)$.

c. Let $\xi_1, \ldots, \xi_\rho$ be a basis of the $\mathbb{Q}$-vector space $\mathbf{N}_0(S)$, and $A_{(\xi)}$ be the matrix $[I(\xi_i, \xi_j)]_{1 \le i, j \le \rho}$. The ordered pair (p, n) consisting of the numbers of positive eigenvalues and negative eigenvalues of $A_{(\xi)}$ is independent of the choice of basis, and is called the *signature* of the pairing I. The difference $p - n$ is called the *index* of I.

Lemma 8.3. *Let Z be an ample divisor, let C be a curve, and let D be a divisor with $D \neq \varnothing$ on S. Then*

(i) $(C, Z)_S > 0$.
(ii) $(D, Z)_S \geq 0$. *Furthermore,* $(D, Z)_S = 0$ *if and only if* $D \sim 0$.
(iii) $Z^2 > 0$.

PROOF. Since $\mathcal{O}(Z)|_C$ is ample, one has deg $\mathcal{O}(Z)|_C > 0$. On the other hand, $(C, Z)_S = \deg(\mathcal{O}(Z)|_C)$ by §8.1.**d**. Hence, (i) follows. (ii) and (iii) follow immediately from (i). □

Theorem 8.3 (Hodge's Index Theorem). *Let ρ denote $\rho(S)$. Then the signature of I is $(1, \rho - 1)$; i.e., the index is $2 - \rho$.*

PROOF. Let ζ be an element of $N_0(S)$ represented by an ample divisor Z on S, and let M be $\mathbb{Q}\zeta$. $M^\perp = \{\eta \in N_0(S) \mid I(\eta, \zeta) = 0\}$ satisfies $M \cap M^\perp = 0$ and $M + M^\perp = N_0(S)$, since $\zeta^2 = Z^2$ and $Z^2 > 0$ by Lemma 8.3.(iii). We claim that I is negative definite on $M^\perp$. Suppose that there exists $\xi \in M^\perp$ with $\xi^2 > 0$. Some multiple $r\xi$ $(r > 0)$ is represented by $D \in \mathbb{G}(S)$. By Corollary 2 to Theorem 8.2, one has

$$l(mD) + l(K(S) - mD) \geq \frac{D^2}{2} \cdot m^2 - \frac{(K(S), D)_S}{2} \cdot m + \chi_S$$

for any m. The right-hand side is positive for sufficiently large m. Hence, $|mD| \neq \varnothing$ or $|K(S) - mD| \neq \varnothing$. Note $(mD, Z)_S = I(mr\xi, \zeta) = 0$ by the choice of ξ. If $|mD| \neq \varnothing$, then $mD \sim 0$ by the last lemma. The relation $mD \sim 0$ implies that $\xi^2 = 0$, which contradicts the choice of ξ. Thus $|K(S) - mD| \neq \varnothing$ for $m \gg 0$. By Proposition 7.15, there exists $v > 0$ such that $D + vZ$ is ample. Hence, again by the last lemma, $(K(S) - mD, D + vZ)_S \geq 0$. But $(K(S) - mD, D + vZ)_S = (K(S), D)_S + v \cdot (K(S), Z)_S - mr^2\xi^2$ is negative for sufficiently large m. Thus we have arrived at a contradiction. Therefore, I is negative semidefinite and hence negative definite on $M^\perp$. □

Corollary. *Let $D, \Delta \in \mathbb{G}(S)$ with $D^2 > 0$. Then $((D, \Delta)_S)^2 \geq D^2 \cdot \Delta^2$. Furthermore, equality holds if and only if $\alpha_2 \Delta \equiv \alpha_1 D$ mod $\mathbb{G}(S)^\perp$ for some α_1, $\alpha_2 \in \mathbb{Z}$ with $\alpha_2 \neq 0$.*

PROOF. Let ξ and η be elements of $N_0(S)$ defined by D and Δ, respectively. Suppose that $\eta \notin \mathbb{Q}\xi$. Then $M = \mathbb{Q}\xi + \mathbb{Q}\eta$ is two-dimensional and so by the last theorem, the pairing I restricted to M is not definite. Hence, there exists $a\xi + \eta \in M$ with $(a\xi + \eta)^2 = 0$. Thus, the quadratic equation in t, $f(t) = \xi^2 t^2 + 2I(\xi, \eta)t + \eta^2 = 0$, has the real root a, i.e., the discriminant of $f(t)$ is not negative. Hence, $I(\xi, \eta)^2 - \xi^2 \cdot \eta^2 \geq 0$. If equality holds, then $\xi^2 a + I(\xi, \eta) = 0$; hence $I(a\xi + \eta, \xi) = 0$. This implies that $a\xi + \eta \in (\mathbb{Q}\xi)^\perp$, i.e., $(a\xi + \eta)^2 \leq 0$, and $(a\xi + \eta)^2 = 0$ if and only if $a\xi + \eta = 0$ in $N_0(S)$ by the last theorem. Finally, note that if $\eta \in \mathbb{Q}\xi$, then $(I(\eta, \xi))^2 = \xi^2 \cdot \eta^2$. □

d. Let C and Γ be complete nonsingular curves, which are projective by Corollary 1 to Theorem 6.6. Thus $S = C \times \Gamma$ is projective by Proposition 3.7. Taking closed points $p_1 \in C$ and $p_2 \in \Gamma$, define $C_1 = C \times p_2$ and $\Gamma_1 = p_1 \times \Gamma$, which are prime divisors on S. Then $C_1^2 = \Gamma_1^2 = 0$ and

$(C_1, \Gamma_1)_S = 1$. For a divisor D on S, define $d(D) = (C_1, D)_S$ and $\delta(D) = (\Gamma_1, D)_S$. $H = \Gamma_1 + C_1$ satisfies $H^2 = 2$ and so applying the last corollary to H and $mD + nC_1$, one has

$$(H, mD + nC_1)_S^2 \geq 2 \cdot (mD + nC_1)^2 = 2m^2D^2 + 2mn \cdot d(D).$$

Then, letting $t = m/n$, one obtains for all $t \in \mathbb{Q}$ the inequality

$$\{2D^2 - (d(D) + \delta(D))^2\}t^2 - 2(d(D) - \delta(D))t - 1 \leq 0.$$

The discriminant of the above quadratic form is negative, i.e.,

$$2d(D) \cdot \delta(D) \geq D^2.$$

By this, we shall prove the next result.

Theorem 8.4. *For any D and $E \in \mathbb{G}(C \times \Gamma)$, let $d = d(D)$, $\delta = \delta(D)$, $e = d(E)$, and $\varepsilon = \delta(E)$. Then*

(i) $2d\delta - D^2 \geq 0$, $2e\varepsilon - E^2 \geq 0$.
(ii) $|\delta e + d\varepsilon - (D, E)_S|^2 \leq (2d\delta - D^2) \cdot (2e\varepsilon - E^2)$.
(iii) *If $C = \Gamma$, and $E = \Delta_C$, then*

$$|d + \delta - (D, \Delta_C)_S|^2 \leq 4g(d\delta + (g - 1)(d + \delta) - \pi(D) + 1),$$

where $g = g(C)$.
(iv) *If the hypothesis of* (iii) *holds, and if $D = \Gamma_\varphi$ for some $\varphi \in \mathrm{Aut}(C)\backslash\{1\}$, then*

$$v^*(\varphi) = (\Gamma_\varphi, \Delta_C)_S \leq 2g + 2.$$

PROOF. (i) This has been proved.

(ii) This follows from (i) applied to $mD + nE$. In fact, one has the following inequality for all m, n:

$$(2d\delta - D^2)t^2 + 2(d\varepsilon + \delta e - (D, E)_S)t + (2e\varepsilon - E^2) \leq 0,$$

where $t = m/n$. Hence, the discriminant is not positive, i.e.,

$$(d\varepsilon + \delta e - (D, E)_S)^2 - (2d\delta - D^2)(2e\varepsilon - E^2) \leq 0.$$

(iii) and (iv) These follow from (ii) by noting the next lemma. □

Lemma 8.4. *Let $D \in \mathbb{G}(C \times C)$. Then $D^2 = 2\pi(D) - 2 - (2g - 2)(d(D) + \delta(D))$. In particular, if $D = \Gamma_\varphi$ for some $\varphi \in \mathrm{Aut}(C)$, then $\Gamma_\varphi^2 = 2 - 2g$.*

PROOF. Since $K(C \times C) \sim K(C) \times C + C \times K(C)$ (see Exercise 5.2.(v)), it follows that $(K(C \times C), D)_S = (2g - 2)(d(D) + \delta(D))$. Apply the adjunction formula. Furthermore, note that $C \cong \Gamma_\varphi$; hence $\pi(\Gamma_\varphi) = g$, and so $\Gamma_\varphi^2 = 2 - 2g$. □

Remark. Assertion (iv) was also proved in §6.8 (Theorem 6.14.(i)).

§8.3 Intersection Matrix of a Divisor

a. For a reduced divisor $D = \sum_{i=1}^{s} E_i$, i.e., the E_i are curves and $E_i \neq E_j$ if $i \neq j$, define a matrix $[(E_i, E_j)_S]$, which is said to be the *intersection matrix* of D.

Now, let S' be a normal projective surface and $f: S \to S'$ be a birational morphism, where S is assumed to be a projective nonsingular surface. If p is a closed point of S', then $f^{-1}(p)$ is connected by Theorem 2.20. Hence, if $f^{-1}(p)$ is one-dimensional, then $f^{-1}(p)$ is a connected reduced divisor $\sum_{i=1}^{l} E_i$, where the E_i are the curves.

Theorem 8.5. *The intersection matrix of $f^{-1}(p)$ is negative definite.*

Proof. Let Σ be $S' \backslash \mathrm{dom}(f^{-1})$ which is a finite closed subset $\{p = p_1, \ldots, p_r\}$. Take a very ample divisor H of S'. Then a general member H_1 of $|H|$ does not pass through $p_1, \ldots, p_r$. Since $f|_{S\backslash f^{-1}(\Sigma)} : S\backslash f^{-1}(\Sigma) \cong S'\backslash\Sigma$ is an isomorphism, $f^*H_1 \cong H_1$ and so $(f^*H_1)^2 = (f^*H_1, f^*H_2)_S = \sum_{p \in H_1 \cap H_2} I_p(H_1, H_2)$, where $H_2 \in |H|$ is a general member. Since H is very ample, one has $H_1 \cap H_2 \neq \varnothing$; hence $(f^*H_1)^2 > 0$. Thus, $D = f^*H_1$ satisfies $D^2 > 0$ and $D \cap f^{-1}(p) = \varnothing$. Hence, each component E_i of $f^{-1}(p)$ satisfies $(D, E_i)_S = 0$. Therefore, the intersection matrix of $\sum_{i=1}^{l} E_i$ is negative definite by Theorem 8.3. □

b. Proposition 8.5. *Let $E_1, \ldots, E_l$ be curves on S such that the intersection matrix $[(E_i, E_j)_S]$ is negative definite. Then $|\sum_{i=1}^{l} m_i E_i| = \sum_{i=1}^{l} m_i E_i$ for any $m_i \geq 0$.*

Proof. Let D be a member of $|\sum_{i=1}^{l} m_i E_i|$. Then $D^2 = \sum_{i=1}^{l} m_i(D, E_i)_S = (\sum_{i=1}^{l} m_i E_i)^2 < 0$, if $(m_1, \ldots, m_l) \neq 0$. Hence there exists j with $(D, E_j)_S < 0$ and $m_j \geq 1$. Thus by Corollary (ii) to Proposition 8.1, $D_j = D - E_j \geq 0$ and $D_j \in |\sum_{i=1}^{l} m_i E_i - E_j|$. By induction on $\sum_{i=1}^{l} m_i$, we have $D_j = \sum_{i=1}^{l} m_i E_i - E_j$; thus $D = \sum_{i=1}^{l} m_i E_i$. □

c. Definition. For an effective divisor D on S, define $\kappa(D, S)$ to be $\max\{\dim \Phi_{mD}(S) \mid m = 1, 2, \ldots\}$.

By definition, $\kappa(ND, S) = \kappa(D, S)$ for any $N > 0$. Thus, the assertion of Proposition 8.5 can be restated as $\kappa(\sum_{i=1}^{l} E_i, S) = 0$.

Lemma 8.5. *If D is an effective divisor on S with $D^2 > 0$, then $\kappa(D, S) = 2$.*

PROOF. Take a very ample divisor H. Then for any m, by Corollary 2 to Theorem 8.2, one has

$$l(mD - H) + l(K(S) + H - mD) \geq \frac{(mD - H)^2}{2} + \frac{(K(S), mD - H)_S}{2} + \chi_S$$

$$= \frac{D^2}{2} m^2 + \text{a linear term in } m.$$

Since $D^2 > 0$ and $l(K(S) + H) \geq l(K(S) + H - mD)$ for any $m > 0$, one concludes that $|mD - H| \neq \varnothing$ for $m \gg 0$. Thus,

$$\kappa(D, S) = \kappa(mD, S) = \kappa(mD - H + H, S) \geq \kappa(H, S) = 2. \qquad \square$$

Proposition 8.6. *If $\kappa(D, S) \leq 1$ for an effective divisor D, then the intersection matrix of D is negative semidefinite.*

PROOF. Let $\Gamma_i (1 \leq i \leq r)$ be the irreducible components of D. For $F = \sum_{i=1}^r a_i \Gamma_i$ with $a_i \in \mathbb{Z}$, one has effective divisors F_+ and F_- such that $F = F_+ - F_-$. $\kappa(F_\pm, S) \leq \kappa(D, S) \leq 1$ implies $F_\pm^2 \leq 0$ by the last lemma. Hence, $F^2 = F_+^2 - 2(F_+, F_-)_S + F_-^2 \leq F_+^2 + F_-^2 \leq 0$. $\square$

The next important result is a special case of the fibering theorem for D-dimension (cf. Theorem 10.3).

Theorem 8.6 (Pencil Theorem). *If $\kappa(D, S) = 1$, then Φ_{mD} is a morphism for any $m \geq 1$. If C_u is a connected component of a general member of $|mD|_{\text{red}}$, then $(C_u, D)_S = 0$ and $C_u \cap D = \varnothing$.*

PROOF. If $|mD|_{\text{red}} \neq \varnothing$ and has base points, then any general members Δ_1, Δ_2 of $|mD|_{\text{red}}$ have a finite intersection. Thus $(\Delta_1, \Delta_2)_S > 0$ and $\Delta_1^2 = (\Delta_1, \Delta_2)_S > 0$. By Lemma 8.5, $\kappa(\Delta_1, S) = 2$; hence $2 = \kappa(\Delta_1, S) \leq \kappa(mD, S) = \kappa(D, S)$. This contradicts the hypothesis. Hence, $\Delta_1^2 = 0$, and so $\text{Bs}|\Delta_1| = \varnothing$.

Suppose that $(D, \Delta_1)_S > 0$. Then $(D + \nu\Delta_1)^2 = D^2 + 2\nu(D, \Delta_1)_S > 0$ for $\nu \gg 0$. This is again absurd, by Lemma 8.5. Hence, $(D, \Delta_1)_S = 0$; thus $\Delta_1 \cap D = \varnothing$, since $\text{Bs}|\Delta_1| = \varnothing$. $\square$

Note that for $m \gg 0$, $f = \Phi_{mD}: S \to \Gamma_{(m)} = \Phi_{mD}(S)$ has connected fibers and Δ_1 can be written as $\sum_{i=1}^r f_1^*(p_i)$ for some $p_i \in \Gamma'_{(m)}$, where $f_1: S \to \Gamma'_m$ is the normalization of f (see §2.14.**c**).

Corollary. *If $\kappa(K(S), S) = 1$ and $m \gg 0$, then $f_1: S \to \Gamma'_m$ is an elliptic surface, i.e., any general fiber is an elliptic curve.*

Proof. Since $C_u^2 = 0$, one has $2\pi(C_u) - 2 = C_u^2 + (C_u, K(S))_S = 0$; hence $\pi(C_u) = 1$. C_u is nonsingular for a general point u by Theorem 7.18. Thus C_u is an elliptic curve. □

$\kappa(K(S), S)$ is denoted by $\kappa(S)$ (see §10.5).

§8.4 Intersection Numbers of Invertible Sheaves

a. Even on a complete normal surface, one cannot define an intersection number of two arbitrary divisors satisfying (i), (ii), (iii), and (iv) in §8.1. But if we consider only locally principal divisors, we have an intersection theory for such divisors on a complete scheme over a field k, as will be seen below.

Definition. Let V be a k-complete scheme with $n \geq \dim V$ and let $\mathscr{L}_1, \ldots, \mathscr{L}_n$ be invertible sheaves on V. Define the *intersection number* $(\mathscr{L}_1, \ldots, \mathscr{L}_n)_V$ to be

$$\chi_V - \sum_{i=1}^{n} \chi_V(\mathscr{L}_i^{-1}) + \sum_{i<j} \chi_V(\mathscr{L}_i^{-1} \otimes \mathscr{L}_j^{-1}) - \cdots + (-1)^n \chi_V(\mathscr{L}_1^{-1} \otimes \cdots \otimes \mathscr{L}_n^{-1}).$$

As in the case of nonsingular surfaces, one can prove that $(\mathscr{L}_1, \ldots, \mathscr{L}_n)_V$ is a symmetric multilinear form with respect to $\mathscr{L}_1, \ldots, \mathscr{L}_n$, i.e., $(\mathscr{L}_1^{\otimes a} \otimes \mathscr{M}_1^{\otimes b}, \mathscr{L}_2, \ldots, \mathscr{L}_n)_V = a(\mathscr{L}_1, \ldots, \mathscr{L}_n)_V + b(\mathscr{M}_1, \mathscr{L}_2, \ldots, \mathscr{L}_n)_V$ and $(\mathscr{L}_{\sigma(1)}, \ldots, \mathscr{L}_{\sigma(n)})_V = (\mathscr{L}_1, \ldots, \mathscr{L}_n)_V$ for any permutation σ, where $\mathscr{M}_1$ is invertible, and $a, b \in \mathbb{Z}$.

b. By applying Theorem 7.13 (the unscrewing lemma), we shall prove this. First, we generalize the notion of intersection number.

Definition. Let $\mathscr{F}$ be a coherent $\mathscr{O}_V$-module and let $\mathscr{L}_1, \ldots, \mathscr{L}_n$ be invertible sheaves on V. Define $(\mathscr{L}_1, \ldots, \mathscr{L}_n; \mathscr{F})_V$ to be

$$\chi_V(\mathscr{F}) - \sum_{i=1}^{n} \chi_V(\mathscr{F} \otimes \mathscr{L}_i^{-1}) + \sum_{i<j} \chi_V(\mathscr{F} \otimes \mathscr{L}_i^{-1} \otimes \mathscr{L}_j^{-1}) - \cdots + (-1)^n \chi_V(\mathscr{F} \otimes \mathscr{L}_1^{-1} \otimes \cdots \otimes \mathscr{L}_n^{-1}).$$

Theorem 8.7. *Let the notation be as above. Then for all $m \geq 0$:*

I(m). *If $\dim V \leq n \leq m$, then $(\mathscr{L}_1, \ldots, \mathscr{L}_n; \mathscr{F})_V$ is multilinear with respect to $\mathscr{L}_1, \ldots, \mathscr{L}_n$.*

II(m). *If $\dim V \leq n \leq m$ and $\mathscr{M}$ is an invertible sheaf, then $(\mathscr{L}_1, \ldots, \mathscr{L}_n; \mathscr{F})_V = (\mathscr{L}_1, \ldots, \mathscr{L}_n; \mathscr{F} \otimes_{\mathscr{O}_V} \mathscr{M})_V$.*

Proof. I(0) and II(0) being clear, it suffices to show the implication (I($m-1$) and II($m-1$)) $\Rightarrow$ (I(m) and II(m)), for all $m \geq 1$. For V with $\dim V = m$, define $\mathbb{K}_{\mathrm{I}}$ to be $\{$coherent $\mathcal{O}_V$-modules $\mathcal{F} \mid (\mathcal{L}_1, \ldots, \mathcal{L}_n; \mathcal{F})_V$ is multilinear with respect to $\mathcal{L}_1, \ldots, \mathcal{L}_n\}$ and $\mathbb{K}_{\mathrm{II}}$ to be $\{$coherent $\mathcal{O}_V$-modules $\mathcal{F} \mid (\mathcal{L}_1, \ldots, \mathcal{L}_n; \mathcal{F})_V = (\mathcal{L}_1, \ldots, \mathcal{L}_n; \mathcal{F} \otimes \mathcal{M})_V$ for any invertible sheaves $\mathcal{L}_1, \ldots, \mathcal{L}_n, \mathcal{M}\}$.

We claim that both $\mathbb{K}_{\mathrm{I}}$ and $\mathbb{K}_{\mathrm{II}}$ satisfy conditions (1), (2), (3), and (4) of Theorem 7.13. Checking conditions (1), (2), and (4) is easy, and left to the reader. We shall verify condition (3) for $\mathbb{K}_{\mathrm{I}}$; in other words, we shall prove that $\mathcal{O}_V \in \mathbb{K}_{\mathrm{I}}$ if V is a complete variety. Applying Lemma 7.14 to $\mathcal{L}_n^{-1}$, one has closed subschemes Y and Z and exact sequences

$$\begin{array}{c} 0 \to \mathcal{S} \to \mathcal{O}_V \to \mathcal{O}_Y \to 0 \\ \wr\| \\ 0 \to \mathcal{J} \otimes \mathcal{L}_n^{-1} \to \mathcal{L}_n^{-1} \to \mathcal{L}_n^{-1}|_Z \to 0. \end{array}$$

Hence for any $i_1, \ldots, i_s$, the following sequences are exact:

$$0 \to \mathcal{S} \otimes \mathcal{L}_{i_1}^{-1} \otimes \cdots \otimes \mathcal{L}_{i_s}^{-1} \to \mathcal{L}_{i_1}^{-1} \otimes \cdots \otimes \mathcal{L}_{i_s}^{-1} \to (\mathcal{L}_{i_1}|_Y)^{-1} \otimes \cdots \otimes (\mathcal{L}_{i_s}|_Y)^{-1} \to 0,$$

$$0 \to \mathcal{J} \otimes \mathcal{L}_{i_1}^{-1} \otimes \cdots \otimes \mathcal{L}_{i_s}^{-1} \otimes \mathcal{L}_n^{-1} \to \mathcal{L}_{i_1}^{-1} \otimes \cdots \otimes \mathcal{L}_{i_s}^{-1} \otimes \mathcal{L}_n^{-1} \to (\mathcal{L}_{i_1}|_Z)^{-1} \otimes \cdots \otimes (\mathcal{L}_{i_s}|_Z)^{-1} \otimes (\mathcal{L}_n|_Z)^{-1} \to 0$$

From these, one has

$$(\mathcal{L}_1, \ldots, \mathcal{L}_n)_V = (\mathcal{L}_1|_Y, \ldots, \mathcal{L}_{n-1}|_Y)_Y - (\mathcal{L}_1|_Z, \ldots, \mathcal{L}_{n-1}|_Z; \mathcal{L}_n^{-1}|_Z)_Z.$$

Since $\dim Z < \dim V = m$, it follows that $(\mathcal{L}_1|_Z, \ldots, \mathcal{L}_{n-1}|_Z; \mathcal{L}_n^{-1}|_Z)_Z = (\mathcal{L}_1|_Z, \ldots, \mathcal{L}_{n-1}|_Z)_Z$ by II($m-1$). Furthermore, one has $\dim Y < \dim V = m$ and so, by applying I($m-1$) to Z and Y, we conclude that $(\mathcal{L}_1, \ldots, \mathcal{L}_n)_V$ is multilinear.

Next, we prove that $\mathcal{O}_V \in \mathbb{K}_{\mathrm{II}}$ when V is again a complete variety. Then by Lemma 7.14, one has

$$(\mathcal{L}_1, \ldots, \mathcal{L}_n; \mathcal{M})_V = (\mathcal{L}_1|_Y, \ldots, \mathcal{L}_{n-1}|_Y; \mathcal{M}|_Y)_Y - (\mathcal{L}_1|_Z, \ldots, \mathcal{L}_{n-1}|_Z; (\mathcal{M} \otimes \mathcal{L}_n^{-1})|_Z)_Z,$$

that is $(\mathcal{L}_1|_Y, \ldots, \mathcal{L}_{n-1}|_Y)_Y - (\mathcal{L}_1|_Z, \ldots, \mathcal{L}_{n-1}|_Z)_Z$ by II($m-1$). Using this for $\mathcal{M} = \mathcal{O}_V$, one has

$$(\mathcal{L}_1, \ldots, \mathcal{L}_n)_V = (\mathcal{L}_1|_Y, \ldots, \mathcal{L}_{n-1}|_Y)_Y - (\mathcal{L}_1|_Z, \ldots, \mathcal{L}_{n-1}|_Z)_Z;$$

hence

$$(\mathcal{L}_1, \ldots, \mathcal{L}_n; \mathcal{M})_V = (\mathcal{L}_1, \ldots, \mathcal{L}_n)_V. \qquad \square$$

Corollary. *If* $\dim V \leq n$, *then* $(\mathcal{L}_1, \ldots, \mathcal{L}_n, \mathcal{L}_{n+1}; \mathcal{F})_V = 0$.

PROOF. Rearranging the terms appropriately, one has

$$\begin{aligned}(\mathscr{L}_1, \ldots, \mathscr{L}_{n+1}; \mathscr{F})_V = \chi_V(\mathscr{F}) - \sum_{i=1}^{n} \chi_V(\mathscr{F} \otimes \mathscr{L}_i^{-1}) + \cdots \\ + (-1)^n \chi_V(\mathscr{F} \otimes \mathscr{L}_1^{-1} \otimes \cdots \otimes \mathscr{L}_n^{-1}) \\ - \chi_V(\mathscr{F} \otimes \mathscr{L}_{n+1}^{-1}) + \sum_{i=1}^{n} \chi_V(\mathscr{F} \otimes \mathscr{L}_{n+1}^{-1} \otimes \mathscr{L}_i^{-1}) - \cdots \\ + (-1)^{n+1} \chi_V(\mathscr{F} \otimes \mathscr{L}_{n+1}^{-1} \otimes \mathscr{L}_i^{-1}, \otimes \cdots \otimes \mathscr{L}_n^{-1}) \\ = (\mathscr{L}_1, \ldots, \mathscr{L}_n; \mathscr{F})_V - (\mathscr{L}_1, \ldots, \mathscr{L}_n; \mathscr{F} \otimes \mathscr{L}_{n+1}^{-1})_V,\end{aligned}$$

which is zero by II(n). □

c. Lemma 8.6. *Let V be a complete scheme and let V', V'' be closed subschemes of V with $V' \cup V'' = V$ and $\dim(V' \cap V'') < \dim V = n$. If $\mathscr{L}_1, \ldots, \mathscr{L}_n$ are invertible sheaves on V, and $\mathscr{F}$ is a coherent $\mathscr{O}_V$-module, then $(\mathscr{L}_1, \ldots, \mathscr{L}_n; \mathscr{F})_V = (\mathscr{L}'_1, \ldots, \mathscr{L}'_n; \mathscr{F}')_{V'} + (\mathscr{L}''_1, \ldots, \mathscr{L}''_n; \mathscr{F}'')_{V''}$, where $\mathscr{E}' = \mathscr{E}|_{V'}$ and $\mathscr{E}'' = \mathscr{E}|_{V''}$ for any coherent sheaf $\mathscr{E}$ on V.*

PROOF. By using the notation in Lemma 7.12, one has the $\mathscr{O}_V$-homomorphism $\xi: \mathscr{F} \to \mathscr{F}' \oplus \mathscr{F}''$. Thus, $(\mathscr{L}_1, \ldots, \mathscr{L}_n; \mathscr{F})_V = (\mathscr{L}_1, \ldots, \mathscr{L}_n; \operatorname{Ker} \xi)_V + (\mathscr{L}_1, \ldots, \mathscr{L}_n; \operatorname{Im} \xi)_V$ and $(\mathscr{L}_1, \ldots, \mathscr{L}_n; \operatorname{Im} \xi)_V = (\mathscr{L}_1, \ldots, \mathscr{L}_n; \mathscr{F}' \oplus \mathscr{F}'')_V + (\mathscr{L}_1, \ldots, \mathscr{L}_n; \operatorname{Coker} \xi)_V$. By the above Corollary, one obtains $(\mathscr{L}_1, \ldots, \mathscr{L}_n; \mathscr{F})_V = (\mathscr{L}_1, \ldots, \mathscr{L}_n; \mathscr{F}' \oplus \mathscr{F}'')_V$. Since

$$\begin{aligned}(\mathscr{L}_{i_1}^{-1} \otimes \cdots \otimes \mathscr{L}_{i_s}^{-1}) \otimes (\mathscr{F}' \oplus \mathscr{F}'') \\ = (\mathscr{L}'_{i_1})^{-1} \otimes \cdots \otimes (\mathscr{L}'_{i_s})^{-1} \otimes \mathscr{F}' \oplus (\mathscr{L}''_{i_1})^{-1} \otimes \cdots \otimes (\mathscr{L}''_{i_s})^{-1} \otimes \mathscr{F}'',\end{aligned}$$

by Proposition 2.19 it follows that

$$(\mathscr{L}_1, \ldots, \mathscr{L}_n; \mathscr{F}' \oplus \mathscr{F}'')_V = (\mathscr{L}'_1, \ldots, \mathscr{L}'_n; \mathscr{F}')_{V'} + (\mathscr{L}''_1, \ldots, \mathscr{L}''_n; \mathscr{F}'')_{V''}.$$

□

Proposition 8.7. *Let V be a complete variety and let $\mathscr{F}$ be a coherent $\mathscr{O}_V$-module with $r = \dim_{k(*)} \mathscr{F}_*$, where $*$ is the generic point of V. Then $(\mathscr{L}_1, \ldots, \mathscr{L}_n; \mathscr{F})_V = r \cdot (\mathscr{L}_1, \ldots, \mathscr{L}_n)_V$.*

PROOF. Since $\mathscr{F}_* \cong \mathscr{O}_{V,*}^r$, one has an $\mathscr{O}_V$-module $\mathscr{H}$ and $\mathscr{O}_V$-homomorphisms $\theta: \mathscr{H} \to \mathscr{F}$ and $\eta: \mathscr{H} \to \mathscr{O}_V^r$ such that $\eta_* \circ \theta_*^{-1}$ is an isomorphism by Lemma 7.11. Hence by the Corollary to Theorem 8.7, one has $(\mathscr{L}_1, \ldots, \mathscr{L}_n; \mathscr{H})_V = (\mathscr{L}_1, \ldots, \mathscr{L}_n; \mathscr{F})_V$ and $(\mathscr{L}_1, \ldots, \mathscr{L}_n; \mathscr{H})_V = r \cdot (\mathscr{L}_1, \ldots, \mathscr{L}_n)_V$. □

d. Let W be a closed subscheme of V defined by an invertible $\mathscr{O}_V$-ideal $\mathscr{I}$. One has the exact sequence

$$0 \to \mathscr{I} \to \mathscr{O}_V \to \mathscr{O}_W \to 0.$$

For invertible sheaves $\mathscr{L}_1, \ldots, \mathscr{L}_{n-1}$ with $n \geq \dim V$, one has

$$-(\mathscr{L}_1, \ldots, \mathscr{L}_{n-1}, \mathscr{S})_V = (\mathscr{L}_1|_W, \ldots, \mathscr{L}_{n-1}|_W)_W.$$

If $\mathscr{L}$ is an invertible sheaf on V, then $\mathscr{L}^n$ denotes $(\mathscr{L}, \ldots, \mathscr{L})_V$ for $n \geq \dim V$.

Definition. Let V be a closed subscheme of $\mathbf{P}_k^N$. Define the *degree* of V in $\mathbf{P}_k^N$ to be $(\mathscr{O}_{\mathbf{P}_k^N}(1)|_V)^n$, where $n = \dim V$.

Let $\tilde{H}$ be a general hyperplane of $\mathbf{P}_k^N$. $W = V \cap \tilde{H}$ is a closed subscheme of dimension $n-1$. Then $(\mathscr{O}_{\mathbf{P}_k^N}(1)|_W)^{n-1} = (\mathscr{O}_{\mathbf{P}_k^N}(1)|_V)^n$, i.e., $\deg W = \deg V$. Repeating this, one has $\deg V > 0$.

§8.5 Nakai's Criterion on Ample Sheaves

a. Proposition 8.8. *Let $\mathscr{L}$ be an invertible sheaf on a complete scheme V of dimension n. Then $\chi_V(\mathscr{L}^{\otimes m}) = (\mathscr{L}^n/n!)\cdot m^n + Q(m)$, where $Q(m)$ is a polynomial with* $\deg Q \leq n-1$.

PROOF. By definition,

$$\begin{aligned}(\mathscr{L}^{-1})^n &= (\mathscr{L}^{-1}, \ldots, \mathscr{L}^{-1})_V \\ &= \chi_V - n\chi_V(\mathscr{L}) + \tbinom{n}{2}\chi_V(\mathscr{L}^{\otimes 2}) - \cdots + (-1)^n\chi_V(\mathscr{L}^{\otimes n}).\end{aligned}$$

Hence, by the multilinearity of the intersection number, one has for any m

$$\begin{aligned}(-1)^n\mathscr{L}^n\cdot m^n &= \chi_V - n\chi_V(\mathscr{L}^{\otimes m}) \\ &\quad + \tbinom{n}{2}\chi_V(\mathscr{L}^{\otimes 2m}) - \cdots + (-1)^n\chi_V(\mathscr{L}^{\otimes nm}) \\ &= (-n + \tbinom{n}{2}2^n - \cdots + (-1)^n n^n)\frac{\alpha_0 m^n}{n!} + \sum_{i=0}^{n}(-1)^i\tbinom{n}{i}Q(im),\end{aligned}$$

where $\chi_V(\mathscr{L}^{\otimes m}) = \alpha_0 m^n/n! + Q(m)$ with $\deg Q \leq n-1$. From this, $\alpha_0 = (-n + \binom{n}{2}2^n - \cdots + (-1)^n n^n)^{-1}(-1)^n n!\cdot\mathscr{L}^n$. Applying this to the case of $V = \mathbf{P}_k^n$ and $\mathscr{L} = \mathscr{O}_{\mathbf{P}_k}(1)$, one has

$$\chi_V(\mathscr{L}^{\otimes m}) = \binom{n+m-1}{m-1} = \frac{m^n}{n!} + \cdots;$$

hence $\alpha_0 = 1$ and $(\mathscr{O}_{\mathbf{P}_k}(1))^n = 1$. Thus,

$$1 = (-n + \tbinom{n}{2}2^n - \cdots + (-1)^n n^n)^{-1}(-1)^n\cdot n!.$$

Therefore, in the general case, one has $\alpha_0 = \mathscr{L}^n$. □

Corollary. *In addition, if $\mathscr{L}$ is ample, then*

$$\dim_k \Gamma(V, \mathscr{L}^{\otimes m}) = \frac{\mathscr{L}^n}{n!} m^n + Q(m) \quad \textit{for } m \gg 0,$$

where $\deg Q \leq n - 1$.

Proof. This follows from Theorem 7.6. □

If $\mathscr{L}$ is ample, $\mathscr{L}^{\otimes r}$ is very ample for some r. Hence, V can be embedded into $\mathbf{P}(\mathscr{L}^{\otimes r})$ by the morphism defined by $\mathscr{L}^{\otimes r}$. Then $\deg V$ agrees with $(\mathscr{L}^{\otimes r})^n$, which is $(\mathscr{L}^n) \cdot r^n > 0$. Hence, $\mathscr{L}^n > 0$.

b. Theorem 8.8 (Nakai's Criterion). *Let $\mathscr{L}$ be an invertible sheaf on V. $\mathscr{L}$ is ample, if and only if $(\mathscr{L}|_W)^s > 0$ for any $s \geq 0$ and any closed subscheme W with $s = \dim W \geq 0$.*

Proof. If $\mathscr{L}$ is ample, then so is $\mathscr{L}|_W$. Hence, $(\mathscr{L}|_W)^s > 0$. Now, we shall prove the converse by the Noetherian induction on V (cf. Lemma 7.10). Thus, one can assume that $\mathscr{L}|_W$ is ample for any closed subscheme W on V with $W \neq V$. If V is reduced but reducible, i.e., $V = V' \cup V''$ for closed subschemes V', V'' with $V' \neq V$, $V'' \neq V$, then from the hypothesis that $\mathscr{L}|_{V'}$ and $\mathscr{L}|_{V''}$ are ample, one can conclude that $\mathscr{L}$ is ample by applying Theorem 7.8 (see Exercise 7.1.(ii)). Next, we can show that $\mathscr{L}$ is ample if and only if $\mathscr{L}/\mathscr{N}\mathscr{L}$ is ample, where $\mathscr{N}$ is the nilradical of $\mathscr{O}_V$. This can be done easily by applying Theorem 7.8 again (Exercise 7.1.(i)). Thus, one can assume that V is a complete variety. Furthermore, by Exercise 7.2, one can assume V to be normal. By Lemma 7.14 applied to V and $\mathscr{L}$, one has closed subschemes Y and Z and the exact sequences

$$\begin{array}{l} 0 \to \mathscr{S} \otimes \mathscr{L}^{\otimes m} \to \mathscr{L}^{\otimes m} \to \mathscr{L}^{\otimes m}|_Y \to 0 \\ \quad\ \ \wr\| \\ 0 \to \mathscr{J} \otimes \mathscr{L}^{\otimes (m+1)} \to \mathscr{L}^{\otimes (m+1)} \to \mathscr{L}^{\otimes (m+1)}|_Z \to 0 \end{array}$$

for any $m > 0$. Since $\mathscr{L}^{\otimes m}|_Y = (\mathscr{L}|_Y)^{\otimes m}$ and $\mathscr{L}|_Y$ is ample, $H^q(Y, \mathscr{L}^{\otimes m}|_Y) = 0$ for $q > 0$ and $m \gg 0$ by Theorem 7.6. Hence, $H^{q+1}(V, \mathscr{S} \otimes \mathscr{L}^{\otimes m}) \cong H^{q+1}(V, \mathscr{L}^{\otimes m})$. Similarly, $H^{q+1}(V, \mathscr{J} \otimes \mathscr{L}^{\otimes (m+1)}) \cong H^{q+1}(V, \mathscr{L}^{\otimes (m+1)})$; thus $H^{q+1}(V, \mathscr{L}^{\otimes m}) \cong H^{q+1}(V, \mathscr{L}^{\otimes (m+1)})$ for any $q > 0$ and $m \gg 0$. This implies that

$$\chi_V(\mathscr{L}^{\otimes m}) = \dim_k \Gamma(V, \mathscr{L}^{\otimes m}) - \dim_k H^1(V, \mathscr{L}^{\otimes m}) + \text{const.}$$

for $m \gg 0$. But since $\chi_V(\mathscr{L}^{\otimes m}) = (\mathscr{L}^n/n!) \cdot m^n + Q(m)$ with $\deg Q \leq n - 1$ and $\mathscr{L}^n > 0$ by hypothesis, $\dim_k \Gamma(V, \mathscr{L}^{\otimes m}) > 0$ holds. Taking a nonzero section $\alpha \in \Gamma(V, \mathscr{L}^{\otimes m})$, one has a closed subscheme D defined by α, i.e.,

$\mathcal{O}(D) \cong \mathcal{L}^{\otimes m}$. By the hypothesis, $\mathcal{L}|_D$ is ample. Hence $H^1(D, \mathcal{L}^{\otimes r}|_D) = 0$ for $r \gg 0$. Then from the exact sequence

$$0 \to \mathcal{O}((v-1)D) \to \mathcal{O}(vD) \to \mathcal{O}(vD)|_D \to 0,$$

one has the surjection $H^1(V, \mathcal{O}((v-1)D)) \to H^1(V, \mathcal{O}(vD))$, since $H^1(D, \mathcal{O}(vD)|_D) = 0$ for $v \gg 0$. Hence, $\dim_k H^1(V, \mathcal{O}((v-1)D)) \geq \dim_k H^1(V, \mathcal{O}(vD)) \geq \cdots$, i.e., there exists $v_0 \gg 0$ such that $\dim_k H^1(V, \mathcal{O}(vD)) = \dim_k H^1(V, \mathcal{O}(v_0 D))$ for all $v \geq v_0$. Since $H^1(V, \mathcal{O}((v-1)D)) \to H^1(V, \mathcal{O}(vD))$ is surjective, $H^1(V, \mathcal{O}((v-1)D)) \cong H^1(V, \mathcal{O}(vD))$ for all $v \gg 0$. Thus, one obtains the surjection $\Gamma(V, \mathcal{O}(vD)) \to \Gamma(V, \mathcal{O}(vD)|_D)$ derived from the above exact sequence. One can suppose that $\mathcal{O}(vD)|_D$ is very ample for some $v > 0$. Hence, letting $\Sigma = \{\varphi_0, \ldots, \varphi_l\}$ be a basis of $\Gamma(V, \mathcal{O}(vD))$, $V_\Sigma = V$ and one has the morphism $\Phi\colon V \to \mathbf{P}(\mathcal{O}(vD))$ associated with $\mathcal{O}(vD) \cong \mathcal{L}^{\otimes mv}$ (cf. §7.4.**a**). We claim that all fibers $\Phi^{-1}(p)$ are finite. In fact, if C is a curve contained in a fiber $\Phi^{-1}(p)$, then $\mathcal{L}^{\otimes mv}|_C \cong \Phi^*(\mathcal{O}_{\mathbf{P}}(1))|_C = \Phi_1^*(\mathcal{O}_{\mathbf{P}}(1)|_p) \cong \mathcal{O}_C$, where Φ_1 is $\Phi|_C\colon C \to p$. This implies $\deg(\mathcal{L}^{\otimes mv}|_C) = 0$, which contradicts the hypothesis. Then Φ is a finite morphism by Theorem 2.27. Thus, $\mathcal{L}^{\otimes mv}$ is ample by the corollary to Theorem 7.8. □

c. As an application of Nakai's criterion, we shall prove that any complete nonsingular surface is projective. On a complete nonsingular surface S, we have an intersection theory of divisors D and Δ, i.e., $(D, \Delta)_S = (\mathcal{O}(D), \mathcal{O}(\Delta))_S$, which was defined in §8.4.**b**. This is bilinear by Theorem 8.7. The proofs of Propositions 8.1, 8.2, and 8.3 remain valid in this case.

Theorem 8.9. *Any complete nonsingular surface S is projective.*

PROOF. Take a nonsingular projective model (W, μ) of S (by the Corollary to Theorem 7.20). Fix a very ample divisor H of W and take a general member H_1 of $|H|$. Then H_1 is irreducible and so $\mu(H_1)$ is a curve C. $\mu^*C - H_1$ is an exceptional divisor E with respect to μ, i.e., $\mu(E)$ is a finite set. Thus, for any curve Γ on S, one has $(\Gamma, C)_S = (\mu^*\Gamma, \mu^*C)_W$ and $(\mu^*\Gamma, \mu^*C)_W = (\mu^*\Gamma, H)_W + (\mu^*\Gamma, E)_W = (\mu^*\Gamma, H)_W > 0$, since H is ample. By Theorem 8.8, C is ample; thus S is projective. □

EXERCISES

8.1. Let H be a line on $\mathbf{P}_k^2$ and let p be a point on $\mathbf{P}_k^1$. $L = H \times \mathbf{P}_k^1$ and $M = \mathbf{P}_k^2 \times p$ are divisors on $V = \mathbf{P}_k^2 \times \mathbf{P}_k^1$.
   (i) Show that $L^3 = 0$, $(L^2, M)_V = 1$, and $(D', M^2)_V = 0$ for any divisor D' on V.
   (ii) $K(V) \sim -3L - 2M$.
   (iii) If a and $b \geq 1$, prove that $aL + bM$ is a very ample divisor.

(iv) Take a member $S \in |aL + bM|$ such that S is an irreducible and nonsingular surface. Show that

$$K(S) \sim (a - 3)D + (b - 2)\Delta,$$

where $D = L|_S$ and $\Delta = M|_S$.

(v) Furthermore, show that $D^2 = b$, $(D, \Delta) = a$, $\Delta^2 = 0$ and

$$K(S)^2 = (a - 3)(3ab - 3b - 4).$$

(vi) Show that $D + \Delta$ is very ample and that $(D + \Delta)^2 = b^2 + 2a$.

(vii) If $a \leq 2$ or $b \leq 1$, show that S is a rational surface or a ruled surface.

(viii) In the case when $a \geq 3$ and $b \geq 2$, compute $P_m(S)$ for all $m \geq 1$, and determine $\kappa(S)$.

8.2. Let S be a nonsingular surface on $\mathbf{P}_k^1 \times \mathbf{P}_k^1 \times \mathbf{P}_k^1$. Then compute $P_m(S)$ and $\kappa(S)$.

8.3. Let C be a curve on a nonsingular complete surface S.

(i) If $\dim|C| \geq 1$ and any general member of $|C|$ is singular, then show that $C^2 \geq 4$.

(ii) Suppose that every member of $|C|$ is irreducible. If $\dim|C| \geq m(m - 1)/2$, then show that $C^2 \geq m^2$.

(iii) If $\mathrm{Bs}|C|$ is a finite set $\{p_1, \ldots, p_s\}$, then show that $C^2 \geq s$.

8.4. Let S be a complete normal surface which has a unique singular point. Prove that S is projective.

Chapter 9

Curves on a Nonsingular Surface

§9.1 Quadric Transformations

a. Throughout this chapter, let S denote a complete nonsingular surface defined over an algebraically closed field k with characteristic zero. Any monoidal transformation whose center is a point is also said to be a *quadric transformation.*

Proposition 9.1. *Let C be a curve on S and let p be a closed point of C with $v = e(p, C)$. Let $\mu\colon W = Q_p(S) \to S$ be a quadric transformation and let C^* be the strict transform of C by μ. If $E = \mu^{-1}(p)$ and $\{p_1, \dots, p_s\} = C^* \cap E$, then*

(i) $v = (C^*, E)_W = \sum_{j=1}^{s} I_{pj}(C^*, E)$.
(ii) $I_{pj}(C^*, E) \geq e(p_j, C^*)$ *and* $v \geq e(p_j, C^*)$ *for all* j.
(iii) *If* $e(p_1, C^*) = v$, *then* $s = 1$.

PROOF. Since $\mu^*C = C^* + vE$ by Proposition 7.20.(i), one has $0 = (\mu^*C, E)_W = (C^*, E)_W + vE^2 = (C^*, E)_W - v$. Furthermore, by Proposition 8.1, one obtains (i). (ii) follows from (i), and (iii) is a consequence of (i) and (ii). □

b. Points $p_1, \dots, p_s$ as in the above theorem are said to be *infinitely near points on C of the first order*. Inductively, infinitely near points on C^* of the j-th order are said to be *infinitely near points on C of the $(j + 1)$-th order* for all $j \geq 1$.

Thus, we have a sequence of quadric transformations $\{\mu_j\colon S_j \to S_{j-1}$ with center $p_{(j)}\}$ and $\{C_{(j)} \mid$ the strict transform of $C_{(j-1)}$ by $\mu_j\}$ for $1 \leq j \leq l$ such that $S_0 = S$, $p_{(1)} = p$, $C_{(0)} = C$. $C_{(i)} \backslash \mu_i^{-1}(p_{(i)})$ is identified with $C_{(i-1)} \backslash \{p_{(i)}\}$ for each i. With these identifications in mind, we say that points q on $C_{(i)}$ are *infinitely near points* on C. We define $e(q, C)$ to be $e(q, C_{(i)})$. If an infinitely near point q on C has $e(q, C) \geq 2$, then it is said to be an *infinitely near singular point*. Infinitely near singular points and singular points are referred to as *(infinitely near) singular points*.

We shall prove that there exist at most a finite number of infinitely near singular points on C (see Theorem 9.1).

EXAMPLE 9.1. Let $C = V_+(\mathsf{T}_2^5 - \mathsf{T}_1^2\mathsf{T}_0^3)$ on $\mathbf{P}_k^2$. Then C has two singular points and two infinitely near singular points as will be illustrated below: Putting $p_{(1)} = (1:0:0)$, $p_{(3)} = (0:1:0)$ and indicating multiplicities by numbers lying over points, one has the following figure.

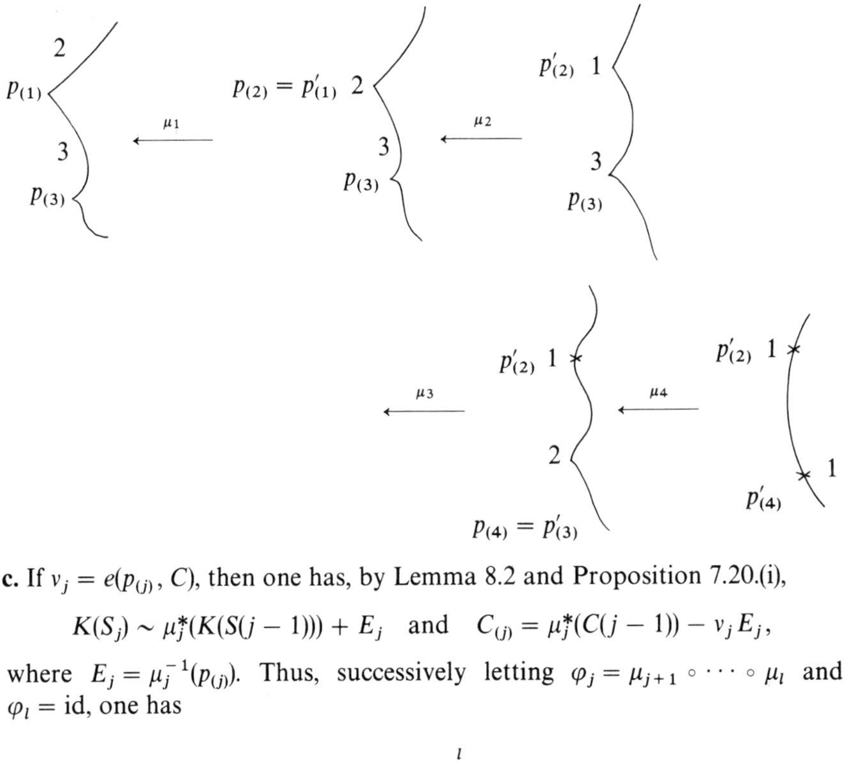

c. If $v_j = e(p_{(j)}, C)$, then one has, by Lemma 8.2 and Proposition 7.20.(i),

$$K(S_j) \sim \mu_j^*(K(S(j-1))) + E_j \quad \text{and} \quad C_{(j)} = \mu_j^*(C(j-1)) - v_j E_j,$$

where $E_j = \mu_j^{-1}(p_{(j)})$. Thus, successively letting $\varphi_j = \mu_{j+1} \circ \cdots \circ \mu_l$ and $\varphi_l = \text{id}$, one has

$$K(S_l) \sim \varphi_0^*(K(S)) + \sum_{j=0}^{l} \varphi_j^*(E_j)$$

and

$$C_{(l)} = \varphi_0^*(C) - \sum_{j=0}^{l} v_j \varphi_j^*(E_j).$$

Hence, one obtains the formula

$$\pi(C_{(l)}) = \pi(C) - \sum_{j=1}^{l} \frac{v_j(v_j - 1)}{2}.$$

In particular,

$$\pi(C) \geq \sum_{j=1}^{l} \frac{\nu_j(\nu_j - 1)}{2}.$$

This implies that there exist at most $\pi(C)$ (infinitely near) singular points on C.

Theorem 9.1

(i) *There exist at most a finite number of (infinitely near) singular points* $p_{(1)}, \ldots, p_{(l)}$ *on* C.
(ii) *The genus of* C *is*

$$g = \pi(C) - \sum_{j=1}^{l} \frac{\nu_j(\nu_j - 1)}{2},$$

where $\nu_j = e(p_{(j)}, C)$.

PROOF. (i) follows from the above discussion. Since $C_{(l)}$ is nonsingular, $(C_{(l)}, \varphi_0)$ is a nonsingular model of C; hence (ii) follows. □

This is a special case of Theorem 7.20. However, this proof uses the global properties of C and S. One can prove the result by local theory, i.e., by using formal power series (cf. [M1, p. 160]).

d. Let q be an infinitely near singular point on C, i.e., there exists j with $q \in C_{(j)}$. Putting $p = \mu_1 \circ \cdots \circ \mu_j(q)$, one says that q *lies over* p. These q and p will be referred to as *(infinitely near) singular points lying over* p.

Definition. Let ε_p denote $\sum \nu_q(\nu_q - 1)/2$, where the q are the (infinitely near) singular points lying over p.

Then one has the next formula, called the *genus formula of Clebsch*:

$$g(C) = \pi(C) - \sum_{p \in C} \varepsilon_p.$$

Definition. If $\nu = e(p, C)$ and $\mu^{-1}(p) \cap C^*$ consists of ν points, then p is said to be an *ordinary singular point* of C.

In this case, if $\{p_1, \ldots, p_\nu\} = \mu^{-1}(p) \cap C^*$, then by Proposition 9.1,

$$\nu = \sum_{j=1}^{\nu} I_{p_j}(C^*, \mu^{-1}(p)) \geq \sum_{j=1}^{\nu} e(p_j, C^*) \geq \nu;$$

hence $e(p_j, C^*) = 1$ for all j. Thus $\varepsilon_p = \nu(\nu - 1)/2$.

Definition. Let $(C^{\#}, \rho)$ be a nonsingular model of C. The point $p \in C$ is said to be a *cusp* of C, if it is a singular point and $\rho^{-1}(p)$ is a set consisting of single point. Furthermore, if μ: $Q_p(S) \to S$ is a quadric transformation and $\mu^{-1}(p) = \{p_1\}$ is a nonsingular point of $C_{(1)}$, then p is said to be a *simple cusp*.

Clearly, if p is a simple cusp with $v = e(p, C)$, then $\varepsilon_p = v(v-1)/2$.
The genus formula of Clebsch says that the genus of C can be computed as if C had only ordinary singular points or simple cusps.

§9.2 Local Properties of Singular Points

a. Let C be a curve on S and let ρ: $C^{\#} \to C$ be the morphism of the nonsingular model as usual. Then by Theorem 5.6, and Example 5.8, one has

$$K(C^{\#}) \sim \rho^*((K(S) + C)|_C) - \mathrm{cond}_\rho.$$

Hence, $g(C) = \pi(C^{\#}) = \pi(C) - \frac{1}{2}\deg(\mathrm{cond}_\rho)$.

Definition. Let $\sigma_p = \sum_{\rho(q)=p} \sigma_q$ for $p \in C$.

Then one obtains the formula:

Theorem 9.2. $g(C) = \pi(C) - \frac{1}{2}\sum_{p \in C} \sigma_p$.

Corollary. $\sigma_p = 2\varepsilon_p$ *for any* $p \in C$.

Proof. We show this by induction on $u = \#(\mathrm{Sing}\ C)$. If $u = 0$, the assertion is clear. If $u > 0$, take $p \in \mathrm{Sing}\ C$ and consider quadric transformations $\{\mu_j\colon S_j \to S_{j-1}|_{1 \le j \le m}\}$ whose centers are (infinitely near) singular points on C lying over p such that the singularities over p are resolved. Then $\#(\mathrm{Sing}\ C_{(m)}) = u - 1$ and $\sigma_q = 2\varepsilon_q$ if $q \neq p$ by the induction hypothesis. By Theorem 9.2, $g(C) = \pi(C) - \frac{1}{2}\sum_{q \neq p} \sigma_q - \frac{1}{2}\sigma_p$; and by the genus formula of Clebsch, one has $g(C) = \pi(C) - \sum_{q \neq p} \varepsilon_q - \varepsilon_p$. Thus, one obtains $2\varepsilon_p = \sigma_p$. □

Remark. By the similar method, one can prove that $\delta_p = \dim_k(\mathcal{O}_{C,\,p})'/(\mathcal{O}_{C,\,p}) = \varepsilon_p$, replacing Theorem 9.2 by Theorem 4.10.(i).

b. We recall the formula for σ_p in Example 5.8. Letting $\rho^{-1}(p)$ be $\{q_1, \ldots, q_s\}$ and ord_j denote the valuation defined by $\mathcal{O}_{C^{\#},\,q_j}$, one has

$$\sigma_p = -\sum_{j=1}^{s} \mathrm{ord}_j\left(\rho^*\left(\frac{dx}{\partial_y f}\right)\right)$$

where (x, y) is a local coordinate system around p of S and C is defined by f around p. Then,

$$\rho^*\left(\frac{dx}{\partial_y f}\right) = \frac{d\rho^*(x)}{dt_j} \cdot \frac{dt_j}{\rho^*(\partial_y f)}$$

and so

$$\begin{aligned}\operatorname{ord}_j\left(\rho^*\left(\frac{dx}{\partial_y f}\right)\right) &= \operatorname{ord}_j\left(\frac{d\rho^*(x)}{dt_j} \cdot \frac{1}{\rho^*(\partial_y f)}\right)\\ &= \operatorname{ord}_j\left(\rho^*\left(\frac{x}{\partial_y f}\right)\right) - 1 \quad \text{for all } j.\end{aligned}$$

Hence,

$$\begin{aligned}\sigma_p &= s - \sum_{j=1}^{s} \operatorname{ord}_j\left(\rho^*\left(\frac{x}{\partial_y f}\right)\right)\\ &= s - \sum_{j=1}^{s} \operatorname{ord}_j(\rho^*(x)) + \sum_{j=1}^{s} \operatorname{ord}_j(\rho^*(\partial_y f)).\end{aligned}$$

But by the formula of Exercise 9.2, one has

$$I_0(x, f) = \sum_{j=1}^{s} \operatorname{ord}_j(\rho^* x) \quad \text{and} \quad I_0(\partial_y f, f) = \sum_{j=1}^{s} \operatorname{ord}_j(\rho^* \partial_y f).$$

Thus one obtains

$$\sigma_p = -I_0(x, f) + I_0(\partial_y f, f) + s.$$

EXAMPLE 9.2. Let $f = y^\nu - x^n$, where $\nu < n$ and $GCD(\nu, n) = 1$. Then $I_0(x, f) = \nu$ and $I_0(\partial_y f, f) = n(\nu - 1)$; hence $\sigma_p = -\nu + n(\nu - 1) + 1 = (n-1)(\nu-1)$. In this case, by definition,

$$\begin{aligned}\delta_p &= \dim_k k[t]/k[t^\nu, t^n]\\ &= \#\{m \in \mathbb{N} \mid m \neq \alpha\nu + \beta n\\ &\qquad \text{for any nonnegative integers } \alpha, \beta\}.\end{aligned}$$

Hence, as a by-product, one has the formula: $\#\{m \in \mathbb{N} \mid m \neq \alpha\nu + \beta n\} = (n-1)(\nu-1)$.

c. Let p be a cusp of C, i.e., $\rho^{-1}(p) = \{q\}$ is a set consisting of a single point. Choosing an affine open neighborhood Spec A of p such that Sing(Spec A) $= \{p\}$, and letting A' denote the integral closure of A in $Q(A)$, one sees that Spec $A' = \rho^{-1}$(Spec A) and Spec$(A' \otimes_A \mathcal{O}_{C,p}) \cong$ Spec$(\mathcal{O}_{C,p})'$ has the unique closed point q. Thus, $(\mathcal{O}_{C,p})' \cong \mathcal{O}_{C^\#,q}$, which is finitely generated as an $\mathcal{O}_{C,p}$-module, since A' is finitely generated as an A-module by Theorem 2.1. Let $\mathfrak{m}$ be the maximal ideal of $\mathcal{O}_{C,p}$ and let $\mathfrak{n}$ be the ideal generated by $\mathfrak{m}$ in $\mathcal{O}_{C^\#,q}$, i.e., $\mathfrak{n} = \mathfrak{m}\mathcal{O}_{C^\#,q}$, which becomes primary. Thus $\hat{\mathcal{O}}_{C^\#,q} \cong \operatorname{projlim}_s \mathcal{O}_{C^\#,q}/\mathfrak{n}^s \cong \operatorname{projlim}_s(\mathcal{O}_{C,p}/\mathfrak{m}^s) \otimes_{\mathcal{O}_{C,p}} \mathcal{O}_{C^\#,q} =$

$\hat{\mathcal{O}}_{C,p} \otimes_{\mathcal{O}_{C,p}} \mathcal{O}_{C^{\#},q}$, which is the formal power series ring, since q is a non-singular point of $C^{\#}$. From $\mathcal{O}_{C,p} \subseteq \mathcal{O}_{C^{\#},q}$, one derives that $\hat{\mathcal{O}}_{C,p} = \mathcal{O}_{C,p} \otimes_{\mathcal{O}_{C,p}} \hat{\mathcal{O}}_{C,p} \subseteq \mathcal{O}_{C^{\#},q} \otimes_{\mathcal{O}_{C,p}} \hat{\mathcal{O}}_{C,p} \cong \hat{\mathcal{O}}_{C^{\#},q}$, which is written as $k[[t]]$, where t denotes a regular parameter of $\mathcal{O}_{C^{\#},q}$. Thus, $\hat{\mathcal{O}}_{C,p}$ has no zero divisor.

Lemma 9.1. *Let p be a cusp of C. Then*

(i) *$\hat{\mathcal{O}}_{C,p}$ is an integral domain.*
(ii) *$\hat{\mathcal{O}}_{C^{\#},q}$ is the normalization of $\hat{\mathcal{O}}_{C,p}$.*

PROOF. (i) This was proved in the above observation.

(ii) Since t is a regular parameter of $\mathcal{O}_{C^{\#},q}$, $t \in \mathrm{Rat}(C^{\#}) = \mathrm{Rat}(C) = Q(\mathcal{O}_{C,p})$ and $\hat{\mathcal{O}}_{C^{\#},q} = k[[t]]$. Furthermore, it was proved that $\hat{\mathcal{O}}_{C^{\#},q}$ is finitely generated as an $\hat{\mathcal{O}}_{C,p}$-module; thus $\hat{\mathcal{O}}_{C^{\#},q}$ is an integral extension of $\hat{\mathcal{O}}_{C,p}$ and $Q(\hat{\mathcal{O}}_{C^{\#},q}) = k((t)) = Q(\hat{\mathcal{O}}_{C,p})$. □

Letting (x, y) be a local coordinate system around p, one has $\hat{\mathcal{O}}_{S,p} \cong k[[x, y]]$ and $\hat{\mathcal{O}}_{C,p} \cong k[[x, y]]/(f)$. Since $v(f)$ is $v = e(p, C)$, f can be written as a sum of the homogeneous parts $f_{(j)}$; i.e., $f = f_{(v)} + f_{(v+1)} + \cdots$ with $f_{(v)} \neq 0$. Thus letting x be $\xi + \alpha y$ for a suitable α, one has $f_{(v)} = \xi g_{(v-1)} + (\sum_{i+j=v} c_{ij}\alpha^i)y^v$ with $\alpha^* = \sum_{i+j=v} c_{ij} \cdot \alpha^i \neq 0$. Furthermore, replacing f by f/α^*, and ξ by x, one can assume that $f_{(v)} = y^v + \cdots$. Then by Corollary 4 to Theorem 2.6, there exist a unit ε, and $A_1(x), \ldots, A_v(x) \in k[[x]]$ such that

$$f = \varepsilon \cdot (y^v + A_1(x)y^{v-1} + \cdots + A_v(x)).$$

The ring $\mathfrak{o} = k[[x, y]]/(f)$ is an integral domain by Lemma 9.1.(i). Now since f is irreducible, one has $[Q(\mathfrak{o})\colon k((x))] = v$. By (ii), the integral closure $\mathfrak{o}'$ is $k[[t]]$; hence $Q(\mathfrak{o}) = Q(\mathfrak{o}') = k((t))$ and $[k((t))\colon k((x))] = v$.

Since $x \in k[[x]] \subseteq \mathfrak{o} \subseteq \mathfrak{o}' = k[[t]]$, x can be written as a power series $\psi(t)$. If $e = \mathrm{ord}_t(\psi)$, then $\psi = t^e\varepsilon$ for some unit ε. $\varepsilon_1 = \varepsilon^{1/e} \in k[[t]]$ and so $t\varepsilon_1 = \tau \in k[[t]]$ with $\tau^e = x$. Clearly, $\varepsilon_1 \in k[[\tau]]$; hence $k[[t]] = k[[\tau]]$. Thus, $e = [k((\tau))\colon k((x))] = [k((t))\colon k((x))] = v$. Hence, $k((\tau))/k((x))$ is a Galois extension whose Galois group is generated by γ with $\gamma(\tau) = \zeta\tau$, where ζ is a primitive e-th root of 1. Since $k[[x, y]]/(f) \subset k[[\tau]]$, there exists $Q(\tau) \in k[[\tau]]$ such that $f(\tau^v, Q(\tau)) = 0$. Hence, letting $h(\mathsf{T})$ be $\mathsf{T}^v + A_1(\tau^v)\mathsf{T}^{v-1} + \cdots + A_v(\tau^v) \in k((\tau^v))[\mathsf{T}]$, $h(\mathsf{T}) = 0$ has a solution $Q(\tau)$. Furthermore, since $\gamma^i(Q(\tau)) = Q(\zeta^i\tau)$, $\{Q(\tau), Q(\zeta\tau), \ldots, Q(\zeta^{v-1}\tau)\}$ is a complete system of solutions of $h(\mathsf{T}) = 0$.

Therefore, one has the factorization

$$h(\mathsf{T}) = (\mathsf{T} - Q(\tau))(\mathsf{T} - Q(\zeta\tau)) \cdots (\mathsf{T} - Q(\zeta^{v-1}\tau)).$$

If $Q(\tau) = \alpha_1\tau^{\lambda_1}$ + terms of higher degree, with $\alpha_1 \in k^*$, then letting $\delta = GCD(\lambda_1, v)$, $m_1 = \lambda_1/\delta$, $n_1 = v/\delta$, one has

$$h(\mathsf{T}) = (\mathsf{T}^{n_1} - \alpha_1^{n_1}x^{m_1})^\delta + \text{terms of higher degree}.$$

Since $f(x, y) = \varepsilon \cdot h(y)$, one obtains the next result.

Proposition 9.2. *If p is a cusp of C defined by a formal power series $f = \sum_{i,j} c_{ij} x^i y^j$ with $\nu = \nu(f)$ and $f_{(\nu)} = y^\nu + \cdots$, and if λ_1 is defined to be* $\min\{i \mid c_{i0} \neq 0\}$, *then*

(i) *$i/\lambda_1 + j/\nu \geq 1$ holds, for (i, j) with $c_{i,j} \neq 0$.*
(ii) *Letting $(i, j) \in \mathbb{L}$ be defined by $i/\lambda_1 + j/\nu = 1$, $\sum_{(i,j) \in \mathbb{L}} c_{i,j} x^i y^j$ equals $(y^{n_1} - \alpha_1^{n_1} x^{m_1})^\delta$ where*

$$\delta = GCD(\lambda_1, \nu),\ \lambda_1 = m_1 \delta,\ \nu = n_1 \delta.$$

Remark. This holds for any irreducible formal power series, including those not obtained from a cusp.

d. In the last proposition, if $n_1 = 1$, then putting $y_1 = y - \alpha_1 x^{m_1}$, we repeat the above argument for x and y_1. Thus, after such procedures, finally we arrive at the case where $n_1 > 1$, whenever $\nu \geq 2$.

Letting $Q(\tau) = \sum_{m=1}^{\infty} \alpha_m \tau^{\lambda_m}$ where $\alpha_m \neq 0$ and $\lambda_1 < \lambda_2 < \cdots$, we define $a(1)$ to be $\min\{a \mid \lambda_a/\nu \text{ is not an integer}\}$. Thus

$$Q(\tau) = \alpha_1 x^{\rho_1} + \cdots + \alpha_{a(1)-1} x^{\rho_{a(1)-1}} + \alpha_{a(1)} \tau^{\lambda_{a(1)}} + \cdots,$$

where $\rho_i = \lambda_i/\nu$, and

$$h(\mathsf{T}) = \prod_{i=0}^{\nu-1} (\mathsf{T} - Q(\zeta^i \tau)) = \prod_{i=0}^{\nu-1} (\mathsf{T} - \alpha_1 \tau^{\lambda_1} - \cdots - \alpha_{a(1)-1} \tau^{\lambda_{a(1)-1}}$$
$$- \alpha_{a(1)} \zeta^{i\lambda_{a(1)}} \tau^{\lambda_{a(1)}}) + \text{terms of higher degree.}$$

By this observation, we can prove the next result.

Proposition 9.3. *If p is a cusp, then*

$$\lambda_{a(1)} = \max\{I_p(C, L) \mid L \text{ is nonsingular at } p\}.$$

PROOF. Left to the reader as an exercise. □

Remark. A curve L satisfying $\lambda_{a(1)} = I_p(C, L)$ is said to be a *maximal contact curve* at p, if L is nonsingular at p.

When p is a cusp of C, one has

$$\sigma_p = -\mathrm{ord}_t(\rho^*(x/\partial_y f)) + 1 = \mathrm{ord}_t(\rho^*(\partial_y f)) - \nu + 1$$

by using the notation in §9.2.**b**, where t is a regular parameter. Letting h be $h(y) = y^\nu + A_1(x) y^{\nu-1} + \cdots + A_\nu(x) = f/\varepsilon$, one has

$$\mathrm{ord}_t(\rho^*(\partial_y f)) = \mathrm{ord}_t(\rho^*(\partial_y h)) = \mathrm{ord}_\tau(\partial_y h)(Q(\tau)),$$

which is computed as follows.

From $h(y) = \prod_{i=0}^{\nu-1} (y - Q(\zeta^i \tau))$, it follows that

$$\partial_y h(y) = \sum_{j=0}^{\nu-1} \prod_{i \neq j} (y - Q(\zeta^i \tau)).$$

Hence, $\text{ord}_\tau(\partial_y h)(Q(\tau)) = \text{ord}_\tau(\prod_{i=1}^{\nu-1} (Q(\tau) - Q(\zeta^i\tau)))$. Since $Q(\tau) - Q(\zeta^i\tau) = \sum_{m=1}^{\infty} \alpha_m(1 - \zeta^{i\lambda_m})\tau^{\lambda_m}$, letting $m(i) = \min\{m \mid \lambda_m i \not\equiv 0 \mod \nu\}$ for $1 \leq i \leq \nu - 1$, one has $\text{ord}_\tau(Q(\tau) - Q(\zeta^i\tau)) = m(i)$ and so

$$\text{ord}_\tau(\partial_y h)(Q(\tau)) = \sum_{i=1}^{\nu-1} \lambda_{m(i)}.$$

Thus, it suffices to compute $\{m(i) \mid 1 \leq i \leq \nu - 1\}$ from the given sequence $\{\lambda_1, \lambda_2, \ldots\}$ of integers. First of all, note that $a(1) = \min\{m(i) \mid 1 \leq i \leq \nu - 1\}$. If one writes $\lambda_{a(1)}/\nu = m_1/n_1$ as a simple fraction (i.e., $GCD(m_1, n_1) = 1$), then

$$a(1) = m(i) \Leftrightarrow \lambda_{a(1)}\, i \not\equiv 0 \mod \nu \Leftrightarrow m_1\, i \not\equiv 0 \mod n_1 \Leftrightarrow i \not\equiv 0 \mod n_1.$$

Thus $\#\{i \mid m(i) = a(1)\} = (n_1 - 1)\nu/n_1 = \nu - \nu/n_1$. Next, suppose that $\{i \mid m(i) > a(1)\}$ is not empty and define $a(2)$ to be $\min\{m(i) \mid m(i) > a(1)\}$. If one writes $\lambda_{a(2)}\, n_1/\nu = m_2/n_2$ as a simple fraction, then $n_2 > 1$ and

$$\begin{aligned} a(2) = m(j) &\Leftrightarrow \lambda_{a(1)}\, j \equiv 0 \mod \nu, \qquad \lambda_{a(2)}\, j \not\equiv 0 \mod \nu \\ &\Leftrightarrow j m_1 \equiv 0 \mod n_1, \qquad j m_2 \not\equiv 0 \mod n_1 n_2 \\ &\Leftrightarrow j = l n_1, \qquad l \not\equiv 0 \mod n_2 \end{aligned}$$

Thus

$$\#\{j \mid m(j) = a(2)\} = \frac{\nu}{n_1} - \frac{\nu}{n_1 n_2} = \frac{\nu}{n_1}\left(1 - \frac{1}{n_2}\right).$$

Further, inductively define $a(t)$ to be $\min\{m(i) \mid m(i) > a(t-1)\}$ and write $(\lambda_{a(t)}\, n_1 \cdots n_{t-1})/\nu = m_t/n_t$ as a simple fraction, provided that the set $\{i \mid m(i) > a(t-1)\}$ is not empty. Then $n_t > 1$ and $u_t = \#\{i \mid m(i) = a(t)\}$ equals $(\nu/n_1 \cdots n_{t-1}) \cdot (1 - 1/n_t)$.

Letting ω denote the maximal number of these t, one has a system of pairs $(m_1, n_1), \ldots, (m_\omega, n_\omega)$ such that $\lambda_m n_1 \cdots n_\omega/\nu$ is an integer for all m. The integer $\nu/(n_1 \cdots n_\omega)$ is a common divisor of $\nu, \lambda_1, \lambda_2, \ldots$, which is 1, since $k((\tau^\nu, \sum_{i=1}^{\infty} \alpha_i \tau^{\lambda_i})) = k((\tau))$. Thus,

$$\sigma_p = \sum_{t=1}^{\omega} m_t(n_t - 1)(n_{t+1} \cdots n_\omega)^2 + 1 - \nu,$$

when p is a cusp of multiplicity ν.

If $\omega = 1$, then $\sigma_p = (\lambda_{a(1)} - 1)(\nu - 1)$ (cf. Example 9.2).

Remark. The system of pairs $(m_1, n_1), \ldots, (m_\omega, n_\omega)$ is determined by the local ring $k[[x, y]]/(f)$, and called the *characteristic pairs* of *the cusp* p (cf. [Z, p. 7]). (m_1, n_1) is said to be *the first characteristic pair*, which could be obtained as a simple fraction, i.e., $m_1/n_1 = I_p(C, L)/\nu$, where L is one of the maximal contact curves of C at p (see Proposition 9.3).

e. Now, we shall investigate the behavior of cusps under quadric transformations. Using the previous notation, if $\lambda_1 = \nu$, then one can replace $y - \alpha_1 x$ by y. Thus, one may assume $\lambda_1 > \nu$ and then $f_{(\nu)}$ is y^ν. In this case, $V(y)$ ($\subset U = \text{Spec } R$) is said to be the *local tangent* to C at p with respect to (x, y). Actually, $I_0(f, y) = \lambda_1$, $I_0(f, y - ax) = \nu$ for any $\alpha \neq 0$, and $I_0(f, x) = \nu$. Let L be the closure of $V(y)$ in S. Then $I_p(C, L) = \lambda_1$. Denoting by C^* and L^* the strict transforms of C and L, respectively, one obtains the following result.

Proposition 9.4. *Let* $E = \mu^{-1}(p)$ *and* $\{q\} = E \cap C^*$. *Then*

(*i*) $I_q(C^*, E) = \nu$,
(*ii*) $I_q(L^*, E) = 1$,
(*iii*) $I_q(C^*, L^*) = \lambda_1 - \nu$.

PROOF. (i) and (ii) follow from $e(p, C) = \nu$ and $e(p, L) = 1$ by Proposition 9.1. (iii) follows from Theorem 8.1 and Proposition 8.1. $\square$

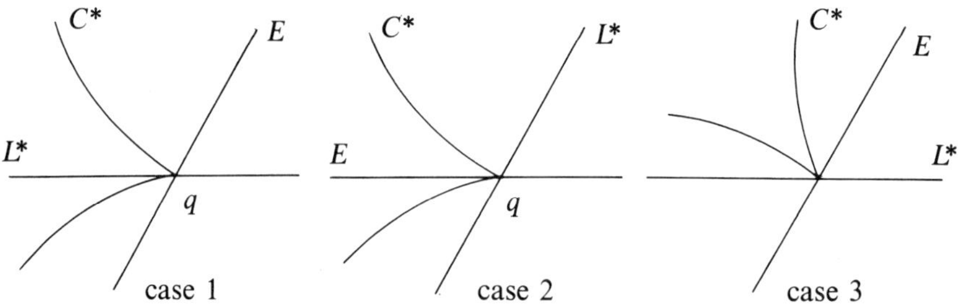

Since q is a cusp or a non-singular point of C^*, C^* has a unique local tangent. Thus, if $\nu^* = e(q, C^*)$, we have the following three cases.

Case 1. $\lambda_1 - \nu > \nu$. Then L^* becomes the local tangent to C^* at q, since $I_q(C^*, E) = \nu$ and $\nu^* = \nu$.

Case 2. $\lambda_1 - \nu < \nu$. Then E becomes the local tangent at q and $\nu^* = \lambda_1 - \nu$, since $\lambda_1 - \nu = I_q(C^*, L^*)$.

Case 3. $\lambda_1 - \nu = \nu$. Then both E and L^* do not give rise to local tangents and so $\nu^* = \nu$.

Thus, if $\lambda_1 > \nu$, then one has $\lambda_1 = q_0\nu + \nu_1$ with $0 \leq \nu_1 < \nu$. In the sequence of successive quadric transformations whose centers are corresponding cusps, one has case 1 repeated $(q_0 - 1)$ times and case 2 (if $\nu_1 > 0$) or case 3 (if $\nu_1 = 0$). Repeating this, we have the Euclidean algorithm:

$$\begin{aligned}
\lambda_1 &= q_0\nu + \nu_1, & 0 < \nu_1 < \nu \\
\nu &= q_1\nu_1 + \nu_2, & 0 < \nu_2 < \nu_1 \\
&\cdots\cdots & \cdots\cdots \\
\nu_{h-1} &= q_h\nu_h,
\end{aligned} \qquad (\#)$$

where $\nu_h = GCD(\lambda_1, \nu)$.

In this case one has case 1 repeated $(q_0 - 1)$ times, case 2 once, case 1 $(q_1 - 1)$ times, ..., and finally case 3.

If $\nu_h = 1$, then one has the infinitely near points lying over p as follows, with the numbers indicating multiplicities.

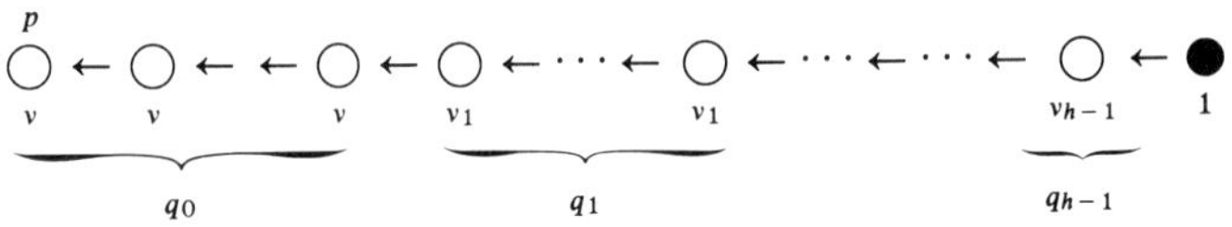

Hence,

$$\varepsilon_p = q_0 \frac{\nu(\nu - 1)}{2} + q_1 \frac{\nu_1(\nu_1 - 1)}{2} + \cdots + q_{h-1} \frac{\nu_{n-1}(\nu_{h-1} - 1)}{2}.$$

From (#), one has

$$\sum_{i=0}^{h-1} q_i \nu_i = \lambda_1 + \nu - \nu_{h-1},$$

where $\nu_0 = \nu$, and

$$\sum_{i=0}^{h-1} q_i \nu_i^2 = \lambda_1 \nu - \nu_{h-1} + 1.$$

Hence,

$$\varepsilon_p = \frac{\lambda_1 \nu - \lambda_1 - \nu + 1}{2} = \frac{(\lambda_1 - 1)(\nu - 1)}{2}.$$

This formula agrees with the one obtained in §9.2.**d**.

It is not hard to carry out successive quadric transformations for more general cusps in order to compute ε_p.

§9.3 Linear Pencil Theorem

a. If p is a singular point of a curve C such that $C^0 = C \backslash \{p\}$ is isomorphic to $\mathbf{A}_k^1$, then p is a cusp of C. To see this, let $(C^\#, \rho)$ be the nonsingular model of C. Since C is birationally equivalent to $\mathbf{A}_k^1$, $C^\#$ is $\mathbf{P}_k^1$ and $\rho^{-1}(C^0) = \mathbf{P}_k^1 \backslash \rho^{-1}(p)$ is isomorphic to $\mathbf{A}_k^1$. Thus $\rho^{-1}(p)$ is a set consisting of a single point.

Let Λ be a linear pencil, i.e., a linear system of dimension 1, on $\mathbf{P}_k^2 = \text{Proj}\, k[\mathsf{T}_0, \mathsf{T}_1, \mathsf{T}_2]$ and let $H = V_+(\mathsf{T}_2)$.

Theorem 9.3 (Linear Pencil Theorem of Jung, Gutwirth, Nagata). *Suppose that a general member C_u of Λ satisfies*

(1) $C_u \cap H = \{p_u\}$,
(2) $C_u \backslash \{p_u\} \cong \mathbf{A}_k^1$.

Then, letting $d = \deg C_u$ and $\nu = e(p_u, C_u)$, $d - \nu$ divides d.

PROOF. If p_u is not a base point of Λ, then C_u is nonsingular. Hence, $d = 1$ or 2, since C_u is a rational curve. Thus, the assertion is obvious. We can suppose that p_u is a singular point. Thus p_u is a base point of Λ, which is independent of u. Hence, we just write p_u as p. By (1), $d = (C_u, H)_{\mathbf{P}^2} = I_p(C_u, H)$. Since $v = I_p(C_u, V_+(\mathsf{T}_1))$, letting $\lambda_1 = d$, one uses the Euclidean algorithm (#) in §9.2.**e**.

If $v_h = 1$, then there exist only $q_0 - 1 + q_1 + \cdots + q_{h-1}$ infinitely near singular points lying over p, as was indicated in §9.2.**e**. If $v_h > 1$, then there exist $q_h + s$ more infinitely near singular points. The multiplicity of the infinitely near singular point of $(\sum_{i=0}^{h} q_i + j)$-th order is indicated by v_{h+j}. Thus, one has the following figure:

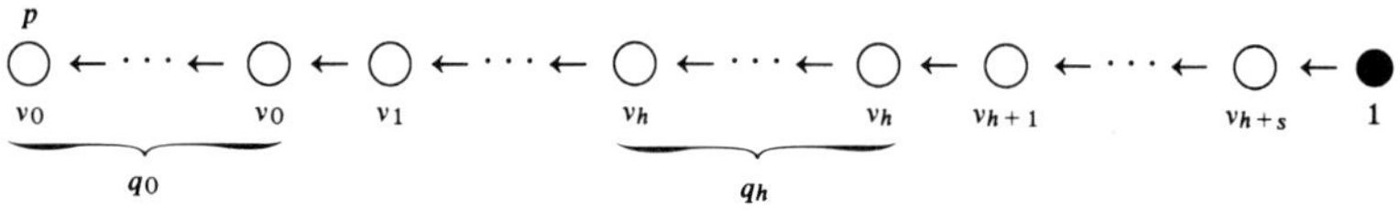

By the genus formula of Clebsch, one has

$$0 = 2g(C_u) = (d-1)(d-2) - \sum_{i=0}^{h} q_i v_i(v_i - 1) - \sum_{m=h+1}^{h+s} v_m(v_m - 1). \qquad (*)$$

In view of (#), one obtains

$$\sum_{i=0}^{h} q_i v_i(v_i - 1) = (d-1)(v-1) + v_h - 1;$$

hence $(*)$ can be rewritten as

$$0 = (d-1)(d-v-1) - (v_h - 1) - \sum_{m=h+1}^{h+s} v_m(v_m - 1),$$

in which the right-hand side is not smaller than $(d-1)(d-v-1) - (v_h - 1) \cdot (1 + \sum_{m=h+1}^{h+s} v_m)$, because $v_h \geq v_m$ for all $m > h$.

Now, if $v_h = 1$, then $0 \geq (d-1)(d-v-1)$. Hence $d - 1 \leq v$, i.e., $d - 1 = v$ and so $1 = d - v$ divides d.

In the case of $v_h > 1$, the inequality above implies

$$\frac{(d-1)(d-v-1)}{v_h - 1} \leq 1 + \sum_{m=h+1}^{h+s} v_m. \qquad (**)$$

On the other hand, since C_u is a general member of Λ, by Theorem 8.1 applied to successive quadric transformations, one obtains

$$d^2 \geq \sum_{i=0}^{h} q_i v_i^2 + \sum_{m=h+1}^{h+s} v_m^2.$$

Combining this with $(*)$, one has

$$d^2 \geq \sum_{i=0}^{h} q_i v_i^2 + \sum_{m=h+1}^{h+s} v_m^2 = (d-1)(d-2) + \sum_{i=0}^{h} q_i v_i + \sum_{m=h+1}^{h+s} v_m.$$

Hence,

$$3d - 2 \geq \sum_{i=0}^{h} q_i v_i + \sum_{m=h+1}^{h+s} v_m.$$

Since $\sum_{i=0}^{h} q_i v_i = d + v - v_h$, it follows that

$$2d - 2 + v_h - v \geq \sum_{m=h+1}^{h+s} v_m.$$

Then in view of (**), one has

$$\frac{(d-1)(d-v-1)}{v_h - 1} \leq 1 + 2d - 2 + v_h - v = d + v_h + d - v - 1;$$

hence

$$(d-1)(d-v-1) \leq (v_h - 1)(d + v_h) + (v_h - 1)(d - v - 1).$$

Thus

$$(d - v_h)(d - v - 1) \leq (v_h - 1)(d + v_h).$$

Since $v_h = GCD(d, v)$, $\upsilon = (d - v)/v_h$ and $w = d/v_h$ are integers with $1 \leq \upsilon < w$. Therefore, one has

$$(w-1)(\upsilon v_h - 1) \leq (v_h - 1)(w + 1) = (v_h - 1)(w - 1) + 2(v_h - 1),$$

and so

$$(w-1)(\upsilon - 1)v_h \leq 2(v_h - 1) < 2v_h.$$

Hence, $(w-1)(\upsilon-1) < 2$. This implies $(w-1)(\upsilon-1) \leq 1$ and $(\upsilon-1)^2 < 1$, i.e., $\upsilon = 1$. Therefore, $d - v = v_h$ divides d. □

b. As an application of the linear pencil theorem, one can prove a theorem of Jung concerning the structure of the automorphism group of $\mathbf{A}_k^2 = \operatorname{Spec} k[\mathsf{X}, \mathsf{Y}]$.

First, we recall two subgroups $\mathit{Aff}_2(k)$ and $J_2(k)$ of $G = \operatorname{Aut}_k(\mathbf{A}_k^2)$ which was introduced in §1.19.**c**. Note that these groups depend on the choice of affine coordinate system (X, Y). Elements of $\mathit{Aff}_2(k)$ are said to be *linear transformations*; while elements of $J_2(k)$ are called *de Jonquièrre transformations.* We shall prove that any automorphism of $\mathbf{A}_k^2$ is a composition of linear transformations and de Jonquièrre transformations.

We begin by proving a more general result. Let φ and ψ be polynomials in X and Y such that $V(\varphi) \cap V(\psi) = \varnothing$. For a general point λ of k, we suppose that $V(\varphi + \lambda\psi) \cong \mathbf{A}_k^1$. Define $d_{(\varphi,\psi)}$ to be $\max\{\deg \varphi, \deg \psi\}$. Choose a homogeneous coordinate system $(\mathsf{T}_0, \mathsf{T}_1, \mathsf{T}_2)$ such that $\mathsf{T}_1/\mathsf{T}_0 = \mathsf{X}$ and $\mathsf{T}_2/\mathsf{T}_0 = \mathsf{Y}$. Assuming $d = d_{(\varphi,\psi)} = \deg \varphi \geq \deg \psi$, we let $F_1 = \mathsf{T}_0^d \cdot \varphi$ and $F_2 = \mathsf{T}_0^d \cdot \psi$. Then F_1 and F_2 define a linear system Λ on $\mathbf{P}_k^2$, which satisfies the hypothesis of the last theorem. We use the notation $L_i = V_+(\mathsf{T}_i)$ for $0 \leq i \leq 2$, $f = \varphi + \lambda\psi$ for a general point of k, and

$C_\lambda = V_+(F_1 + \lambda F_2)$. Suppose $d \geq 2$. Then Λ has a base point p, which lies on L_0, by hypothesis. Since $L_0 \cap C_\lambda = \{p\}$, L_0 is the unique tangent to C at p. After a suitable linear transformation with respect to X and Y, we can assume that $p = (0 : 0 : 1)$. $I_p(C_\lambda, L_0)$ is equal to the multiplicity of C at p, denoted by v. By the last theorem, $d - v$ divides d, i.e., $d = e(d - v)$ for some integer e. One has

$$d - v = \sum_{q \notin L_0} I_q(C_\lambda, L_1) = \dim_k k[\mathsf{X}, \mathsf{Y}]/(f, \mathsf{X})$$

by Corollary (iii) to Proposition 8.1. Furthermore, since $p \notin L_2 = V_+(\mathsf{T}_2)$, one has

$$d = \sum_q I_q(C_\lambda, L_2) = \sum_{q \notin L_0} I_q(C_\lambda, L_2) = \dim_k k[\mathsf{X}, \mathsf{Y}]/(f, \mathsf{Y}).$$

Furthermore, since $V(f) \cong \mathbf{A}_k^1$, one has an isomorphism

$$\Psi\colon k[\mathsf{X}, \mathsf{Y}]/(f) \cong k[\mathsf{T}].$$

Thus $F(\mathsf{T}) = \Psi(\mathsf{X})$ and $G(\mathsf{T}) = \Psi(\mathsf{Y})$ satisfy

$$k[\mathsf{X}, \mathsf{Y}]/(f, \mathsf{X}) \cong k[\mathsf{T}]/(F(\mathsf{T}))$$

and

$$k[\mathsf{X}, \mathsf{Y}]/(f, \mathsf{Y}) \cong k[\mathsf{T}]/(G(\mathsf{T})).$$

Therefore, deg $F(\mathsf{T}) = d - v$ and deg $G(\mathsf{T}) = d$. Hence, there exists $\gamma \in k$ such that $\deg(G(\mathsf{T}) - \gamma F(\mathsf{T})^e) < d$. Hence, letting $\mathsf{X}_1 = \mathsf{X}$, $\mathsf{Y}_1 = \mathsf{Y} - \gamma \mathsf{X}^e$, $\varphi_1(\mathsf{X}_1, \mathsf{Y}_1) = \varphi(\mathsf{X}, \mathsf{Y})$, $\psi_1(\mathsf{X}_1, \mathsf{Y}_1) = \psi(\mathsf{X}, \mathsf{Y})$, and $f_1(\mathsf{X}_1, \mathsf{Y}_1) = f(\mathsf{X}, \mathsf{Y}) = \varphi_1 + \lambda\psi_1$, we obtain

$$k[\mathsf{X}_1, \mathsf{Y}_1]/(f_1, \mathsf{X}_1) \cong k[\mathsf{T}]/(F(\mathsf{T}))$$

and

$$k[\mathsf{X}_1, \mathsf{Y}_1]/(f_1, \mathsf{Y}_1) \cong k[\mathsf{T}]/(G(\mathsf{T}) - \gamma F(\mathsf{T})^e).$$

Hence, $d_{(\varphi_1, \psi_1)} = \max\{\deg F(\mathsf{T}), \deg(G(\mathsf{T}) - \gamma F(\mathsf{T})^e)\} < d = d_{(\varphi, \psi)}$. Using this result, we prove the following theorem.

Theorem 9.4. *Let $g\colon \mathbf{A}_k^2 \to \mathbf{A}_k^1$ be a dominating morphism such that a general fiber $g^{-1}(\lambda)$ is isomorphic to $\mathbf{A}_k^1$. Then there exists a composition f of linear transformations and of de Jonquièrre transformations such that $g \circ f$ is the projection morphism of $\mathbf{A}_k^2$ to $\mathbf{A}_k^1$. In particular, $\mathbf{A}_k^2 \backslash g^{-1}(a) \cong \mathbf{A}_k^1 \times (\mathbf{A}_k^1 \backslash \{a\})$* for any $a \in k$.

Proof. Fix an affine coordinate system (X, Y) of $\mathbf{A}_k^2$ and take two general points p and q of $g(\mathbf{A}^2)$. Define polynomials φ and ψ of $V(\varphi) = g^{-1}(p)$ and $V(\psi) = g^{-1}(q)$. Since $d_{(\varphi, \psi)} = \deg g^{-1}(\lambda)$ for a general point λ of k, we complete the proof by induction of $d_{(\varphi, \psi)}$. □

Corollary (Jung). *Any automorphism of $k[\mathsf{X}, \mathsf{Y}]$ is a composition of linear transformations and de Jonquièrre transformations.*

PROOF. Let h be a morphism associated to a given automorphism of $k[\mathsf{X}, \mathsf{Y}]$. Define $g: \mathbf{A}_k^2 \to \mathbf{A}_k^1$ to be the composition of h and the projection of $\mathbf{A}_k^2$. Applying the last theorem to g, we complete the proof. $\square$

EXAMPLE 9.3. Let $\varphi = \mathsf{X} - (\mathsf{Y} - \mathsf{X}^2)^2$ and $\psi = \mathsf{Y} - \mathsf{X}^2$. These satisfy the hypothesis of the theorem. $\mathsf{T}_0^4 \varphi = \mathsf{T}_1 \mathsf{T}_0^3 - (\mathsf{T}_2 \mathsf{T}_0 - \mathsf{T}_1^2)^2$ and $C = V(\mathsf{T}_0^4 \varphi)$ has a double point p and $C - \{p\} \cong \mathbf{A}^1$.

§9.4 Dual Curves and Plücker Relations

a. A curve L on $\mathbf{P}_k^2 = \operatorname{Proj} k[\mathsf{T}_0, \mathsf{T}_1, \mathsf{T}_2]$ is said to be a *line*, if $\deg L = 1$, i.e., $L = V_+(a_0 \mathsf{T}_0 + a_1 \mathsf{T}_1 + a_2 \mathsf{T}_2)$ for some $(a_0, a_1, a_2) \in k^3 \backslash \{0\}$. Clearly,

$$V_+(a_0 \mathsf{T}_0 + a_1 \mathsf{T}_1 + a_2 \mathsf{T}_2) = V_+(b_0 \mathsf{T}_0 + b_1 \mathsf{T}_1 + b_2 \mathsf{T}_2)$$
$$\Leftrightarrow a_i = \lambda b_i, \quad i = 0, 1, 2, \quad \text{for some} \quad \lambda \in k.$$

Therefore, one has the bijection

$$\mathscr{T}: \{\text{lines on } \mathbf{P}_k^2\} \to \mathbf{P}_k^2(k)$$

defined by $\mathscr{T}(V_+(a_0 \mathsf{T}_0 + a_1 \mathsf{T}_1 + a_2 \mathsf{T}_2)) = (a_0 : a_1 : a_2)$.

Take a curve $C = V_+(F)$ with $r = \deg C = \deg F \geq 2$. For a nonsingular point p of C, let l_p denote the tangent to C at p. When $p = (\alpha_0 : \alpha_1 : \alpha_2)$, one writes $\partial_i F(\alpha_0, \alpha_1, \alpha_2)$ as $(\partial_i F)(p)$. Then, clearly, $l_p = V_+((\partial_0 F)(p)\mathsf{T}_0 + (\partial_1 F)(p)\mathsf{T}_1 + (\partial_2 F)(p)\mathsf{T}_2)$, i.e., $\mathscr{T}(l_p) = ((\partial_0 F)(p) : (\partial_1 F)(p) : (\partial_2 F)(p))$. The rational map defined by $\partial_0 F, \partial_1 F, \partial_2 F$ is denoted by τ_C, i.e., $\tau_C = (\partial_0 F : \partial_1 F : \partial_2 F)$. Then $\mathbf{P}_k^2 \backslash \operatorname{dom}(\tau_C) = V_+(\partial_0 F, \partial_1 F, \partial_2 F)$ and $\tau_C(p) = \mathscr{T}(l_p)$.

Lemma 9.2. *$\tau_C(C)$ is a curve.*

PROOF. Suppose that $\tau_C(C)$ is a point $(a_0 : a_1 : a_2)$. Then $(\partial_0 F(p) : \partial_1 F(p) : \partial_2 F(p)) = (a_0 : a_1 : a_2)$ for any nonsingular point p of C; hence $(\partial_i F)(p) = \gamma a_i$, $i = 0, 1, 2$ for some $\gamma \in k$. By Euler's identity, $\alpha_0(\partial_0 F)(p) + \alpha_1(\partial_1 F)(p) + \alpha_2(\partial_2 F)(p) = r \cdot F(p) = 0$ where $p = (\alpha_0 : \alpha_1 : \alpha_2)$; hence $\alpha_0 a_0 + \alpha_1 a_1 + \alpha_2 a_2 = 0$. This implies that $\operatorname{Reg} C \subseteq V_+(a_0 \mathsf{T}_0 + a_1 \mathsf{T}_1 + a_2 \mathsf{T}_2)$; hence $C = V_+(a_0 \mathsf{T}_0 + a_1 \mathsf{T}_1 + a_2 \mathsf{T}_2)$. Thus $r = \deg C$ must be 1. This contradicts the assumption $r \geq 2$. $\square$

Definition. $\tau_C(C)$ is said to be the *dual curve* of C. If $\tau_C(C)$ is expressed as $V_+(G)$ by an irreducible homogeneous polynomial G, then G is also said to be a *dual homogeneous polynomial* of F.

b. Let Γ be the dual curve of C and let $*$ be the generic point of C. Then $\tau_C(*)$ is the generic point of Γ. Let $f(\mathsf{X}, \mathsf{Y}) = F(1, \mathsf{X}, \mathsf{Y}) = F(\mathsf{T}_0, \mathsf{T}_1, \mathsf{T}_2)/\mathsf{T}_0^r$

and $x = \mathsf{X} \bmod f$, $y = \mathsf{Y} \bmod f$, where r is deg F. One has $f(x, y) = 0$ and puts

$$u = (\partial_1 F)(1, x, y)/(\partial_0 F)(1, x, y) = -f_x/\Delta f,$$

$$v = (\partial_2 F)(1, x, y)/(\partial_0 F)(1, x, y) = -f_y/\Delta f,$$

where $f_x = (\partial f/\partial \mathsf{X})(x, y), f_y = (\partial f/\partial \mathsf{Y})(x, y)$ and $\Delta f = xf_x + yf_y$.

From $f(x, y) = 0$, it follows that $f_x\, dx + f_y\, dy = 0$ and $u\, dx = -dy/\psi$, $v\, dx = dx/\psi$, where ψ is defined by $x\, dy - y\, dx = \psi\, dx$. Here, note that $\psi \neq 0$, because x and y are linearly independent (see §6.7). Clearly, one has

$$ux + vy + 1 = 0 \quad \text{and} \quad u\, dx + v\, dy = 0.$$

Thus, $du \cdot x + u \cdot dx + dv \cdot y + v \cdot dy = 0$ and so $x\, du + y\, dv = 0$. Therefore, if $u \in k$, then $du = 0$ and $dv = 0$, i.e., Γ is a point, which contradicts Lemma 9.2. Hence, $u \notin k$ and so $du \neq 0$. For any $\varphi \in \mathrm{Rat}(C)$, define φ' and φ'' by $d\varphi = \varphi'\, du$ and $d\varphi' = \varphi''\, du$. From $ux' + vy' = 0$ and $x + v'y = 0$, we have

$$x' + ux'' + vy'' + v'y' = 0 \quad \text{and} \quad x' + v'y' + v''y = 0.$$

But $ux'' + vy'' = (y'x'' - y''x')/\psi \neq 0$, since $\deg C \geq 2$. Thus $v''y = -(x' + v'y') = ux'' + vy'' \neq 0$, which implies v' is not a constant; hence $1, u, v$ are linearly independent. So, $\tau_\Gamma \colon \mathbf{P}_k^2 \to \mathbf{P}_k^2$ is defined and $\tau_\Gamma(\Gamma) = Z$ is a curve by Lemma 9.2. Letting $(1 : \xi : \eta)$ be $(\partial_0 G : \partial_1 G : \partial_2 G)(1, u, v)$, one has again $\xi u + \eta v + 1 = 0$ and $\xi\, du + \eta\, dv = 0$. On the other hand, we have had the relations:

$$xu + yv + 1 = 0 \quad \text{and} \quad x\, du + y\, dv = 0.$$

Since $u\, dv - v\, du \neq 0$, it follows that $\xi = x$ and $\eta = y$ from these relations. Therefore, $Z = C$ has been established. Thus we obtain the next theorem.

Theorem 9.5. *Let Γ be the dual curve of C. Then the dual curve of Γ coincides with C. In particular, $\tau_C|_C \colon C \to \Gamma$ is birational.*

c. Take a point $0 \in \mathbf{P}_k^2 \backslash C$ and define a linear system Λ_1 on a nonsingular model $C^{\#}$ of C by

$$\Lambda_1 = \{\rho^* L \mid L \text{ is a line through } 0 \text{ on } \mathbf{P}_k^2\},$$

where $\rho \colon C^{\#} \to C \subset \mathbf{P}_k^2$. Since $\dim \Lambda_1 = 1$, the degree of the Weierstrass divisor $W(C^{\#}, \Lambda_1)$ is $2r + 2g - 2$, where $r = \deg C$ and $g = g(C)$ by Theorem 6.11.

For a closed point p on $C^{\#}$, we shall compute the Λ_1-gap sequence $\{a_1, a_2\}$ at p as follows: Since $\mathrm{Bs}\, \Lambda_1 = \varnothing$, $a_1 = 0$. If $q = \rho(p)$ is a nonsingular point, then letting L_{q0} be the line connecting q and 0, one has $a_2 = I_q(L_{q0}, C)$ by definition. Suppose that q is a singular point and $\rho^{-1}(q) = \{p = p_1, p_2, \ldots, p_s\}$. $\rho^* L_{q0}$ is a divisor $\sum_{i=1}^{s} \mu_i p_i$ and $a_2 = \mu_1$. To study the μ_i, we assume q is $(1 : 0 : 0)$ and let $v = e(q, C)$. Setting $A = k[\mathsf{X},$

$\mathsf{Y}]/(f)$ and $S = A\backslash(\mathsf{X}, \mathsf{Y})$, one has $S^{-1}A = \mathcal{O}_{C,q}$, denoted $\mathfrak{o}$. Let $C_0 = \operatorname{Spec} A$ and $C_0^{\#} = \operatorname{Spec} A'$, where A' is the integral closure of A in $Q(A)$. Then $C_0^{\#} = \rho^{-1}(C_0)$ and $\rho|_{C_0^{\#}}$ is associated with the inclusion $A \subseteq A'$. Since $\mathfrak{o}' = S^{-1}(A')$ by Proposition 2.1, and since A' is finitely generated as an A-module by Theorem 2.1, $\mathfrak{o}'$ is finitely generated as an $\mathfrak{o}$-module. Since $\rho^{-1}(q) = \{p_1, \ldots, p_s\}$, $\mathfrak{o}'$ has finitely many maximal ideals $\mathfrak{m}_1, \ldots, \mathfrak{m}_s$ such that $\mathcal{O}_{C^{\#}, p_j}$ is obtained as the localization of $\mathfrak{o}'$ by $\mathfrak{m}_j$, denoted by $\mathfrak{o}_j'$. Let $J = \mathfrak{m}_1 \cap \cdots \cap \mathfrak{m}_s$, which can also be written as $\mathfrak{m}_1 \cdots \mathfrak{m}_s$. For any $n > 0$, $\mathfrak{m}_i^n + \prod_{j \neq i} \mathfrak{m}_j^n = \mathfrak{o}'$. Hence, $\mathfrak{o}'/J^n \cong \bigoplus_{i=1}^s \mathfrak{o}'/\mathfrak{m}_i^n$. Noting that $\mathfrak{o}'/\mathfrak{m}_i^n \cong \mathfrak{o}_i'/\mathfrak{m}_i^n \mathfrak{o}_i'$ for all i, n, one has

$$(\mathfrak{o}')^\wedge = \operatorname*{projlim}_n \mathfrak{o}'/J^n \cong \bigoplus_{i=1}^{s} (\mathfrak{o}_i')^\wedge.$$

Since $\mathfrak{o}_i'$ is a DVR with $\mathfrak{o}_i'/\mathfrak{m}_i \mathfrak{o}_i' \cong k$, $(\mathfrak{o}_i')^\wedge$ is isomorphic to $k[[t_i]]$ by Corollary 1 to Theorem 2.6. Hence, $(\mathfrak{o}')^\wedge$ has no nonzero nilpotent elements and is integrally closed in its total quotient ring $Q((\mathfrak{o}')^\wedge)$.

Referring to [N2, Theorem 17.8], one knows that $(\mathfrak{o}')^\wedge \cong \mathfrak{o}' \otimes_{\mathfrak{o}} \hat{\mathfrak{o}}$, which is integral over $\hat{\mathfrak{o}}$. Since $Q(\mathfrak{o}' \otimes_{\mathfrak{o}} \hat{\mathfrak{o}}) = Q(\hat{\mathfrak{o}})$, $(\mathfrak{o}')^\wedge$ is the integral closure of $\hat{\mathfrak{o}}$ in $Q(\hat{\mathfrak{o}})$. Thus, $(\mathfrak{o}')^\wedge \cong (\hat{\mathfrak{o}})'$.

Let $\mathfrak{p}_1, \ldots, \mathfrak{p}_u$ be the minimal prime ideals of $\hat{\mathfrak{o}}$. Then since $(\hat{\mathfrak{o}})'$ is integral over $\hat{\mathfrak{o}}$, every $\mathfrak{p}_i$ corresponds to some minimal prime ideal $\mathfrak{q}_i$ of $\hat{\mathfrak{o}}'$ by Theorem 1.7; hence $u = s$. If $\mathfrak{q}_1$ corresponds to $\mathfrak{p}_1$, then $(\hat{\mathfrak{o}})'/\mathfrak{p}_1 \cong (\mathfrak{o}_1')^\wedge$ is integral over $\hat{\mathfrak{o}}/\mathfrak{p}_1$; thus $(\mathfrak{o}_1')^\wedge$ is the integral closure of $\hat{\mathfrak{o}}/\mathfrak{p}_1$ in its field of fractions. Therefore, since $k[\mathsf{X}, \mathsf{Y}] \subseteq \mathcal{O}_{S,q} \subset \hat{\mathcal{O}}_{S,q} = k[[\mathsf{X}, \mathsf{Y}]]$, f is regarded as a power series, and is a product of s distinct irreducible power series $f_1, \ldots, f_s$, where $(f_i) = \mathfrak{p}_i$ and $\hat{\mathfrak{o}} = k[[\mathsf{X}, \mathsf{Y}]]/(f)$. Furthermore, $\hat{\mathfrak{o}}/\mathfrak{p}_i = k[[\mathsf{X}, \mathsf{Y}]]/(f_i)$ has $k[[t_i]]$ as an integral closure. If ord_i denotes the valuation at p_i, then $\operatorname{ord}_i(t_j) = \delta_{ij}$. Thus for any $\psi \in \mathcal{O}_{S,q}$,

$$I_0(f, \psi) = \sum_{i=1}^{s} \operatorname{ord}_i(\rho^*\psi).$$

After a suitable linear transformation with respect to X, Y, one can assume that the v-th homogeneous part of f is a product of $\mathsf{Y} - \delta_1\mathsf{X}, \ldots, \mathsf{Y} - \delta_v\mathsf{X}$ for some $\delta_i \in k$. Then, since each f_i is irreducible, letting $v_i = v(f_i)$, one knows that the v_i-th homogeneous part of f_i is of the form $\gamma(\mathsf{Y} - \delta_t\mathsf{X})^{v_i}$ for some t and γ. δ_t is characterized by the inequality:

$$I_0(f_i, \mathsf{Y} - \delta_t\mathsf{X}) > v_i.$$

$I_0(f_i, \mathsf{Y} - \delta_t\mathsf{X})$ is denoted $\lambda_1(f_i)$ or $\lambda_{1,i}(f)$.

Now, if $\rho^*(L_{q0})$ is expressed as $\sum_{i=1}^s \mu_i p_i$, then $\mu_i = \operatorname{ord}_i(\rho^*\psi) = I_0(f_i, \psi)$, where ψ is the homogeneous polynomial of degree one defining L_{q0}. In particular, when 0 is a general point of $\mathbf{P}_k^2$, $\mu_i = v_i$ and so $\sum_{i=1}^s \rho_{p_i}(C^{\#}, \Lambda_1) = \sum_{i=1}^s (v_i - 1) = v - s$.

Therefore, letting $v_q = e(q, C)$ and letting s_q denote the number of points on $C^\#$ lying over q, one obtains

$$2r + 2g - 2 = \sum_{q \in \operatorname{Reg} C} (I_q(L_{q0}, C) - 1) + \sum_{q \in \operatorname{Sing} C} (v_q - s_q).$$

Now, it is easy to see that $\{\mathscr{T}(L) \mid L \text{ is a line passing through } 0\}$ gives rise to a line M on the dual plane $\mathbf{P}_k^{\check{2}}$ and furthermore that $\sum_{q \in \operatorname{Reg} C} (I_q(L_{q0}, C) - 1) = (M, \Gamma)_{\mathbf{P}^2}$ if 0 is a general point.

Definition. deg Γ is said to be the *class* of C, denoted by c. Furthermore, define R to be $\sum_q (v_q - 1)$ where the q ranges over all simple cusps, and define α to be $\sum (v_q - s_q)$ where the q ranges over singular points which are not simple cusps.

Using these, one establishes

$$c = 2r + 2g - 2 - R - \alpha. \tag{I}$$

d. Now, let Λ_2 be the linear system $\{\rho^* L \mid L \text{ is a line on } \mathbf{P}_k^{\check{2}}\}$. Then $\dim \Lambda_2 = 2$ and so $\deg W(C^\#, \Lambda_2) = 3r + 6g - 6$ by Theorem 6.11. At $p \in C^\#$, we have the Λ_2-gap sequence $\{a_1, a_2, a_3\}$. Since Bs $\Lambda_2 = \varnothing$, it follows that $a_1 = 0$. If $q = \rho(p)$ is a nonsingular point, then $a_2 = 1$ and $a_3 = I_q(C, l_q)$, where l_q denote *the tangent to* C at q. If q is a singular point, one chooses the coordinate system in such a way that q is $(1:0:0)$. Using the notation in §9.4.c, one has $\rho^{-1}(q) = \{p = p_1, p_2, \ldots, p_s\}$ and furthermore $a_2 = v_1$, $a_3 = \lambda_{1,1}(f)$. Hence, $\sum_{i=1}^s \rho_{p_i}(C^\#, \Lambda_2) = v - s + \sum_{i=1}^s \lambda_{1,i}(f) - 2s$.

Definition. If q is a nonsingular point, define w_q to be $I_q(C, l_q) - 2$. Furthermore, if q is a singular point, then define w_q to be $\sum_{i=1}^s \lambda_{1,i}(f) - 2s$.

If q is a simple cusp, then $w_q = v_q - 1$.

Definition. If q is a nonsingular point with $w_q > 0$, then q is said to be a *flex* (or a *point of inflection*) of order w_q.

Definition. Define W to be $\sum_q w_q$ where the q ranges over all flexes on C and β to be $\sum_q w_q$ where the q ranges over all singular points which are not simple cusps.

By using these R and α, one obtains

$$3r + 6g - 6 = \deg W(C^\#, \Lambda_2) = W + \alpha + R + \beta + R.$$

Rewriting this, one has

$$W = 3r + 6g - 6 - 2R - \alpha - \beta. \tag{II}$$

One can derive the following relations from (I) and (II)

$$r = 2c + 2g - 2 - W + \alpha - \beta. \tag{I*}$$

$$R = 3c + 6g - 6 - 2W - 2\beta + \alpha. \tag{II*}$$

By applying (I) to Γ, one can also derive (I*). Thus (I*) is the dual form of (I). Similarly, (II*) is regarded as the dual form of (II). These relations (I), (I*), (II), (II*) are said to be the *Plücker relations of* C and Γ.

§9.5 Decomposition of Birational Maps

a. Let S be a nonsingular complete surface and let $\mu: Q_p(S) \to S$ be a quadric transformation with center p. $E = \mu^{-1}(p)$ has the following properties.

(1) $Q_p(S)\backslash E \cong S\backslash\{p\}$, and E is irreducible (see Proposition 7.18.(ii)).
(2) $E \cong \mathbf{P}_k^1$ and $E^2 = -1$ (see Lemma 8.2.).

We shall show that either (1) or (2) characterizes $\mu^{-1}(p)$.

First, let $f: W \to S$ be a birational morphism between nonsingular complete surfaces W and S. Suppose that $f^{-1}(p)$ is one-dimensional, which is connected by Theorem 2.21. Then let $\mu: Q_p(S) \to S$ be a quadric transformation and define a rational map g to be $\mu^{-1} \circ f: W \to Q_p(S)$. We claim that g is a morphism. To prove this claim, we suppose that g is not a morphism, i.e., $g(p_1)$ is one-dimensional for some point p_1. Then $\mu(g(p_1))$ is a point $f(p_1)$ and $g(p_1) \subseteq \mu^{-1}(f(p_1))$. Since $Q_p(S)\backslash\mu^{-1}(p) \cong S\backslash\{p\}$, $f(p_1)$ is p and $g(p_1) = \mu^{-1}(p)$. Let $h = g^{-1}: Q_p(S) \to W$, and choose a point q from $\mu^{-1}(p) \cap \operatorname{dom}(h)$. Then $\mu = f \circ h$ and $\mathcal{O}_{Q_p(S),q} \supset \mathcal{O}_{W,p_1} \supset \mathcal{O}_{S,p}$, where we identify $\operatorname{Rat}(W)$ and $\operatorname{Rat}(Q_p(S))$ with $\operatorname{Rat}(S)$. One can choose regular systems of parameters (x_1, y_1), (u, v), and (x, y) of $\mathcal{O}_{W,p_1}$, $\mathcal{O}_{Q_p(S),q}$, and $\mathcal{O}_{S,p}$, respectively such that $x = u$ and $y = uv$. Since $f^{-1}(p)$ is one-dimensional, one has $x = \xi\varphi_1$ and $y = \xi\varphi_2$ for some $\xi, \varphi_1, \varphi_2 \in \mathcal{O}_{W,p_1}$ where ξ defines $f^{-1}(p)$; hence $\xi(p_1) = 0$. Furthermore, $x_1 = u\psi_1$ and $y_1 = u\psi_2$ for some $\psi_1, \psi_2 \in \mathcal{O}_{Q_p(S),q}$, since $h(U \cap E) = p_1$ for some open neighborhood U of q. Letting $\bar{\mathfrak{o}} = \mathcal{O}_{Q_p(S),q}/(u)$, we define a k-derivation $\mathfrak{d}$ of $\mathcal{O}_{Q_p(S),q}$ into $\bar{\mathfrak{o}}$ by $\mathfrak{d}(\eta) = \partial\eta/\partial u \bmod (u)$. From $u = \xi\varphi_1$, one derives $1 = \mathfrak{d}(\xi)\bar{\varphi}_1(p_1)$, since $\varphi_1 \bmod (u) = \bar{\varphi}_1(p_1)$ and $\xi \bmod (u) = 0$. Furthermore $uv = \xi\varphi_2$ implies that $v = \mathfrak{d}(\xi)\bar{\varphi}_2(p_1)$; thus $v\bar{\varphi}_1(p_1) = \bar{\varphi}_2(p_1)$ in $\bar{\mathfrak{o}}$, which is absurd, since $\bar{\varphi}_1(p_1) \neq 0$ and $\bar{\varphi}_2(p_1) \in k$.

Thus, we establish the following result due to Zariski and Hopf, independently.

Theorem 9.6. *Let $f: W \to S$ be a birational morphism between nonsingular complete surfaces. Suppose that $f^{-1}(p)$ is one-dimensional. Then $g = \mu^{-1} \circ f: W \to Q_p(S)$ is a birational morphism.*

Furthermore, if $g: W \to Q_p(S)$ is not an isomorphism, in other words, if $g^{-1}(p')$ is one-dimensional for some p' (see Theorem 2.22), then we repeat quadric transformations. We note the next result.

Lemma 9.3. *Let $f: W \to S$ be a birational morphism. Then $K(W)^2 < K(S)^2$, whenever f is not an isomorphism. Furthermore, if $W = Q_p(S)$, then $K(W)^2 = K(S)^2 - 1$.*

PROOF. One has the ramification formula: $K(W) = f^*(K(S)) + R_f$ by Theorem 5.5. On the other hand, R_f is exceptional with respect to f by Proposition 5.8, i.e., $R_f = 0$ if and only if $R_f^2 = 0$. Hence, if f is not an isomorphism, then $R_f^2 < 0$ by Theorem 8.5; hence $K(W)^2 = K(S)^2 + R_f^2 < K(S)^2$. Furthermore, if $\mu: Q_p(S) \to S$ is a quadric transformation, then $R_\mu = \mu^{-1}(p)$; hence $R_\mu^2 = -1$ by Lemma 8.2. □

Given a birational morphism $f: W \to S$ which is not an isomorphism, one has a quadric transformation $\mu_1: S_1 \to S$ such that $g = \mu_1^{-1} \circ f: W \to S_1$ is a birational morphism. Hence, $K(W)^2 \leq K(S_1)^2$, and equality holds if and only if g is an isomorphism. Thus, one has a finite sequence of quadric transformations $\mu_l: S_l \to S_{l-1}, \ldots, \mu_1: S_1 \to S$ such that $\mu_l^{-1} \circ \mu_{l-1}^{-1} \circ \cdots \circ \mu_1^{-1} \circ f: W \to S_l$ is an isomorphism. Note that $K(W)^2 = K(S)^2 - l$ by the last lemma. Hence, one obtains the next result.

Proposition 9.5. *Any birational morphism is a composition of an isomorphism and a morphism which is the composition of a finite number of quadric transformations. In particular, if $f: W \to S$ has the property that $f^{-1}(p)$ is irreducible and $W \setminus f^{-1}(p) \cong S \setminus \{p\}$, then $W \cong Q_p(S)$.*

b. Let Λ be a reduced linear system on S. Suppose that $\operatorname{Bs} \Lambda \neq \varnothing$. Then $\operatorname{Bs} \Lambda$ is a finite set. Take $p \in \operatorname{Bs} \Lambda$ and consider a general member D_λ of Λ. Then $e(p, D_\lambda)$ does not depend on the choice of a general member D_λ. Let $m_p = e(p, D_\lambda)$ and let $\mu: Q_p(S) \to S$ be the quadric transformation. Then $D_\lambda^* = \mu^*(D_\lambda) - m_p E_p$ is the strict transform of D_λ, where E_p is $\mu^{-1}(p)$. Furthermore, $\mu^*(D) - m_p E_p \geq 0$ for any $D \in \Lambda$ and these form a linear system Λ_1, which is also reduced. Λ_1 is said to be the *weak transform* of Λ by μ. Then it is clear that $\Phi_{\Lambda_1} = \Phi_\Lambda \circ \mu: Q_p(S) \to \mathbf{P}|\Lambda| = \mathbf{P}|\Lambda_1|$.

Now, define $\sigma(\Lambda)$ to be D^2, where $D \in \Lambda$. Then $\sigma(\Lambda) \geq 0$, since Λ is reduced, and $\sigma(\Lambda_1) = (D_\lambda^*)^2 = (\mu^*(D_\lambda) - m_p E_p)^2 = D_\lambda^2 - m_p^2 = \sigma(\Lambda) - m_p^2 < \sigma(\Lambda)$. Thus, one obtains the next result.

Proposition 9.6. *If Λ is a reduced linear system on S, then there exist a sequence of quadric transformations $\mu_l: S_l \to S_{l-1}, \ldots, \mu_1: S_1 \to S_0 = S$ and a sequence of reduced linear systems Λ_j on S_j such that each Λ_j is the weak transform of Λ_{j-1} by μ_j for all j, $\Lambda_0 = \Lambda$, and $\operatorname{Bs}(\Lambda_l) = \varnothing$.*

Corollary. *If $f\colon S \to Z$ is a rational map into a projective variety Z, then there exists a composition $\varphi\colon S_l \to S$ of quadric transformations such that $f \circ \varphi$ is a morphism.*

PROOF. First recall a result in §2.11.**f** which asserts that f is obtained as Φ_Λ for some reduced linear system Λ. Then applying the above observation to this Φ_Λ, one obtains the result. □

Proposition 9.7 (Castelnuovo). *Let E be a curve on S such that $E^2 = -1$ and $E \cong \mathbf{P}_k^1$. Then there exist a nonsingular projective surface S_1 and a point p on S_1 such that $S \cong Q_p(S_1)$ which induces an isomorphism $E \cong \mu^{-1}(p)$, where $\mu\colon Q_p(S_1) \to S_1$ is a quadric transformation.*

PROOF. We refer the reader to [G–H, p. 476]. □

EXERCISES

9.1. The following affine plane curves have cusp singularities at the origin. Compute the characteristic pairs of these cusps.
(a) $V((\mathsf{Y}^{n-1} - \mathsf{X}^n)^m - \mathsf{Y}^{nm-1})$ for $n, m \geq 2$,
(b) $V((\mathsf{Y} - \mathsf{X}^2)(\mathsf{Y} - \mathsf{X}^2 + 2\mathsf{Y}\mathsf{X}^2) + \mathsf{Y}^5)$,
(c) $V((\mathsf{Y} - \mathsf{X}^2)^2 + \mathsf{Y}\mathsf{X}^3)$,
(d) $V((\mathsf{Y} - \mathsf{X}^2)^2 + \mathsf{Y}^3\mathsf{X})$.

9.2. Let C be a curve on a complete nonsingular surface and D be an effective divisor with $D \not\geq C$ on S. For a closed point $p \in C$, we choose a local coordinate system (x, y) at p on some coordinate neighborhood in S. Let $g(x, y)$ be a formal power series defining D around p, i.e., $\hat{\mathcal{O}}_{D,p} \cong \hat{\mathcal{O}}_{S,p}/(g)$. As in §9.2.**a**, choose a nonsingular model $\rho\colon C^\# \to C$ and let $\{q_1, \ldots, q_s\}$ be the set $\rho^{-1}(p)$. Letting ord_j denote the valuation of $\mathrm{Rat}(C)$ at q_j, prove the formula:

$$I_p(C, D) = \sum_{j=1}^{s} \mathrm{ord}_j(\rho^*(g(x, y))).$$

9.3. Let $h(\mathsf{T})$ be a polynomial of degree n and let C^0 be the graph of h. Let C be the closure of C^0 in $\mathbf{P}_k^2$. Then compute the class c, α, β, R and W of C (see §9.4.**d**).

9.4. Let C be a projective plane curve defined by $\mathsf{Y}^2 - (\mathsf{X} - a_1)(\mathsf{X} - a_2) \cdots (\mathsf{X} - a_m)$, where $a_i \neq a_j$ for $i \neq j$, and $m \geq 5$. Compute the class of C.

9.5. Let C be a projective plane curve with $\deg(C) \geq 2$ and let Γ be its dual curve. One has the canonical birational map $f\colon C \to \Gamma$ and the normalization of f becomes an isomorphism h between the nonsingular models $(C^\#, \rho)$ and $(\Gamma^\#, \rho_1)$ of C and Γ, respectively. Suppose that $h(p_1) = q_1$ and that $p = \rho(p_1)$ and $q = \rho_1(q_1)$ are both cusps. If C is defined around p by the Puiseux expansion: $x = \tau^\nu$, $y = \sum_{i=1}^\infty a_i \tau^{\lambda_i}$ with $\nu < \lambda_1$, then show that the characteristic pair of the cusp q is obtained as the characteristic pair of a cusp defined by a Puiseux expansion $\xi = \tau^{\lambda_1 - \nu}$, $\eta = \sum_{j=1}^\infty b_j \cdot \tau^{\lambda_j}$.

Chapter 10

D-Dimension and Kodaira Dimension of Varieties

§10.1 D-Dimension

a. Let V be a complete normal variety and let D be a divisor on V. Define $\mathbb{N}(D)$ to be the set $\{m \in \mathbb{N} \mid l_V(mD) > 0\}$, which is a semigroup, if it is not empty. In this case, the subgroup of $\mathbb{Z}$ generated by $\mathbb{N}(D)$ is of the form $m_0 \mathbb{Z}$ for some $m_0 > 0$, denoted by $m_0(D)$. Since $GCD\{m \in \mathbb{N} \mid mm_0 \in \mathbb{N}(D)\} = 1$, there exists $m' > 0$ such that if $m \geq m'$, then $mm_0 \in \mathbb{N}(D)$.

Definition. The *D-dimension* $\kappa(D, V)$ of V is defined to be $\max\{\dim \Phi_{mD}(V) \mid m \in \mathbb{N}(D)\}$ if $\mathbb{N}(D) \neq \varnothing$ and $\kappa(D, V) = -\infty$ otherwise.

b. Let D be an effective divisor and define V_0 to be $V \backslash D$.

The closed image of V under Φ_D is denoted by W_D, and so $\Phi_D: V \to W_D$ is a dominating rational map. The field $\text{Rat}(W_D)$ is denoted by $Q(D)$ which is therefore the field generated by $\mathbf{L}(D)$ in $\text{Rat}(V)$. From the sequence $\mathbf{L}(D) \subseteq \mathbf{L}(2D) \subseteq \cdots \subseteq A(V_0) \subseteq \text{Rat}(V)$ one obtains $Q(D) \subseteq Q(2D) \subseteq \cdots \subseteq QA(V_0) \subseteq \text{Rat}(V)$.

For any $\varphi \in A(V_0) = \Gamma(V \backslash D, \mathcal{O}_V)$, one has j such that $\text{div}(\varphi)_- = (\varphi)_\infty \leq jD$; hence $A(V_0) = \bigcup_{j=1}^{\infty} \mathbf{L}(jD)$. Thus, $QA(V_0) = \bigcup_{j=1}^{\infty} Q(jD)$. We note the next lemma.

Lemma 10.1. *Let $\mathfrak{K}/L$ be a finitely generated extension of fields and $L \subseteq L_1 \subseteq \cdots \subseteq L_m \subseteq \cdots \subset \mathfrak{K}$ be a sequence of intermediate fields. Then $L_{m_1} = L_{m_1+1} = \cdots = \bigcup_{m=1}^{\infty} L_m$, for some m_1.*

Proof. Left to the reader. □

Therefore, one has m_1 such that $QA(V_0) = Q(m_1D) = \text{Rat}(W_{m_1D})$. Clearly, $\kappa(D, V) = \text{tr. deg}_k\, Q(m_1D)$, and $\gamma(V_0) = \text{tr. deg}_k\, QA(V_0)$ by definition (see §1.24.**d**). Hence, one obtains the next result due to Tsunoda.

Proposition 10.1. *If D is effective, then $\gamma(V\backslash D) = \kappa(D, V)$.*

Corollary. *If $\kappa(D, V) = 0$, then $l(jD) \leqq 1$ for all $j > 0$.*

Proof. By definition, $\kappa(D, V) = \kappa(iD, V)$ for any $i \geq 1$. Hence, if $l(jD) \geq 1$, one can apply the last proposition to $D' \in |jD|$ and obtain $\gamma(V\backslash D') = 0$. Hence, $l(iD') = l(ijD) = 1$ for all $i > 0$. □

Proposition 10.2. *If $\kappa(D, V) = 0$, then $\mathbb{N}(D) = m_0(D)\mathbb{N}$.*

Proof. By the Corollary to Proposition 10.1, $l_V(mD) \leq 1$ for all $m \geq 1$. Define m^* to be $\min\{m \mid l_V(mD) = 1\}$. Then letting $m_0 = m_0(D)$, one has $m^* = em_0$ for some $e \geq 1$. There exists a such that $(ae + 1)m_0 \in \mathbb{N}(D)$. Hence, we may choose effective divisors E_1 and E_2 from $|em_0 D|$ and $|(ae + 1)m_0 D|$, respectively. Then

$$eE_2 \sim (ae + 1)em_0 D \sim (ae + 1)E_1.$$

But since $l_V(e(ae + 1)m_0 D) = 1$, one has $eE_2 = (ae + 1)E_1$, and so $E_1 = eF$ and $E_2 = (ae + 1)F$ for some $F \geq 0$. Furthermore,

$$aE_1 = aeF \sim aem_0 D \text{ and } E_2 = (ae + 1)F \sim (ae + 1)m_0 D.$$

Hence, $m_0 D \sim (ae + 1)F - aeF = F$, i.e., $m_0 \in \mathbb{N}(D)$; thus $m^* = m_0$. □

Corollary

(i) *If $\kappa(D, V) = 0$ and $\alpha, \beta \in \mathbb{N}(D)$, then $GCD(\alpha, \beta) \in \mathbb{N}(D)$.*
(ii) *If $\alpha, \beta \in \mathbb{N}(D)$ and $GCD(\alpha, \beta) \notin \mathbb{N}(D)$, then $l(\delta D) \geq 2$, where δ is $LCM(\alpha, \beta)$.*

c. To prove and state the fundamental theorems for D-dimension, we need the following notion.

Definition. A morphism $f\colon V \to W$ together with V and W is said to be a *fibered variety* if f is dominating and the extension of fields $\text{Rat}(V)/\text{Rat}(W)$ is algebraically closed.

If f is proper, a fibered variety $f\colon V \to W$ is said to be *proper*. Furthermore, if V and W are nonsingular, f is proper, and if $R_f = \varnothing$, then f is said to be a *proper smooth fibered variety*.

Given a proper fibered variety $f\colon V \to W$, where V is nonsingular, first we let $W_0 = \text{Reg}\, W$, $V_0 = f^{-1}(W_0)$, $f_0 = f|_{V_0}$, and then let $W_1 = W_0 \backslash f_0(R_{f_0})$,

$V_1 = f^{-1}(W_1)$, and $f_1 = f_0|_{V_1}$. Thus, a proper smooth fibered variety f_1: $V_1 \to W_1$ is obtained. We say that f_1: $V_1 \to W_1$ is the *smooth part* of f: $V \to W$.

Note that for any closed point x of W_1, $f^{-1}(x)$ is a nonsingular complete variety of dimension dim V − dim W.

The following result is not hard to prove.

Theorem 10.1. *Let f: $V \to W$ be a proper smooth fibered variety and let $\mathscr{L}$ be a locally free $\mathscr{O}_V$-module of finite type. Then the function $\delta_{\mathscr{L}, f}(x) = \dim_{k(x)} \Gamma(f^{-1}(x), \mathscr{L}_x)$ is upper semicontinuous, where $\mathscr{L}_x$ is defined to be the pullback of $\mathscr{L}$ over $f^{-1}(x)$; i.e., $\mathscr{L}_x = \mathscr{L}|_{f^{-1}(x)}$.*

PROOF. We refer the reader to [H, III. Theorem 12.8]. □

Thus, letting $*$ be the generic point of W, the set $\{x \in W \mid \delta_{\mathscr{L}, f}(x) = \delta_{\mathscr{L}, f}(*)\}$ is an open dense subset of W.

Definition. Let f: $V \to W$ be a proper fibered variety, where V is a normal variety, and let $D = \sum_{i=1}^{r} m_i \Gamma_i$ be an effective divisor on V. Then define D_{hor} to be $\sum m_i \Gamma_i$ where $f(\Gamma_i) = W$ and D_{ver} to be $\sum m_i \Gamma_i$ where $f(\Gamma_i) \neq W$.

With this notation, $D = D_{\text{hor}} + D_{\text{ver}}$. If $D_{\text{hor}} = 0$, (respectively $D_{\text{ver}} = 0$), D is said to be a *vertical* (respectively *horizontal*) *divisor with respect to* f.

§10.2 The Asymptotic Estimate for $l(mD)$

a. Let V be a nonsingular complete variety and let D be an effective divisor. We choose $m_1 > 0$ such that $QA(V \backslash D) = Q(m_1 D)$ ($= \text{Rat}(\Phi_{m_1 D}(V))$ as in §10.1.**a**. Then there exist a complete nonsingular variety $V^\#$ and a birational morphism μ: $V^\# \to V$ such that $\Phi_{m_1 D} \circ \mu$: $V^\# \to V \to \mathbf{P}|m_1 D|$ is a morphism. Letting W be $\Phi_{m_1 D}(V)$ and $f = \Phi_{m_1 D} \circ \mu$: $V^\# \to W$, one obtains a proper fibered variety f: $V^\# \to W$. Let $D^\#$ be $\mu^* D$. Then $\Phi_{m_1 D^\#} = \Phi_{m_1 D} \circ \mu$ (cf. §2.20.**e**).

Proposition 10.3. *For any $m > 0$, and any member X_m of $|mD^\#|$, one has $(X_m)_{\text{hor}} = m(X_1)_{\text{hor}}$. Hence, $m(X_1)_{\text{hor}}$ is a fixed component of $|mD^\#|$ and $l_V(mD) = l_{V^\#}(mD^\#) = l_{V^\#}(m(D^\#_{\text{ver}}))$.*

PROOF. One has $\varphi \in \mathbf{L}(mD^\#) \subseteq A(V^\# \backslash D^\#) = A(V \backslash D)$ such that $X_m = mD^\# + \text{div}(\varphi)$. Since $A(V \backslash D) \subseteq QA(V \backslash D) = f^*(\text{Rat}(W))$, one has $\varphi = f^*(\psi)$ for some $\psi \in \text{Rat}(W)$. We claim that if Γ is a prime divisor on $V^\#$ horizontal with respect to f, then $\text{ord}_\Gamma(f^*(\psi)) = 0$. In fact, letting y be the generic point of Γ, $f(y)$ is the generic point $*$ of W, since $f(\Gamma) = W$. Then f^*:

$\mathcal{O}_{W, *} = \mathrm{Rat}(W) \to \mathcal{O}_{V^\#, y}$ and so the image of a nonzero element under f^* is a unit; hence $\mathrm{ord}_\Gamma(f^*(\psi)) = 0$. This implies that $\mathrm{div}(\varphi) = \mathrm{div}(f^*(\psi))$ is a vertical divisor, and so $(X_m)_{\mathrm{hor}} = (mD^*)_{\mathrm{hor}} = m(X_1)_{\mathrm{hor}}$. □

b. Theorem 10.2. *If D is a divisor on V and $\kappa = \kappa(D, V) \geq 0$, then there exist $\alpha, \beta > 0$ such that*

$$\alpha m^\kappa \leq l_V(mm_0(D)D) \leq \beta m^\kappa$$

for sufficiently large m (which may also be written as $m \gg 0$).

Proof. First assume D is effective. Since $l_V(mD) = l_{V^\#}(mD^\#)$ for any $m > 0$, one can replace V by $V^\#$, i.e., we may assume that $f = \Phi_{m_1 D}\colon V \to W$ is a morphism and $\dim W = \kappa$. Letting H be a hyperplane of $\mathbf{P}|m_1 D|$, and F be the fixed part of $|m_1 D|$, one has $m_1 D \sim \Phi_{m_1 D}{}^*(H) + F$ and so by Proposition 2.24 $l(mm_1 D) \geq l(\Phi_{m_1 D}{}^*(mH)) = l_W(\mathcal{O}_W(m))$, where $\mathcal{O}_W(1) = \mathcal{O}_{\mathbf{P}|m_1 D|}(H)|_W$. Since $l_W(\mathcal{O}_W(m))$ is a polynomial of degree κ for $m \gg 0$, there exists α' such that $\alpha' m^\kappa \leq l_W(\mathcal{O}_W(m))$ for $m \gg 0$, i.e., $\alpha' m^\kappa \leq l_V(mm_1 D)$.

On the other hand, by Proposition 10.3, one can assume D to be vertical. Now, take a nonsingular model $(W^\#, \lambda)$ of W, i.e., $\lambda\colon W^\# \to W$ is birational and $W^\#$ is a nonsingular complete variety, and choose a proper birational morphism $\mu\colon V^\# \to V$ such that $h = \lambda^{-1} \circ f \circ \mu$ is a morphism. $\mu^* D$ is also vertical with respect to h and so there exists an effective divisor Y on $W^\#$ such that $h(\mu^* D) \subseteq Y$ and there is $s > 0$ with $m_1 \mu^*(D) \leq h^*(sY)$. Thus, $l_V(mm_1 D) = l_{V^\#}(mm_1 \mu^*(D)) \leq l_{V^\#}(mh^*(sY)) = l_{W^\#}(msY)$. Since $\dim W^\# = \kappa$, there exists β' such that $l_{W^\#}(msY) \leq \beta' m^\kappa$ for $m \gg 0$.

Now, if $m \gg 0$, then $m = rm_1 + q$, $0 \leq q < m_1$, and from

$$\alpha' r^\kappa \leq l(rm_1 D) \leq \beta' r^\kappa,$$

one obtains

$$\alpha'\left(\frac{m}{m_1} - 1\right)^\kappa \leq l(mD) \leq l((r+1)m_1 D) \leq \beta'\left(\frac{m}{m_1} + 1\right)^\kappa,$$

i.e.,

$$\alpha m^\kappa \leq l(mD) \leq \beta m^\kappa \quad \text{for certain} \quad \alpha, \beta > 0.$$

By the same consideration, one can prove the asymptotic estimate for general D. □

Remark. In the case $\kappa(D, V) = -\infty$, by interpreting $m^{-\infty}$ to be 0, the asymptotic estimate still holds for $l_V(mD)$.

$$\begin{array}{ccc} V & \xrightarrow{f} & W \subset \mathbf{P}|m_1 D| \\ {\scriptstyle\mu}\uparrow & & \uparrow{\scriptstyle\lambda} \\ V^\# & \xrightarrow{h} & W^\# \end{array}$$

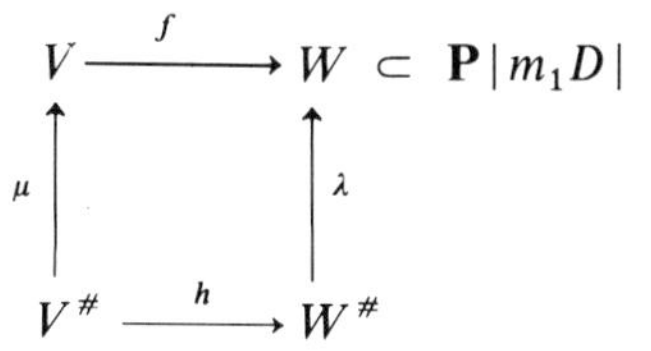

§10.3 Fundamental Theorems for D-Dimension

a. Let D be an effective divisor on a normal complete variety V, and let V_0 be $V \backslash D$. We choose m_1 and $\mu: V^{\#} \to V$ as in the previous section. Then $f = \Phi_{m_1 D^{\#}}: V^{\#} \to W$ is a proper fibered variety.

Proposition 10.4. *Let* $V_0^{\#} = V^{\#} \backslash D^{\#}$ *and* $f_0 = f|_{V_0^{\#}}: V_0^{\#} \to W$. *Then* $A(f_0^{-1}(*)) = k(*) \cong \mathrm{Rat}(W)$, *where* $*$ *is the generic point of* W.

Proof. Since $\mu: V_0^{\#} \to V_0$ is a proper birational morphism, $A(V_0^{\#})$ is isomorphic to $A(V_0)$ (see Exercise 2.10). Now, since $V_0^{\#} = V^{\#} \backslash m_1 D^{\#}$, one can assume D is $m_1 D$, i.e., $m_1 = 1$. Thus, letting W_0 be $W \backslash H$ (where H is a hyperplane of $\mathbf{P}|m_1 D|$), which is an affine variety, one has $f(V_0^{\#}) \subseteq W_0$. Hence by the universal property of the canonical morphism $\Psi: V_0^{\#} \to \mathrm{Spec}\, A(V_0^{\#})$ ($\cong \mathrm{Spec}\, A(V_0)$), one has a morphism $\rho: \mathrm{Spec}\, A(V_0^{\#}) \to W_0$ such that $\rho \circ \Psi = f|_{V_0^{\#}}$. If $*_1$ is the generic point of $\mathrm{Spec}\, A(V_0^{\#})$, then ρ induces an isomorphism $k(*_1) \cong k(*)$, since $k(*_1) \cong QA(V_0) = \mathrm{Rat}(W)$ and $k(*) = \mathrm{Rat}(W)$; hence $\rho^{-1}(*) = \{*_1\}$.

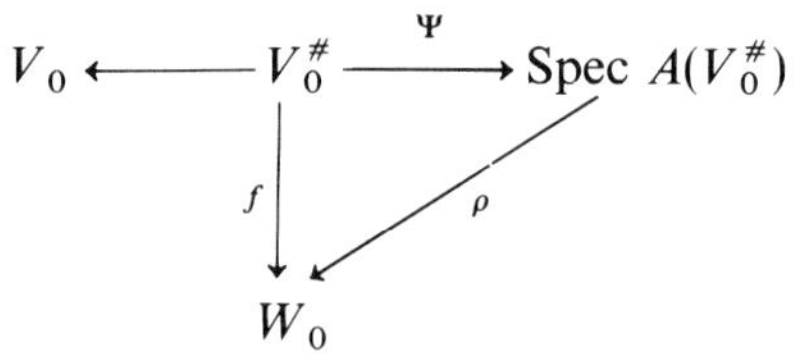

Since $A(\Psi^{-1}(*_1)) = k(*_1)$ by Theorem 1.18, one obtains the result. □

b. Now, we state the first fundamental theorem for D-dimension.

Theorem 10.3 (Fibering Theorem). *If D is a divisor on a nonsingular complete variety V such that $\kappa = \kappa(D, V) \geq 0$, then there exist a nonsingular complete variety $V^{\#}$, a birational morphism $\mu: V^{\#} \to V$, and a proper fibered variety f: $V^{\#} \to W$ such that*

(1) $\dim W = \kappa$,
(2) *If* $l(m_0 D) \geqslant 1$ *and* $f_1: V_1^{\#} \to W_1$ *is the smooth part of* $f: V^{\#} \to W$, *and* m *is a positive integer, then there is an open dense subset* $W_{(m)}$ *of* W_1 *with the property that* $l(mm_0 D|_{f^{-1}(x)}) = 1$ *for all* $x \in W_{(m)}$.

Proof. Clearly, we may assume D is effective. We use the same m_1, μ as in §10.2. Letting $\mathscr{L} = \mathcal{O}(mD^{\#})$ and $D_x = D|_{f^{-1}(x)}$, one has $\mathscr{L}_* = \mathcal{O}(mD_*^{\#})$ and $\Gamma(f^{-1}(*), \mathscr{L}_*) \subseteq A(f^{-1}(*) \backslash D_*^{\#}) = A(f_0^{-1}(*)) = \mathrm{Rat}(W)$, i.e., $\delta_{\mathscr{L}, f}(*) = 1$. Hence, applying Theorem 10.1 to the smooth part $f_1: V_1^{\#} \to W_1$, one has $W_{(m)} \subseteq W_1$ with the property that $\delta_{\mathscr{L}, f}(x) = 1$ for all $x \in W_{(m)}$. □

Remarks. (1) The statement (2) in the last theorem is restated as follows:
(2)′ $\kappa(D_x, f^{-1}(x)) = 0$ for "general" points x of W, i.e., for $x \in \bigcap_{m=1}^{\infty} W_{(m)}$.
(2) The case $\dim V = 2$ has been proved in §8.3.**c**.
(3) $f\colon V^{\#} \to W$ is said to be a *D-canonical fibered variety.*

c. Now, to prove the second fundamental theorem, we consider a simple result about the invariant γ.

Lemma 10.2. *Let V be a normal variety and let $f\colon V \to W$ be a fibered variety Then $\gamma(V) \le \gamma_*(f^{-1}(*)) + \dim W$, where $*$ is the generic point of W. In particular, if W is affine, then equality holds, i.e., $\gamma(V) = \gamma_*(f^{-1}(*)) + \gamma(W) = \gamma_*(f^{-1}(*)) + \dim W$.*

Proof. If W is an affine variety Spec R, then by Lemma 1.44, $A(f^{-1}(*)) = A(V) \otimes_R Q(R)$. Hence, $QA(f^{-1}(*)) = QA(V) \supseteq Q(R)$, since $A(V) \supseteq R$. Thus $\gamma_*(f^{-1}(*) = \text{tr.}\deg_{Q(R)} QA(V) = \text{tr.}\deg_k QA(V) - \text{tr.}\deg_k Q(R) = \gamma(V) - \gamma(W) = \gamma(V) - \dim W$.

For a general variety W, let W_α be an affine open dense subset of W. Then $V_\alpha = f^{-1}(W_\alpha) \subseteq V$ and so

$$\gamma(V) \le \gamma(V_\alpha) = \gamma_*(f^{-1}(*)) + \dim W_\alpha = \gamma_*(f^{-1}(*)) + \dim W. \qquad \square$$

d. Theorem 10.4 (Easy Addition Theorem). *Let $f\colon V \to W$ be a proper fibered variety and let D be a divisor on V. If V is a nonsingular complete variety and $f_1\colon V_1 \to W_1$ is a smooth part of $f\colon V \to W$, then*

$$\kappa(D, V) \le \kappa(D_x, f^{-1}(x)) + \dim W \quad \textit{for all} \quad x \in W_1.$$

Proof. If $\kappa(D, V) = -\infty$, the assertion is obvious. Now, if $\kappa(D, V) \ge 0$, there exists $m > 0$ with $|mD| \ne \emptyset$. Replacing D by a member of $|mD|$, we assume D is effective. Recall that $\kappa(D, V) = \gamma(V \setminus D)$ by Lemma 10.1. Letting $f_0 = f|_{V \setminus D}$, one has by Lemma 10.2,

$$\gamma(V \setminus D) \le \gamma_*(f_0^{-1}(*)) + \dim W.$$

We claim $\gamma_*(f_0^{-1}(*)) \le \kappa(D_x, f^{-1}(x))$ for any $x \in W_1$.

In fact, for any $m > 0$, let $\mathscr{L}$ be $\mathscr{O}(mD)$. Then

$$\delta_{\mathscr{L}, f}(*) \le \delta_{\mathscr{L}, f}(x)$$

by Theorem 10.1.

Hence, $l(mD_*) = \dim_{k(*)} \Gamma(f^{-1}(*), \mathscr{O}(mD_*)) \le l(mD_x)$, if $x \in W_1$. Letting $\kappa(*) = \kappa(D_*, f^{-1}(*))$ and $\kappa(x) = \kappa(D_x, f^{-1}(x))$, there exist $\alpha_*, \beta_*, \alpha_x, \beta_x$ such that by Theorem 10.2

$$\alpha_* m^{\kappa(*)} \le l(mD_*) \le \beta_* m^{\kappa(*)} \quad \text{for} \quad m \ge m_*,$$

$$\alpha_x m^{\kappa(x)} \le l(mD_x) \le \beta_x m^{\kappa(x)} \quad \text{for} \quad m \ge m_x.$$

where m_* and m_x are constants depending on x. Thus

$$\alpha_x m^{\kappa(*)} \leq \beta_x m^{\kappa(x)} \quad \text{for all } m \geq \max(m_*, m_x).$$

On the other hand $\gamma_*(f_0^{-1}(*)) = \kappa(*)$; hence $\gamma_*(f_0^{-1}(*)) \leq \kappa(x)$. □

e. Lemma 10.3. *Let $f\colon V \to W$ be a finite surjective morphism and let D be a divisor on W. If V and W are locally factorial and complete, then $\kappa(f^*D, V) = \kappa(D, W)$.*

PROOF. If $\kappa(D, W) \geq 0$, then one can assume D is effective. By Lemma 10.1, $\gamma(W \backslash D) = \kappa(D, W)$ and $\gamma(V \backslash f^{-1}(D)) = \kappa(f^{-1}(D), V)$, where $f^{-1}(D)$ is the *reduced inverse image* of D by f, i.e., if $f^*(D) = \sum_{i=1}^{s} m_i \Gamma_i$, where the Γ_i are prime divisors with $\Gamma_i \neq \Gamma_j$ for $i \neq j$, then $f^{-1}(D)$ is defined to be $\sum_{i=1}^{s} \Gamma_i$. Since $f^{-1}(D) = \sum_{i=1}^{s} \Gamma_i \leq f^*(D) \leq m \sum_{i=1}^{s} \Gamma_i$ for $m = \max\{m_1, \ldots, m_s\}$, one has $\kappa(f^{-1}(D), V) \leq \kappa(f^*D, V) \leq \kappa(mf^{-1}(D), V) = \kappa(f^{-1}(D), V)$; hence $\kappa(f^{-1}(D), V) = \kappa(f^*D, V)$. Thus, one obtains $\kappa(f^*D, V) = \gamma(V \backslash f^{-1}(D))$. On the other hand, by the Corollary to Theorem 2.30, $\gamma(V \backslash f^{-1}(D)) = \gamma(W \backslash D)$, since $f_1 = f|_{V \backslash f^{-1}(D)}\colon V \backslash f^{-1}(D) \to W \backslash D$ is proper surjective. Thus $\kappa(f^*D, V) = \kappa(D, W)$.

If $\kappa(D, W) = -\infty$, it suffices to derive a contradiction from the assumption $l(rf^*D) \geq 1$ for some $r \geq 1$. Take a finite affine open cover $\{W_\lambda \mid \lambda \in \Lambda\}$ of W such that $\mathcal{O}(D)|_{W_\lambda} = \mathcal{O}_{W_\lambda} \cdot a_\lambda^{-1}$, where $a_\lambda \in \mathrm{Rat}(W)$ for all λ. Then $a_{\lambda\mu} = a_\lambda \cdot a_\mu^{-1} \in \Gamma(W_\lambda \cap W_\mu, \mathcal{O}_W^*)$, and if $W_\lambda = \mathrm{Spec}\, A_\lambda$, then $\{U_\lambda = f^{-1}(W_\lambda) = \mathrm{Spec}\, B_\lambda \mid \lambda \in \Lambda\}$ is an affine open cover of V and each B_λ is finite over each A_λ. For a nonzero $\beta \in \Gamma(V, \mathcal{O}(rf^*D))$, one has $b_\lambda \in B_\lambda$ with $\beta|_{U_\lambda} = b_\lambda a_\lambda^{-r}$. One has a monic minimal polynomial $\Phi_\lambda(\mathrm{X}) \in A_\lambda[\mathrm{X}]$ with $\Phi_\lambda(b_\lambda) = 0$. By Lemma 2.31 in §2.17.**b**, $\Phi_\lambda(\mathrm{X})$ is irreducible as a polynomial over $\mathrm{Rat}(W)$. Also $b_\lambda \cdot b_\mu^{-1} = a_{\lambda\mu}^r \in \Gamma(W_\lambda \cap W_\mu, \mathcal{O}_W^*)$ and so $\deg \Phi_\lambda$ does not depend on λ. If $m = \deg \Phi_\lambda$ and $\alpha_\lambda = \Phi_\lambda(0) \in A_\lambda$, then $b_\lambda^m + \cdots + \alpha_\lambda = 0$ and so $a_{\lambda\mu}^{mr} b_\mu^m + \cdots + \alpha_\lambda = 0$, i.e., $\alpha_\mu = \alpha_\lambda \cdot a_{\mu\lambda}^{mr}$. Corresponding to $\{\alpha_\lambda\}$, one therefore has a nonzero $\alpha \in \Gamma(W, \mathcal{O}(mrD))$. □

From the above result, we can derive the third fundamental theorem.

Theorem 10.5 (Covering Theorem). *Let $f\colon V \to W$ be a proper surjective morphism between nonsingular complete varieties. If D is a divisor on W and E is an effective divisor on V that is exceptional with respect to f, then $\kappa(f^*D + E, V) = \kappa(D, W)$.*

PROOF. Let $\mu \circ g = f$ be the Stein factorization of f, in which $g\colon V \to Z$ is a proper surjective morphism onto a complete normal variety Z. Applying Proposition 2.25 to $\mu^*(\mathcal{O}(D))$, one has $l_V(mf^*D + mE) = l_Z(m\mu^*D)$ for all $m \geq 1$. Since $\kappa(\mu^*D, Z) = \kappa(D, W)$ by the last lemma, one obtains the assertion. □

f. Finally we have to prove the uniqueness theorem for the D-canonical fibered variety. We begin by showing the uniqueness theorem for γ.

Lemma 10.4. *Let $f: V \to W_\alpha$ be a fibered variety. Suppose that V is normal, W_α is affine, and $\gamma(V) = \dim W_\alpha$. Then there exists a morphism ρ: $\operatorname{Spec} A(V) \to W_\alpha$ with $f = \rho \circ \Psi$, such that ρ is birational, i.e., ρ^*: $\operatorname{Rat}(W_\alpha) \cong QA(V)$.*

Proof. By the universal property of Ψ: $V \to \operatorname{Spec} A(V)$, one has a morphism ρ with $f = \rho \circ \Psi$. The morphism ρ is dominating, because $f: V \to W_\alpha$ is a fibered variety. Thus, $\rho^*: \operatorname{Rat}(W_\alpha) \to QA(V) \subset \operatorname{Rat}(V)$ gives rise to the finite extension $QA(V)/\operatorname{Rat}(W_\alpha)$. But since $\operatorname{Rat}(V)/\operatorname{Rat}(W_\alpha)$ is algebraically closed, one sees that $QA(V) = \rho^* \operatorname{Rat}(W_\alpha)$. □

Remark. If $*$ denotes the generic point of W_α, then $\gamma(V) = \dim W$ is equivalent to $\gamma_*(f^{-1}(*)) = 0$ by Lemma 10.2.

Theorem 10.6. *Let D be a divisor on a nonsingular complete variety V and let $f: V^\# \to W$ be a D-canonical fibered variety with $\mu: V^\# \to V$ a birational morphism. Suppose that a fibered variety $\varphi: V_1^\# \to Z$ has the following properties*

(1) *$V_1^\#$ is a nonsingular complete variety birationally equivalent to V via a birational morphism $\lambda: V_1^\# \to V$,*
(2) *$\dim Z = \kappa(D, V)$, and*
(3) *$\kappa(\lambda^*(D)_x, \varphi^{-1}(x)) = 0$ for any "general" points $x \in Z$.*

Then there exists a birational map $\rho: W \to Z$ such that $\rho \circ f \circ \mu^{-1} = \varphi \circ \lambda$.

Proof. Clearly, one can assume D is effective. Taking an affine open dense subset Z_α of Z, let $V_\alpha = \varphi^{-1}(Z_\alpha) \setminus \lambda^{-1}(D)$ and $\psi_\alpha = \varphi|_{V_\alpha}: V_\alpha \to Z_\alpha$. By the universality of Ψ, one has ρ_α: $\operatorname{Spec} A(V_\alpha) \to Z_\alpha$ and σ, which make the following diagram commutative:

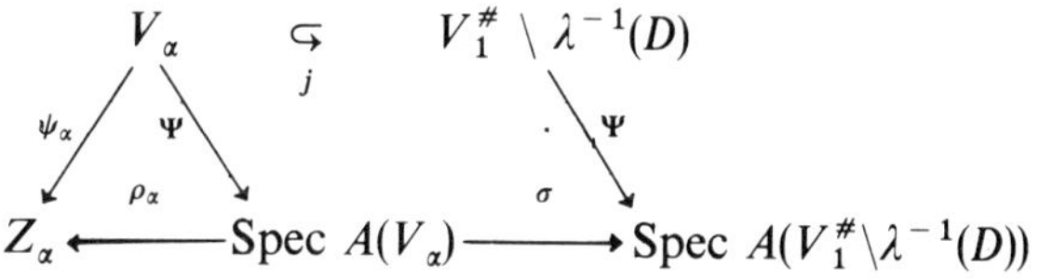

Here, σ is the morphism ${}^a\Gamma(j)$, where j is the inclusion. By hypothesis (3), one can show that $\gamma_*(\psi_\alpha^{-1}(*)) = 0$ with $*$ the generic point of Z. Hence, by Lemma 10.4, ρ_α is birational. Since Ψ is dominating, σ is also dominating. Thus,

$$QA(V_1^\#\setminus\lambda^{-1}(D)) \xrightarrow{\sigma^*} QA(V_\alpha) \cong \operatorname{Rat}(Z_\alpha) = \operatorname{Rat}(Z)$$

is derived. Furthermore, recall that $QA(V\setminus D) \cong QA(V_1^\#\setminus\lambda^{-1}(D))$.

Composing these homomorphisms, one obtains a homomorphism $\tau\colon QA(V\backslash D) \to \mathrm{Rat}(Z)$. By hypothesis (2), one has $\dim Z = \kappa(D, V)$, i.e., $\mathrm{tr.\,deg}_k\, QA(V\backslash D)$. Hence, τ is finite, i.e., $\mathrm{Rat}(Z)/QA(V\backslash D)$ is finite, and so $\mathrm{Rat}(Z) = QA(V\backslash D) \cong \mathrm{Rat}(W)$. Therefore, $\mathrm{Rat}(Z) \cong \mathrm{Rat}(W)$ gives rise to a birational map $\rho\colon W \to Z$ which has the required property. □

g. Let $\mathscr{L}$ be an invertible sheaf on a complete variety V. Define $\mathbb{N}(\mathscr{L}) = \{m \in \mathbb{N} \mid l_V(\mathscr{L}^{\otimes m}) \geq 1\}$ and Φ_m to be the rational map associated with $\lambda^*(\mathscr{L}^{\otimes m})$, where $\lambda\colon V' \to V$ is the normalization. Furthermore, define $\kappa(\mathscr{L}, V)$ to be $\max\{\dim \Phi_m(V') \mid m \geq 1\}$.

Take a nonsingular model (V^*, μ) of V'. Then $\mu^*(\lambda^*\mathscr{L}) \cong \mathscr{O}(D)$ for some divisor D on V^* and $l_{V^*}(mD) = l_V(\mathscr{L}^{\otimes m})$, $\Phi_{mD} = \Phi_m \circ \mu$. Hence $\kappa(D, V^*) = \kappa(\mathscr{L}, V)$. Thus, applying Theorems 10.3, 10.4, 10.5, and 10.6 to D on V^*, we obtain corresponding theorems for $\mathscr{L}$ on V.

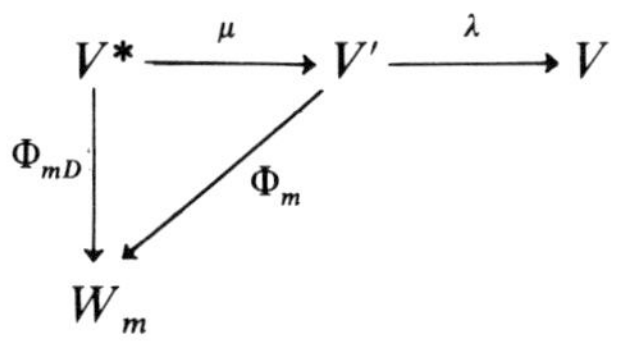

§10.4 D-Dimensions of a $K3$ Surface and an Abelian Variety

a. Definition. A nonsingular complete surface S is said to be a $K3$ *surface* if $K(S) \sim 0$ and $q(S) = 0$.

For example, a nonsingular quartic surface in $\mathbf{P}_k^3$ is a $K3$ surface (see Example 5.7).

Let S be a $K3$ surface and let D be an effective divisor. We shall compute the D-dimension of S.

First note that $\dim_k H^1(S, \mathscr{O}_S) = q(S) = 0$ and $\dim_k H^2(S, \mathscr{O}_S) = p_g(S) = 1$ by the Remark in §5.4.f. Thus, for any divisor Δ on S, letting K be $K(S)$, one has by Theorem 8.2,

$$\chi_S(\mathscr{O}(\Delta)) = \frac{\Delta^2 - (K, \Delta)_S}{2} + 1 + p_g - q = \frac{\Delta^2}{2} + 2.$$

If Δ is effective and nonzero, then by the Corollary to Theorem 5.7,

$$\dim_k H^2(S, \mathscr{O}(\Delta)) = l_S(K - \Delta) = l_S(-\Delta) = 0.$$

Thus, one obtains

$$l_S(\Delta) \geq \frac{\Delta^2}{2} + 2.$$

To compute the D-dimension, the following simple facts are very helpful.

Lemma 10.5 (κ-calculus). *Let V be a complete normal variety and let D, D_1, $\ldots$, D_r be divisors on V.*

(*i*) *For any $m_1 > 0, \ldots, m_r > 0$, if $\kappa(D_1, V) \geq 0, \ldots, \kappa(D_r, V) \geq 0$, then*

$$\kappa\left(\sum_{j=1}^{r} D_j, V\right) = \kappa\left(\sum_{j=1}^{r} m_j D_j, V\right).$$

(*ii*) *If $\kappa(D_1, V) \geq 0$, then $\kappa(D, V) \leq \kappa(D + D_1, V)$.*

PROOF. (i) Clearly, one can suppose that these D_i are effective. If $m = \max\{m_1, \ldots, m_r\}$ and $\mathfrak{D} = \sum_{j=1}^{r} D_j$, then

$$\mathfrak{D} \leq \sum_{j=1}^{r} m_j D_j \leq m\mathfrak{D}.$$

Hence

$$\kappa(\mathfrak{D}, V) \leq \kappa\left(\sum_{j=1}^{r} m_j D_j, V\right) \leq \kappa(m\mathfrak{D}, V) = \kappa(\mathfrak{D}, V).$$

Thus, the equalities hold.

(ii) is obvious. □

Now, we compute $\kappa(D, S)$ by examining the following cases separately.

Case (i) $\kappa(D, S) = 0$. Let $D = \sum_{i=1}^{r} C_i$ be the sum of the curves C_i. Then one has $l(\sum_{i=1}^{r} m_i C_i) = 1$ for any $m_i \geq 0$ by Lemma 10.5.(i). Thus

$$1 = l(C_i) \geq C_i^2/2 + 2,$$

hence $C_i^2 \leq -2$. By the adjunction formula (Corollary 1 to Theorem 8.2),

$$-2 \leq 2\pi(C_i) - 2 = C_i^2 + (K, C_i) = C_i^2 \leq -2.$$

Hence, $C_i^2 = -2$. Furthermore, for any $m_1, \ldots, m_r \geq 0$ with $(m_1, \ldots, m_r) \neq (0, \ldots, 0)$, one has

$$1 = l\left(\sum_{i=1}^{r} m_i C_i\right) \geq \sum_{i,j} m_i m_j (C_i, C_j)/2 + 2;$$

hence

$$\sum_{i,j} m_i m_j (C_i, C_j) \leq -2.$$

This implies that the matrix $[(C_i, C_j)]$ is negative definite. Since $C_i^2 = -2$ and $(C_i, C_j) \geq 0$ for any $i \neq j$, it follows that $(C_i, C_j) = 0$ or 1. Such a matrix is associated to a (direct sum) of Dynkin diagrams. In particular, $\sum_{i=1}^{r} C_i$ is said to be a *reducible curve of Dynkin type*.

Given a $K3$ surface, it is known that there exist at most 16 curves of Dynkin type on that surface.

Case (ii) $\kappa(D, S) = 1$. Applying Theorem 8.6 to D, one has a proper fibered variety $f\colon S \to W$ whose general fiber $C_x = f^{-1}(x)$ satisfies $(C_x, D) = 0$.

By the adjunction formula, one has

$$2\pi(C_x) - 2 = (K, C_x) + C_x^2 = 0.$$

Hence, $\pi(C_x) = 1$, i.e., $f: S \to W$ is an elliptic surface and

$$\text{supp } D \subseteq \bigcup_{i=1}^{s} f^{-1}(a_i)$$

for some $a_1, \ldots, a_s \in W$. Thus, the matrix $[(C_i, C_j)]$ becomes negative semi-definite.

Case (iii) $\kappa(D, S) = 2$. Neither of the above two cases hold.

b. Definition. A complete variety which admits the structure of an algebraic group is said to be an *Abelian variety*.

Proposition 10.5. *Let D be an effective divisor on an Abelian variety V. Then*

(i) *If $D \neq 0$, then $l_V(2D) \geq 2$ and $\mathrm{Bs}\,|2D| = \varnothing$; i.e., $\kappa(D, V) > 0$.*
(ii) *D is ample if and only if $D^n > 0$, where n is* dim *V.*

PROOF. Left to the reader (cf. [M2, Chapter II]). □

Proposition 10.6. *Let D be a divisor on an Abelian variety V.*

(i) *If $\kappa(D, V) = 0$, then $iD \sim 0$ for some $i > 0$.*
(ii) *If $\kappa = \kappa(D, V) > 0$ and $m \geq 3$, then $\Phi_{mD}: V \to \mathbf{P}|mD|$ gives rise to a proper fibered variety $f = \Phi_{mD}: V \to W = \Phi_{mD}(V)$, where W is an Abelian variety of dimension κ.*
(iii) *$f: V \to W$ is smooth and any fiber $f^{-1}(x)$ is an Abelian variety.*
(iv) *There exists an ample divisor Δ on W such that $D = f^*(\Delta)$.*

PROOF. Left to the reader. □

c. Let V, W be nonsingular complete varieties and let D, Δ be divisors on V, W, respectively. Then $\mathfrak{D} = D \times W + V \times \Delta$ is a divisor on $V \times W$. Letting $p: V \times W \to V$ and $q: V \times W \to W$ be projections, one has an isomorphism $\mathcal{O}(\mathfrak{D}) \cong p^*\mathcal{O}(D) \otimes q^*\mathcal{O}(\Delta)$.

Proposition 10.7. *If $m > 0$, then $l(m\mathfrak{D}) = l(mD) \cdot l(m\Delta)$. Furthermore, $\kappa(\mathfrak{D}, V \times W) = \kappa(D, V) + \kappa(\Delta, W)$.*

PROOF. By Exercise 5.1, we have

$$\begin{aligned} l(m\mathfrak{D}) &= \dim_k H^0(V \times W, p^*\mathcal{O}(mD) \otimes q^*\mathcal{O}(m\Delta)) \\ &= \dim_k H^0(V, \mathcal{O}(mD)) \cdot \dim_k H^0(W, \mathcal{O}(m\Delta)) \\ &= l(mD) \cdot l(m\Delta). \end{aligned}$$

Then by applying Theorem 10.2, one obtains the equality. □

§10.5 Kodaira Dimension

a. Definition. Let V be a nonsingular complete variety. The *Kodaira dimension* $\kappa(V)$ of V is defined to be $\kappa(K(V), V)$.

For any $m > 0$, $P_m(V) = l_V(mK(V))$; hence by Theorem 10.2, there exist α, $\beta > 0$ such that

$$\alpha m^{\kappa(V)} \leq P_{mm_0}(V) \leq \beta m^{\kappa(V)} \quad \text{for } m \gg 0,$$

where $m_0 = m_0(K(V))$.

Since the $P_m(V)$ are birational invariants, $\kappa(V)$ is also a birational invariant.

Definition. Given any variety V, define $\kappa(V)$ to be $\kappa(V^*)$, where V^* is a nonsingular complete variety birationally equivalent to V.

b. Proposition 10.8. *If V and W are varieties, then*

$$P_m(V \times W) = P_m(V) \cdot P_m(W) \quad \textit{and} \quad \kappa(V \times W) = \kappa(V) + \kappa(W).$$

PROOF. One can assume both V and W are nonsingular complete. Then $K(V \times W) \sim K(V) \times W + V \times K(W)$. Hence, by Proposition 10.7, one obtains the result. □

Definition. A variety V is said to be a *ruled variety*, if V is birationally equivalent to $\mathbf{P}_k^1 \times W$ for some variety W.

Thus, if V is ruled, then $\kappa(V) = \kappa(\mathbf{P}_k^1) + \kappa(W) = -\infty$.

EXAMPLE 10.1. Let V be a nonsingular hypersurface of degree r in $\mathbf{P}_k^{n+1}$. Then by Example 5.7,

$$r \leq n + 1 \Leftrightarrow \kappa(V) = -\infty,$$

$$r = n + 2 \Leftrightarrow \kappa(V) = 0,$$

$$r \geq n + 3 \Leftrightarrow \kappa(V) = n.$$

EXAMPLE 10.2. If V is an Abelian variety of dimension n, then $\Omega_V \cong \mathcal{O}_V^n$; hence $\Omega_V^q \cong \mathcal{O}_V^{\binom{n}{q}}$, $\Omega_V^n \cong \mathcal{O}_V$. Thus $K(V) \sim 0$, $P_m(V) = 1$ and $\kappa(V) = 0$.

Proposition 10.9. *Let C be a nonsingular complete curve of genus g. Then*

(i) $\kappa(C) = -\infty$ *if and only if* $g(C) = 0$, *i.e.*, $C \cong \mathbf{P}_k^1$.
(ii) $\kappa(C) = 0$ *if and only if* $g(C) = 1$, *i.e.*, C *is an elliptic curve.*
(iii) $\kappa(C) = 1$ *if and only if* $g(C) \geq 2$. *In this case* $K(C)$ *is ample.*

PROOF. Left to the reader. □

c. By Theorems 10.3, 10.4, 10.5, and 10.6, one obtains the fundamental theorems for Kodaira dimension.

Theorem 10.7 (Fibering Theorem). *Given a variety V with $\kappa(V) \geq 0$, there exists a proper fibered variety $f: V^{\#} \to W$ with the following properties:*

(0) *$V^{\#}$ is a nonsingular complete variety birationally equivalent to V,*
(1) $\dim W = \kappa(V)$,
(2) $\kappa(f^{-1}(x)) = 0$ *for "general" points x of W.*

Proof. Clearly, one can assume V is nonsingular complete. Applying Theorem 10.3 to $K(V)$, one obtains a $K(V)$-canonical fibered variety $f: V^{\#} \to W$. Then the result follows from $K(V^{\#})|_{f^{-1}(x)} \sim K(f^{-1}(x))$ which will be proved by the next lemma. □

Lemma 10.6. *Let $f: V \to W$ be a proper smooth fibered variety. Then $K(V)|_F \sim K(F)$, where $F = f^{-1}(x)$, for a general point x of W.*

Proof. We may assume that W is a coordinate neighborhood of x with $(w^1, \ldots, w^r)$ a local coordinate system. Then since f is smooth, we can also assume that V has a coordinate cover $\{U_\lambda | \lambda \in \Lambda\}$ with $(z_\lambda^1, \ldots, z_\lambda^n, w^1, \ldots, w^r)$ a local coordinate system on each U_λ, where $\mathrm{Rat}(W)$ is identified with the subfield $f^*\mathrm{Rat}(W)$ of $\mathrm{Rat}(V)$.

Take a rational n-form ω such that $\omega|_F$ is defined and is not identically equal to zero (for example, $\omega = dz_1^1 \wedge \cdots \wedge dz_1^n$, for some $1 \in \Lambda$). Then $\omega|_{U_\lambda} = \varphi_\lambda dz_\lambda + \psi_\lambda^1 \wedge dw^1 + \cdots + \psi_\lambda^r \wedge dw^r$, where dz_λ is $dz_\lambda^1 \wedge \cdots \wedge dz_\lambda^n$, φ_λ is a rational function and where the ψ_λ^i are rational $(n-1)$-forms. Thus,

$$K(V) \sim \mathrm{div}(\omega \wedge dw^1 \wedge \cdots \wedge dw^r) = \mathrm{div}\{\varphi_\lambda\}$$

and so

$$K(V)|_F \sim \mathrm{div}\{\varphi_\lambda|_F\}.$$

On the other hand, $\omega|_{F \cap U_\lambda} = (\varphi_\lambda|_F)dz_\lambda$ is a rational n-form on F, and hence

$$K(F) \sim \mathrm{div}(\omega|_F) = \mathrm{div}\{\varphi_\lambda|_F\} \sim K(V)|_F.$$ □

Remark. Since $l(mK(V^{\#})) = l(mK(V))$, when V is a nonsingular complete variety, one has $\Phi_{mK(V^{\#})} = \Phi_{mK(V)} \circ \mu$ for any $m \geq 1$. Thus, the canonical fibered variety $f: V^{\#} \to W$ is obtained from a pluricanonical morphism of $V^{\#}$.

Theorem 10.8 (Easy Addition Theorem). *Let $f: V \to W$ be a fibered variety. Then*

$$\kappa(V) \leq \kappa(f^{-1}(x)) + \dim W$$

for general points x of W.

PROOF. Clearly, we may assume V is nonsingular complete. By Theorem 10.4 and Lemma 10.6, one obtains the result. □

Corollary. *Let $f: V \to W$ be a fibered variety.*

(i) *If $\kappa(f^{-1}(x)) = -\infty$ for general points x of W, then $\kappa(V) = -\infty$.*
(ii) *If $\kappa(f^{-1}(x)) = 0$ for general points x of W, then $\kappa(V) \leq \dim W$.*

Theorem 10.9 (Covering Theorem). *Let $f: V_1 \to V_2$ be an étale covering between nonsingular complete varieties V_1 and V_2. Then $\kappa(V_1) = \kappa(V_2)$.*

PROOF. By Theorem 5.5, $K(V_1) \sim f^*K(V_2) + R_f = f^*K(V_2)$. Hence, by Theorem 10.5, one obtains the result. □

Theorem 10.10 (Uniqueness Theorem). *Let $\varphi: V \to Z$ be a fibered variety such that $\dim Z = \kappa(V)$ and $\kappa(\varphi^{-1}(x)) = 0$ for "general" points x of Z. Then for a canonical fibered variety $f: V^\# \to W$, where $V^\#$ is birationally equivalent to V via a birational map $\mu: V^\# \to V$, there exists a birational map $\rho: W \to Z$ such that $\rho \circ f = \varphi \circ \mu$.*

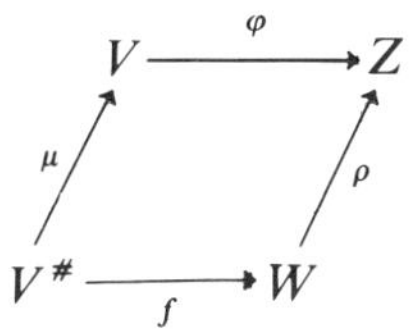

PROOF. This follows immediately from Theorem 10.6. □

Thus any fibered variety $\varphi: V \to Z$ may be said to be a *canonical fibered variety* of V_1, if it satisfies the condition of the last theorem and if V is birationally equivalent to V_1.

§10.6 Types of Varieties

a. Let V be a variety of dimension n. V is said to be of *elliptic type* or of *type* I if $\kappa(V) = -\infty$. Ruled varieties are of elliptic type. In general if there is a fibered variety $f: V \to W$ whose general fibers are of elliptic type, then so is V; in particular, if $f^{-1}(x)$ are rational curves, then $\kappa(V) = -\infty$.

V is said to be of *parabolic type* or of *type* II if $\kappa(V) = 0$. Abelian varieties and nonsingular hypersurfaces $V_+(F)$ of $\mathbf{P}_k^{n+1}$ with $\deg F = n + 2$ are of parabolic type.

V is said to be of *hyperbolic type* or of *type* III if $\kappa(V) = n$. Traditionally, varieties of hyperbolic type are called of *general type*. Nonsingular hypersurfaces $V_+(F)$ with $\deg F > n + 2$ are of hyperbolic type. Roughly speaking, almost all varieties are of hyperbolic type, so they are called of general type.

Finally, V is said to be of *fiber type* or *type* $\mathrm{II}\frac{1}{2}$, if $0 < \kappa(V) < n$. Then, by Theorem 10.7 there is a canonical fibered variety $f\colon V^{\#} \to W$ such that $V^{\#}$ is birationally equivalent to V, $\dim W = \kappa(V)$, and "general" fibers are of parabolic type. But, $\kappa(W)$ may take one of values $-\infty, 0, \ldots, \kappa(V)$.

$\kappa(V)$	V	
$-\infty$	of elliptic type	type I
0	of parabolic type	type II
$1, \ldots, \dim V - 1$	of fiber type	type $\mathrm{II}\frac{1}{2}$
$\dim V$	of hyperbolic type (or of general type)	type III

b. If V is a variety of hyperbolic type, then the f in the canonical fibered variety $f\colon V^{\#} \to W$ is a birational morphism.

Proposition 10.10. *If V is a variety of hyperbolic type, then any dominating rational map $\varphi\colon V \to V$ is birational.*

Proof. Clearly, we can assume V is $V^{\#}$. Then by Proposition 5.6 and its corollary, φ induces an isomorphism $p_m(\varphi)\colon W \to W$ such that $p_m(\varphi) \circ f = f \circ \varphi$ i.e., $\varphi = f^{-1} \circ p_m(\varphi) \circ f$, which is a birational map. □

The following result is a special case of Theorem 11.12 which will be proved in the next chapter.

Theorem 10.11. *If V is a variety of hyperbolic type, then* Bir(V) *is a finite group.*

§10.7 Subvarieties of an Abelian Variety

a. Here, we assume the base field k is the field of complex numbers. We shall study closed subvarieties of an Abelian variety by making use of canonical fibered varieties. A key lemma in this section is the following result due to Ueno. However, the proof uses the analytic theory of Abelian varieties, which is omitted in this book.

Theorem 10.12 (K. Ueno). *Let V be a closed subvariety of an Abelian variety $\mathscr{A}$ with $n = \dim V$ and $N = \dim \mathscr{A}$. Then,*

(*i*) $\kappa(V) \geq 0$, $p_g(V) \geq 1$, *and* $q(V) \geq n$.

(ii) *The following conditions are equivalent*:

(a) V *is a translation of an Abelian subvariety* (i.e., *a connected closed algebraic subgroup*) *of* $\mathscr{A}$.

(b) $\kappa(V) = 0$.

(c) $p_g(V) = 1$.

(d) $q(V) = n$.

PROOF. (i) Since the base field is the field $\mathbb{C}$ of complex numbers, we can consider $\mathscr{A}$ as a complex torus and V as a closed complex subspace of $\mathscr{A}$. Then $T_1(\mathscr{A})$ has a basis $\omega_1, \ldots, \omega_N$ and the universal covering manifold $\mathscr{A}^{\text{univ}}$ of $\mathscr{A}$ is (complex analytically) isomorphic to a complex affine space $\mathbb{C}^N$ which admits an affine coordinate system $(u_1, \ldots, u_N)$ such that $du_1 = \omega_1, \ldots, du_N = \omega_N$. Let $\pi: \mathscr{A}^{\text{univ}} \to \mathscr{A}$ be the covering map.

Take a general closed point x of V and choose a point x_a from $\pi^{-1}(x)$. Replacing $u_i - u_i(x_a)$ by u_i, we may assume that $u_i(x_a) = 0$ for $1 \le i \le N$. The replacement corresponds to a translation of the zero element of $\mathscr{A}$ as the group variety.

Since π is a local isomorphism, $(u_1, \ldots, u_N)$ can be considered as a local analytic coordinate system around x of $\mathscr{A}$. Since V is nonsingular at x, there exists a local coordinate system $(z_1, \ldots, z_N)$ around x of $\mathscr{A}$ such that $z_j|_V = 0$ for all $j \ge n + 1$. Since both $(z_1, \ldots, z_N)$ and $(u_1, \ldots, u_N)$ are local analytic coordinate systems, one has $a_{ij} \in \mathbb{C}$ and convergent power series $\varphi_i(u_1, \ldots, u_N)$ with $v(\varphi_i) \ge 2$ such that

$$z_i = \sum_{j=1}^{N} a_{ij} u_j + \varphi_i(u_1, \ldots, u_N) \quad \text{for } 1 \le i \le N,$$

where $v(\varphi)$ denotes the order of φ with respect to the maximal ideal generated by $u_1, \ldots, u_N$. Note that φ_i may be zero.

After a suitable linear transformation of $u_1, \ldots, u_N$, one can assume that $a_{ij} = \delta_{ij}$; hence $z_i = u_i + \varphi_i$ for $1 \le i \le N$. From these equalities, one can express $u_1, \ldots, u_N$ as power series in $z_1, \ldots, z_N$, i.e., there exist convergent power series $\psi_1, \ldots, \psi_N$ in $z_1, \ldots, z_N$ with $v(\psi_i) \ge 2$ such that $u_i = z_i + \psi_i$ for $1 \le i \le N$.

We have a small neighborhood U of x in $\mathscr{A}$ as a complex manifold where the ψ_i are convergent, and define W to be $V \cap U$. Let $\zeta_i = z_i|_W$ and $v_i = u_i|_W$ for all i. Then $\zeta_{n+1} = \cdots = \zeta_N = 0$ and $\zeta_1, \ldots, \zeta_n$ form a local analytic coordinate system of W. Hence $d\zeta_1 \wedge \cdots \wedge d\zeta_n \ne 0$ and so $d\zeta_1, \ldots, d\zeta_n$ are independent. Let (V^*, λ) be a nonsingular model of V such that λ is an isomorphism on a neighborhood of W. Then the $\lambda^*(\omega_i|_V)$ are regular 1-forms and $\lambda^*(\omega_i|_V)|_W = dv_i$. Therefore, $\lambda^*(\omega_i|_V)$ can be considered as an analytic continuation of dv_i on V^*. For simplicity, we denote $\lambda^*(\omega_i|_V)$ by dv_i.

Since $dv_i = d\zeta_i + \sum_{j=1}^{n} (\partial\bar{\psi}_i/\partial\zeta_j)d\zeta_j$ on W and $(\partial\bar{\psi}_i/\partial\zeta_j)(0) = 0$ where $\bar{\psi}_i = \psi_i|_W = \psi_i(\zeta_1, \ldots, \zeta_n, 0, \ldots, 0)$, it follows that $dv_1 \wedge \cdots \wedge dv_n \ne 0$; thus $dv_1, \ldots, dv_n$ are linearly independent. Therefore, $p_g(V) \ge 1$ and $q(V) \ge n$.

(ii) Since $p_g(V) \geq 1$, $\kappa(V) = 0$ implies $p_g(V) = 1$. Thus, we shall prove that (c) $\Rightarrow$ (a). Fix $j \geq n + 1$ and define $\tilde{\omega}_j$ to be $dv_1 \wedge \cdots \wedge dv_{n-1} \wedge dv_j$, which is an element of $T_n(V^*)$. Since $p_g(V) = 1$ by hypothesis, letting $\tilde{\omega}_n$ be $dv_1 \wedge \cdots \wedge dv_n$, which is not zero, one has $T_n(V^*) = \mathbb{C}\tilde{\omega}_n$ and so $\tilde{\omega}_j = \lambda_j \tilde{\omega}_n$ for some $\lambda_j \in \mathbb{C}$.

Since $d\zeta_j = 0$ for $n + 1 \leq j \leq N$, one has

$$dv_j + \sum_{l=n+1}^{N} \left.\frac{\partial \varphi_j}{\partial u_l}\right|_W \cdot dv_l = -\sum_{i=1}^{n} \left.\frac{\partial \varphi_j}{\partial u_i}\right|_W \cdot dv_i$$

Recalling that $(\partial \varphi_j / \partial u_i)(0) = 0$, one can form an inverse system of the above system, i.e., there exist convergent power series A_{ji} in $v_1, \ldots, v_N$ with $A_{ji}(0) = 0$ such that

$$dv_j = \sum_{i=1}^{n} A_{ji}\, dv_i \quad \text{for} \quad n + 1 \leq j \leq N.$$

Then $\tilde{\omega}_j = A_{jn} \tilde{\omega}_n$ and so $A_{jn} = \lambda_j$, since $\tilde{\omega}_j = \lambda_j \tilde{\omega}_n$. However, one has $A_{jn}(0) = 0$; thus $\lambda_j = 0$ and $A_{jn} = 0$. Similarly, one can obtain $A_{ji} = 0$ for all $n + 1 \leq j \leq N$ and $1 \leq i \leq n$. Hence, $v_j = 0$ for $n + 1 \leq j \leq N$, which implies that $\pi^{-1}(V)$ is contained in $\mathbb{L} = \{u_{n+1} = \cdots = u_N = 0\}$ locally around x_a. Since both have dimension n, a connected component of $\pi^{-1}(V)$ is $\mathbb{L}$ and hence $V^{\text{univ}} \cong \mathbb{L}$. This implies that V is nonsingular and an Abelian subvariety of $\mathscr{A}$. By the same way, one can derive (a) from (d). Furthermore, (b) $\Rightarrow$ (c) is clear by the first statement. (a) $\Rightarrow$ (c) and (a) $\Rightarrow$ (d) are obvious. □

Corollary to Theorem 10.12.(i). *Let U be a nonsingular variety and let $f: U \to \mathscr{A}$ be a rational map, where $\mathscr{A}$ is an Abelian variety. Then f is a morphism.*

Proof. It is known that if $p \in U \backslash \mathrm{dom}(f)$, then $f(p)$ would contain a rational curve Γ (see Exercise 7.5), and so $\kappa(\Gamma) = -\infty$. However by Theorem 10.12.(i), $\kappa(\Gamma) \geq 0$, since $\Gamma \subseteq \mathscr{A}$. Hence $U = \mathrm{dom}(f)$. □

b. To study general subvarieties, we use the following lemmas.

Lemma 10.7 (Kawamata). *Let $f: V \to Y$ and $g: V \to W$ be morphisms of varieties. Suppose that there exists a dense subset S of* $\mathrm{spm}(Y)$ *such that $g(f^{-1}(x))$ is a set consisting of single point for every $x \in S$. Then g can be written as $h \circ f$ for some rational map $h: Y \to W$.*

Proof. Let Φ be $(f, g)_k: V \to Y \times W$ and let G be the closed image of V by Φ. One has the projection morphism $p: Y \times W \to Y$ and defines φ to be $p|_G: G \to Y$. Since $\varphi(G) \supseteq p(\Phi(V)) \supseteq S$, φ is dominating. G has a dense open subset U (see Exercise 3.1) such that $U \subseteq \Phi(V)$ and $\psi = \varphi|_U: U \to Y$ satisfies the hypothesis of the following lemma; hence ψ is birational and so $h = q|_G \circ \varphi^{-1}$ has the required property, where $q: Y \times W \to W$ is the projection morphism.

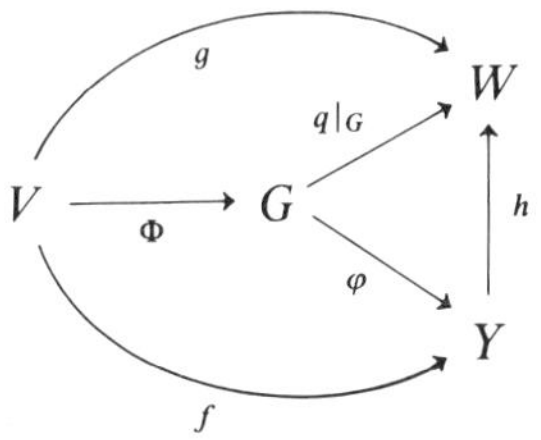

□

Lemma 10.8. *Let $\psi: U \to Y$ be a dominating morphism of varieties. Suppose that there exists a dense subset S of Y such that $\psi^{-1}(y)$ is either $\varnothing$ or a single point set for $y \in S$. Then ψ is birational.*

PROOF. Clearly, one may assume that both U and Y are affine varieties, i.e., $U = \operatorname{Spec} B$ and $Y = \operatorname{Spec} A$ such that the inclusion $A \subseteq B$ corresponds to ψ. Since B is finitely generated as an A-algebra, one has $\beta_1, \ldots, \beta_m \in B$ such that $B = A[\beta_1, \ldots, \beta_m]$. Applying Theorem 1.3 to $B \otimes_A Q(A)$ as a $Q(A)$-algebra, one obtains $y_1, \ldots, y_r \in B$ which are algebraically independent over $Q(A)$ such that $B \otimes_A Q(A)$ is an integral extension of $Q(A)[y_1, \ldots, y_r]$. Thus the β_i are integral over $Q(A)[y_1, \ldots, y_r]$ and so one has an element $a \in A\setminus\{0\}$ such that all the $a\beta_i$ are integral over $A[y_1, \ldots, y_r]$. Therefore, B_a is an integral extension of $A_a[y_1, \ldots, y_r]$. This implies that letting Y_0 be $D(a)$ and U_0 be $\psi^{-1}(Y_0)$, $\psi_0 = \psi|_{U_0}: U_0 \to Y_0$ factors through a finite surjective morphism: $U_0 \to Y_0 \times \mathbf{A}_k^r$. Hence, even for $y \in S \cap Y_0$, $\psi^{-1}(y)$ has dimension r; thus by hypothesis, $r = 0$ follows. Therefore, $\psi_0: U_0 \to Y_0$ is a finite surjective morphism. Choose a smooth part of ψ_0, which becomes an étale cover $\psi_1: U_1 \to Y_1$ (see §10.1.**c**). For any closed point y of Y_1, $\psi_1^{-1}(y) = \psi^{-1}(y)$ consists of n distinct points, where $n = \deg \psi$. Since $S \cap Y_1 \neq \varnothing$, it follows that $n = 1$, i.e., ψ is birational. □

Lemma 10.9. *Let U be a variety and let $\{U_j \mid j = 1, 2, \ldots\}$ be a countable collection of subsets of U. Suppose that the base field k is algebraically closed and has uncountably many elements, and that* $\operatorname{spm}(U) \subseteq \bigcup_{j=1}^{\infty} U_j$. *Then U_j is dense for some j.*

PROOF. Replacing the U_j by their closures, one can assume that the U_j are closed sets. Further, without any loss of generality, we can assume that the U_j are irreducible. Supposing that all $U_j \neq U$, we shall derive a contradiction by induction on $\dim U$. If $\dim U = 1$, then the U_j are sets consisting of a single point and so $\operatorname{spm} U$ is a countable set. This contradicts the cardinality hypothesis for k. If $r = \dim U \geq 2$, one can replace U by an affine open dense subset. Hence, U is assumed to be an affine variety. Applying

Theorem 1.3 to U, one has a finite surjective morphism $U \to \mathbf{A}_k^r$. Then U can be replaced by A_k^r and so there exists a hyperplane H of U such that $H \neq U_j$ for all j. Thus $\text{spm}(H) \subseteq \bigcup_{j=1}^{\infty} (H \cap U_j)$ and $\dim H \cap U_j < \dim U_j$. This contradicts the induction hypothesis. □

Furthermore, we shall have to use the following elementary results concerning Abelian varieties.

Lemma 10.10. *An Abelian variety has at most countably many Abelian subvarieties.*

PROOF. We refer the reader to [M2, IV, Theorem 1]. □

Lemma 10.11. *Let B be an Abelian subvariety of an Abelian variety $\mathscr{A}$. Then the quotient variety $\mathscr{A}/B$ is an Abelian variety. Moreover, there exists an étale cover $C \to \mathscr{A}/B$ such that the fiber product of $\mathscr{A}$ and C over $\mathscr{A}/B$ is isomorphic to $B \times C$.*

PROOF. We refer the reader to [M2, IV, Theorem 1 on p. 173]. □

Remark. C is an Abelian variety.

c. Let V be a closed subvariety of an Abelian variety $\mathscr{A}$ with $n = \dim V$, $N = \dim \mathscr{A}$, and suppose that $\kappa(V) > 0$. By Theorem 10.3, there exist a proper birational morphism $\mu: V^{\#} \to V$ and a proper fibered variety f: $V^{\#} \to W$ such that

(1) W is a nonsingular projective variety with $\dim W = \kappa(V)$,
(2) $\kappa(f^{-1}(x)) = 0$ for "general" points x of W.

Thus, there exists an affine open dense subset $W_{(1)}$ of W such that all $f^{-1}(x)$ (where $x \in W_{(1)}$) are nonsingular varieties with $p_g(f^{-1}(x)) \leq 1$ and the restrictions of μ to $f^{-1}(x)$, $\mu|_{f^{-1}(x)}: f^{-1}(x) \to \mu(f^{-1}(x))$ are birational. Let V_x be $\mu(f^{-1}(x))$, which is a closed subvariety of $\mathscr{A}$. Then $p_g(V_x) \geq 1$ by Theorem 10.12.(i). However, one notes that $p_g(V_x) = p_g(f^{-1}(x)) \leq 1$; thus $p_g(V_x) = 1$ follows. By the implication (c) $\Rightarrow$ (a) of Theorem 10.12.(ii), V_x is a translation of some Abelian subvariety $B\langle x \rangle$ of $\mathscr{A}$. $x \mapsto B\langle x \rangle$ gives rise to a map of $\text{spm}(W_{(1)})$ into the set of all Abelian subvarieties of $\mathscr{A}$, which is a countable set by Lemma 10.9. Thus by Lemma 10.8, there exists an Abelian subvariety B such that the set $S_B = \{x \in \text{spm}(W_{(1)}) \mid B\langle x \rangle = B\}$ is dense.

Letting $\mathscr{A}_0$ be the quotient variety $\mathscr{A}/B$, one has the canonical morphism $\pi: \mathscr{A} \to \mathscr{A}_0$ (as a quotient variety). For $x \in S_B$, V_x is a translation of B; hence $\pi(V_x)$ is a set consisting of a single point. Thus one can apply Lemma 10.7 to $V^{\#} \to V \subseteq \mathscr{A} \to \mathscr{A}_0$ and f: $V^{\#} \to W$, and one obtains a rational map g: $W \to \mathscr{A}_0$ such that $g \circ f = \pi|_V \circ \mu$. However, g is a morphism, since $\mathscr{A}_0$ is also an Abelian variety (cf. Corollary to Theorem 10.12). Let Z be $g(W)$ and so $\pi(V) = Z$. Hence, letting $\varphi = \pi|_V$, one has a surjective morphism $\varphi: V \to Z$ such that $\varphi \circ \mu = g \circ f$.

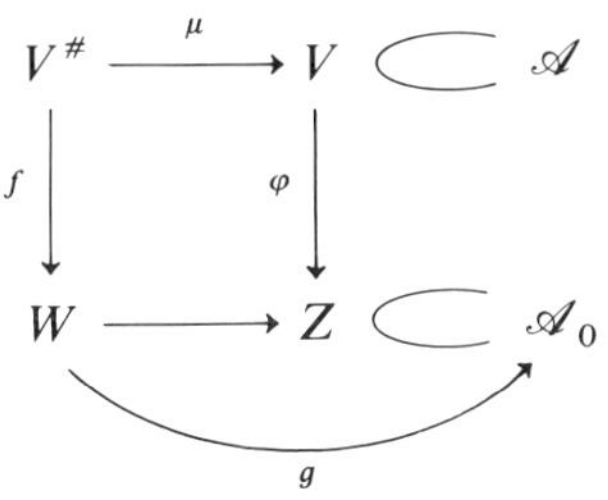

For $x \in S_B$, one has $f^{-1}(g^{-1}(g(x))) = \mu^{-1}(\varphi^{-1}(g(x)))$, which contains $f^{-1}(x)$; hence $\dim \varphi^{-1}(g(x)) \geq \dim f^{-1}(x) = n - \kappa(V)$. On the other hand, $\varphi^{-1}(g(x)) = V \cap (B + \tau(x))$ where $\tau(x)$ is a closed point of $\mathscr{A}$ such that $\pi(\tau(x)) = g(x)$. Thus $\dim(\varphi^{-1}(g(x))) \leq \dim B = n - \kappa(V)$. Therefore, combining this with the previous inequality, one obtains $\dim \varphi^{-1}(g(x)) = n - \kappa(V)$ and so $\varphi^{-1}(g(x)) = B + \tau(x) \subseteq V$. Hence, if x is a general point, $f^{-1}(g^{-1}(g(x))) = \mu^{-1}(\varphi^{-1}(g(x)))$ is irreducible; hence $f^{-1}(g^{-1}(g(x))) = f^{-1}(x)$ and so $g^{-1}(g(x)) = x$. This implies that $g: W \to Z$ is birational. Thus $\varphi: V \to Z$ is a canonical fibered variety.

We shall prove that $\kappa(Z) = \dim Z$, i.e., Z is of hyperbolic type. Applying Lemma 10.10, one has an étale cover $\rho: C \to \mathscr{A}_0$ such that the fiber product of $\mathscr{A}$ and C over $\mathscr{A}_0$ is isomorphic to a product $B \times C$. The projection morphisms $B \times C \to \mathscr{A}$ and $B \times C \to C$ are denoted by ρ_1 and π_1, respectively. Let Z_1 be a connected component of $\rho^{-1}(Z)$. Then one has a connected component of $\rho_1^{-1}(V)$, which is a fiber product of V and Z_1 over Z, denoted by V_1. The projection morphism $V_1 \to Z_1$ is denoted by φ_1.

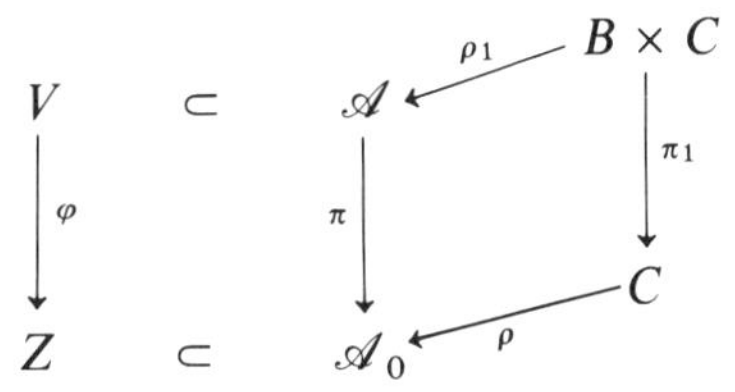

Since $\varphi_1(V_1) \subseteq Z_1$, V_1 is contained in $B \times Z_1$, where $\dim(B \times Z_1) = n - \kappa(V) + \kappa(V) = n$; hence $V_1 = B \times Z_1$. By Proposition 10.8, one has $\kappa(V_1) = \kappa(B) + \kappa(Z_1) = \kappa(Z_1)$. Applying Theorem 10.5 to $V_1 \to V$ and $Z_1 \to Z$, one obtains $\kappa(V) = \kappa(V_1)$ and $\kappa(Z_1) = \kappa(Z)$. Therefore, we conclude that $\kappa(Z) = \dim Z$, and this establishes the following theorem due to Ueno.

Theorem 10.13 (Ueno). *Let V be a closed subvariety of $\mathscr{A}$. Then there exists a canonical fibered variety $\varphi: V \to Z$ such that Z is of hyperbolic type and its* *general fibers are isomorphic to some Abelian subvariety B. Moreover, there exists an étale cover V_1 of V such that $V_1 = B \times Z_1$, where Z_1 is a closed subvariety of some Abelian variety with* $\dim Z_1 = \kappa(Z_1)$.

Exercises

10.1. (Fujita) Let $f\colon V \to W$ be a fibered variety, where V and W are complete nonsingular varieties. If H is a divisor on W with $\kappa(H, W) = \dim W$ and if there exist a and $b > 0$ with $\kappa(aD - bf^*(H), V) \geq 0$, then prove the addition formula $\kappa(D, V) = \kappa(D_y, V_y) + \kappa(H, W)$ for general points y of W.

10.2. Let D be a divisor on a complete normal variety V. If $\kappa(D, V) = \dim V$, then show that $m_0(D) = 1$; i.e. $l(mD) > 0$ for any $m >$ some integer m_1. Moreover, if $\kappa(D, V) = \dim V$ and Δ is an arbitrary divisor on V, then show that there exists m such that $|mD - \Delta| \neq \emptyset$.

10.3. (Tsunoda) Let V be a normal variety. Show that there exists a nonzero $f \in A(V)$ such that $A(V)_f$ is finitely generated as a k-algebra.

10.4. Let V be a complete nonsingular variety and let D be a divisor with normal crossings on V. Suppose that there exists $m > 0$ such that $\kappa(mK(V) + (m - 1)D, V) \geq 0$ and that $\kappa(D, V) = \dim V$. Prove that

$$\kappa(K(V) + D, V) = \max\{\kappa(mK(V) + (m - 1)D, V) \mid m \geq 2\} = \dim V.$$

10.5. Let (V, D) denote an ordered pair consisting of a complete normal variety V and a divisor D on V. A rational map $f\colon V_1 \to V_2$ is said to be a *D-rational map* from (V_1, D_1) to (V_2, D_2) if there exists a normal reduction model (Z, μ) of f such that $\mu^*D_1 \sim g^*D_2$, where g is a morphism $f \circ \mu\colon Z \to V_2$. Show that $f^*D_2 \sim D_1$, where f^*D_2 is defined to be $\mu_*g^*D_2$.

(i) If $f\colon (V_1, D_1) \to (V_2, D_2)$ and $g\colon (V_2, D_2) \to (V_3, D_3)$ are D-rational maps and the composition $g \circ f$ exists as a rational map, then show that $g \circ f$ is also a D-rational map and that $(g \circ f)^*D_1 \sim f^*(g^*D_1)$.

(ii) Let $\mathrm{Bir}(V, D)$ be the subgroup of $\mathrm{Bir}(V)$ generated by birational and D-rational maps from (V, D) into itself. Show that if $\mu\colon V^* \to V$ is a birational morphism, then $\mathrm{Bir}(V, D) \cong \mathrm{Bir}(V^*, \mu^*D)$.

(iii) If D is ample then show that $\mathrm{Bir}(V, D) \subseteq \mathrm{Aut}(V)$.

(iv) Suppose that a D-rational map $f\colon (V_1, D_1) \to (V_2, D_2)$ is dominating and that $\kappa(D_2, V_2) = \dim V_1 = \dim V_2$. Show that f is birational.

10.6. Let V be a variety over $\mathbb{C}$ and let $W_{(1)}, \ldots, W_{(m)}, \ldots$ be a sequence of open subsets such that $W_{(1)} \supseteq W_{(2)} \supseteq \cdots$ and let $W_\infty = \bigcap_{i=1}^\infty W_{(i)}$, which is nonempty. If there exists a countable set of subsets S_j of $\mathrm{spm}(V)$ such that $W_\infty \cap \mathrm{spm}(V) \subseteq \bigcup_{j=1}^\infty S_j$, then show that there exists some S_j which is dense in V.

10.7. Let $f\colon V \to W$ be a proper fibered variety with both V and W nonsingular. Define $\omega_{V/W}$ to be $\Omega_V^n \otimes f^*(\Omega_W^r)^{-1}$, where $n = \dim V$ and $r = \dim W$. If $g\colon Y \to W$ is a proper fibered variety, Y is nonsingular, and $\psi\colon V \to Y$ is a birational map such that $g \circ \psi = f$, then show that $f_*(\omega_{V/W}^{\otimes m}) \cong g_*(\omega_{Y/W}^{\otimes m})$ for any $m \geq 1$.

10.8. (Tsunoda) Let V and W be complete nonsingular varieties and let $f\colon V \to W$ be a dominating rational map. Suppose that $\dim V > \dim W$. Then for any prime divisor Γ on V with $\kappa(\Gamma, V) > \dim W$, show that $f(\Gamma)$ is dense in W.

10.9. (Maehara) Let D and Δ be effective divisors on a complete normal variety V such that $\mathrm{Bs}\,|mD| = \mathrm{Bs}\,|m\Delta| = \emptyset$ for some $m > 0$. Then show that

$$\kappa(D, V) + \kappa(\Delta, V) \geq \kappa(D + \Delta, V).$$

Chapter 11

Logarithmic Kodaira Dimension of Varieties

§11.1 Logarithmic Forms

a. In this chapter we also fix an algebraically closed field k with characteristic zero and assume that all varieties are over k.

Definition. If D is a divisor with simple normal crossings on a complete nonsingular variety $\bar{V}$, then $\bar{V}$ is said to be a *smooth completion of* $V = \bar{V} \backslash D$ *with smooth boundary* D.

By virtue of Theorem 7.21, any nonsingular variety V has a smooth completion $\bar{V}$ with smooth boundary D (cf. §7.10.f).

Lemma 11.1. *Let D be a reduced divisor $\sum_{j=1}^{r} \Gamma_j$ on a complete nonsingular variety $\bar{V}$ where the Γ_j are prime divisors and $\Gamma_i \neq \Gamma_j$ for $i \neq j$. The divisor D has only simple normal crossings if and only if $D_J = \bigcap_{j \in J} \Gamma_j$ are disjoint unions of nonsingular subvarieties for all subsets $J \subseteq \{1, \ldots, r\}$ with* $\operatorname{codim}(D_J) = \#(J)$ *or* $D_J = \varnothing$.

Proof. Assume that D has only simple normal crossings. For a closed point p of D_J, one can choose a coordinate neighborhood U_λ with a local coordinate system $(z_\lambda^1, \ldots, z_\lambda^n)$ such that $\Gamma_{j_i} \cap U_\lambda = V(z_\lambda^i)$ for $1 \leq i \leq s$ where $J = \{j_1, \ldots, j_s\}$. $V(z_\lambda^1, \ldots, z_\lambda^s) = D_J \cap U_\lambda$ is nonsingular and hence p is a nonsingular point of D_J. The converse is easily proved. □

Proposition 11.1. *Let $f\colon V \to W$ be a smooth fibered variety and let D be a divisor with simple normal crossings on V. Then for a general point x of W, $D_x = f^{-1}(x)$ is also a divisor with simple normal crossings.*

PROOF. Let $\sum_{j=1}^{r} \Gamma_j$ be the irreducible decomposition of D. For all subsets J of $\{1, \ldots, r\}$, $D_J \cap f^{-1}(x)$ is nonsingular for a general point x of W by Theorem 7.17. Since $(D \cap f^{-1}(x))_J = D_J \cap f^{-1}(x)$, $D \cap f^{-1}(x)$ has only simple normal crossings by Lemma 11.1.

b. Let $\bar{V}$ be a smooth completion of a nonsingular variety V with a smooth boundary D. At any point $p \in D$, let $\Gamma_1, \ldots, \Gamma_s$ be the irreducible components of D containing p. By the definition of a divisor with simple normal crossings at p, there exists a coordinate neighborhood U_λ with a local coordinate system $(z_\lambda^1, \ldots, z_\lambda^n)$ such that $\Gamma_i \cap U_\lambda = V(z_\lambda^i)$ for all $1 \le i \le s$, where $n = \dim V$. Such a local coordinate system is said to be a *logarithmic coordinate system along D at p*.

Definition. The *$\mathcal{O}_{\bar{V}}$-module of logarithmic 1-forms of $\bar{V}$ along D* is defined to be an $\mathcal{O}_{\bar{V}}$-submodule $\Omega_{\bar{V}}^1(\log D)$ of $\Omega_{\bar{V}}^1(D) = \Omega_{\bar{V}}^1 \otimes_{\mathcal{O}_{\bar{V}}} \mathcal{O}_{\bar{V}}(D)$ such that

(i) $\Omega_{\bar{V}}^1(\log D)|_V = \Omega_V^1$.
(ii) At any closed point p of D,

$$\omega_p \in \Omega_{\bar{V}}^1(\log D)_p \Leftrightarrow \omega_p \in \Omega_{\bar{V}}^1(D)_p$$

and

$$\omega_p = \sum_{i=1}^{s} a_i \frac{dz_\lambda^i}{z_\lambda^i} + \sum_{j=s+1}^{n} b_j\, dz_\lambda^j,$$

where $(z_\lambda^1, \ldots, z_\lambda^n)$ is a logarithmic coordinate system along D at p and the $a_i, b_j \in \mathcal{O}_{\bar{V},p}$.

Clearly, $\Omega_{\bar{V}}^1(\log D)_p = \sum_{i=1}^{s} \mathcal{O}_{\bar{V},p} \cdot (dz_\lambda^i/z_\lambda^i) + \sum_{j=s+1}^{n} \mathcal{O}_{\bar{V},p}\, dz_\lambda^j \cong \mathcal{O}_{\bar{V},p}^n$. Hence, $\Omega_{\bar{V}}^1(\log D)$ is locally free of rank n.

$\Omega_{\bar{V}}^q(\log D)$, which is defined to be $\bigwedge^q(\Omega_{\bar{V}}^1 \log D)$, is also locally free of rank $\binom{n}{q}$ for any $0 \le q \le n$.

Note that $\Omega_{\bar{V}}^n(\log D)$ coincides with $\Omega_{\bar{V}}^n(D) \cong \mathcal{O}(K(\bar{V}) + D)$, since

$$\frac{dz_\lambda^1}{z_\lambda^1} \wedge \cdots \wedge \frac{dz_\lambda^s}{z_\lambda^s} \wedge dz_\lambda^{s+1} \wedge \cdots \wedge dz_\lambda^n = \frac{1}{z_\lambda^1 \cdots z_\lambda^s}\, dz_\lambda^1 \wedge \cdots \wedge dz_\lambda^n.$$

Definition. For any n-tuple $M = (\nu_1, \ldots, \nu_n)$ of nonnegative integers, define $\Omega_{\bar{V}}(\log D)^M$ to be $(\Omega_{\bar{V}}^1 \log D)^{\otimes \nu_1} \otimes \cdots \otimes (\Omega_{\bar{V}}^n \log D)^{\otimes \nu_n}$.

Write $\Gamma(\bar{V}, \Omega_{\bar{V}}(\log D)^M)$ as $T_M(\bar{V}, D)$. Thus, in particular, $T_{(0,\ldots,0,m)}(\bar{V}, D) = \Gamma(\bar{V}, (\Omega_{\bar{V}}^n(D))^{\otimes m}) \cong \mathbf{L}_V(m(K(\bar{V}) + D))$.

Definition. An element of $T_M(\bar{V}, D)$ is said to be a *logarithmic M-form* of $\bar{V}$ along D.

c. We shall consider pullbacks of logarithmic forms. Let $\bar{W}$ be a smooth completion of a nonsingular variety W with smooth boundary B. Let $\bar{f}$:

$\bar{V} \to \bar{W}$ be a morphism such that $\bar{f}(V) \subseteq W$. For a logarithmic 1-form ω of $\bar{W}$ along B, the pullback $\bar{f}^*\omega$ is a rational 1-form (cf. §5.4.**b**). We claim that $\bar{f}^*\omega$ is a logarithmic 1-form of $\bar{V}$ along D. Since $\bar{f}(V) \subseteq W$, $(\bar{f}^*\omega)|_V = (\bar{f}|_V)^*\omega$ is regular on V. In order to check condition (ii), take $p \in D$ and let $p_1 = \bar{f}(p) \in \bar{W}$. Choose a coordinate neighborhood U_λ around p and a coordinate neighborhood W_α around p_1 such that (1) $\bar{f}(U_\lambda) \subseteq W_\alpha$ and (2) there exist logarithmic coordinate systems $(z_\lambda^1, \ldots, z_\lambda^n)$ on U_λ and $(w_\alpha^1, \ldots, w_\alpha^m)$ on W_α. Then since $D \cap U_\lambda = V(z_\lambda^1) + \cdots + V(z_\lambda^s) \supseteq \bar{f}^{-1}(B \cap W_\alpha) \cap U_\lambda$ and $B \cap W_\alpha = V(w_\alpha^1) + \cdots + V(w_\alpha^t)$, one has

$$\bar{f}^*(V(w_\alpha^i)) \cap U_\lambda = \sum_{j=1}^{s} n_{ij} V(z_\lambda^j) \quad \text{where } n_{ij} \geq 0 \quad \text{and } 1 \leq i \leq t.$$

These are rewritten in terms of local coordinates as

$$\bar{f}^*(w_\alpha^i) = \varepsilon_i \prod_{j=1}^{s} (z_\lambda^j)^{n_{ij}} \quad \text{for} \quad 1 \leq i \leq t, \tag{L}$$

where $\varepsilon_i \in \mathcal{O}_{\bar{V},p}^*$. From these, one obtains

$$\bar{f}^*\left(\frac{dw_\alpha^i}{w_\alpha^i}\right) = \sum_{j=1}^{s} n_{ij} \frac{dz_\lambda^j}{z_\lambda^j} + \frac{d\varepsilon_i}{\varepsilon_i}. \tag{$\frac{dL}{L}$}$$

On the other hand,

$$\omega \Big|_{W_\alpha} = \sum_{i=1}^{t} a_i \frac{dw_\alpha^i}{w_\alpha^i} + \sum_{j=t+1}^{m} b_j \, dw_\alpha^j,$$

where $a_i, b_j \in \mathcal{O}_{\bar{W},p_1}$. Thus,

$$\bar{f}^*(\omega)\Big|_{U_\lambda} = \sum_{i=1}^{t} \sum_{j=1}^{s} \bar{f}^*(a_i) n_{ij} \frac{dz_\lambda^j}{z_\lambda^j} + \sum_{i=1}^{t} \bar{f}^*(a_i) \frac{d\varepsilon_i}{\varepsilon_i} + \sum_{j=t+1}^{m} \bar{f}^*(b_j) \bar{f}^*(dw_\alpha^j),$$

which clearly belongs to $\Omega_{\bar{V}}^1(\log D)_p$. Therefore, it has been proved that $\bar{f}^*(\omega)$ is a logarithmic 1-form of $\bar{V}$ along D.

In the same way, one can prove that if ω is a logarithmic $(v_1, \ldots, v_m)$-form of $\bar{W}$ along B, then $\bar{f}^*(\omega)$ is a logarithmic $(v_1, \ldots, v_m, 0, \ldots, 0)$-form of $\bar{V}$ along D.

Note that if $n < m$ and $v_b \neq 0$ for some $b > n$, then $\bar{f}^*(\omega)$ must be zero. To simplify the notation, we identify $(v_1, \ldots, v_a)$ with $(v_1, \ldots, v_a, 0, \ldots, 0)$. Thus $\bar{f}^*$ gives rise to a linear map from $T_M(\bar{W}, B)$ to $T_M(\bar{V}, D)$ for any $M = (v_1, \ldots, v_m)$.

Proposition 11.2. *If $\bar{f}$: $\bar{V} \to \bar{W}$ is dominating and $\bar{f}^{-1}(B) \subseteq D$, then $\bar{f}^*$: $T_M(\bar{W}, B) \to T_M(\bar{V}, D)$ is one-to-one.*

PROOF. This is clear from Proposition 5.5. □

d. With the notation as above, we shall prove the proper birational invariance of $T_M(\bar{V}, D)$.

Lemma 11.2. *Let $f = \bar{f}|_V$: $V \to W$. Then f is proper if and only if $\bar{f}^{-1}(B) = D$.*

Proof. Let $V_1 = \bar{V} \backslash \bar{f}^{-1}(B)$, which contains V. Since $\varphi = \bar{f}|_{V_1}$ is $(\bar{f})_W$ by Lemma 1.36, φ is a proper morphism by Theorem 1.22.(iii). f is proper if and only if $V = V_1$, since f is a composition of φ with the open immersion $V \to V_1$. □

Theorem 11.1. *If f: $V \to W$ is a proper birational morphism, then $\bar{f}^*$: $T_M(\bar{W}, B) \to T_M(\bar{V}, D)$ is an isomorphism.*

Proof. In view of the last proposition, it suffices to prove that $\bar{f}^*$ is surjective. Let $W_0 = \mathrm{dom}(\bar{f}^{-1})$ and $B_0 = B \cap W_0$. Then $g = \bar{f}^{-1}|_{W_0}$: $W_0 \to \bar{V}$ is a morphism. For any $\omega \in T_M(\bar{V}, D)$, $g^*\omega$ belongs to $T_M(W_0, g^{-1}D)$. However, since $D = \bar{f}^{-1}(B)$, $g^{-1}(D) = g^{-1}\bar{f}^{-1}(B) = B_0$. Hence, $g^*\omega \in \Gamma(W_0, \Omega_{\bar{W}}(\log B)^M)$. Since $\Omega_{\bar{W}}(\log B)^M$ is locally free and $\mathrm{codim}(W \backslash W_0) \geq 2$, one sees that

$$T_M(\bar{W}, B) = \Gamma(\bar{W}, \Omega_{\bar{W}}(\log B)^M) = \Gamma(W_0, \Omega_{\bar{W}}(\log B)^M)$$

by Lemma 2.32. Hence, $g^*\omega \in T_M(\bar{W}, B)$, and $\bar{f}^*(g^*\omega) = \omega$. □

e. Given a nonsingular variety V, take two smooth completions $\bar{V}_1$ and $\bar{V}_2$ of V with smooth boundaries D_1 and D_2, respectively. Then one can choose a smooth completion $\bar{V}_3$ of V with smooth boundary D_3 such that the inclusions $V \subseteq \bar{V}_1$ and $V \subseteq \bar{V}_2$ determine birational morphisms λ: $\bar{V}_3 \to \bar{V}_1$ and μ: $\bar{V}_3 \to \bar{V}_2$, respectively. Then $\lambda|_V = \mathrm{id}$ and $\mu|_V = \mathrm{id}$; hence λ^*: $T_M(\bar{V}_1, D_1) \cong T_M(\bar{V}_3, D_3)$ and μ^*: $T_M(\bar{V}_2, D_2) \cong T_M(\bar{V}_3, D_3)$ by Theorem 11.1. $\mu^{*-1} \circ \lambda^*$: $T_M(\bar{V}_1, D_1) \xrightarrow{\sim} T_M(\bar{V}_2, D_2)$ does not depend on the choice of $\bar{V}_3$; actually $\mu^{*-1} \circ \lambda^*$ can be written as $(\lambda \circ \mu^{-1})^*$, which was defined in §5.4.**c**.

Definition. Define $T_M(V)$ to be $T_M(\bar{V}, D)$, where $\bar{V}$ is a nonsingular completion of V with smooth boundary D (§7.10.**d**). When V is a singular variety, define $T_M(V)$ to be $T_M(V^*)$, where V^* is a nonsingular model of V.

Note that by the above observation, $T_M(V)$ does not depend on the choice of nonsingular model V^* of V.

Definition. Elements of $T_M(V)$ are said to be *logarithmic M-forms* on V.

Given any morphism f: $V \to W$ of varieties, choose completions $\bar{V}$ and $\bar{W}$ such that f defines a morphism $\bar{f}$: $\bar{V} \to \bar{W}$. Furthermore, choose nonsingular models $(\bar{V}^*, \lambda)$ and $(\bar{W}^*, \mu)$ of $\bar{V}$ and $\bar{W}$, respectively, such that the rational map $\bar{\varphi} = \mu^{-1} \circ \bar{f} \circ \lambda$ is a morphism.

$$\begin{array}{ccc} V & \xrightarrow{f} & W \\ \cap & & \cap \\ \bar{V} & \xrightarrow{\bar{f}} & \bar{W} \\ \uparrow{\scriptstyle\lambda} & & \uparrow{\scriptstyle\mu} \\ \bar{V}^* & \xrightarrow{\bar{\varphi}} & \bar{W}^* \end{array}$$

Then since $\bar{f}(V) \subseteq W$, one has $\bar{\varphi}(\lambda^{-1}(V)) \subseteq \mu^{-1}(W)$, and so $\bar{\varphi}^*$: $T_M(\mu^{-1}(W)) \to T_M(\lambda^{-1}(V))$ is induced for every M. Now, define f^* to be $\bar{\varphi}^*$. Thus, one has the linear map f^*: $T_M(W) \to T_M(V)$. By Proposition 11.2 and Theorem 11.1, one obtains the next result.

Proposition 11.3

(i) *If f is dominating, then f^*: $T_M(W) \to T_M(V)$ is one-to-one.*
(ii) *If f is a proper birational morphism, then f^*: $T_M(W) \cong T_M(V)$ is an isomorphism.*

EXAMPLE 11.1 Let $a_1, \ldots, a_r \in k$ such that $a_i \neq a_j$ for $i \neq j$. Let $C = \mathbf{A}_k^1 \backslash \{a_1, \ldots, a_r\} = \mathbf{P}_k^1 \backslash \{\infty, a_1, \ldots, a_r\}$, and let $z = \mathsf{T}_1/\mathsf{T}_0$. Then $T_{(1)}(C)$ is spanned by $dz/(z - a_1), \ldots, dz/(z - a_r)$.

EXAMPLE 11.2. Let $\mathbf{P}_k^n = \operatorname{Proj} k[\mathsf{T}_0, \ldots, \mathsf{T}_n]$, $G_m^n = \operatorname{Spec}(k[\mathsf{X}_1, \ldots, \mathsf{X}_n, \mathsf{X}_1^{-1}, \ldots, \mathsf{X}_n^{-1}])$, and let $\Gamma_j = V_+(\mathsf{T}_j)$ for all j. Then $\mathbf{P}_k^n$ is a smooth completion of G_m^n with smooth boundary $D = \Gamma_0 + \cdots + \Gamma_n$. Letting $\mathsf{X}_i = \mathsf{T}_i/\mathsf{T}_0$ for $1 \leq i \leq n$, $T_{(1)}(G_m^n)$ is spanned by $d\mathsf{X}_1/\mathsf{X}_1, \ldots, d\mathsf{X}_n/\mathsf{X}_n$. If φ is an automorphism of G_m^n, then $\varphi^*(d\mathsf{X}_i/\mathsf{X}_i) = \sum_{j=1}^n n_{ij}\, d\mathsf{X}_j/\mathsf{X}_j$ for some $n_{ij} \in \mathbb{Z}$ such that $[n_{ij}]$ is a unimodular matrix. Furthermore, note that $K(\mathbf{P}_k^n) + D \sim 0$.

EXAMPLE 11.3. $\mathbf{P}_k^n$ is a smooth completion of $\mathbf{A}_k^n$ with smooth boundary Γ_0. Thus $T_M(\mathbf{A}_k^n) = 0$ for any $M \neq (0, \ldots, 0)$.

§11.2 Logarithmic Genera

a. Let f: $V \to W$ be a strictly rational map, i.e., there exists a reduction model (Z, μ) of f such that μ is proper. Then μ^*: $T_M(Z) \to T_M(V)$ is an isomorphism; hence a k-linear map $(\mu^*)^{-1} \circ (f \circ \mu)^*$: $T_M(W) \to T_M(V)$ is defined, and is denoted by f^*.

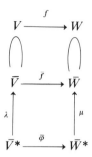

Theorem 11.2. *Let $f: V \to W$ be a strictly rational map.*

(i) *Pullback of forms induces a k-linear map $f^*: T_M(W) \to T_M(V)$ for every M.*
(ii) *If f is dominating, then f^* is one-to-one.*
(iii) *If $h: W \to U$ is a strictly rational map such that the composition $h \circ f$ is defined, then $(h \circ f)^* = f^* \circ h^*: T_M(U) \to T_M(W) \to T_M(V)$.*
(iv) *If f is a proper birational map, then $f^*: T_M(W) \to T_M(V)$ is an isomorphism.*

PROOF. These follow immediately from Proposition 11.3. □

Remark. In Theorem 5.3, the characteristic of k was not assumed to be zero. Thus the condition of being a separable map is necessary in §5.4.

b. If $\varphi \in \mathrm{PBir}(V)$, then $\varphi^*: T_M(V) \to T_M(V)$ is an isomorphism, i.e., $\varphi^* \in GL(T_M(V))$. Since $(\varphi \circ \psi)^* = \psi^* \circ \varphi^*$ for all $\varphi, \psi \in \mathrm{PBir}(V)$, $\varphi \mapsto {}^t\varphi^* \in GL(T_M(V))$ gives rise to a group representation of $\mathrm{PBir}(V)$ with $T_M(V)$ the representation space.

Let $\varphi: V \to V$ be a birational and strictly rational map. Then by Theorem 11.2, $\varphi^*: T_M(V) \to T_M(V)$ is one-to-one. Since $T_M(V)$ is finite-dimensional, φ^* is an isomorphism. Thus, letting $\mathrm{SBir}(V)$ be the group generated by all birational and strictly rational maps of V into itself, $\varphi^* \mapsto {}^t\varphi^* \in GL(T_M(V))$ gives rise to a group homomorphism $\rho_M: \mathrm{SBir}(V) \to GL(T_M(V))$.

Definition. The group $\mathrm{SBir}(V)$ is said to be the *strictly birational automorphism group* of V.

Clearly, one has

$$\mathrm{Aut}(V) \subseteq \mathrm{PBir}(V) \subseteq \mathrm{SBir}(V) \subseteq \mathrm{Bir}(V).$$

c. Definition. Given a variety V of dimension n and an n-tuple M of nonnegative integers, $\bar{P}_M(V) = \dim_k T_M(V)$ is said to be the *logarithmic M-genus* of V.

By Theorem 11.2.(ii), these $\bar{P}_M(V)$ are proper birational invariants.

If $\bar{V}$ is a completion of V, then $\bar{P}_M(V) \geq \bar{P}_M(\bar{V}) = P_M(V)$.

As in §5.4.e, if $M = (0, \ldots, 0, 1, 0, \ldots, 0)$ (where the i-th component is 1), then $T_M(V)$ is denoted by $T_i(V)$ and $\dim_k T_i(V)$ is denoted by $\bar{q}_i(V)$. $\bar{q}_1(V)$ is said to be the *logarithmic irregularity of V*, often denoted by $\bar{q}(V)$. Furthermore, if $M = (0, \ldots, 0, m)$ where the n-th component is m ($n = \dim V$), then $T_M(V)$ is written as $T_{\{m\}}(V)$. $\dim_k T_{\{m\}}(V)$ is denoted by $\bar{P}_m(V)$, which is said to be the *logarithmic m-genus* of V. In particular, $\bar{p}_g(V) = \bar{P}_1(V)$ is often called the *logarithmic geometric genus* of V.

When $\bar{V}$ is a smooth completion of a nonsingular variety V with smooth boundary D, the logarithmic m-genus $\bar{P}_m(V) = l_{\bar{V}}(m(K(\bar{V}) + D))$ for each m.

Definition. The *logarithmic Kodaira dimension* of V is defined to be $\kappa(K(V) + D, \bar{V})$, and is denoted by $\bar{\kappa}(V)$.

If $\bar{\kappa}(V) \geq 0$, then one has m_0 with $\bar{P}_{m_0}(V) > 0$, and $\alpha, \beta > 0$ such that

$$\alpha m^{\bar{\kappa}(V)} \leq \bar{P}_{mm_0}(V) \leq \beta m^{\bar{\kappa}(V)} \quad \text{for} \quad m \gg 0$$

by Theorem 10.2. Thus $\bar{\kappa}(V)$ is a proper birational invariant.

As a corollary to Theorem 11.2.(i), one obtains the following result.

Proposition 11.4

(i) *If f: $V \to W$ is a strictly rational dominating morphism, then $\bar{P}_M(V) \geq \bar{P}_M(W)$ and $\bar{q}_i(V) \geq \bar{q}_i(W)$ for all M and i.*
(ii) *In addition, if* dim V = dim W, *then $\bar{P}_m(V) \geq \bar{P}_m(W)$ for all m and so $\bar{\kappa}(V) \geq \bar{\kappa}(W)$.*
(iii) *If V_0 is a dense open subset of V, then*

$$\bar{q}_i(V_0) \geq \bar{q}_i(V), \quad \bar{P}_m(V_0) \geq \bar{P}_m(V), \quad \textit{and} \quad \bar{\kappa}(V_0) \geq \bar{\kappa}(V).$$

d. If C is a curve, then $\bar{q}(C) = \bar{p}_g(C)$, which is denoted by $\bar{g}(C)$. Let $\bar{C}$ be a nonsingular complete curve with $g = g(\bar{C})$, and let D be $p_0 + \cdots + p_t$ on $\bar{C}$ such that $C = \bar{C} \backslash D$. Then

$$\bar{g}(C) = l(K(\bar{C}) + D) = 1 - g + 2g - 2 + t + 1 - l(-D).$$

If $D \neq 0$, i.e., if C is not complete, then $\bar{g}(C) = g + t$. Thus

$$\bar{g}(C) = 0 \Leftrightarrow g = t = 0 \Leftrightarrow C \cong \mathbf{A}_k^1 \Leftrightarrow \bar{\kappa}(C) = -\infty.$$

Furthermore, if $\bar{g}(C) = 1$, then $g = 1$, $t = 0$ or $g = 0$, $t = 1$. If $g = 1$, one has $\bar{P}_2(C) = 2(t + 1)$. Hence,

$$\bar{g}(C) = \bar{P}_2(C) = 1 \Leftrightarrow g = 0, \quad t = 1 \Leftrightarrow C \cong G_m \Leftrightarrow \bar{\kappa}(C) = 0.$$

Here, G_m denotes Spec $k[\mathsf{X}, \mathsf{X}^{-1}]$, called the 1-dimensional *algebraic torus* over k. Thus, one obtains the next table.

type	$\bar{\kappa}(C)$	C
I	$-\infty$	$\mathbf{P}_k^1$, $\mathbf{A}_k^1$
II	0	elliptic curves, G_m
III	1	the others

EXAMPLE 11.4. Let F be a nonconstant polynomial in n variables $\mathsf{X}_1, \ldots, \mathsf{X}_n$. If R is the integral closure of $k[F]$ in $k(\mathsf{X}_1, \ldots, \mathsf{X}_n)$, then $R = k[\psi]$ for

some polynomial ψ and F can be written as a polynomial in ψ. In particular, $k(X_1, \ldots, X_n)/k(\psi)$ is an algebraically closed extension of fields.

This can be easily shown as follows: F determines a dominating morphism f from $\mathbf{A}_k^n$ to $\mathbf{A}_k^1 = C$. Then one has the Stein factorization of f in the form: $\mathbf{A}_k^n \to \operatorname{Spec} R \to C$. Spec R is a nonsingular affine curve and $\bar{g}(\operatorname{Spec} R) \leq \bar{q}(\mathbf{A}_k^n) = 0$ by Proposition 11.3.(i). Thus $\bar{g}(\operatorname{Spec} R) = 0$, which implies $\operatorname{Spec} R \cong \mathbf{A}_k^1$; hence $R = k[\psi]$.

e. Proposition 11.5. *Let V_0 be an open subset of a nonsingular variety V such that* $\operatorname{codim}(V \backslash V_0) \geq 2$. *Then $\bar{P}_M(V_0) = \bar{P}_M(V)$ for all M.*

Proof. Let $j\colon V_0 \to V$ be the inclusion map. Take smooth completions $\bar{V}_0$ and $\bar{V}$ of V_0 and V with smooth boundaries D_0 and D, respectively, such that j defines a morphism $\mu\colon \bar{V}_0 \to \bar{V}$. Let F be the closure of $V \backslash V_0$ in $\bar{V}$. Then $\operatorname{codim}(F) \geq 2$ and so by Lemma 2.32 one has

$$\Gamma(\bar{V} \backslash F, \Omega(\log D)^M) = \Gamma(\bar{V}, \Omega(\log D)^M) = T_M(V).$$

Since $\mu|_{\bar{V}_0 \backslash \mu^{-1}(F)}\colon \bar{V}_0 \backslash \mu^{-1}(F) \to \bar{V} \backslash F$ is a proper birational morphism, μ^* gives rise to an isomorphism:

$$\Gamma(\bar{V} \backslash F, \Omega(\log D)^M) \cong \Gamma(\bar{V}_0 \backslash \mu^{-1}(F), \Omega(\log D)^M).$$

However, since $\Gamma(\bar{V}_0 \backslash \mu^{-1}(F), \Omega(\log D_0)^M) \supseteq \Gamma(\bar{V}_0, \Omega(\log D_0)^M) = T_M(V_0)$, one has $\bar{P}_M(V) \geq \bar{P}_M(V_0)$ and hence $\bar{P}_M(V) = \bar{P}_M(V_0)$ by Proposition 11.4.(iii). □

Theorem 11.3. *Let V and W be varieties with* $\dim V = n$ *and* $\dim W = r$. *Then $\bar{q}(V \times W) = \bar{q}(V) + \bar{q}(W)$, $\bar{P}_m(V \times W) = \bar{P}_m(V) \cdot \bar{P}_m(W)$ for all $m \geq 1$, and $\bar{\kappa}(V \times W) = \bar{\kappa}(V) + \bar{\kappa}(W)$.*

Proof. We can assume that both V and W are nonsingular. As usual, take smooth completions $\bar{V}$ and $\bar{W}$ of V and W with smooth boundaries D and B, respectively. Then $\mathfrak{D} = \bar{V} \times B + D \times \bar{W}$ has only simple normal crossings and $\bar{Z} = \bar{V} \times \bar{W}$ is a smooth completion of $V \times W$ with smooth boundary $\mathfrak{D}$. Letting $p\colon \bar{V} \times \bar{W} \to \bar{V}$ and $q\colon \bar{V} \times \bar{W} \to \bar{W}$ be projections, one has $\Omega_{\bar{Z}}^1(\log \mathfrak{D}) = p^*(\Omega_{\bar{V}}^1(\log D)) \oplus q^*(\Omega_{\bar{W}}^1(\log B))$ and so $\Omega_{\bar{Z}}^{n+r}(\log \mathfrak{D}) = p^*(\Omega_{\bar{V}}^n(\log D)) \otimes q^*(\Omega_{\bar{W}}^r(\log B))$. Therefore,

$$\begin{aligned}\bar{q}(V \times W) &= \dim_k \Gamma(\bar{Z}, \Omega_{\bar{Z}}^1(\log \mathfrak{D})) \\ &= \dim_k \Gamma(\bar{V}, \Omega_{\bar{V}}^1(\log D)) + \dim_k \Gamma(\bar{W}, \Omega_{\bar{W}}^1(\log B)) \\ &= \bar{q}(V) + \bar{q}(W).\end{aligned}$$

Furthermore, recalling Proposition 10.7, one has

$$\begin{aligned}\bar{P}_m(V \times W) &= \dim_k \Gamma(\bar{Z}, (\Omega_{\bar{Z}}^{n+r}(\log \mathfrak{D}))^{\otimes m}) \\ &= \dim_k \Gamma(\bar{V}, (\Omega_{\bar{V}}^{n}(\log D))^{\otimes m}) \\ &\quad \times \dim_k \Gamma(\bar{W}, (\Omega_{\bar{W}}^{r}(\log B))^{\otimes m}) \\ &= \bar{P}_m(V) \cdot \bar{P}_m(W).\end{aligned}$$

Therefore, $\bar{\kappa}(V \times W) = \bar{\kappa}(V) + \bar{\kappa}(W)$. □

§11.3 Reduced Divisor as a Boundary

a. We shall study noncomplete surfaces which are realized as complements of families of lines on $\mathbf{P}_k^2$. Let $L_0, \ldots, L_r$ be distinct lines on $\mathbf{P}_k^2$, and let S be $\mathbf{P}_k^2 \backslash (L_0 \cup \cdots \cup L_r)$.

First, we consider the following family of lines, called type I.

type I

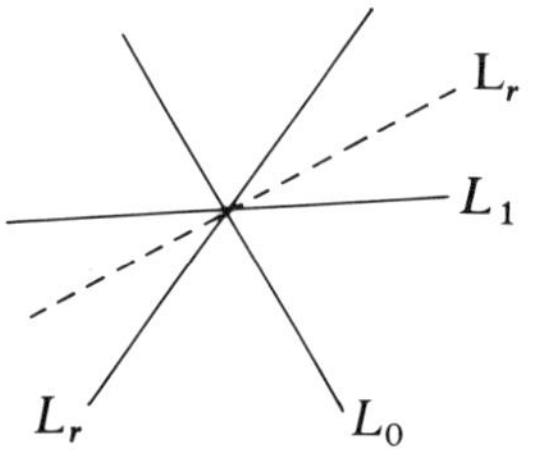

type II

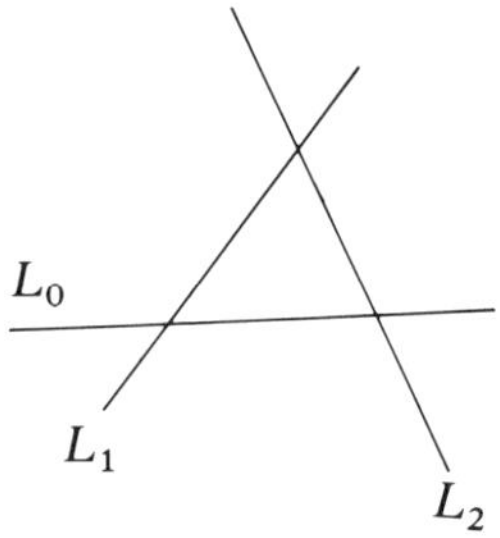

Then obviously, $S \cong \mathbf{A}_k^1 \times (\mathbf{A}_k^1 - \{a_1, \ldots, a_r\})$; hence $\bar{\kappa}(S) = -\infty$.

Second, if L_0, L_1, L_2 form a projective coordinate system of $\mathbf{P}_k^2$ (in this case, the family is called type II), then $S = G_m^2$ and so $\bar{\kappa}(S) = 0$.

Third, we consider the following family of lines, called type II$\frac{1}{2}$ where $r \geq 3$.

type II$\frac{1}{2}$,

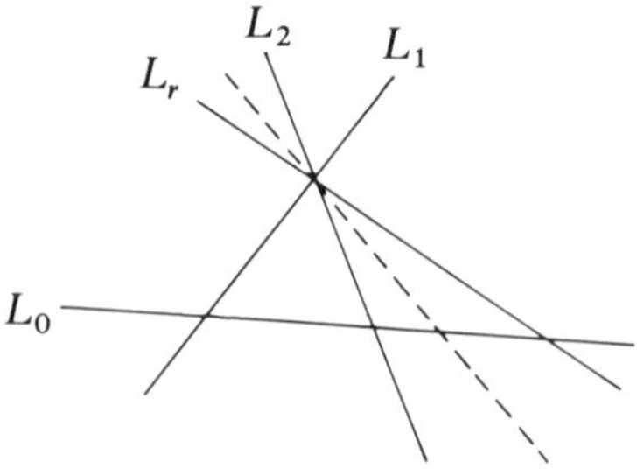

Then, $S \cong G_m \times (G_m \backslash \{a_3, \ldots, a_r\})$; hence $\bar{\kappa}(S) = \bar{\kappa}(G_m) + \bar{\kappa}(G_m \backslash \{a_3, \ldots, a_r\}) = 1$.

Finally, consider the following family of lines, called type a.

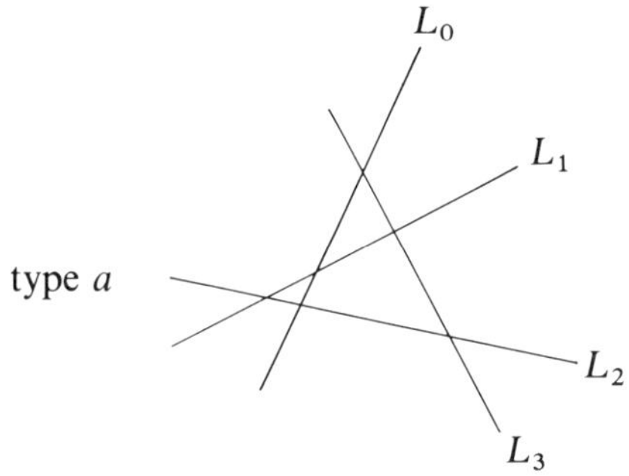

Then, since $L_0 + L_1 + L_2 + L_3$ has only simple normal crossings, $\bar{\kappa}(S)$ can be computed as $\kappa(K(\mathbf{P}_k^2) + 4L_0, \mathbf{P}_k^2)$, which is 2.

Lemma 11.3. *A family of lines is of type I, or type II, or type $II\frac{1}{2}$ or it contains a subfamily of type a.*

Proof. Left to the reader as an exercise. □

Propositon 11.6. *$S = \mathbf{P}_k^2 \backslash (L_0 \cup \cdots \cup L_r)$ can be classified as follows:*

$\bar{\kappa}(S)$	type of $\{L_0, \ldots, L_r\}$	S
$-\infty$	I	$\mathbf{A}_k^1 \times \Gamma$
0	II	$G_m \times G_m$
1	$\mathrm{II}\frac{1}{2}$	$G_m \times \Gamma_1$ with $\bar{\kappa}(\Gamma_1) = 1$
2		$S \subseteq \mathbf{P}_k^2 \backslash (\text{type } a)$

b. Now, let S be a nonsingular complete surface and let D be a reduced divisor on S.

Definition. D is *smooth* at p if D has only simple normal crossings at p.

Let p be a point at which D is not smooth. Let $\mu_1 : S_1 \to S$ be a quadric transformation with center p and define $D_{[1]}$ to be $\mu_1^{-1}(D)$. Then, letting $m_1 = e(p, D)$ and $E_1 = \mu_1^{-1}(p)$, one has

$$K(S_1) \sim \mu_1^* K(S) + E_1, \quad \text{and} \quad D_{[1]} = \mu_1^*(D) - (m_1 - 1)E_1.$$

Thus, $K(S_1) + D_{[1]} \sim \mu_1^*(K(S) + D) - (m_1 - 2)E_1$.

When $D_{[1]}$ is not a smooth boundary, we repeat the above process. Finally, we obtain a sequence of quadric transformations $\{\mu_j : S_j \to S_{j-1}$, with center $p_{(j)} \in D_{[j-1]} \mid 1 \le j \le l\}$, and $\{D_{[j]} \mid$ the reduced transform of $D_{[j-1]}$ by μ_j, i.e., $D_{[j]} = \mu_j^{-1}(D_{[j-1]})\}$, such that $S_0 = S$, $p_{(1)} = p$, and $D_{[0]} = D$, where each

$D_{[j-1]}$ is not smooth at $p_{(j)}$. Then, letting $m_j = e(p_{(j)}, D_{[j-1]})$, $E_j = \mu_j^{-1}(p_{(j)})$, and $\varphi_j = \mu_{j+1} \circ \cdots \circ \mu_l$, we have the formula:

$$K(S_l) + D_{[l]} = \varphi_0^*(K(S) + D) - \sum_{j=1}^{l} (m_j - 2)\varphi_j^*(E_j).$$

Thus,

$$\pi(D_{[l]}) = \pi(D) - \sum_{j=1}^{l} (m_j - 1)(m_j - 2)/2.$$

By making use of this, one can give another proof of Theorem 7.21, in this case. We begin by showing the following simple result.

Lemma 11.4. *Let D be a reduced divisor on S with s connected components. Then* $\dim_k H^1(S, \mathcal{O}(-D)) \le q(S) + s - 1$.

Proof. From the exact sequence

$$0 \to \mathcal{O}(-D) \to \mathcal{O}_S \to \mathcal{O}_D \to 0,$$

one obtains the exact sequence

$$0 \to \Gamma(S, \mathcal{O}_S) \to \Gamma(D, \mathcal{O}_D) \to H^1(S, \mathcal{O}(-D)) \to H^1(S, \mathcal{O}_S).$$

Hence, $\dim_k H^1(S, \mathcal{O}(-D)) \le \dim_k H^1(S, \mathcal{O}_S) + s - 1$. Applying the formula $q(S) = \dim_k H^1(S, \mathcal{O}_S)$ in §5.4.f., one obtains the required inequality. □

By Theorem 5.7, one has $\dim_k H^1(S, \mathcal{O}(-D)) = \dim_k H^1(S, \mathcal{O}(K(S) + D))$. Then in view of Theorem 8.2,

$$\chi(\mathcal{O}(K(S_l) + D_{[l]})) = \pi(D_{[l]}) + p_g(S_l) - q(S_l) + 1.$$

Since S_l is birationally equivalent to S, one has $p_g(S_l) = p_g(S)$ and $q(S_l) = q(S)$. Combining these with the formula, we obtain

$$0 \le l(K(S_l) + D_{[l]}) \le \pi(D) - \sum_{j=1}^{l} \frac{(m_j - 1)(m_j - 2)}{2} + p_g(S) + s.$$

Thus,

$$\sum_{j=1}^{l} (m_j - 1)(m_j - 2)/2 \le \pi(D) + p_g(S) + s.$$

Therefore, we know that after a finite number of quadric transformations, $D_{[l]}$ becomes a reduced divisor B with $e(p, B) = 1$ or 2 for all $p \in B$. □

Lemma 11.5. *Let $p \in B$ satisfy $e(p, B) = 2$. If p is a singular point of an irreducible component Γ of B, then*

(1) *p is an ordinary double point, or*
(2) *there exists $p_2 \in B_{[1]} = \mu_1^{-1}(B)$ such that $e(p_2, B_{[1]}) = 3$, or*

(3) *there exist* $p_1 \in B_{[1]}$ *with* $e(p_1, B_{[1]}) = 2$ *and* $p_2 \in B_{[2]} = \mu_2^{-1}(B_{[1]})$ *such that* $e(p_2, B_{[2]}) = 3$,

where μ_i *is a quadric transformation with center* p_i *for each* p_i.

Moreover, if p *is a point of* $\Gamma_1 \cap \Gamma_2$, *where* Γ_1 *and* Γ_2 *are irreducible components of* B, *then*

(4) B *is smooth at* p, *or*
(5) *there exists* $p_1 \in B_{[1]}$ *such that* $e(p_1, B_{[1]}) = 3$.

PROOF. This is clearly seen by observing the following figures.

(1)

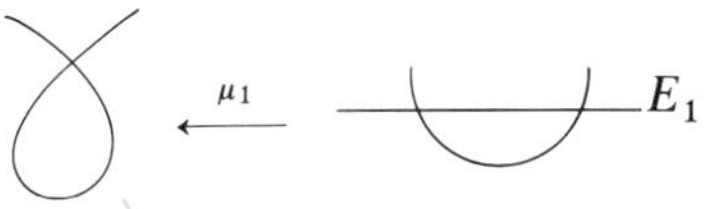

(2)

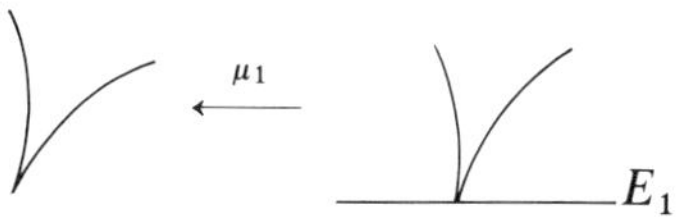

(3)

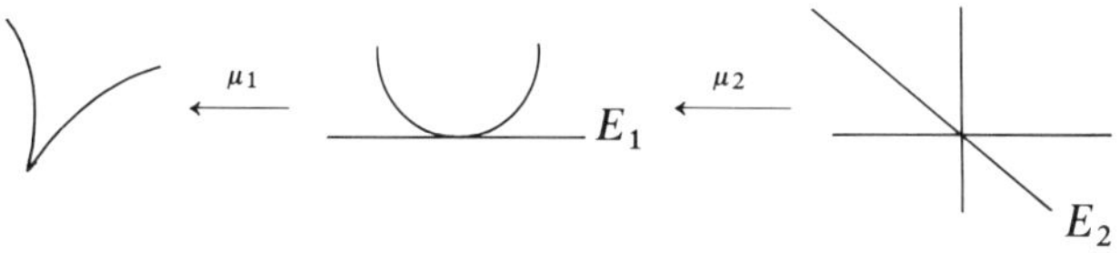

(4)

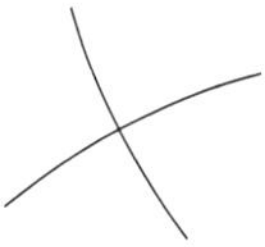

(5)

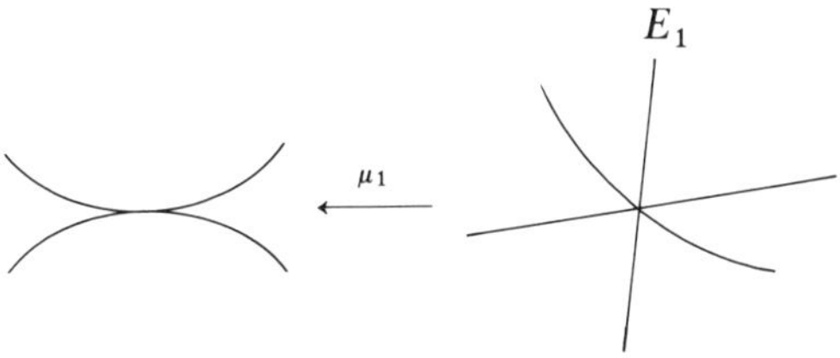

□

Accordingly, we obtain the following result.

Proposition 11.7. *Given a reduced divisor* D *on* S, *there exist a finite number of successive quadric transformations* $\{\mu_j : S_j \to S_{j-1} \mid 1 \le j \le l\}$ *such that* $\varphi_0^{-1}(D)$ *has only simple normal crossings, where* $\varphi_0 = \mu_1 \circ \cdots \circ \mu_l$.

By using the corresponding multiplicities $m_1, \ldots, m_l$, one obtains the following formula:

Proposition 11.8. *If* $q(S) = 0$ *and* D *is connected, then*

$$\bar{p}_g(S \backslash D) - p_g(S) = \pi(D) - \sum_{j=1}^{l} (m_j - 1)(m_j - 2)/2.$$

In particular, if $S = \mathbf{P}_k^2$ *and* D *is a reduced divisor of degree* δ, *then letting* $S_0 = \mathbf{P}_k^2 \backslash D$, *one has*

$$\bar{p}_g(S_0) = \frac{(\delta - 1)(\delta - 2)}{2} - \sum_{j=1}^{l} \frac{(m_j - 1)(m_j - 2)}{2}.$$

c. Let $\bar{V}$ be a nonsingular complete variety and let D be a reduced divisor on $\bar{V}$. In order to compute the logarithmic plurigenera and the logarithmic Kodaira dimension of $V = \bar{V} \backslash D$, one has to smooth the boundary D by a composition of monoidal transformations. Actually, in view of Theorem 7.21, there exists a birational morphism $\mu\colon \bar{V}^{\#} \to \bar{V}$ such that $\bar{V}^{\#}$ is a nonsingular complete variety and $D^{\#} = \mu^{-1}(D)$ has only simple normal crossings.

Using this notation, we shall prove the following results.

Lemma 11.6

(i) $E_1 = \mu^* D - \mu^{-1}(D) = \mu^* D - D^{\#}$ *is an effective divisor exceptional with respect to* μ.
(ii) R_μ *is also exceptional, and all exceptional divisors with respect to* μ *appear in* R_μ.
(iii) *There exists an* N *such that* $E_1 \leq N R_\mu$.

Theorem 11.4

(i) $\bar{P}_m(V) \leq l(m(K(\bar{V}) + D))$ *for all* m; *in particular,* $\bar{\kappa}(V) \leq \kappa(K(\bar{V}) + D, \bar{V})$.
(ii) *If* $\kappa(\bar{V}) \geq 0$, *then* $\bar{\kappa}(V) = \kappa(K(\bar{V}) + D, \bar{V})$.
(iii) *If* $\kappa(\bar{V}) \geq 0$, Δ *is an effective divisor such that* $D^{\#} = \Delta + E_2$, *and* E_2 *is an effective divisor exceptional with respect to* μ, *then* $\bar{\kappa}(\bar{V}^{\#} \backslash \Delta) = \kappa(K(\bar{V}) + D, \bar{V})$.

PROOF. (i) Since $D^{\#} \leq \mu^* D$, one has by Proposition 2.25,

$$\bar{P}_m(V) = l(m(K(\bar{V}^{\#}) + D^{\#})) \leq l(m(\mu^* K(\bar{V}) + \mu^* D) + mR_\mu) = l(m(K(\bar{V}) + D)).$$

(ii) This is a special case of (iii).

(iii) By Lemma 10.5, one has $\bar{\kappa}(\bar{V} \backslash \Delta) = \kappa(K(\bar{V}^{\#}) + \Delta, \bar{V}^{\#}) = \kappa(2K(\bar{V}^{\#}) + \Delta, \bar{V}^{\#})$, since $\kappa(\bar{V}^{\#}) \geq 0$ and $\kappa(\Delta, \bar{V}^{\#}) \geq 0$. Then $\kappa(K(\bar{V}^{\#}), \bar{V}^{\#}) = \kappa(\mu^*(K(\bar{V})) + R_\mu, \bar{V}^{\#}) \geq \kappa(R_\mu, \bar{V}^{\#})$ and hence $\bar{\kappa}(\bar{V} \backslash \Delta) \geq \kappa(K(\bar{V}^{\#}) + R_\mu + \Delta, \bar{V}^{\#}) \geq \kappa(K(\bar{V}^{\#}) + D^{\#}, \bar{V}^{\#})$, which is $\kappa(\mu^*(K(\bar{V})) + R_\mu + \mu^{-1}D - E_1, \bar{V}^{\#}) = \kappa(\mu^*(K(\bar{V})) + NR_\mu + \mu^{-1}D - E_1, \bar{V}^{\#}) \geq \kappa(\mu^*(K(\bar{V}) + D), \bar{V}^{\#}) = \kappa(K(\bar{V}) + D, \bar{V})$, by Lemmas 10.5.(i) and 11.6.(iii), and by Theorem 10.5.

Therefore, $\bar{\kappa}(V^{\#}\backslash\Delta) \geq \kappa(K(V) + D, \bar{V})$. However, by (i) and the inclusion $V \subseteq \bar{V}^{\#}\backslash\Delta$, one has

$$\bar{\kappa}(\bar{V}^{\#}\backslash\Delta) \leq \bar{\kappa}(V) \leq \kappa(K(\bar{V}) + D, \bar{V}).$$

Thus the equality has been established. □

EXAMPLE 11.5. Let $\mu: \bar{S} \to \bar{S}_0$ be a birational morphism between nonsingular complete surfaces. Suppose that $\bar{S}_0$ is a $K3$ surface and D is a reduced divisor on $\bar{S}$. Let $S = \bar{S}\backslash D$ and let D_* be the direct image of D by μ, i.e., if $D = \sum_{i=1}^{r} m_i \Gamma_i$, where the Γ_i are curves, then $D_* = \sum^* m_j \Delta_j$ where the Δ_j are 0 or curves obtained from Γ_j; in other words, $\Delta_j = \mu_*(\Gamma_j)$ (cf. §11.5.**a**). By Theorem 11.4, $\bar{\kappa}(S) = \kappa(D_*, \bar{S}_0)$. Hence, if $\bar{\kappa}(S) = 0$, then $D_* = 0$ or it consists of curves of Dynkin type (see §10.4.**a**).

§11.4 Logarithmic Ramification Formula

a. We use the notation in §11.1.**c**, i.e., $\bar{f}: \bar{V} \to \bar{W}$ is a morphism and $f = \bar{f}|_V$ is a morphism into W. We assume, furthermore that $\dim V = \dim W = n$ and f is dominating.

At a closed point $p \in \bar{V}$, we choose a coordinate neighborhood U_λ around p admitting a logarithmic coordinate system $(z_\lambda^1, \ldots, z_\lambda^n)$ along D at p such that $\bar{f}(p)$ has a coordinate neighborhood W_α admitting a logarithmic coordinate system $(w_\alpha^1, \ldots, w_\alpha^n)$ along B with the property that $\bar{f}(U_\lambda) \subseteq W_\alpha$. Then one has $\psi_{\alpha\lambda} \in \Gamma(U_\lambda, \mathcal{O}_V)$ such that

$$f^*\left(\frac{dw_\alpha^1}{w_\alpha^1} \wedge \cdots \wedge \frac{dw_\alpha^t}{w_\alpha^t} \wedge dw_\alpha^{t+1} \wedge \cdots \wedge dw_\alpha^n\right) = \psi_{\alpha\lambda} \frac{dz_\lambda^1}{z_\lambda^1} \wedge \cdots \wedge \frac{dz_\lambda^s}{z_\lambda^s} \wedge \cdots \wedge dz_\lambda^n.$$

$\{U_\lambda\}$ is an open cover and $\psi_{\alpha\lambda} \cdot \psi_{\beta\mu}^{-1} \in \Gamma(U_\lambda \cap U_\mu, \mathcal{O}_{\bar{V}}^*)$ for all λ, μ. Thus one has an effective divisor D such that $D|_{U_\lambda} = \operatorname{div}(\psi_{\alpha,\lambda})$. Clearly, D depends only on $\bar{f}$ and f.

Definition. D is said to be the *logarithmic ramification divisor*, denoted by $\bar{R}_f$.

By definition, $\bar{R}_f \cap V$ is the ramification divisor for $f: V \to W$, i.e., $\bar{R}_f \cap V = R_f$. But note that $\bar{R}_f$ may not agree with the closure divisor of R_f in $\bar{V}$.

By the same argument as in the proof of Theorem 5.5, one can establish the following formula.

Theorem 11.5 (Logarithmic Ramification Formula).

$$K(\bar{V}) + D \sim \bar{f}^*(K(\bar{W}) + B) + \bar{R}_f,$$

where $\bar{R}_f$ is the logarithmic ramification divisor.

b. Example 11.6. Let $\bar{W}$ be a smooth completion of W with a smooth boundary B, and C be a nonsingular closed subvariety of $\bar{W}$ contained in $\Gamma_1, \ldots, \Gamma_s$, where the Γ_j are the irreducible components of B. We let μ: $\bar{V} = Q_C(\bar{W}) \to \bar{W}$ be the monoidal transformation and let D be $\mu^{-1}(B)$. Then letting $V = \bar{V} \backslash D$, $E = \mu^{-1}(C)$, $f = \mu|_V \colon V \to W$, and $n = \dim V$, $b = \dim C$, one has $\bar{R}_f = (n - s - b)E$. In particular, $\bar{R}_f = 0$ if and only if C is an irreducible component of $\Gamma_1 \cap \cdots \cap \Gamma_s$ with $b = n - s$.

c. Let $\bar{V}$ be a smooth completion of V with a smooth boundary D. Suppose that $\bar{P}_m(V) \neq 0$. Then one has the rational map $\Phi_{m,V} \colon \bar{V} \to \mathbf{P}(m(K(\bar{V}) + D))$ associated with $|m(K(\bar{V}) + D)|$. Let $\bar{W}$ be a smooth completion of W with a smooth boundary B and let $f \colon V \to W$ be a dominating morphism which defines a rational map $\bar{f} \colon \bar{V} \to \bar{W}$. Suppose that $\dim V = \dim W = n$ and $\bar{P}_m(V) = \bar{P}_m(W)$. Then $\bar{f}^* \colon T_{\{m\}}(W) \to T_{\{m\}}(V)$ is an isomorphism, which induces an isomorphism $\mathbf{P}(m(K(\bar{W}) + B)) \to \mathbf{P}(m(K(\bar{V}) + D))$, denoted by $p_m(f)$. Since

$$m(K(\bar{V}) + D) \sim f^*(m(K(\bar{W}) + B)) + m\bar{R}_f,$$

one has

$$|m(K(\bar{V}) + D)| \supseteq \bar{f}^* |m(K(\bar{W}) + B)| + m\bar{R}_f.$$

Since $\bar{P}_m(V) = \bar{P}_m(W)$, one obtains

$$|m(K(\bar{V}) + D)| = \bar{f}^* |m(K(\bar{W}) + B)| + m\bar{R}_f.$$

Thus, the following square is commutative:

$$\begin{array}{ccc}
\bar{V} & \xrightarrow{\bar{f}} & \bar{W} \\
\Big\downarrow {\scriptstyle \Phi_{m,V}} & \circlearrowleft & \Big\downarrow {\scriptstyle \Phi_{m,W}} \\
 & {\scriptstyle p_m(f)} & \\
\mathbf{P}(m(K(\bar{V}) + D)) & \simeq & \mathbf{P}(m(K(\bar{W}) + B))
\end{array}$$

Accordingly, one obtains the following result.

Proposition 11.9. *If* $\dim V = \dim W$ *and* $\bar{P}_m(V) = \bar{P}_m(W) > 0$, *then*

$$p_m(f) \circ \Phi_{m,V} = \Phi_{m,W} \circ \bar{f}.$$

Now, let $f \colon V \to W$ be a strictly rational map, which defines a rational map $\bar{f} \colon \bar{V} \to \bar{W}$. Then, there exist a nonsingular complete variety $\bar{Z}$, and a proper birational morphism $\bar{\mu} \colon \bar{Z} \to \bar{V}$ such that (1) $\Delta = \bar{\mu}^{-1}(\bar{V} - V)$ is a divisor with simple normal crossings, and (2) $\bar{\varphi} = \bar{f} \circ \bar{\mu} \colon \bar{Z} \to \bar{V}$ is a morphism.

Since f is a strictly rational map, $\bar{\varphi}(\bar{\mu}^{-1}(V))$ is contained in W; hence $\varphi = \bar{\varphi}|_{\bar{\mu}^{-1}(V)} \colon \bar{\mu}^{-1}(V) \to W$ is a morphism. Applying the previous argument to $\mu = \bar{\mu}|_{\bar{\mu}^{-1}(V)}$ and φ, one has the following commutative diagram:

$$\begin{array}{ccccc} \bar{V} & \xleftarrow{\bar{\mu}} & \bar{Z} & \xrightarrow{\bar{\varphi}} & \bar{W} \\ \downarrow & & \downarrow & & \downarrow \\ \mathbf{P}_k^l & \xleftarrow{p_m(\mu)} & \mathbf{P}_k^l & \xrightarrow{p_m(\varphi)} & \mathbf{P}_k^l \end{array}$$

Here l denotes $\bar{P}_m(V) - 1$.

Putting $F = \Phi_{m,V}(D)$, $F' = \Phi_{m,W}(B)$, and $F'' = \Phi_{m,Z}(\Delta)$, where $Z = \bar{Z}\backslash\Delta$, one has $B \subseteq \bar{\varphi}(\Delta)$ and so $F' = \Phi_{m,W}(B) \subseteq \Phi_{m,W} \circ \bar{\varphi}(\Delta) = p_m(\varphi) \circ \Phi_{m,Z}(\Delta) = p_m(\varphi)(F'')$. Furthermore, since $\bar{\mu}(\Delta) = D$, one has $F = \Phi_{m,V}(D) = p_m(\mu)(F'')$. $p_m(f) = p_m(\varphi) \circ p_m(\mu)^{-1}$ is an isomorphism and $p_m(f)(F) = p_m(\varphi)(F'') \supseteq F'$.

Now, suppose that $V = W$ and $\bar{V} = \bar{W}$. Then $F = F'$; hence $p_m(f)(F) \supseteq F$.

Lemma 11.7. *Let X be a Noetherian space and let F be a closed subset of X. If a homeomorphism $h: X \to X$ satisfies $h(F) \supseteq F$, then $h(F) = F$.*

Proof. Let $h_1 = h^{-1}$. Then $F \supseteq h_1(F)$; hence one has an infinite sequence:

$$F \supseteq h_1(F) \supseteq h_1^2(F) \supseteq \cdots .$$

Since X is Noetherian, there exists $r > 0$ such that $h_1^r(F) = h_1^{r+1}(F)$, and so $h_1(F) = F$, i.e., $h(F) = F$. □

By this lemma, one obtains $p_m(f)(F) = F$, which implies that $p_m(f)$ gives rise to an automorphism of $\Phi_{m,V}(\bar{V})\backslash F$.

When $\bar{\kappa}(V) = \dim V$, letting $\bar{Y}_m$ be $\Phi_{m,V}(\bar{V})$, one has $m > 0$ such that $\Phi_{m,V}: \bar{V} \to \bar{Y}_m$ is birational. Then $F \not\supseteq \bar{Y}_m$; hence $Y_m = \bar{Y}_m\backslash F$ is not empty. By the above observation, $p_m(f)|_{\bar{Y}_m}$ is an automorphism denoted by $\beta_m(f)$. β_m gives rise to a group homomorphism from SBir(V) into $L\langle Y_m\rangle$, which is defined to be $\{\sigma \in \mathrm{PGL}(T_{\{m\}}(V)) \mid \sigma(\bar{Y}_m) = \bar{Y}_m \text{ and } \sigma(F) = F\}$.

d. Theorem 11.6. *If $\bar{\kappa}(V) = \dim V = n$, then any dominating strictly rational map $f: V \to V$ is birational.*

Proof. One can assume that V is nonsingular, and admits a smooth completion $\bar{V}$ with a smooth boundary D. Applying the last proposition, one has the isomorphism $p_m(f)$: $\mathbf{P}(m(K(\bar{V}) + D)) \to \mathbf{P}(m(K(\bar{V}) + D))$ such that $\Phi_{m,V} \circ \bar{f} = p_m(f) \circ \Phi_{m,V}$ for any $m > 0$. Choose $m \gg 0$ such that $\bar{V} \to \Phi_{m,V}(\bar{V})$ is birational. Since $p_m(f)$ gives rise to an isomorphism between $\Phi_{m,V}(\bar{V})$ and $\Phi_{m,V}(\bar{V})$, one concludes that $\bar{f}$ is also birational. □

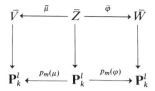

§11.5 Étale Endomorphisms

a. Let $\mu: Z \to V$ be a birational morphism between normal complete varieties. For a prime divisor Γ on Z, define $\mu_*(\Gamma) = \mu(\Gamma)$ if $\mu(\Gamma)$ is a divisor, and $\mu_*(\Gamma) = 0$ if $\mu(\Gamma)$ is not a divisor, i.e., if $\operatorname{codim}(\mu(\Gamma)) \geq 2$. For a general divisor $D = \sum_{i=1}^{s} m_i \Gamma_i$, where the Γ_i are prime divisors and the $m_i \in \mathbb{Z}$, we define $\mu_* D$ to be $\sum_{i=1}^{s} m_i \mu_*(\Gamma_i)$.

Lemma 11.8. *If $D_1 \sim D_2$, then $\mu_* D_1 \sim \mu_* D_2$.*

PROOF. We can assume that $D_2 = 0$. Then $D_1 = \operatorname{div}(\varphi)$ for some $\varphi \in \operatorname{Rat}(Z)$. Since $\mu: Z \to V$ is birational, one has $\psi \in \operatorname{Rat}(V)$ with $\varphi = \mu^*(\psi)$. If $\mu(\Gamma)$ is a divisor Δ, then $\Delta \cap \operatorname{dom}(\mu^{-1}) \neq \varnothing$ and hence $\mathcal{O}_{Z,v} \cong \mathcal{O}_{V,\mu(v)}$ where v is the generic point of Γ. Therefore, $\operatorname{ord}_\Gamma(\mu^*(\psi)) = \operatorname{ord}_\Delta(\psi)$. Since $\operatorname{div}(\mu^*(\psi)) = \sum_{\mu_*(\Gamma) \neq 0} \operatorname{ord}_\Gamma(\mu^*(\psi))\Gamma + \sum_{\mu_*(E)=0} \operatorname{ord}_E(\mu^*(\psi))E$, one has

$$\mu_*(\operatorname{div}(\mu^*(\psi))) = \sum_\Gamma \operatorname{ord}_\Gamma(\mu^*\psi)\Delta = \sum_\Delta \operatorname{ord}_\Delta(\psi)\Delta = \operatorname{div}(\psi),$$

i.e., $\mu_* D_1 \sim 0$. □

Lemma 11.9. *If Z and V are nonsingular complete varieties and $\mu: Z \to V$ is a birational morphism, then $\mu_*(K(Z)) \sim K(V)$.*

PROOF. This follows immediately from the ramification formula (Theorem 5.5). □

Definition. Let $f: V \to W$ be a strictly rational map between normal complete varieties. Take a normal reduction model (Z, μ) of f such that μ is proper. Given a divisor D on W, define f^*D to be $\mu_*(f \circ \mu)^*D$.

Then if $D_1 \sim D_2$ and W is nonsingular, one has $f^*D_1 \sim f^*D_2$ by Lemma 11.8.

Remark. Let $\mu: Z \to V$ be a birational morphism between nonsingular complete varieties. If Γ is a prime divisor on V and $\mu(\Gamma)$ is a divisor Δ, i.e., $\Delta = \mu_*(\Gamma)$, then $\mu_*(\mathcal{O}(\Gamma)) \subseteq \mathcal{O}(\Delta)$. In general, $\mu_*(\mathcal{O}(\Gamma)) \neq \mathcal{O}(\Delta)$. If E is an effective divisor exceptional with respect to μ, then $\mu_*(\mathcal{O}(\mu^*\Delta + E)) = \mathcal{O}(\Delta)$ by §2.20.**f**.

b. Let V be a nonsingular variety and let $f: V \to V$ be a dominating morphism. Take a smooth completion $\bar{V}$ of V with smooth boundary D and let $\bar{f}$: $\bar{V} \to \bar{V}$ be a rational map represented by $f: V \to V \subseteq \bar{V}$. One has a nonsingular complete variety $\bar{Z}$ and a proper birational morphism $\mu: \bar{Z} \to \bar{V}$ such that (1) $\mu|_{\mu^{-1}(V)}$ is the identity, (2) $\Delta = \mu^{-1}(D)$ has only simple normal crossings, and (3) $\bar{\varphi} = \bar{f} \circ \mu: \bar{Z} \to \bar{V}$ is a morphism.

Letting $Z = \mu^{-1}(V)$ and $\Delta = \mu^{-1}(D)$, one has the logarithmic ramification formula for $\varphi = \bar{\varphi}|_Z: Z = V \to V$; i.e.,

$$K(\bar{Z}) + \Delta \sim \bar{\varphi}^*(K(\bar{V}) + D) + \bar{R}_\varphi.$$

Since $\mu_* K(\bar{Z}) \sim K(\bar{V})$ by Lemma 11.9 and $\mu_*(\Delta) = D$, letting $\bar{R}_f$ be $\mu_* \bar{R}_\varphi$, one has the formula:

$$K(\bar{V}) + D \sim \bar{f}^*(K(\bar{V}) + D) + \bar{R}_f.$$

Proposition 11.10. *If there exists $m > 0$ such that $m(K(\bar{V}) + D) \sim 0$, then $R_f = 0$, i.e., f is étale.*

PROOF. By the above formula, one has $m\bar{R}_f \sim 0$. But since $\bar{R}_f$ is effective, this implies $m\bar{R}_f = 0$; hence $R_f = \bar{R}_f|_V = 0$. □

Theorem 11.7. *Suppose that $f: V \to V$ is a proper dominating morphism and $\bar{\kappa}(V) \geq 0$. Then f is an étale covering.*

PROOF. One has m such that $|m(K(\bar{V}) + D)| \neq \emptyset$. Then choosing Γ_m from this linear system, one has

$$\Gamma_m \sim \bar{f}^*\Gamma_m + m\bar{R}_f.$$

Therefore,

$$\Gamma_m \sim \bar{f}^{*2}\Gamma_m + m\bar{f}^*\bar{R}_f + m\bar{R}_f \sim \cdots \sim \bar{f}^{*s}\Gamma_m + m\left(\sum_{i=0}^{s-1} \bar{f}^{*i}\bar{R}_f\right)$$

for any $s \geq 2$. Thus

$$|\Gamma_m| = |\bar{f}^{*s}\Gamma_m + m \sum_{i=0}^{s-1} \bar{f}^{*i}\bar{R}_f| \supseteq \bar{f}^{*s}|\Gamma_m| + m \sum_{i=0}^{s-1} \bar{f}^{*i}\bar{R}_f.$$

Since $\dim(\bar{f}^*|\Gamma_m|) = \dim|\Gamma_m|$, one has

$$|\Gamma_m| = \bar{f}^{*s}|\Gamma_m| + m \sum_{i=0}^{s-1} \bar{f}^{*i}\bar{R}_f.$$

This implies $|\Gamma_m|_{\text{fix}} \geq m \sum_{i=0}^{s-1} \bar{f}^{*i}\bar{R}_f$. Thus, there exists s_0 such that $\bar{f}^{*s}\bar{R}_f = 0$ for all $s \geq s_0$. Since $f = \bar{f}|_V: V \to V$ is proper, $f: V \to V$ is surjective. Since $\bar{f}^{*s}\bar{R}_f|_V = f^{*s}R_f = 0$, one has $R_f = 0$ and hence f is étale. □

EXAMPLE 11.7. Let $A = k[\mathsf{X}, \mathsf{Y}, (\mathsf{XY} - 1)^{-1}]$ and let $\varphi: A \to A$ be a k-homomorphism. $f = {}^a\varphi$ is dominating if and only if φ is one-to-one. In this case $R_f = 0$, since Spec A has a smooth completion $\mathbf{P}_k^2$ with smooth boundary $D = V_+(\mathsf{T}_0) + V_+(\mathsf{T}_0^2 - \mathsf{T}_1\mathsf{T}_2)$ which satisfies $K(\mathbf{P}_k^2) + D \sim 0$. Note that $(\mathsf{X} - a, \mathsf{Y} - b)$ is a coordinate system around (a, b). Thus R_f is defined by $\delta = \partial(\varphi(\mathsf{X}), \varphi(\mathsf{Y}))/\partial(\mathsf{X}, \mathsf{Y})$ as $V(\delta)$. Since $R_f = 0$, one has $V(\delta) = \emptyset$ in Spec A, which implies that δ is a unit of A, i.e., $\delta = c(\mathsf{XY} - 1)^m$ for some $m \in Z$ and $c \in k\backslash\{0\}$.

§11.6 Logarithmic Canonical Fibered Varieties

a. Given a variety V with $\bar{\kappa}(V) \geq 0$, we let $\bar{V}$ be a completion of V and let $(\bar{V}^{\#}, \mu)$ be a nonsingular model of $\bar{V}$ such that $D^{\#} = \mu^{-1}(\bar{V} \backslash V)$ has only simple normal crossings. Applying Theorem 10.3 to $K(\bar{V}^{\#}) + D^{\#}$, one has a nonsingular complete variety $\bar{Z}$ and a birational morphism $\lambda: \bar{Z} \to \bar{V}^{\#}$ such that (1) $\Delta = \lambda^{-1}(D^{\#})$ has only simple normal crossings, and (2) $\bar{f} = \Phi_{m, Z}: \bar{Z} \to \bar{Y}_m = \Phi_{m, Z}(\bar{Z})$ is a $(K(\bar{Z}) + \Delta)$-canonical fibered variety for $m = rm_0$, where $r \gg 0$, $\bar{P}_{m_0}(V) > 0$, and $Z = \bar{Z} \backslash \Delta$. By Proposition 11.1, for a general point x of $\bar{Y}_m$, letting $\bar{Z}_x = \bar{f}^{-1}(x)$, $\Delta_x = \Delta \cap \bar{Z}_x$ is a divisor with simple normal crossings on $\bar{Z}_x$, i.e., $Z_x = \bar{Z}_x \backslash \Delta_x$ has a smooth completion $\bar{Z}_x$ with smooth boundary Δ_x. Thus by Lemma 10.6, $\bar{P}_m(Z_x) = \dim_k \Gamma(\bar{Z}_x, \mathcal{O}(m(K(\bar{Z}) + \Delta))|_{\bar{Z}_x})$. Hence, if x is a "general" point of $\bar{Y}_m$, then

$$\bar{\kappa}(Z_x) = \kappa((K(\bar{Z}) + \Delta)|_{\bar{Z}_x}, \bar{Z}_x) = 0$$

by Theorem 10.3. Letting $V^* = Z, f = \bar{f}|_{V*}, \mu_1 = \mu \circ \lambda|_Z$, and $W = \bar{Y}_m$, one has a fibered variety $f: V^* \to W$ such that $\bar{\kappa}(f^{-1}(x)) = 0$ for any "general" points x of W.

Theorem 11.8. *Given V with $\bar{\kappa}(V) \geq 0$, there exist V^*, a proper birational morphism $\mu_1: V^* \to V$, and a fibered variety $f: V^* \to W$ such that*

(1) $\dim W = \bar{\kappa}(V)$,
(2) $\bar{\kappa}(f^{-1}(x)) = 0$ *for any "general" points x of W.*

Such an $f: V^* \to W$ is said to be a *logarithmic canonical fibered variety.*

b. Similarly one can prove the following "easy addition" for logarithmic Kodaira dimension.

Theorem 11.9. *Let $f: V \to W$ be a fibered variety. Then for a general point x of W,*

$$\bar{\kappa}(V) \leq \bar{\kappa}(f^{-1}(x)) + \dim W.$$

The covering theorem can be formulated as follows.

Theorem 11.10. *Let $f: V \to W$ be an étale covering between nonsingular varieties. Then $\bar{\kappa}(V) = \bar{\kappa}(W)$.*

Proof. One can choose smooth completions $\bar{V}$ of V with a smooth boundary D and $\bar{W}$ of W with a smooth boundary B such that $\bar{f}: \bar{V} \to \bar{W}$ defined by f is a morphism. Then, by Lemma 11.2, $\bar{f}^{-1}(B)$ coincides with D. Since $\bar{R}_f \cap V = R_f = 0$, $\mathrm{supp}(\bar{R}_f)$ is contained in D. We claim that $\bar{R}_f$ is exceptional with respect to $\bar{f}$. Let Γ be an irreducible component of D which is not exceptional with respect to $\bar{f}$, i.e., $\bar{f}(\Gamma) = \Delta$ is a divisor contained in B. $h = \bar{f}|_\Gamma$ is étale at a general point p of Γ by Theorem 7.17. Γ and Δ are nonsingular at

p and $p_1 = \bar{f}(p)$, respectively. Choose local coordinate systems $(z_\lambda^1, \ldots, z_\lambda^n)$ and $(w_\alpha^1, \ldots, w_\alpha^n)$ around p and p_1, respectively, in such a way that $U_\lambda \cap V(z_\lambda^1) = U_\lambda \cap \Gamma$ and $W_\alpha \cap V(w_\alpha^1) = W_\alpha \cap \Delta$ where U_λ and W_α are the coordinate neighborhoods. Since $\bar{f}^{-1}(\Delta) \cap U_\lambda = \Gamma \cap U_\lambda$, $\bar{f}^*(w_\alpha^1) = \varepsilon(z_\lambda^1)^\nu$ for some $\nu > 0$ and $\varepsilon \in \Gamma(U_\lambda, \mathcal{O}_{\bar{V}}^*)$. Since $U_\lambda \cap V(z_\lambda^1) \to W_\alpha \cap V(w_\alpha^1)$ is étale, letting $\bar{z}_\lambda^i = z_\lambda^i|_\Gamma$ and $\bar{w}_\alpha^j = w_\alpha^j|_\Delta$, one has $h^*(d\bar{w}_\alpha^2 \wedge \cdots \wedge d\bar{w}_\alpha^n) = \bar{\eta}\, d\bar{z}_\lambda^2 \wedge \cdots \wedge d\bar{z}_\lambda^n$ for some nonvanishing $\bar{\eta} \in \Gamma(U_\lambda \cap \Gamma, \mathcal{O}^*)$. Hence,

$$\begin{aligned}\bar{f}^*\left(\frac{dw_\alpha^1}{w_\alpha^1} \wedge dw_\alpha^2 \wedge \cdots \wedge dw_\alpha^n\right) &= \left(\nu\,\frac{dz_\lambda^1}{z_\lambda^1} + \frac{d\varepsilon}{\varepsilon}\right) \wedge \bar{f}^*(dw_\alpha^2 \wedge \cdots \wedge dw_\alpha^n) \\ &= \nu\eta\,\frac{dz_\lambda^1}{z_\lambda^1} \wedge \cdots \wedge dz_\lambda^n,\end{aligned}$$

where $\eta \in \Gamma(U_\lambda, \mathcal{O})$ is such that $\eta|_\Gamma = \bar{\eta}$, which is nonvanishing at p. Thus $p \notin \bar{R}_f$, i.e., $\Gamma \not\subseteq \bar{R}_f$. By Theorem 11.5, one now has

$$K(\bar{V}) + D \sim \bar{f}^*(K(\bar{W}) + B) + \bar{R}_f.$$

Since $\bar{R}_f$ is exceptional with respect to $\bar{f}$, one has by Theorem 10.5,

$$\begin{aligned}\bar{\kappa}(V) &= \kappa(K(\bar{V}) + D, \bar{V}) = \kappa(\bar{f}^*(K(\bar{W}) + B) + \bar{R}_f, \bar{V}) \\ &= \kappa(K(\bar{W}) + B, \bar{W}) = \bar{\kappa}(W).\end{aligned}$$

□

The following result is the uniqueness theorem for $\bar{\kappa}$.

Theorem 11.11. *Let $\psi: V^\# \to Y$ be a fibered variety such that*

(1) *$V^\#$ is proper birationally equivalent to V,*
(2) *$\dim Y = \bar{\kappa}(V)$,*
(3) *for any "general" points x of Y, $\bar{\kappa}(\psi^{-1}(x)) = 0$.*

If $f: V^* \to W$ is a logarithmic canonical fibered variety of V, then there exist a proper birational map $\mu: V^\# \to V^*$ and a birational morphism $\rho: Y \to W$ such that $\rho \circ \psi = f \circ \mu$.

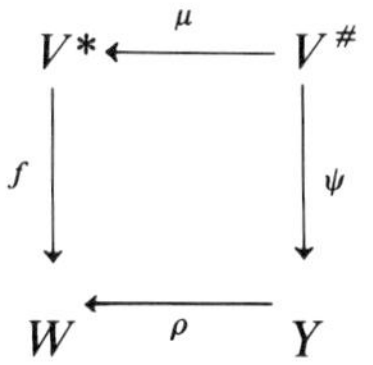

Proof. Left to the reader. □

§11.7 Finiteness of the Group SBir(V)

a. An algebraic group G is said to be *affine* if G is an affine variety. A connected affine algebraic group G of dimension 1 is either $\mathbf{A}_k^1$ or G_m (cf. [B, Theorem 10.9]). Here, we shall use the following simple results on algebraic groups.

Lemma 11.10. *Any nontrivial connected affine algebraic group contains* $\mathbf{A}_k^1$ *or* G_m.

PROOF. Left to the reader. □

Lemma 11.11. *If an algebraic group G of dimension 1 acts on a variety V, there exists a G-admissible open dense subset V_0 of V such that the quotient variety V_0/G exists.*

PROOF. Left to the reader. □

Using these, we have the following result.

Proposition 11.11. *If a nontrivial connected affine algebraic group G acts effectively on a variety V, then $\bar{\kappa}(V) < \dim V$.*

PROOF. By Lemma 11.10, one can assume that G is $\mathbf{A}_k^1$ or G_m. Then one has an open dense set V^0 and a quotient variety $W = V^0/G$ by Lemma 11.11. If x is a general point of V, then the orbit $G \cdot x$ is a general fiber of $V^0 \to W$. However, $\bar{\kappa}(G \cdot x) \leq \bar{\kappa}(G) = -\infty$ or 0. Hence by Theorem 11.9, one has $\bar{\kappa}(V^0) \leq \bar{\kappa}(G \cdot x) + \dim W \leq \dim W < \dim V$. Since $V^0 \subseteq V$, it follows that $\bar{\kappa}(V) \leq \bar{\kappa}(V^0)$ and hence $\bar{\kappa}(V) < \dim V$. □

b. The next result is one of the main results.

Theorem 11.12. *If $\bar{\kappa}(V) = \dim V$, then* SBir(V) *is a finite group.*

PROOF. Since SBir(V) is a proper birational invariant, one can assume V is nonsingular, and that it admits a smooth completion $\bar{V}$ with smooth boundary D. Using the notation in §11.4, choose $m \gg 0$ such that $\Phi_{m,V}: \bar{V} \to \bar{Y}_m$ is birational. Then, one can also assume that $\Phi_{m,V}$ is a morphism $\bar{f}$. Hence, by §11.4.c, one has a group homomorphism β_m: SBir(V) $\to L\langle Y_m\rangle$. The set $V_0 = \bar{V}\backslash\bar{f}^{-1}(F)$ is contained in V and $V_0 \to Y_m = \bar{Y}_m - F$ is proper birational. Hence, $\bar{\kappa}(V) \leq \bar{\kappa}(V_0) = \bar{\kappa}(Y_m) \leq \dim Y_m = \dim V$, which implies $\bar{\kappa}(Y) = \dim Y_m$.

$L\langle Y_m\rangle$ is clearly an affine algebraic group, and let G denote its connected component. Then the quotient group $L\langle Y_m\rangle/G$ is finite. Suppose that G is not trivial. Then $\bar{\kappa}(Y) < \dim Y_m$ by Proposition 11.11, which contradicts the above result. Therefore, $G = \{1\}$ and so $L\langle Y_m \rangle$ is a finite group. Since $\bar{f}: \bar{V} \to \bar{Y}_m$ is birational, Ker β_m is trivial and hence $\text{SBir}(V) \subseteq L\langle Y_m \rangle$. □

Corollary. *If* $\bar{\kappa}(V) = \dim V$, *then* SBir(V) *coincides with* PBir(V).

Proof. Let $f: V \to V$ be a birational and strictly rational map. Then $f \in$ SBir(V) and f has finite order, since SBir(V) is a finite group by the last theorem. There exists $r > 0$ such that $f^r = \text{id}$ and so $f^{-1} = f^{r-1}$, which is a strictly rational map. Hence by Proposition 2.17.(i), f is proper. □

c. Let A be an integral domain finitely generated over k.

Theorem 11.13. *Suppose that A is normal and* $\bar{\kappa}(\text{Spec } A) = \dim A$. *Then any one-to-one k-endomorphism φ of A is an isomorphism.*

Proof. By Theorem 11.6, $f = {}^a\varphi: \text{Spec } A \to \text{Spec } A$ is a birational morphism. Hence $f \in \text{PBir}(\text{Spec } A)$ by the last corollary and then $f \in \text{Aut}_k(A)$ by the Corollary to Theorem 2.22.

§11.8 Some Applications

a. As an application of Theorem 11.10 (the covering theorem), we shall prove the following result.

Theorem 11.14. *Let $f: \bar{V} \to \mathbf{P}_k^n$ be a surjective morphism from a nonsingular complete variety $\bar{V}$ of dimension n and let B be the codimension 1 part of* $f(R_f)$. If $\bar{\kappa}(\mathbf{P}_k^n \backslash B) < n$, *then* $\kappa(\bar{V}) = -\infty$.

Proof. Let $D = f^{-1}(f(R_f))$. Then $V = \bar{V} \backslash D \to \mathbf{P}_k^n \backslash f(R_f)$ is an étale covering and hence by Theorem 11.10 $\bar{\kappa}(V) = \bar{\kappa}(\mathbf{P}_k^n \backslash f(R_f))$, which is $\bar{\kappa}(\mathbf{P}_k^n \backslash B) < n$.

Choose a nonsingular complete variety $\bar{V}^{\#}$ and a birational morphism μ: $\bar{V}^{\#} \to \bar{V}$ such that $D^{\#} = \mu^{-1}(D)$ has only simple normal crossings. Then $\bar{\kappa}(V) = \kappa(K(\bar{V}^{\#}) + D^{\#}, \bar{V}^{\#})$. Let $D_0 = f^{-1}(B)$ and $D_0^{\#} = \mu^{-1}(D_0)$. Then there exists $N > 0$ such that $(f \circ \mu)^* B \leq N D_0^{\#}$. Define $D_1^{\#}$ to be $D^{\#} - D_0^{\#}$, which is exceptional with respect to $f \circ \mu$. Therefore, supposing $\kappa(\bar{V}) \geq 0$, one has $\bar{\kappa}(V) = \kappa(K(\bar{V}^{\#}) + D^{\#}, \bar{V}^{\#}) = \kappa(K(\bar{V}^{\#}) + N D_0^{\#} + D_1^{\#}, \bar{V}^{\#}) \geq \kappa((f \circ \mu)^*(B) + D_1^{\#}, \bar{V}^{\#}) \geq \kappa(B, \mathbf{P}_k^n) = n$, if $B \neq 0$. Assume $B = 0$. Then codim $f(R_f) \geq 2$ and so $\bar{\kappa}(V) = \bar{\kappa}(\mathbf{P}_k^n \backslash f(R_f)) = -\infty$. This contradicts the hypothesis $\bar{\kappa}(V) \geq \kappa(\bar{V}) \geq 0$. □

Let $F(\mathsf{X}_1, \ldots, \mathsf{X}_n)$ be a polynomial in variables $\mathsf{X}_1, \ldots, \mathsf{X}_n$.

Corollary. *If* $\bar{\kappa}(\text{Spec } k[\mathsf{X}_1, \ldots, \mathsf{X}_n, F^{-1}]) < n$, *then the field* $Q(k[\mathsf{X}_1, \ldots, \mathsf{X}_n, \mathsf{Y}]/(\mathsf{Y}^m - F))$ *has Kodaira dimension* $-\infty$ *for any* m.

PROOF. This can be proved easily using the last theorem. Details are left to the reader. □

b. The following theorem has been proved by Kawamata.

Theorem 11.15. *Let $f: V \to W$ be a fibered variety such that a general fiber $f^{-1}(x)$ is a curve. If x is a "general" point, then $\bar{\kappa}(V) \geq \bar{\kappa}(f^{-1}(x)) + \bar{\kappa}(W)$.*

PROOF. We refer the reader to [Kawamata(1)]. □

Now, let F be a non-constant polynomial in variables X, Y, and let S be $D(F)$ in $\mathbf{A}_k^2$, i.e., $S = \operatorname{Spec} k[\mathsf{X}, \mathsf{Y}, F^{-1}]$. F gives rise to a morphism f: $S \to G_m \subset \mathbf{A}^1$ such that $f^*(\lambda) = V(F - \lambda)$ in S, for all $\lambda \in G_m$. Consider the case when a general fiber $f^{-1}(\lambda)$ is irreducible.

First, suppose that $\bar{\kappa}(S) = -\infty$. Then since $\bar{\kappa}(G_m) = 0$, one has by Theorem 11.15,

$$-\infty = \bar{\kappa}(S) \geq \bar{\kappa}(f^{-1}(\lambda)) + \bar{\kappa}(G_m) = \bar{\kappa}(f^{-1}(\lambda)).$$

Hence, $f^{-1}(\lambda) \cong \mathbf{A}_k^1$. Therefore $\{V(F - \lambda) \mid \lambda \in G_m\}$ is a linear system whose general members $V(F - \lambda) \cong \mathbf{A}^1$. Hence, applying Theorem 9.4, one has a new affine coordinate system ξ, η, i.e., $k[\mathsf{X}, \mathsf{Y}] = k[\xi, \eta]$, such that $F = \eta$, which implies that $S = \mathbf{A}_k^1 \times G_m$.

Next, suppose that $\bar{\kappa}(S) = 0$. Then by a similar argument, one has $\bar{\kappa}(f^{-1}(\lambda)) = 0$ and hence $f^{-1}(\lambda) \cong G_m$. In [Iitaka (13)] it is proved that a new affine coordinate system (ξ, η) exists in which F can be written in the form $\xi^l\eta + a_0 + a_1\xi + \cdots + a_s\xi^s$, where $l > 0$, $s \geq 0$, and $a_i \in k$.

Finally, suppose $\bar{\kappa}(S) = 1$. By Theorem 11.8, one has the logarithmic canonical fibered variety $\psi: S \to \Gamma$. One can assume ψ is surjective. Then the general fibers $\psi^{-1}(x)$ are isomorphic to G_m. Thus, there exists a polynomial H such that ψ corresponds to H/F^m for some $m \geq 0$. It is very desirable to seek a new affine coordinate system (ξ, η) by which F and H could be expressed in a simple form.

If a general fiber $f^{-1}(\lambda)$ is reducible, then there exist an irreducible polynomial $F_1 \in k[\mathsf{X}, \mathsf{Y}]$ and a polynomial Ψ such that $F = \Psi(F_1)$ (see Example 11.4). For a general $\lambda \in k$, one has $F - \lambda = \prod_{i=1}^r (F_1 - \lambda_i)$ where $\Psi(\mathsf{T}) - \lambda = \prod_{i=1}^r (\mathsf{T} - \lambda_i)$. Thus $f^{-1}(\lambda) = \sum_{i=1}^r f_1^{-1}(\lambda_i)$, where f_1 is the morphism: $\mathbf{A}^2 \to \mathbf{A}^1$ corresponding to F_1. In particular, if $\bar{\kappa}(\operatorname{Spec} k[\mathsf{X}, \mathsf{Y}, F^{-1}]) = -\infty$, then $F = \prod_{i=1}^r (\eta - \lambda_i)$, where (ξ, η) is an affine coordinate system.

EXERCISES

11.1. Let V be a closed subvariety of G_m^N over $\mathbb{C}$ with $\dim V = n$. Show that
   (1) $\bar{\kappa}(V) \geq 0$, $\bar{p}_g(V) \geq 1$, and $\bar{q}(V) \geq n$.
   (2) The following conditions are equivalent.
      (a) $V \cong G_m^n$ and $V \times G_m^{N-n} \cong G_m^N$.
      (b) $\bar{\kappa}(V) = 0$.
      (c) $\bar{p}_g(V) = 1$.
      (d) $\bar{q}(V) = n$.

(3) If $\bar{\kappa}(V) = m > 0$, then V is isomorphic to a product $W \times G_m^{n-m}$, where W is a closed subvariety of some G_m^{N+m-n} with $\bar{\kappa}(W) = m = \dim W$.

11.2. Let D be a reduced divisor on G_m^n and let V be $G_m^n \backslash D$. Show that
(1) $\bar{\kappa}(V) \geq 0$, $\bar{p}_g(V) \geq 1$, and $\bar{q}(V) \geq n$.
(2) The following conditions are equivalent:
(a) $D = \varnothing$, i.e., $V = G_m^n$.
(b) $\bar{\kappa}(V) = 0$.
(c) $\bar{p}_g(V) = 1$.
(d) $\bar{q}(V) = n$.

11.3. Let $H_0, \ldots, H_t$ be hyperplanes on $\mathbf{P}_k^n$ and let V be $\mathbf{P}_k^n \backslash \bigcup_{j=0}^t H_j$. Show that $\bar{q}(V) = t$. In addition, show that $V \cong \mathbf{A}^\alpha \times G_m^\beta \times W$, where W is a variety of hyperbolic type, realized as the complement of a finite union of hyperplanes in $\mathbf{P}_k^{n+\alpha-\beta}$.

11.4. Let V be a variety which has an open subset isomorphic to G_m^n, and suppose that $\bar{q}(V) = n$. Show that $V \cong G_m^n$.

11.5. (Ueno) Let V be a variety with $\dim V = n$, $\bar{p}_g(V) = 1$, and $\bar{q}(V) = n$. If there exists a dominating morphism $f: V \to G_m^n$, then show that $\bar{q}_i(V) = \binom{n}{i}$ for all i.

11.6. Let $L_0, \ldots, L_t$ be lines on $\mathbf{P}_k^2$ and let $\sum$ be a set of points where $\sum_{i=0}^t L_i$ does not have normal crossings, i.e., $p \in \sum$ if and only if $\#\{j \mid L_j \ni p\} \geq 3$. Let $\mu: S \to \mathbf{P}_k^2$ be the composition of all quadric transformations with centers $p \in \sum$. Then show that
(i) $D = \mu^{-1}(\sum_{j=0}^t L_j)$ has only simple normal crossings.
(ii) $K(S) + D$ is ample if and only if S is not a product of curves.

11.7 Let $l_1, \ldots, l_t$ be linear polynomials in X, Y over $\mathbb{C}$ such that $V(l_i) \neq V(l_j)$ for any $i \neq j$. Then compute $\dim_{\mathbb{C}} \sum_{i \neq j} \mathbb{C}\, dl_i \wedge dl_j / l_i l_j$, denoted p. Letting $S = \mathbf{A}_{\mathbb{C}}^2 \backslash \bigcup_{i=1}^t V(l_i)$, and letting r_m denote the number of m-ple points of $\sum_{i=1}^t V(l_i)$, prove that

$$p = \bar{p}_g(S) \quad \text{and} \quad p = \sum_{m=2}^{\infty} (m-1) r_m.$$

11.8. Suppose that there exists a dominating morphism $G_m^n \to V$. Show that $\bar{\kappa}(V) \leq 0$.

11.9. Let g be an irreducible element of $\mathbb{C}[\mathsf{X}_1, \ldots, \mathsf{X}_n, \mathsf{X}_1^{-1}, \ldots, \mathsf{X}_n^{-1}]$ and let V be $G_m^n \backslash V(g)$. Then $\bar{q}(V) = n + 1$. Supposing that $\bar{\kappa}(V) = n$, prove that $\bar{q}_i(V) \geq (n+1)n \cdots (n-i)/i!$ for $1 \leq i \leq n$, and so, in particular, $\bar{p}_g(V) \geq n + 1$.

In Exercises 11.10–11.15, let $\bar{S}$ denote a complete rational non-singular surface.

11.10. If C is a curve on $\bar{S}$ and S is a surface $\bar{S} \backslash C$, then show that $\bar{p}_g(S) = g(C) + \sum_{p \in C} (s_p - 1)$, where s_p was defined in §9.4.**c**, called the number of analytic branches of C at p.

11.11. Suppose that C is a rational curve on $\bar{S}$ and all singular points on C are cusps. If r is the number of singular points of C, then show that $\bar{P}_2(S) \geq r - 1$, where $S = \bar{S} \backslash C$.

11.12. Let C_1 and C_2 be rational curves on $\bar{S}$ such that $(\mathrm{Sing}\ C_1) \cup (\mathrm{Sing}\ C_2)$ consists of more than one point, and let $S = \bar{S} \backslash (C_1 \cup C_2)$. Show that $\bar{P}_2(S) \geq 1$ and $\bar{\kappa}(S) = 1$ or 2. Furthermore, if $C_1 \cap C_2 \neq \varnothing$, then show that $\bar{\kappa}(S) = 2$.

11.13. (Kuramoto) Let g be a polynomial in X and Y and let $S = \mathbf{A}_k^2 \backslash V(g)$. Suppose that $\bar{P}_2(S) = 0$. Show that $V(g)$ is isomorphic to $\mathbf{A}_k^1$, and that $\bar{\kappa}(S) = -\infty$.

11.14. (Tsunoda) Let D be a reduced divisor on $\bar{S}$ and let S be $\bar{S} \backslash D$. If $\bar{\kappa}(S) = 0$, then show that $\bar{p}_g(S) = 1$ and $\bar{q}(S) \leq 2$. Furthermore, if $\bar{q}(S) = 2$, then prove that S is isomorphic to G_m^2.

11.15. Let V, W and Z be varieties such that $V \times Z \cong W \times Z$. Then for any n-tuple M of nonnegative integers, show that $\bar{P}_M(V) = \bar{P}_M(W)$; in particular, $\bar{\kappa}(V) = \bar{\kappa}(W)$.

11.16. Let C_1 be a cubic curve with a cusp on $\mathbf{P}_k^2$ and take a nonsingular point p which is not of inflexion. Choose a conic C_2 such that $I_p(C_1, C_2) = 5$ and let q be another common point of C_1 and C_2. Perform a quadric transformation $\bar{S} = Q_q(\mathbf{P}_k^2) \to \mathbf{P}_k^2$ with q as center and let C_1^*, C_2^* be the strict transforms of C_1, C_2, respectively. Then, letting S be $\bar{S} \backslash (C_1^* \cup C_2^*)$, show that $\bar{q}(S) = 0$ and $\bar{\kappa}(S) = 2$. Further, show that $S \times V$ cannot be isomorphic to $\mathbf{A}_k^2 \times V$ for an arbitrary variety V.

References

[A–K] Altman, A. and Kleiman, S. 1970. *Introduction to Grothendieck Duality Theory*. Heidelberg: Springer–Verlag.

[B] Borel, A. 1969. *Linear Algebraic Groups*. New York: Benjamin.

[EGA] Grothendieck, A. and Dieudonné, J.
Eléments de Géométrie Algébrique.
EGA I. Le langage des schémas, *Publ. Math. IHES* 4 (1960).
EGA II. Étude globale élémentaire de quelques classes de morphismes, Ibid. 8 (1961).
EGA III. Étude cohomologique des faisceaux cohérents, Ibid. 11 (1961), and 17 (1963).
EGA IV. Étude locale des schémas et des morphismes de schémas, Ibid. 20 (1964), 24 (1965), 28 (1966), 32 (1967).
EGA I. *Eléments de Géométrie Algébrique, I.* Heidelberg: Springer–Verlag (1971).

[G] Godement, R. 1958. *Topologie Algébrique et Théories des Faisceaux*. Paris: Hermann.

[G–H] Griffiths, P. A. and Harris, J. 1978. *Principles of Algebraic Geometry*. New York: John Wiley & Sons.

[H] Hartshorne, R. 1977. *Algebraic Geometry*. New York: Springer–Verlag.

[M1] Mumford, D. B. 1975. *Algebraic Geometry I. Complex Projective Varieties.* Berlin: Springer–Verlag.

[M2] ———. 1974. *Abelian Varieties*. Oxford: Oxford University Press.

[N1] Nagata, M. 1977. *Field Theory*. New York: Marcel Dekker.

[N2] ———. 1962. *Local Rings*. New York: John Wiley & Sons.

[Z] Zariski, O. 1971. *Algebraic Surfaces*. Heidelberg: Springer–Verlag.

Books and Papers Related to Chapters 10 and 11

Deligne, P.

1. Théorie de Hodge II, *Publ. I. H. E. S.* **40,** (1971), 5–57.

Fujita, T.

1. Some remarks on Kodaira dimensions of fiber spaces, *Proc. Japan Acad.* **53**(A), (1977), 28–30.
2. On Kaehler fiber spaces over curves, *J. Math. Soc. Japan* **30,** (1978), 779–794.
3. On Zariski problem, *Proc. Japan Acad.* **55A,** (1979), 106–110.
4. (with Iitaka) Cancellation theorem for algebraic varieties, *J. Fac. Sci. Univ. of Tokyo*, Sec. IA, **24,** (1977), 123–127.

Iitaka, S.

1. On D-dimensions of algebraic varieties, *J. Math. Soc. Japan* **23,** (1971), 356–373.
2. Deformations of compact complex surfaces, II, III, *J. Math. Soc. Japan* **22,** (1970), 247–261; **23,** (1971), 692–705.
3. Logarithmic forms of algebraic varieties, *J. Fac. Sci. Univ. Tokyo* **23,** (1976), 525-544.
4. On logarithmic Kodaira dimension of algebraic varieties, pages 123–127 in *Complex Analysis and Algebraic Geometry*, 1977, Iwanami, Tokyo.
5. Some applications of logarithmic Kodaira dimension, pages 185–206 in *Proc. Int. Symp. Algebraic Geometry, Kyoto*, 1978, Kinokuniya, Tokyo.
6. Geometry on complements of lines in $\mathbf{P}^2$, *Tokyo J. Math.* **1,** (1978), 1–19.
7. Minimal models in proper birational geometry, *Tokyo J. Math.* **2,** (1979), 29–45.
8. Finiteness property of weakly proper birational maps, *J. Fac. Sci. Univ. of Tokyo*, Sec. IA, **24,** (1977), 491–502.
9. On the Diophantine equation $\varphi(X, Y) = \varphi(x, y)$, *J. reine angew. Math.* **298,** (1978), 43–52.
10. Classification of algebraic varieties, *Proc. Japan Acad.* **53,** (1977), 103–105.
11. A numerical criterion of quasi-abelian surfaces, *Nagoya J. Math.* **73,** (1979), 99–115.
12. On the homogeneous Lüroth theorem, *Proc. Japan Acad.* **55,** (1979), 88–91.
13. On logarithmic $K3$ surfaces, *Osaka J. Math.* **16,** (1979), 697–705.
14. The virtual singularity theorem and logarithmic bigenus theorem, *Tohoku J. Math.* **32,** (1980), 337–351.

Kambayashi, T.

1. On Fujita's strong cancellation theorem, *J. Fac. Sci. Univ. Tokyo.*, Sec. IA. **27,** (1980), 535–548.

Kawamata, Y.

1. Addition formula of logarithmic Kodaira dimension for morphisms of relative dimension one, pages 207–217 in *Proc. Int. Symp. Algebraic Geometry, Kyoto*, 1978, Kinokuniya, Tokyo.
2. Deformations of compactifiable complex manifolds, *Math. Ann.* **235,** (1978), 247–265.
3. Classification of non-complete algebraic surfaces, *Proc. Japan Acad.* **54,** (1978), 133–135.

4. (with E. Viehweg) On a characterization of an abelian variety in the classification theory of algebraic varieties, *Compositio Math.*, **41,** (1980), 355–359.
5. Characterization of Abelian varieties, *Compositio Math.*, **43,** (1981), 253–276.

Kuramoto, Y.
1. On the logarithmic plurigenera of algebraic surfaces, *Composito Math.*, **42,** (1981).

Liebermann, D. and Sernesi, E.
1. Semicontinuity of L-dimention, *Math. Ann.* **225,** (1977), 77–88.

Miyanishi, M.
1. An algebraic characterization of the affine plane, *J. Math. Kyoto Univ.* **15,** (1975), 449–456.
2. *Lectures on Curves on Rational and Unirational Surfaces*, 1978, Springer–Verlag, New York.
3. (with Sugie), Affine surfaces containing cylinder-like open subsets, *J. Math. Kyoto Univ.* **20,** (1980), 11–42.
4. Regular subrings of a polynomial ring I and II, *Osaka J. Math.* **17,** (1980), 329–338 to appear.
5. Singularities of normal affine surfaces containing cylinder-like open sets, *J. of Algebra* 68, (1981), 268–275.
6. Theory of non-complete surfaces, 1981, Springer-Verlag, New York.

Norimatsu, Y.
1. Kodaira vanishing theorem and Chern classes for ∂-manifolds, *Proc. Japan Acad.* **54**A, (1978), 107–108.

Popp, H.
1. *Moduli Theory and Classification Theory of Algebraic Varieties*, 1977, Springer–Verlag, New York.

Russel, P.
1. On affine ruled rational surfaces, *Math. Ann.*, **255,** (1981), 287–302.

Sakai, F.
1. Kodaira dimensions of complements of divisors, pages 239–257 in *Complex Analysis and Algebraic Geometry*, 1977, Iwanami, Tokyo.
2. Canonical models of compliments of stable curves, pages 643–661 in *Proc. Int. Symp. Algebraic Geometry, Kyoto* 1978, Kinokuniya, Tokyo.
3. Semi-stable curves on algebraic surfaces and logarithmic pluricanonical maps, *Math. Ann.* **254,** (1980), 89–120.

Ueno, K.
1. Classification of algebraic varieties, I, *Composition Math.* **27,** (1973), 277–342.
2. *Classification Theory of Algebraic Varieties and Compact Complex Spaces*, 1975, Springer–Verlag, New York.
3. Kodaira dimensions for certain fiber spaces, pages 279–292 in *Complex Analysis and Algebraic Geometry*, 1977, Iwanami, Tokyo.
4. On algebraic fiber space of abelian varieties, *Math. Ann.* 237, (1978), 1–22.
5. Classification of algebraic varieties, II, in pages 693–708 in *Proc. Intl. Symp. on Algebraic Geometry, Kyoto*, 1978, Kinokuniya, Tokyo.

Viehweg, E.

1. Canonical divisors and the additivity of the Kodaira dimensions for morphisms of relative dimension one, *Compositio Math.* **35,** (1977), 197–223.

2. Klassifikationstheorie algebraischer Varietaeten der Dimension drei, *Compositio Math.* **41,** (1980), 361–400.

Wakabayashi, I.

1. On the logarithmic dimension of the complement of a curve in $\mathbf{P}^2$, *Proc. Japan Acad.* **54**A, (1978), 157–162.

Wilson, P. M. H.

1. The behaviour of the plurigenera of surfaces under algebraic smooth deformations, *Inv. Math.* **47,** (1978), 289–299.
2. The pluricanonical map on varieties of general type, *Bull. Lond. Math. Soc.* **12,** (1980), 103–107.

Yoshihara, H.

1. On rational plane curves. *Proc. Japan Acad.*, **55**A, (1979), 152–155.

Index